Edited by
Wolfgang Osten and Nadya Reingand

Optical Imaging and Metrology

Related Titles

Dörband, B., Müller, H., Gross, H.

Handbook of Optical Systems
Volume 5: Metrology of Optical Components and Systems

2011
ISBN: 978-3-527-40381-3

Kaufmann, G. H. (ed.)

Advances in Speckle Metrology and Related Techniques

2011
ISBN: 978-3-527-40957-0

Horn, A.

Ultra-fast Material Metrology

2009
ISBN: 978-3-527-40887-0

Ackermann, G. K., Eichler, J.

Holography
A Practical Approach

2007
ISBN: 978-3-527-40663-0

Malacara, D. (ed.)

Optical Shop Testing

2006
ISBN: 978-0-471-48404-2

Edited by Wolfgang Osten and Nadya Reingand

Optical Imaging and Metrology

Advanced Technologies

WILEY-VCH Verlag GmbH & Co. KGaA

The Editors

Prof. Dr. Wolfgang Osten
Universität Stuttgart
Institut für Technische Optik
Pfaffenwaldring 9
70569 Stuttgart

Dr. Nadya Reingand
CeLight Inc.
12200 Tech Rd., Ste. 200
Silver Spring, MD 20904
USA

Frontcover Illustration

Distribution of the electric field intensity behind a subwavelength slit (width = $\lambda/5$) and a double meander-type structure. The slit basically represents a point source, which is then focused twice on the backside of the double meander lens. Since the meander system partly mimicks Pendry's perfect lens, the first focus is a 1:1 nearfield image of the slit. On the other hand, the second focus shows up at a distance of about 2 wavelengths behind the structure. Hence, a subwavelength slit can be observed in the far field using conventional microscopy.

All books published by **Wiley-VCH** are carefully produced. Nevertheless, authors, editors, and publisher do not warrant the information contained in these books, including this book, to be free of errors. Readers are advised to keep in mind that statements, data, illustrations, procedural details or other items may inadvertently be inaccurate.

Library of Congress Card No.: applied for

British Library Cataloguing-in-Publication Data
A catalogue record for this book is available from the British Library.

Bibliographic information published by the Deutsche Nationalbibliothek
The Deutsche Nationalbibliothek lists this publication in the Deutsche Nationalbibliografie; detailed bibliographic-data are available on the Internet at <http://dnb.d-nb.de>.

© 2012 Wiley-VCH Verlag & Co. KGaA, Boschstr. 12, 69469 Weinheim, Germany

All rights reserved (including those of translation into other languages). No part of this book may be reproduced in any form – by photoprinting, microfilm, or any other means – nor transmitted or translated into a machine language without written permission from the publishers. Registered names, trademarks, etc. used in this book, even when not specifically marked as such, are not to be considered unprotected by law.

Print ISBN: 978-3-527-41064-4
ePDF ISBN: 978-3-527-64847-4
ePub ISBN: 978-3-527-64846-7
mobi ISBN: 978-3-527-64845-0
oBook ISBN: 978-3-527-64844-3

Cover Design   Adam-Design, Weinheim
Typesetting   Laserwords Private Limited, Chennai, India
Printing and Binding   Markono Print Media Pte Ltd, Singapore

Printed in Singapore
Printed on acid-free paper

Contents

Preface *XV*
List of Contributors *XVII*

1	**LCOS Spatial Light Modulators: Trends and Applications** *1*	

Grigory Lazarev, Andreas Hermerschmidt, Sven Krüger, and Stefan Osten

1.1 Introduction *1*
1.2 LCOS-Based SLMs *2*
1.2.1 LCOS Technology *2*
1.2.1.1 Manufacturing and Assembly Technologies *2*
1.2.1.2 Signal Flow *4*
1.2.1.3 Drive Schemes and Latency *6*
1.2.2 Operation Modes *8*
1.2.2.1 Amplitude Modulation *9*
1.2.2.2 Phase Modulation *9*
1.2.2.3 Polarization Modulation *10*
1.2.2.4 Complex-Valued Modulation *10*
1.2.3 Performance Evaluation *11*
1.3 Some Applications of Spatial Light Modulators in Optical Imaging and Metrology *12*
1.4 Conclusion *23*
References *23*

2 Three-Dimensional Display and Imaging: Status and Prospects *31*
Byoungho Lee and Youngmin Kim

2.1 Introduction *31*
2.2 Present Status of 3D Displays *32*
2.2.1 Hardware Systems *32*
2.2.1.1 Stereoscopic Displays *32*
2.2.1.2 Autostereoscopic Displays *36*
2.2.1.3 Volumetric Displays *40*
2.2.1.4 Holographic Displays *42*
2.2.1.5 Recent 3D Display Techniques *44*

2.2.2	Software Systems	44
2.3	The Human Visual System	47
2.4	Conclusion	48
	Acknowledgments	50
	References	50

3 Holographic Television: Status and Future 57
Małgorzata Kujawińska and Tomasz Kozacki

- 3.1 Introduction 57
- 3.2 The Concept of Holographic Television System 60
- 3.3 Holographic Display Configuration 63
- 3.3.1 Planar and Circular Configurations 63
- 3.3.2 Real Image Wigner Distribution Analysis of the Holographic Displays in Planar and Circular Configurations 65
- 3.3.2.1 Planar Configuration 66
- 3.3.2.2 Circular Configuration 67
- 3.3.3 Visual Perception Analysis of Holographic Displays in Planar and Circular Configurations 68
- 3.3.4 Comparison of the Display Configurations for Commercially Available SLMs 70
- 3.4 Capture Systems 71
- 3.4.1 The Theory of Wide-Angle Holographic Capture 71
- 3.4.2 The Capture System with Multiple Cameras 72
- 3.4.3 The Capture System with a Single Camera and Rotated Static Object 75
- 3.5 Display System 77
- 3.5.1 1LCoS SLM in Holographic Display 77
- 3.5.2 Experimental Configurations of Holographic Displays 79
- 3.5.2.1 Display Basing on Spherical Illumination 80
- 3.5.2.2 Display Configuration Matching the Capture Geometry 80
- 3.5.2.3 Display Based on Tilted Plane Wave Illumination 81
- 3.6 Linking Capture and Display Systems 85
- 3.6.1 Mismatch in Sampling and Wavelength 86
- 3.6.2 Processing of Holograms to the Display Geometry 86
- 3.7 Optical Reconstructions of Real Scenes in Multi-SLM Display System 88
- 3.8 Conclusions and Future Perspectives 91
- References 92

4 Display Holography – Status and Future 95
Ventseslav Sainov and Elena Stoykova

- 4.1 Introduction 95
- 4.2 Types of Holograms 97
- 4.3 Basic Parameters and Techniques of Holographic Recording 99

4.4	Light-Sensitive Materials for Holographic Recording in Display Holography	*102*
4.4.1	Photoresists	*103*
4.4.2	Dichromate Gelatin	*103*
4.4.3	Photopolymers	*104*
4.4.4	Silver Halide Emulsions	*105*
4.5	Diffraction Efficiency of Discrete Carrier Holograms	*108*
4.6	Multicolor Holographic Recording	*111*
4.7	Digital Holographic Display: Holoprinters	*115*
4.8	Conclusion	*117*
	References	*117*

5	**Incoherent Computer-Generated Holography for 3D Color Imaging and Display**	***121***
	Toyohiko Yatagai and Yusuke Sando	
5.1	Introduction	*121*
5.2	Three-Dimensional Imaging and Display with CGHs	*122*
5.3	Theory of this Method	*123*
5.3.1	Relation between Object Waves and 3D Fourier Spectrum	*123*
5.3.2	Extraction Method for Paraboloid of Revolution	*124*
5.3.3	Extension to Full-Color Reconstruction	*126*
5.4	Imaging System and Resolution	*127*
5.4.1	Size of Object	*127*
5.4.2	Spatial Resolution	*128*
5.4.3	Magnification along the z-direction	*128*
5.5	Experiments	*129*
5.5.1	Computer Simulation and Some Parameters	*129*
5.5.2	Optical Reconstruction	*130*
5.6	Biological Specimen	*131*
5.7	Conclusion	*133*
	Acknowledgments	*133*
	References	*133*

6	**Approaches to Overcome the Resolution Problem in Incoherent Digital Holography**	***135***
	Joseph Rosen, Natan T. Shaked, Barak Katz, and Gary Brooker	
6.1	Introduction	*135*
6.2	Digital Incoherent Protected Correlation Holograms	*136*
6.3	Off-Axis Optical Scanning Holography	*142*
6.4	Synthetic Aperture with Fresnel Elements	*147*
6.5	Summary	*159*
	Acknowledgments	*160*
	References	*160*

7		**Managing Digital Holograms and the Numerical Reconstruction Process for Focus Flexibility** *163*
		Melania Paturzo and Pietro Ferraro
7.1		Introduction *163*
7.2		Fresnel Holograms: Linear Deformation *165*
7.3		Fresnel Holograms: Quadratic and Polynomial Deformation *168*
7.4		Fourier Holograms: Quadratic Deformation *170*
7.5		Simultaneous Multiplane Imaging in DH *172*
7.6		Summary *175*
		References *176*
8		**Three-Dimensional Particle Control by Holographic Optical Tweezers** *179*
		Mike Woerdemann, Christina Alpmann, and Cornelia Denz
8.1		Introduction *179*
8.2		Controlling Matter at the Smallest Scales *180*
8.2.1		Applications of Optical Tweezers *181*
8.2.2		Dynamic Optical Tweezers *181*
8.3		Holographic Optical Tweezers *183*
8.3.1		Diffractive Optical Elements *183*
8.3.2		Iterative Algorithms *184*
8.3.3		Experimental Implementation *185*
8.4		Applications of Holographic Optical Tweezers *187*
8.4.1		Colloidal Sciences *187*
8.4.2		Full Three-Dimensional Control over Rod-Shaped Bacteria *189*
8.4.3		Managing Hierarchical Supramolecular Organization *190*
8.5		Tailored Optical Landscapes *192*
8.5.1		Nondiffracting Beams *193*
8.5.1.1		Mathieu Beams *196*
8.5.2		Self-Similar Beams *197*
8.5.2.1		Ince-Gaussian Beams *198*
8.6		Summary *200*
		References *200*
9		**The Role of Intellectual Property Protection in Creating Business in Optical Metrology** *207*
		Nadya Reingand
9.1		Introduction *207*
9.2		Types of Intellectual Property Relevant to Optical Metrology *208*
9.3		What Kind of Business Does Not Need IP Protection? *210*
9.4		Does IP Protect Your Product from Counterfeiting? *211*
9.5		Where to Protect Your Business? *212*
9.6		International Patent Organizations *212*
9.7		Three Things Need to Be Done Before Creating Business *214*
9.7.1		Prior Art Search *214*

9.7.2	Patent Valuation	*216*
9.8	Ownership Clarification	*217*
9.9	Patent Filing	*219*
9.10	Commercialization	*220*
9.10.1	Licensing to Existing Companies	*220*
9.10.2	Start-Ups	*221*
9.11	Conclusions	*222*
	References	*223*

10 On the Difference between 3D Imaging and 3D Metrology for Computed Tomography *225*
Daniel Weiß and Michael Totzeck

10.1	Introduction	*225*
10.2	General Considerations of 3D Imaging, Inspection, and Metrology	*226*
10.2.1	3D Imaging	*226*
10.2.2	3D Inspection	*227*
10.2.3	3D Metrology	*229*
10.3	Industrial 3D Metrology Based on X-ray Computed Tomography	*229*
10.3.1	X-Ray Cone-Beam Computed Tomography	*230*
10.3.1.1	Two-Dimensional Image Formation	*230*
10.3.1.2	Imaging Geometries and 3D Reconstruction	*231*
10.3.2	X-Ray CT-Based Dimensional Metrology	*232*
10.3.2.1	Why CT Metrology?	*232*
10.3.2.2	Surface Extraction from 3D Absorption Data	*232*
10.3.3	Device Imperfections and Artifacts	*233*
10.3.3.1	Geometrical Alignment of the Components	*233*
10.3.3.2	Beam Hardening	*233*
10.3.4	Standards for X-Ray CT-Based Dimensional Metrology	*235*
10.3.4.1	Length Measurement Error and Scanning Errors of Form and Size	*235*
10.3.4.2	Dependence on Material and Geometry	*237*
10.4	Conclusions	*237*
	References	*238*

11 Coherence Holography: Principle and Applications *239*
Mitsuo Takeda, Wei Wang, and Dinesh N. Naik

11.1	Introduction	*239*
11.2	Principle of Coherence Holography	*240*
11.2.1	Reciprocity in Spatial Coherence and Hologram Recording	*240*
11.2.2	Similarity between the Diffraction Integral and van Cittert–Zernike Theorem	*241*
11.3	Gabor-Type Coherence Holography Using a Fizeau Interferometer	*241*
11.4	Leith-Type Coherence Holography Using a Sagnac Interferometer	*243*
11.5	Phase-Shift Coherence Holography	*246*
11.6	Real-Time Coherence Holography	*247*

11.7	Application of Coherence Holography: Dispersion-Free Depth Sensing with a Spatial Coherence Comb *248*	
11.8	Conclusion *252*	
	Acknowledgments *252*	
	References *252*	

12	**Quantitative Optical Microscopy at the Nanoscale: New Developments and Comparisons** *255*	
	Bernd Bodermann, Egbert Buhr, Zhi Li, and Harald Bosse	
12.1	Introduction *255*	
12.2	Quantitative Optical Microscopy *257*	
12.2.1	Metrological Traceability *257*	
12.2.2	Measurands and Measurement Methods *260*	
12.2.3	Image Signal Modeling *261*	
12.2.4	Experimental Aspects *263*	
12.2.5	Measurement Uncertainty *265*	
12.3	Comparison Measurements *268*	
12.4	Recent Development Trends: DUV Microscopy *271*	
12.4.1	Light Source and Coherence Reduction *274*	
12.4.2	Illumination System *275*	
12.4.3	Imaging Configuration *276*	
12.5	Points to Address for the Further Development of Quantitative Optical Microscopy *278*	
	References *279*	

13	**Model-Based Optical Metrology** *283*	
	Xavier Colonna de Lega	
13.1	Introduction *283*	
13.2	Optical Metrology *283*	
13.3	Modeling Light–Sample Interaction *284*	
13.3.1	From Light Detection to Quantitative Estimate of a Measurand *284*	
13.3.2	Two Types of Light–Sample Interaction Models *285*	
13.4	Forward Models in Optical Metrology *287*	
13.5	Inverse Models in Optical Metrology *290*	
13.5.1	Wave Front Metrology *290*	
13.5.2	Thin-Film Structures Metrology *291*	
13.5.3	Unresolved Structures Metrology *295*	
13.6	Confidence in Inverse Model Metrology *298*	
13.6.1	Modeling Pitfalls *299*	
13.6.2	Sensitivity Analysis *300*	
13.6.3	Validation of the Overall Tool Capability *300*	
13.7	Conclusion and Perspectives *301*	
	References *302*	

14	**Advanced MEMS Inspection by Direct and Indirect Solution Strategies** *305*
	Ryszard J. Pryputniewicz
14.1	Introduction *305*
14.2	ACES Methodology *307*
14.2.1	Computational Solution *308*
14.2.2	Experimental Solution Based on Optoelectronic Methodology *309*
14.2.2.1	The OELIM System *312*
14.3	MEMS Samples Used *314*
14.4	Representative Results *317*
14.4.1	Deformations of a Microgyroscope *317*
14.4.2	Functional Operation of a Microaccelerometer *319*
14.4.3	Thermomechanical Deformations of a Cantilever Microcontact *319*
14.5	Conclusions and Recommendations *322*
	Acknowledgments *323*
	References *323*

15	**Different Ways to Overcome the Resolution Problem in Optical Micro and Nano Metrology** *327*
	Wolfgang Osten
15.1	Introduction *327*
15.2	Physical and Technical Limitations in Optical Metrology *328*
15.2.1	Optical Metrology as an Identification Problem *329*
15.2.2	Diffraction-Limited Lateral Resolution in Optical Imaging *331*
15.2.3	Diffraction-Limited Depth of Focus in Optical Imaging *333*
15.2.4	Space-Bandwidth Product of Optical Imaging Systems *333*
15.3	Methods to Overcome the Resolution Problem in Optical Imaging and Metrology *334*
15.3.1	New Strategies for the Solution of Identification Problems *335*
15.3.1.1	Active Measurement Strategies *335*
15.3.1.2	Model-Based Reconstruction Strategies *336*
15.3.1.3	Sensor Fusion Strategies *337*
15.3.2	Different Approaches for Resolution Enhancement of Imaging Systems *338*
15.3.2.1	Conventional Approaches to Achieve the Resolution Limit *340*
15.3.2.2	Unconventional Approaches to Break the Resolution Limit *340*
15.4	Exemplary Studies on the Performance of Various Inspection Strategies *343*
15.4.1	Model-Based Reconstruction of Sub-λ Features *343*
15.4.1.1	The Application of Scatterometry for CD-Metrology *343*
15.4.1.2	Model-Based and Depth-Sensitive Fourier Scatterometry for the Characterization of Periodic Sub-100 nm Structures *348*
15.4.2	High-Resolution Measurement of Extended Technical Surfaces with Multiscale Sensor Fusion *355*

15.5	Conclusion *360*	
	Acknowledgments *362*	
	References *362*	

16   Interferometry in Harsh Environments *369*
Armando Albertazzi G. Jr

16.1	Introduction *369*	
16.2	Harsh Environments *369*	
16.3	Harsh Agents *370*	
16.3.1	Temperature *370*	
16.3.2	Humidity *372*	
16.3.3	Atmosphere and Pressure *373*	
16.3.4	Shock and Vibration *374*	
16.3.5	Radiation and Background Illumination *374*	
16.4	Requirements for Portable Interferometers *375*	
16.4.1	Robustness *375*	
16.4.2	Flexibility *376*	
16.4.3	Compactness *376*	
16.4.4	Stability *376*	
16.4.5	Friendliness *376*	
16.4.6	Cooperativeness *377*	
16.5	Current Solutions *377*	
16.5.1	Isolation *377*	
16.5.1.1	Atmosphere Isolation *377*	
16.5.1.2	Temperature Isolation *378*	
16.5.1.3	Radiation Isolation *378*	
16.5.1.4	Vibration Isolation *378*	
16.5.2	Robustness *379*	
16.6	Case Studies *381*	
16.6.1	Dantec ESPI Strain Sensor (Q-100) *382*	
16.6.2	Monolitic GI/DSPI/DHI Sensor *382*	
16.6.3	ESPI System for Residual Stresses Measurement *384*	
16.6.4	Pixelated Phase-Mask Dynamic Interferometer *385*	
16.6.5	Digital Holographic Microscope *386*	
16.6.6	Shearography *387*	
16.6.7	Fiber-Optic Sensors *388*	
16.7	Closing Remarks *389*	
16.7.1	Summary *389*	
16.7.2	A Quick Walk into the Future *390*	
	References *390*	

17   Advanced Methods for Optical Nondestructive Testing *393*
Ralf B. Bergmann and Philipp Huke

17.1	Introduction *393*	
17.2	Principles of Optical Nondestructive Testing Techniques (ONDTs) *393*	

17.2.1	Material or Object Properties	394
17.2.2	Application of Thermal or Mechanical Loads for NDT	395
17.2.3	Selected Measurement Techniques Suitable for Optical NDT	396
17.2.4	Comparison of Properties of Selected NDT Techniques	397
17.3	Optical Methods for NDT	399
17.3.1	Thermography	399
17.3.2	Fringe Reflection Technique (FRT)	399
17.3.2.1	Principle of FRT	400
17.3.2.2	Experimental Results	401
17.3.3	Digital Speckle Shearography	402
17.3.3.1	Principle of Shearography	402
17.3.3.2	Experimental Results	402
17.3.4	Laser Ultrasound	404
17.3.4.1	Principle of Operation	404
17.3.4.2	Experimental Results	406
17.4	Conclusions and Perspectives	408
	Acknowledgments	409
	References	409

18 Upgrading Holographic Interferometry for Industrial Application by Digital Holography 413

Zoltán Füzessy, Ferenc Gyímesi, and Venczel Borbély

18.1	Introduction	413
18.2	Representative Applications	414
18.3	Contributions to Industrial Applications by Analog Holography	414
18.3.1	Portable Interferometer in the Days of Analog Holographic Interferometry	414
18.3.2	Difference Holographic Interferometry (DHI) –Technique for Comparison and Fringe Compensation	418
18.3.3	Straightforward Way of Managing Dense Holographic Fringe Systems	423
18.3.3.1	Upper Limit of the Evaluating Camera–Computer System	424
18.3.3.2	Measuring Range of Holographic Interferometry	425
18.3.3.3	PUZZLE Read-Out Extension Technique –for Speckled Interferograms	426
18.4	Contributions to Industrial Applications by Digital Holography	428
18.4.1	Scanning and Magnifying at Hologram Readout	429
18.4.2	Digital Holography for Residual Stress Measurement	431
18.5	Conclusion and a Kind of Wish List	434
	Acknowledgments	434
	References	434

Color Plates 439

Index 475

Preface

This book contains all contributions which were presented on occasion of the last HoloMet workshop "HoloMet 2010" in Balatonfüred, Hungary. The year of HoloMet 2010 closed the first decade of the 21st century. During this period many key technologies such as micro electronics, photonics, informatics, and mechatronics have shown an extremely fast development that perhaps was never seen before. Moreover, in 2010 we celebrated the 50th anniversary of the birth of the laser and the 45th anniversary of the invention of holographic interferometry. And even more, in June 2010 the inventor of holography, Dennis Gabor, had his 110th birthday. Thus it was a welcome opportunity to come together again and to discuss the state of the art, modern trends, and future perspectives in optics. Obviously, optical sciences have experienced dramatic transformations within the last few decades, influenced by revolutionary inventions of many branches such as image processing, digital holography, plasmonics, photonic sources and sensors from which optical imaging and metrology have benefited remarkably.

The HoloMet workshop series was started in May 2000 with the objective to provide an effective forum for the discussion of recent activities and new perspectives in optical technologies. While the first workshops were dedicated to Holography and Optical Metrology (see [1], [2]), the scope of HoloMet 2010 was extended to the broader scope of advanced technologies focused around optical imaging and optical metrology. On one hand, optical imaging and optical metrology are topics with long traditions. On the other hand, the current trend in both disciplines shows an increasing dynamic that is stimulated by many fascinating innovations such as high resolution microscopy, 3D imaging and nanometrology. Consequently, both are getting younger every day and stimulate each other more and more. Thus, the main objective of the workshop was to bring experts from both fields together and to bridge these strongly related and emerging fields.

The editors are very thankful to all authors and really appreciate that they took the time out of their busy schedules to work on the book chapters to provide insights into current and future trends in optical imaging and metrology. Their dedication made it possible to present the complete collection of all contributions that were presented at the workshop. Special thanks go to our co-organizers, Zoltan Füzessy and Ferenc Gyimesi from the Holography Group at the Budapest University of Technology and Economics, and Sven Krüger from HoloEye Photonics AG Berlin.

And finally, we thank the publisher, Wiley-VCH, especially Valerie Molière and Anja Tschörtner, for the wonderful and successful cooperation.

Looking forward to HoloMet 2012 in Japan.

April 2012 *Wolfgang Osten and Nadya Reingand*

The participants of the HoloMet 2010 Workshop in Balatonfüred, Hungary, June 2010 (from left to right: Toyohiko Yatagai, Mrs. Yatagai, Grigory Lazarev, Ryszard J. Pryputniewicz, Mitsuo Takeda, Aladar Czitrovsky, Joseph Rosen, Zoltan Füzessy, Malgorzata Kujawinska, Michael Totzeck, Wolfgang Osten, Armando G. Albertazzi Jr, Ralf B. Bergmann, Cornelia Denz, Nadya Reingand, Byoungho Lee, Pietro Ferraro, Harald Bosse, Xavier Colonna de Lega, Ferenc Gyimesi).

References

1. Osten, W. and Jüptner, W. New Prospects of Holography and 3D-Metrology, Proc of the International Berlin Workshop HoloMet 2000, Strahltechnik Band 14, BIAS, Bremen 2000.

2. Osten, W. and Jüptner, W. New Perspectives for Optical Metrology, Proc of the International Balatonfüred Workshop, HoloMet 2001, BIAS Verlag, Bremen 2001.

List of Contributors

Armando Albertazzi G. Jr.
Universidade Federal de Santa Catarina
Mechanical Engineering Department LabMetro
CEP 88 040-970
Florianopolis, SC
Brazil

Christina Alpmann
Westfälische Wilhelms-Universität Münster
Institut für Angewandte Physik
Corrensstr. 2/4
48149 Münster
Germany

Ralf B. Bergmann
Bremer Institut für angewandte Strahltechnik (BIAS) GmbH
Klagenfurter Strasse 2
D-28359 Bremen
Germany

Bernd Bodermann
Physikalisch-Technische Bundesanstalt
Division Optics
Bundesallee 100
D-38116
Braunschweig
Germany

Venczel Borbély
Budapest University of Technology and Economics
Institute of Physics
Department of Physics
Budafoki út 8.
1111 Budapest
Hungary

Harald Bosse
Physikalisch-Technische Bundesanstalt
Division Precision Engineering
Bundesallee 100
D-38116
Braunschweig
Germany

Gary Brooker
Johns Hopkins University
Department of Biomedical Engineering
Microscopy Center
9605 Medical Center Drive
Rockville
MD 20850
USA

Egbert Buhr
Physikalisch-Technische
Bundesanstalt
Division Optics
Bundesallee 100
D-38116
Braunschweig
Germany

Xavier Colonna de Lega
ZYGO Corporation
Laurel Brook Road
Middlefield
CT 06455
USA

Cornelia Denz
Westfälische
Wilhelms-Universität Münster
Institut für Angewandte Physik
Corrensstr. 2/4
48149 Münster
Germany

Pietro Ferraro
Istituto Nazionale di Ottica (INO)
del CNR
Via Campi Flegrei 34
Pozzuoli (Na) I-80078
Italy

Zoltán Füzessy
Budapest University of
Technology and Economics
Institute of Physics
Department of Physics
Budafoki út 8.
1111 Budapest
Hungary

Ferenc Gyímesi
Budapest University of
Technology and Economics
Institute of Physics
Department of Physics
Budafoki út 8.
1111 Budapest
Hungary

Andreas Hermerschmidt
HOLOEYE Photonics AG
R&D Department
Albert-Einstein-Str. 14
12489 Berlin
Germany

Philipp Huke
Bremer Institut für angewandte
Strahltechnik (BIAS) GmbH
Klagenfurter Strasse 2
D-28359 Bremen
Germany

Barak Katz
Ben-Gurion University of the
Negev
Department of Electrical and
Computer Engineering
P. O. Box 653
Beer-Sheva 84105
Israel

Youngmin Kim
Seoul National University
School of Electrical Engineering
Gwanak-Gu Gwanakro 1
Seoul 151-744
Korea

Tomasz Kozacki
Warsaw University of Technology
Department of Mechatronics
Institute of Micromechanics and
Photonics
8 Sw. A. Boboli Street
02-525 Warsaw
Poland

Sven Krüger
HOLOEYE Photonics AG
R&D Department
Albert-Einstein-Str. 14
12489 Berlin
Germany

Małgorzata Kujawińska
Warsaw University of Technology
Department of Mechatronics
Institute of Micromechanics and
Photonics
8 Sw. A. Boboli Street
02-525 Warsaw
Poland

Grigory Lazarev
HOLOEYE Photonics AG
R&D Department
Albert-Einstein-Str. 14
12489 Berlin
Germany

Byoungho Lee
Seoul National University
School of Electrical Engineering
Gwanak-Gu Gwanakro 1
Seoul 151-744
Korea

Zhi Li
Physikalisch-Technische
Bundesanstalt
Division Precision Engineering
Bundesallee 100
D-38116
Braunschweig
Germany

Dinesh N. Naik
Universität Stuttgart
Institut für Technische
Optik (ITO)
Pfaffenwaldring 9
D-70569 Stuttgart
Germany

Stefan Osten
HOLOEYE Photonics AG
R&D Department
Albert-Einstein-Str. 14
12489 Berlin
Germany

Wolfgang Osten
Universität Stuttgart
Institut für Technische
Optik (ITO)
Pfaffenwaldring 9
D-70569 Stuttgart
Germany

Melania Paturzo
Istituto Nazionale di Ottica (INO)
del CNR
Via Campi Flegrei 34
Pozzuoli (Na) I-80078
Italy

Ryszard J. Pryputniewicz
Department of Mechanical Engineering
CHSLT
School of Engineering
Worcester Polytechnic Institute
100 Institute Road
Worcester
MA 01609
USA

Nadya Reingand
Patent Hatchery
7 Clifton Ct.
Pikesville
MD 21208
USA

Joseph Rosen
Ben-Gurion University of the Negev
Department of Electrical and Computer Engineering
P. O. Box 653
Beer-Sheva 84105
Israel

Ventseslav Sainov
Holography and Optical Metrology
Institute of Optical Materials and Technologies
Bulgarian Academy of Sciences (IOMT-BAS)
Acad. G. Bonchev Str., Bl. 101
Sofia 1113
Bulgaria

Yusuke Sando
Utsunomiya University
Center for Optical Research and Education
Yoto 7-1-2
Utsunomiya Tochigi 321-8585
Japan

Natan T. Shaked
Tel Aviv University
Department of Biomedical Engineering
Tel-Aviv 69978
Israel

Elena Stoykova
Holography and Optical Metrology
Institute of Optical Materials and Technologies
Bulgarian Academy of Sciences (IOMT-BAS)
Acad. G. Bonchev Str., Bl. 101
Sofia 1113
Bulgaria

Michael Totzeck
Carl Zeiss AG
Corporate Research and Development
Carl-Zeiss-Straße 22
73447 Oberkochen
Germany

Mitsuo Takeda
Utsunomiya University
Center for Optical Research and Education
Yoto 7-1-2
Utsunomiya Tochigi 321-8585
Japan

Wei Wang
Heriot-Watt University
Department of Mechanical Engineering
School of Engineering and Physical Sciences
Edinburgh
EH14 4AS
UK

Daniel Weiß
Carl Zeiss Industrielle
Messtechnik GmbH
Development CT systems
Carl-Zeiss-Straße 22
73447 Oberkochen
Germany

Mike Woerdemann
Westfälische
Wilhelms-Universität Münster
Institut für Angewandte Physik
Corrensstr. 2/4
48149 Münster
Germany

Toyohiko Yatagai
Utsunomiya University
Center for Optical Research and
Education
Yoto 7-1-2
Utsunomiya Tochigi 321-8585
Japan

1
LCOS Spatial Light Modulators: Trends and Applications

Grigory Lazarev, Andreas Hermerschmidt, Sven Krüger, and Stefan Osten

1.1
Introduction

Spatial light modulator (SLM) is a general term describing devices that are used to modulate amplitude, phase, or polarization of light waves in space and time. Current SLM–based systems use either optical MEMS (microelectromechanical system, [1]) or LCD technology [2]. In this chapter, we review trends and applications of SLMs with focus on liquid crystal on silicon (LCOS) technology.

Most developments of liquid crystal (LC) microdisplays are driven by consumer electronics industry for rear–projection TVs, front projectors, and picoprojectors. Also, MEMS technologies such as digital micromirror device (DMD, [3]) and grating light valve (GLV, [4]) are driven by these industries, except for membrane mirrors. Some industrial applications have forced MEMS development for scanning, printing technologies, and automotive applications [5]. But the major R&D-related driving force for new SLM technologies is the defense industry.

Technological advances in lithography are the basis for MEMS developments. Phase modulators based on 2D pistonlike mirror arrays [6, 7] or ribbonlike 1D gratings [8] show high performance in frame rate. Unfortunately, the availability of these technologies is limited because they are developed either company-internal or within defence projects. The major advantages of MEMS are frame rate, spectral range, and an efficient use of nonpolarized light. Phase modulators and other optical implementations are still niche markets for the MEMS industry. Even now, customized MEMS developments are quite challenging and expensive.

LC panels still have an advantage out of their projection applications in terms of resolution and minimal pixel size for 2D displays. Only LC-based technology is able to modulate intensity, phase, and/or polarization because of polarization rotation and/or electrically controlled birefringence (ECB).

LCOS technology [9] was developed for front- and rear- (RPTV) projection systems competing with AMLCD (active matrix LCD) and DMD. The reflective arrangement due to silicon backplane allows putting a high number of pixels in a small panel, keeping the fill factor ratio high even for micrometer-sized pixels.

Optical Imaging and Metrology: Advanced Technologies, First Edition.
Edited by Wolfgang Osten and Nadya Reingand.
© 2012 Wiley-VCH Verlag GmbH & Co. KGaA. Published 2012 by Wiley-VCH Verlag GmbH & Co. KGaA.

The history of companies shutting down their LCOS activities (Intel, Philips, etc.) and the downfall of the RPTV market made it difficult to demonstrate the promised LCOS advantages in performance and volume pricing. However, the classic three-panel architectures for high-end front- and rear-projection systems led to the development of high-quality microdisplays, high-performance polarization, and compensation optics as well as sophisticated electronic driving solutions. Single-panel designs for field sequential color systems never really entered the high-end and professional market, but are as good candidates as DMD and laser scanning technologies for small and embedded projection devices, such as cell phone, companion, camera, and toy projectors.

1.2 LCOS-Based SLMs

LCOS SLMs based on "consumer" microdisplay technology inherited features and drawbacks of projection displays. Backplanes of the front-projection microdisplays usually have a diagonal exceeding 0.55 in. or higher and pixel sizes starting from 8 μm. Smaller panels are not popular in front-projection microdisplays because of étendue limitations when used in an incoherent system [10] as well as heat dissipation considerations. However, microdisplays intended for embedded projectors and picoprojectors utilize quite small panels from 0.17 to 0.55 in. with pixel sizes ranging from 6 to 10 μm. Whereas HMD products require a larger panel diagonal for high field of view (FOV) designs. Reduction of the pixel size has a significant economical advantage in customer electronics industry since it allows placing more dies on the wafer. Both types of microdisplays are usually driven color field sequential, which means they offer a higher frame rate when potentially used as SLMs.

1.2.1 LCOS Technology

1.2.1.1 Manufacturing and Assembly Technologies

LCOS technology has different descriptions/brand names with different suppliers; for example, JVC's "D-ILA" and Sony's "SXRD," are basically the same CMOS wafer technology, processed typically using 8 in. silicon wafers (Figure 1.1). Both Sony and JVC introduced 4K LCOS panels (4096 × 2160 and 4096 × 2400) to the market. JVC was also successful in building a prototype of an 8K panel (8192 × 4320) with 5 μm pixel size [11].

In order to get a good reflectivity out of these reflective pixelated arrays, high reflective aluminum mirrors, mostly with a passivation layer, are used. Various techniques for planarization or reduced interpixel gap effects have been developed over the years. At present, foundries offer processes to reduce the interpixel gap in the design down to 200 nm [12].

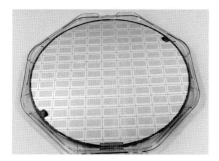

Figure 1.1 Silicon wafer with panel backplanes of Omnivision (OVT).

It is also possible to cover the backplane with a dielectric mirror so that the pixelated structure is not seen any more. It allows to increase the reflectivity of the SLM, so the light utilization efficiency for a low-frequency content is higher. Unfortunately, dielectric layers limit the spectral range of the device typically to an 80–100 nm band. Another disadvantage is that the cell requires higher voltages as it becomes thicker and therefore introduces a higher cross talk between adjacent pixels. As the modulation is then strongly dependent on the addressed spatial frequency, the effective resolution of the microdisplay as well as the achievable diffraction efficiency will decrease. The pixel resolution of such commercially available SLMs (Hamamatsu Photonics, Boulder Nonlinear Systems) does not exceed SVGA, and pixel sizes are typically as large as 16–32 μm [13].

The production of the actual LCOS cell can be done on the wafer level or based on single cell. The LCOS cell production on the wafer level has advantages on the economics side but lacks flexibility. An important process is the implementation of the spacers, defining the cell gap of the LCOS cell. There are spacers designed into the CMOS backplane, spacers distributed into the LC material itself, and also spacers as part of the gasket definition (frame spacer technology). Assembled LCOS cells typically show a dashed shape (Figure 1.2). Wafer-scale-processed parts have a lower curvature than single-cell-assembly-manufactured parts [2].

In the peak of the RPTV and front-projection development phase, the alignment layer technologies became an important factor because of lifetime issues with higher-density illumination and the lower end of the blue spectrum. So, the standard alignment layer material polyimide (PI) was replaced by inorganic SiO_x alignment structures [14]. Besides projection applications various LC modes for photonic applications [15] were also designed and tested, covering twisted nematic, ECB, and VAN (vertically aligned nematic) materials. In order to use the LCOS as an addressable optical component the packaging design and the packaging processes, such as die attach/bonding and wire bonding, as well as device handling are critical. Packaging also significantly influences the total device cost when going into volume production.

Related to HOLOEYE's history in diffractive optics and SLM technologies, we developed phase modulating LCOS SLMs in conjunction with Omnivision Technologies (OVT), see [16]. The phase modulating LCOS SLMs are based on

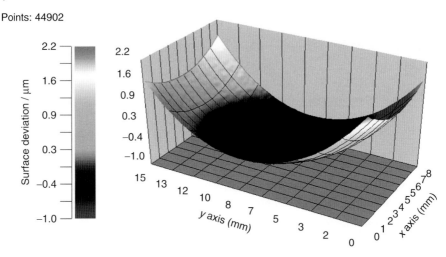

Figure 1.2 Curvature of the microdisplay measured with a Twyman–Green interferometer.

OVT's commercially available 0.7 in. full HD backplane and have been used in numerous adaptive optics applications. HOLOEYE is using nematic LC for their LCOS SLMs with almost analogous phase modulation, because of the higher diffraction efficiency of multilevel hologram structures (256 level, typically 80%). This is one of the main advantages compared with ferroelectric LC (FLC) technology, as FLC can only display binary holograms that always create a symmetrical image leading to a basic 50% loss of light in the system in addition to 50% duty cycle of FLCs. The clear advantage of the FLC is its switching speed, that is, it can be refreshed in the kilohertz range.

1.2.1.2 Signal Flow

In most microdisplay technologies/applications, the signal flow starts with a defined input signal form to be transferred to the microdisplay driver directly in contact with the microdisplay. The schematic representation shown in Figure 1.3 can be used to derive the first batch of parameters leading to an optimized microdisplay drive solution.

Most pixel-based display technologies show a native resolution of a certain display standard (compare, e.g., VESA (Video Electronics Standards Association)), showing optimized performance with input signals representing this native resolution. For higher quality microdisplays, the device can embed the input signal of lower resolution into the physical pixel matrix, filling the whole array of pixels. Pure digital input signals and an EDID (extended display identification data) adapted to the display controllers guarantee the correct choice of timing and addressed display

```
┌─────────────────────────────────────────────────────────────┐
│  Input signal                                                │
│    Analog or digital video input from a graphics processor   │
│    (e.g. based on FPGA) combining video input and            │
│    digital signal generator, digital output                  │
└─────────────────────────────────────────────────────────────┘
                              ↓
┌─────────────────────────────────────────────────────────────┐
│  Signal transfer                                             │
│    High bandwidth on PCB signal data lines (Bit-Parallel) for short distances │
│    Signal transfer by Bit-Serial (e.g. DVI or special nonstandard interface)  │
│    for long distances with differential signal like LVDS     │
└─────────────────────────────────────────────────────────────┘
                              ↓
  ┌───────────────────────────────────────────────────────────┐
  │  Display driver                                            │
  │    Input signal processing depending on display drive scheme │
  │    Signal interface (for digital drive schemes through    │
  │    image frame buffer) to microdisplay                    │
  └───────────────────────────────────────────────────────────┘
                              ↓
  ┌───────────────────────────────────────────────────────────┐
  │  Microdisplay                                              │
  │    Processing of display bus signal depending on drive scheme │
  │    (from simply pixel voltage set to integrated LUT and   │
  │    illumination feedback to controller)                   │
  └───────────────────────────────────────────────────────────┘
                Integrated with some display technologies
```

Figure 1.3 Block diagram of microdisplay driving. LUT, lookup table; PCB, printed circuit board.

resolution. Consumer products are facing the challenge of processing various digital and analog signals, always leading to a compromise in the transferred signal.

Image processors (e.g., based on FPGAs – field-programmable gate array) can combine several input signals and create digital video output according to the capabilities of the display driver technology, leading to a significant design freedom in resolution (aspect ratio), frame rate, and bit depth.

For video-only applications, standard DVI (digital visual interface) and/or HDMI (high-definition multimedia interface) signals provide highest quality, but are rather complex, require a large cable diameter, and cause a relatively high power consumption. Special LVDS (low-voltage differential signal) multiplexer/demultiplexer chipsets are available for custom (frame rate and bit depth) digital signals, such as single pair LVDS transmission solutions, especially for near-to-eye (NTE) applications.

ASIC (application-specific integrated circuit) along with microdisplay combinations (like most HD projection microdisplays) show high flexibility in addressing parameters, but are large and power hungry. Epson and Micron showed more simple approaches, where Epson integrated an LVDS receiver on the microdisplay flex cable (flexible flat cable) and Micron integrated the driver ASIC onto the silicon backplane of the LCOS. The Epson solution can reduce the portion of electronic circuitry right at the microdisplay, so the driver electronics can be separated. The Micron solution (at the moment only up to the resolution of 960×540 pixels) has a fully integrated driver reducing power consumption drastically, but limiting the choice of possible drive schemes.

1.2.1.3 Drive Schemes and Latency

Two major types of drive schemes are used: analog and digital, and this differentiation is based on the actual generated voltage applied to individual pixels.

Analog Drive In analog drive schemes, the microdisplay utilizes analog voltages directly for the representation of a gray level in an individual LC cell. The scheme is well suited for short illumination pulses because of analog voltages on the pixel (there is no specific digital flicker noise, see further). Analog drive uses typically lower clock frequency and hence has lower power consumption. As a result, longer flex cables between driver and microdisplay are possible.

The analogue scheme also has a number of drawbacks. Drift- and channel-depending variations of drive voltages need to be compensated. The ability for that compensation is evidently limited. The display addressing is progressive, that is, pixels are addressed in a consecutive way and not simultaneously. Effectively, the video signal goes through a low-pass filter. The analog signal path affects the slew rate of the video signal and can superimpose ringing, nonlinear distortion, noise, and echoes. Since the frequency of the video signal is relatively low in the vertical direction and relatively high in the horizontal direction, a significant "cross talk" occurs for the latter one, that is, for sequentially written pixels. As a result, a decrease of phase (or amplitude) modulation for high spatial frequencies in the addressed image is observed [17], which corresponds to a decrease in resolution.

The field inversion, which is always required in LCs, in typical analog progressive scan architectures is limited to the single or double frame rate. In the case of the single frame rate field inversion, the DC balancing can fail if the content changes too fast (e.g., due to a specific application). This can cause lifetime issues, that is, destroy the transparent electrode (Indium tin oxide (ITO)). The inversion with the double frame rate requires a frame buffer [18].

Digital Fast Bit-Plane Addressing In a digital drive scheme a pulse width modulation (PWM) encodes a certain gray level into a series of binary pulses in the kilohertz range, referred to as sequence. In principle, every individual pulse interacts with the LC molecule, causing its rotation, that at the end leads to the desired gray level representation. Owing to limited rotational viscosity of the LC material, LC molecules cannot follow individual pulses of the electrical signal in a discrete way. That is why the base addressing frequency cannot be resolved, so that it is possible to achieve an almost analogous LC molecule position representing/resulting in a certain gray level.

The digital scheme is usually more stable than analog and shows a repeatable performance. Field inversion is possible at the kilohertz range (e.g., for each modulation pulse) without image retention. All pixels are addressed simultaneously. The scheme does not suffer from electrical cross talk (i.e., no low path filtering of the signal). However, an electro-optical cross talk for small pixel sizes may still be observable. The electro-optical cross talk is caused by influence of the electrical field between adjacent pixels and can be further compensated [19].

The control electronics of such microdisplays is compact and has low cost. The device itself is highly programmable. The addressed resolution, amplitude or phase bit depth, and frame rate can be changed *in situ* and if required, adapted to environmental changes (e.g., wavelength, temperature).

The advantages of digital addressing are accompanied by limitations. One observes a kind of flicker noise at high frequencies ("supermodulation"). This means the electro-optical response of the SLM is not constant over the frame. In some scenarios (e.g., projection) time averaging can be used to compensate this effect. In other scenarios, in particular, when using phase modulation or short-pulse light sources, time averaging is not possible. Here, special sequences and higher bandwidth to the LCOS panel help to reduce the flicker noise to acceptable level. Another option is synchronization between the light source and the SLM at the frame rate.

OVT Display Driving The OVT technology, implemented in HOLOEYE's phase SLMs, uses a digital PWM technique, based on the idea, that a fast sequence of binary modulation can realize an almost analogous response, in particular, for a low bandwidth detector, for example, the human eye. In this way, LC microdisplays with binary modulating FLC material and MEMS technologies, such as digital light processing (DLP), are operated to deliver gray scale modulation. With nematic LC technology, we also have to consider the larger response time of this LC material leading to an almost analogous optical response because of the convolution of the digital pulse code and the LC response time. With OVT's technology [20], it is actually not the "pulse width" that is varied, whereas the gray scale encoding is done by the sequence of bits. The bits of the sequence have individual durations, selected from the set of fixed values. This pulse code modulation (PCM) is advantageous because of the bandwidth limitation and the digital nature of the drive concept. A typical PCM sequence consists of bits with different weights (duration), which are independently programmable, which is repeated every video frame. With this approach, the sequence design offers a lot of flexibility, enabling the microdisplay to be driven with different frame rates, color bit depth, and color frames. Here, a simple 10 bit interface could be used for RGB (sequentially reproduced red, green, and blue) $3:4:3$ bit depth system, which for technical applications shows a quite reasonable performance. For color systems with mainly monochromatic content (e.g., green) a $2:6:2$ bit depth can also be designed and programed.

The effective latency not only depends on the LC response time but also on the drive scheme. The analog drive is typically operated in an "unbuffered" mode, where the pixel data can be directly addressed to the microdisplay. The analog drive uses a progressive scan approach, where there is a continuous serial refresh of the pixel information.

However, the fast bit-plane addressing needs memory for storing the image content. With the incoming video information, the pulse sequence (encoding the gray value) for the individual pixel is written into a frame buffer. With the next frame, binary information for all pixels (the so-called bit planes) is transferred to all pixels at the same time. The individual gray values are realized with a sufficient number of bit planes.

Most digital drive schemes are designed for video applications, where the human eye is the detector. Any supermodulation (flicker) above 100 Hz is almost not noticeable. For applications using a pulsed light source, such as light-emitting diode (LED) or laser, the PWM of the digital display drive could interfere with the light source driving as it was already mentioned above. It is worth mentioning that the electronic addressing bandwidth is the key in defining the right addressing frequency.

1.2.2 Operation Modes

The use of LC materials in SLMs is based on their optical and electrical anisotropy. Typically, a thin layer of LC material can be described as a birefringent material with two refractive indices. The orientation of the index ellipsoid is dependent on the direction of the molecular axis. This orientation is determined by the alignment layers of the LC cell. The most important cases are twisted, parallel aligned (PA), and vertical aligned (VA) cells. In a twisted cell, the orientation of the molecules differs by typically 90° between the top and the bottom of the LC cell and is arranged in a helix-like structure in between. In both PA and VA cells, the alignment layers are parallel to each other, so the LC molecules have the same orientation.

The effect of an LC cell on a monochromatic, polarized light wave can be described by a Jones matrix. Here, "polarized" does not necessarily refer to linear polarization, but to a fully polarized state in the sense that the Stokes parameters [21] add up to 1. For PA and VA cells, the Jones matrix [22] is given by

$$W_{PN-LC} = \exp(-i\phi) \begin{pmatrix} \exp(-i\beta) & 0 \\ 0 & \exp(i\beta) \end{pmatrix} \qquad (1.1)$$

where the birefringence β and the phase offset ϕ are given by

$$\beta = (n_{eo} - n_o) \frac{\pi d}{\lambda} \qquad (1.2a)$$

$$\phi = (n_{eo} + n_o) \frac{\pi d}{\lambda} \qquad (1.2b)$$

where n_o and n_e are the ordinary and extraordinary indices of refraction of the LC material, respectively, d is the thickness of the cell, and λ is the wavelength of the light field. The possibility of changing the birefringence β as a function of the voltage applied to the LC cell makes this component a switchable waveplate.

The Jones matrix of a TN-LC cell is dependent on the physical parameters twist angle, α, front director orientation, ψ, and birefringence, β. It is given by [23]

$$W_{TN-LC}(f, h, g, j) = \exp -i\phi \begin{pmatrix} f - i \cdot g & h - i \cdot j \\ -h - i \cdot j & f + i \cdot g \end{pmatrix} \qquad (1.3)$$

using parameters f, g, h and j which fulfill $f^2 + g^2 + h^2 + j^2 = 1$ and can be calculated from the physical cell parameters as

$$f = \cos\gamma \cos\alpha + \frac{\alpha}{\gamma} \sin\gamma \sin\alpha \qquad (1.4\text{a})$$

$$h = \cos\gamma \sin\alpha - \frac{\alpha}{\gamma} \sin\gamma \cos\alpha \qquad (1.4\text{b})$$

$$g = \frac{\beta}{\gamma} \sin\gamma \cos(2\psi - \alpha) \qquad (1.4\text{c})$$

$$j = \frac{\beta}{\gamma} \sin\gamma \sin(2\psi - \alpha) \qquad (1.4\text{d})$$

where the parameter γ is given by

$$\gamma = \sqrt{\alpha^2 + \beta^2} \qquad (1.5)$$

The phase factor $\exp(-i\phi)$ can be neglected for most applications. For some commercially available TN-LC-cell-based microdisplays, the parameters α, β, and ψ were not made available by the manufacturers but could be retrieved from optical measurements [24].

It is obvious that the first case of a PA or VA cell is more convenient, at least if we are interested in creating pure phase modulation or polarization modulation.

1.2.2.1 Amplitude Modulation

This modulation type can be achieved with twisted, PA, as well as VA LC cells in a simple optical configuration. The incident light field should be linearly polarized, and after passing through the LC cell, it should be transmitted through a polarizer oriented perpendicular to the incident polarization. For an LCOS-SLM, a polarizing beam-splitter cube is suitable to obtain this. For PA or VA cells, the orientation of the optical axis should be rotated by 45° with respect to the incident polarization. For a phase delay of π introduced by the cell, the polarization appears to be rotated by 90°. The level of attenuation by the second polarizer can be tuned by applying an electric field to the cell, which leads to a change of the birefringence β.

This regime is normally used by projection displays. The three-panel architecture in front projectors can realize very high contrast ratio values above 70 000:1 (JVC DLA-RS series). To get the panels toward fast response time, the VAN mode is used, in which the LC material has a considerable pretilt (also in order to avoid reverse domains), which leads to a residual retardation effect in the dark state, and this needs to be compensated to achieve high contrast ratio. A variety of compensation technologies are available [25, 26], whereas these days, the preferred components are quarter-wave plates (QWPs) and specific retarder plates (see e.g., [22]).

1.2.2.2 Phase Modulation

We have seen that a PA LC cell can be interpreted as a switchable waveplate (Eqs. (1.1) and (1.2a)). It is evident that on transmission through a PA LC cell, light polarized linearly parallel to the extraordinary axis of the LC material is retarded as a function of the voltage-controlled birefringence β (this mode is also known as *ECB*). Therefore, such cell is a convenient phase-only modulator for linearly polarized light.

Obtaining phase-only modulation using twisted cells is significantly more complicated. It has been shown that there are elliptic polarization states that are only subject to phase modulation, with tolerable amplitude modulation introduced by a polarizer behind the SLM [27, 28]. This mode of operation is often referred to as phase-mostly operation. Creating the appropriate elliptic polarization requires a QWP, as well as the conversion to a linearly polarized state behind the SLM.

This regime has many applications ranging from wavefront control (with typically slowly varying phase functions) to dynamic computer-generated holography (with typically fast spatial variation of the phase function). In the latter case, a suitable algorithm for the creation of the phase-only hologram is required. Such computational algorithms have been adapted to match the particular needs of SLM applications in order to deal with fringe field effects, [29], optimize the speed of holographic computations [30], obtain a free choice of diffraction angles [31], and reduce the intensity fluctuations during frame-to-frame SLM update [32].

1.2.2.3 Polarization Modulation

When placing a waveplate of variable retardance ("WP1") between two QWPs, with the optical axes of these QWPs rotated by $+45°$ and $-45°$ with respect to the optical axis of WP1, the Jones matrix of the three waveplates together is a rotation matrix in which the rotation angle is given by the phase shift of WP1. Therefore it is possible to convert a phase modulating SLM into a 2D matrix of phase-rotating pixels by sandwiching it between two such QWPs. Interestingly, the polarization rotating feature is not dependent on the polarization of the incident light wave, only on the degree of polarization, which should be 1, and its wavelength. To give an example, this means that by addressing a vortex phase function to the SLM, a linearly polarized beam can be converted into a beam with radial or azimuthal polarization.

1.2.2.4 Complex-Valued Modulation

A desired mode of operation would be the ability to change both amplitude and phase of an incoming wavefront simultaneously, thereby creating a complex-valued transmittance of the SLM. From the discussion above it is evident that while of course phase and amplitude can be modulated by a single cell, the amplitude and phase values cannot be chosen independently, which makes complex-valued operation using a single cell almost unusable. Following are the options to represent a complex-valued transmittance by using special configurations.

- Stacking two LC cells: This involves sandwiching two LC layers and operating one in phase-only and the the other in amplitude-only mode. A transmissive SLM with a 1D array of pixels is manufactured by Cambridge Research and Instrumentation, but for 2D arrays of pixels the control of the two independent voltages required for each pixel has prevented the realization of such device.
- 4f imaging: In this option two SLM devices, including polarizer(s), with two lenses in 4f configuration are used and one device is operated in phase-only

and the other in amplitude-only mode (or vice versa),. Apart from the spatial frequency bandwidth limitation and inevitable optical aberrations, this would be the straightforward equivalent to physically stacking two LC cells.
- Macro pixel technique: In this, two adjacent pixels can be used to represent the real and imaginary parts of the desired complex transmittance [33] or can be combined together with help of additional thin components, which provides better quality of reconstruction in digital holography [34]; three amplitude pixels can very well represent a complex value [35], and using four pixels, it is possible even to use TN cells with mixed polarization and phase modulation [36].
- Spatial multiplexing: In the special case that the desired complex distribution is the sum of two phase functions it is an option to use a single phase-only SLM and to display each phase function in only every second of the available pixels. The pixel locations used for each phase function can simply be a random pattern [37].

1.2.3 Performance Evaluation

The evaluation of the SLM performance for digital holography applications typically comprises evaluation of phase response versus addressed value (linearity and maximal achievable phase delay) as well as phase response versus time, crosstalk versus spatial frequencies, flatness of the display, response times, and cross modulation.

Phase response can be measured using a Michelson interferometer or also with a common path interferometer [38]. Alternatively, it is possible to get a good estimation indirectly, using amplitude modulation mode. In this case, the incident polarization is oriented at 45° to the slow axis (in phase mode, it is parallel) and the analyzer is set perpendicular to the incident polarization. However, the advantage of a Michelson interferometer is that it is well suited for determining the flatness of the SLM at the same time (Figure 1.2).

Measurement of the phase response requires a high-speed detector, because the specific digital noise has relatively high frequency. One indirect method is to address diffraction gratings to the SLM and to observe intensity of the diffraction orders with a single detector (photodiode), attached to an oscilloscope. This method actually measures the diffraction efficiency over time. Another indirect method is to use an amplitude modulation mode, that is, to measure intensity noise over time. A more direct measurement can be performed using an interferometer with fast acquisition or with a stroboscopic technique [39].

Cross talk can be well evaluated with a simple approach, in which diffraction efficiency is measured versus addressed phase value for different spatial frequencies. The resulting curves can be used to derive actual phase modulation of the addressed grating versus addressed values [38].

Cross modulation is a residual amplitude modulation, which accompanies phase modulation. Residual amplitude modulation means that light coming from the SLM has a certain ellipticity in polarization state. This is simple to measure using an analyzer, which is oriented parallel to the incident polarization, and a power meter. More generally, as mentioned above, a full Jones matrix can be determined and taken into account later in the calculation of holograms [40, 41]. Response times can be basically evaluated with an oscilloscope and photodiode, observing diffraction order intensity of an addressed grating (the grating is then switched on and off). Alternatively, measurements in amplitude modulation mode as well as time-resolved phase measurements can be considered (as already mentioned).

Also of importance is the quality of anti-reflection (AR) coating of the front and back surfaces of the cover glass. Parasite reflections created by one of this surfaces can cause Fabry-Pérot-type interferences in the microdisplay.

For industrial and high-power applications, the behavior of SLM in response to temperature can be important. It is defined mainly by the dependence of the viscosity of the LC material on temperature. Higher temperatures usually decrease response times and increase phase modulation values at the cost of increasing temporal noise.

1.3
Some Applications of Spatial Light Modulators in Optical Imaging and Metrology

SLMs are used in a wide variety of applications mostly as a phase modulator, among which are measurement systems, microscopy, telecommunications, and digital holography. Meeser *et al.* [42] developed a holographic sensor, using an SLM to adapt the reference wave for different object positions as well as a flexible phase shifter. The SLM allows to switch between Fresnel and Fourier holograms and to determine accurately the phase distribution in the CCD plane using phase shifting algorithms (Figures 1.4 and 1.5). Another system from the same group works using

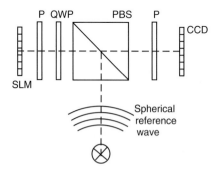

Figure 1.4 Schematic layout of the holographic sensor. P, polarizer; QWP, quarter-wave plate; PBS, polarizing beam splitter; CCD, detector. (Source: Adapted with permission from [42].)

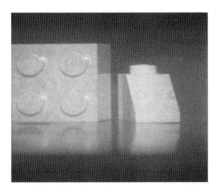

Figure 1.5 Numerical reconstruction of the hologram, captured with the holographic sensor. Source: With permission from [42].

the shearography principle, where the SLM performs the function of a shear [43], see also Chapter 17. Schaal et al. [44] proposed to use a phase SLM in a multipoint dynamic vibrometer, where the vibration is simultaneously measured in a freely selectable set of points. Baumbach et al. [45] demonstrated a digital holography technique, which replaces the holographic interferometry and implements SLM for achieving "analog" optical reconstruction of the master object.

Jenness [46] demonstrated the use of a phase SLM in a holographic lithography system (Figure 1.6), in which he successfully processed micropyramid structures (Figure 1.7) in photoresist [47, 48]. The main limitation of lithography applications with LCOS is the UV absorption of the LC cell, which does not allow the use of short wavelengths and hence affects the increase of the resolution of the lithography. Even though there exist transmission windows in the absorption spectrum [49, 50], the use of these windows does not look feasible. More flexible might be a combination of holographic lithography with two-photon lithography [51], which gives the possibility of working in the visible spectrum and then halving the wavelength due to the two-photon effect in the object plane. A similar principle was used in the realization of scanless two-photon microscopy by Nikolenko et al. [52]. Another maskless lithography application that utilizes polarization modulation in the SLM plane, which is imaged onto the object plane (a photoactive polymer film), is the fabrication of polarization computer-generated holograms (CGHs) [53].

The capability of LCOS to withstand high light intensities (that is actually inherited from its "front projection" origin), permits unusual LC applications. Kuang et al. [54] directly used a high-power laser to make microstructuring with laser ablation parallel at many points of an array. Nevertheless, the SLM function in this case is similar to holographic lithography described before.

Microscopic applications of SLMs have been found in the illumination or in the imaging light path. Examples for SLM usage in the illumination path are structured illumination microscopy [56–58], optical tweezing, and point spread function (PSF)-engineering (discussed below). Use of LC SLMs for optical tweezing was first proposed by Hayasaki et al. [59], followed by a number of publications of Dufresne and Grier [60], Reicherter et al. [61]. Figure 1.8a shows optical tweezer

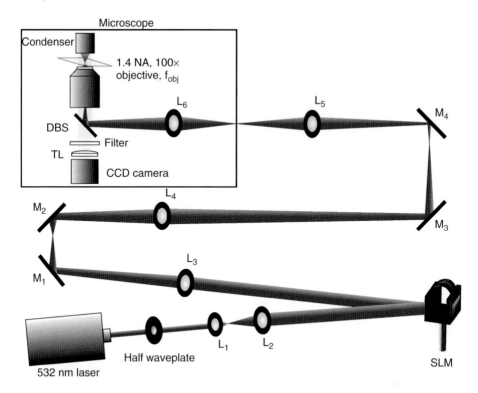

Figure 1.6 Schematic layout of the holographic lithography system. M1–M4, fold mirrors; L1 and L2, beam expander; L3–L6, relay lenses. DBS, dichroic beam splitter; TL, tube lens. Source: With permission from [46].

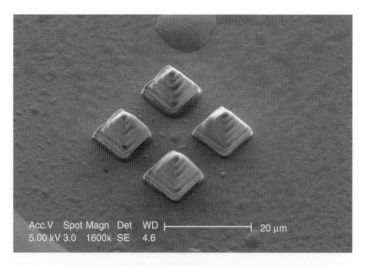

Figure 1.7 Processed structure. Source: With permission from Ref. [46].

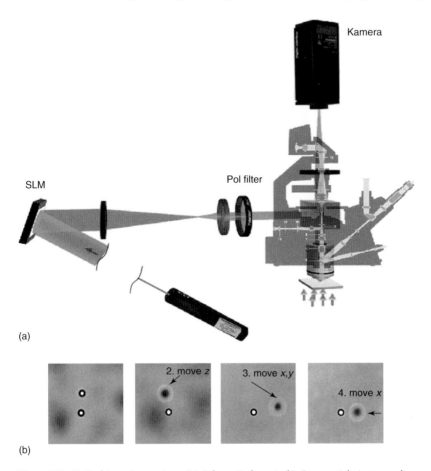

Figure 1.8 Optical tweezing system. (a) Schematic layout. (b) One particle is trapped in x-y-z relative to another particle. Source: With permission from Institut für Technische Optik, Stuttgart. (a) Taken from [55].

setup integrated in a Zeiss Axiovert 200. The SLM is telecentrically imaged into the pupil of the microobjective lens. The object is then illuminated with a pattern, which reconstructs a hologram addressed to the SLM. Moving of optical traps (i.e., addressing holograms to SLM) allows to move one or multiple objects in three dimensions (Figure 1.8b), see [62]. The state of the art in holographic optical tweezing is reviewed in Chapter 8.

For application in the imaging light path, LCOS SLMs were used in implementations of a phase contrast microscope [63–65], where addressing of different phase patterns to the SLM located in the Fourier plane allowed to get phase contrast, DIC (differential interference contrast), and dark field images in the same microscope (Figure 1.9). Figure 1.10 shows a phase bar structure imaged in bright field in panel (a) and SLM-based DIC in panels (b and c). The difference between two DIC images in Figure 1.10 represents two different periods of the gratings used for DIC.

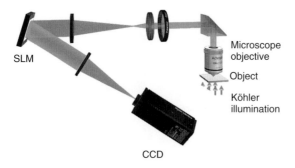

Figure 1.9 Schematic layout of the implementation of phase contrast. Source: With permission from [63].

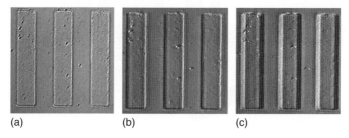

Figure 1.10 Phase object imaged with bright field (a) and DIC (b,c). Source: With permission from [63].

Another attractive application is coherence holography proposed by Takeda *et al.* [66], also see Chapter 11. This principle was originally proposed for microscopy [67], which allows to eliminate dispersion problems [68], see also Chapter 6. The same idea was recently applied for a synthetic aperture telescope [69].

One can find an excellent review of the SLM applications in microscopy in a recent paper by Maurer *et al.* [70].

Very promising are advances in application of phase SLMs in superresolution microscopy. Auksorius *et al.* [72] generated and controlled a doughnut-shaped PSF of a stimulated emission depletion (STED) beam with a phase SLM. They showed rapid switching between different STED imaging modes as well as correction of aberrations in the depletion path of the microscope and achieved significant resolution improvement compared with the standard confocal technique (Figure 1.11). The compensation of aberrations is especially important because of high sensitivity of the STED beam [71]. The standard confocal method can be also improved – as it was proposed recently by Mudry *et al.* [73], confocal microscopes could significantly increase resolution in z-axis, using an SLM for generation of two superimposed spots, which results in an isotropic diffraction-limited spot. These applications are very close to the more general method, called *SLM-based PSF engineering* [74].

PSF engineering is the most common use of LCOS SLMs operated in a mode to provide polarization modulation. Generation of polarization patterns [75–77] already found plenty of applications in microscopy, giving the possibility to reduce

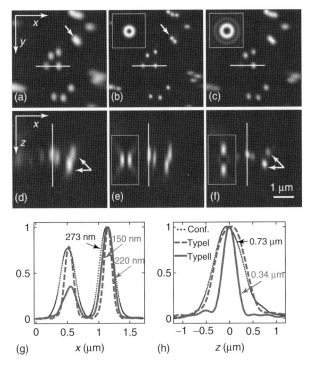

Figure 1.11 (a,b,c) Lateral and (d,e,f) axial fluorescence images of 200 nm beads with (a,d) confocal acquisition, (b,e) acquisition with a doughnut STED beam (type I), and (c,f) acquisition with a "bottle" STED beam (type II). (g,h) Normalized intensity line profiles of lateral and axial images, respectively, with specified FWHM. Insets show the corresponding PSFs. Note the differences in lateral and axial resolution between two STED imaging modes. (Source: With permission from [71]). (Please find a color version of this figure on the color plates.)

a spot size [78], to generate a doughnut, an optical bubble [79], or needle [80]. Generation of the doughnut could also be achieved with spiral phase pattern without additional polarization components (e.g., QWPs) as it was discussed earlier [72, 81]. Beversluis and Stranick [82] showed independent polarization and phase modulation with two SLMs, which succeeded in increasing the contrast of 300 nm polystyrene beads in coherent anti-Stokes Raman spectroscopy (CARS) images. They used independent phase and polarization modulation (Figure 1.12) for PSF engineering (Figure 1.13, see also [74]). SLM 1 in Figure 1.12 performs phase-only modulation of the light wave. SLM 2 with attached QWP is dedicated to polarization-only modulation. These two SLMs with optics form a "mode converter," so that finally phase- and polarization-modulated wavefronts are telecentrically imaged in the pupil plane of the microobjective lenses. The differences between linear, azimuthal, and radial polarizations showed in Figure 1.13 are obvious. Note also the changes in the Z-component of the field.

The methods and applications of PSF engineering were recently reviewed by Zhan [83].

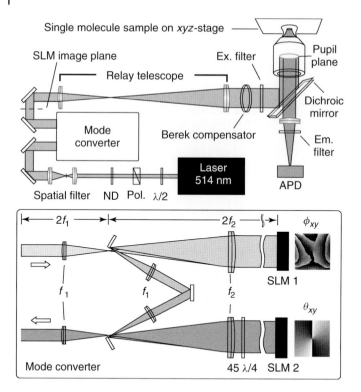

Figure 1.12 Schematic layout of the phase and polarization modulation in a PSF-engineered microscope. Source: With permission from [74].

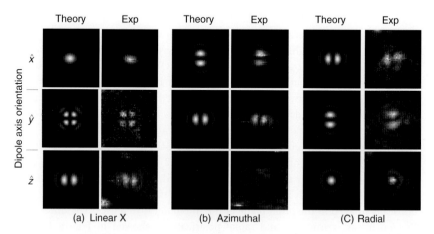

Figure 1.13 Theoretical and experimental images for (a) linear, (b) azimuthal, and (c) radial pupil polarizations (each image has 2 μm side). Source: With permission from [74].

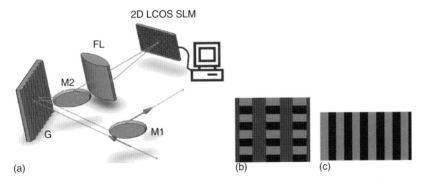

Figure 1.14 Femtosecond pulse shaper. (a) Schematic layout. G, diffraction grating; M1 and M2, mirrors; FL, Fourier lens. (b) Amplitude modulation with SLM. (c) Phase modulation with SLM. Source: With permission from [84].

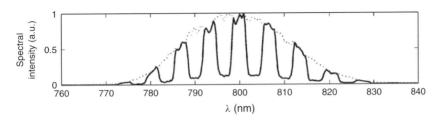

Figure 1.15 Amplitude-only modulation of the spectrum. Source: With permission from [84].

Frumker and Silberberg [84] demonstrated amplitude and phase shaping of femtosecond laser pulse using an LCOS phase modulator (Figure 1.14a). Here illumination femtosecond pulses are split into a spectrum with help of a grating G so that a cylindrical Fourier lens FL focuses each wavelength component onto the SLM in only one direction. The components stay in the same time spread in vertical direction. Then, writing a phase grating in vertical direction (Figure 1.14b) allows to modulate amplitude of the spectral components (Figure 1.15) independent of the phase modulation (the last one uses SLM in horizontal direction – Figure 1.14c). The phase modulation is observed as a correlation function showed at Figure 1.16. A principle capability of LCOS to work in the near-infrared range allowed to develop a telecommunication device based on a similar principle, combining optical switching [85] and pulse shaping capabilities of SLMs [86]. A similar approach was recently patented and commercialized by Finisar [87].

There are several attempts to use SLM for holographic reconstruction in visual systems. A comprehensive review of SLM-based holographic 3D displays is given by Onural et al. [89]. SeeReal Technologies demonstrated an 8 in. full color holographic projection 3D display (Figures 1.17 and 1.18) using an amplitude SLM and observation through a concave mirror – "display" [88]. The color was achieved using color field sequential technology. The holograms were a kind of Fresnel hologram, where complex values were converted to amplitude values using well-known

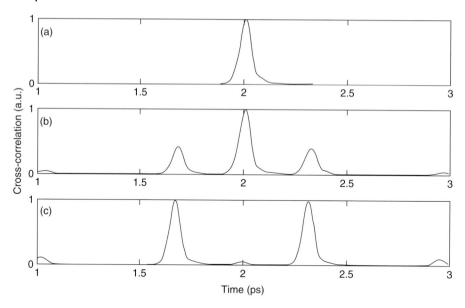

Figure 1.16 Cross-correlation. (a) No phase is applied. (b) Periodic phase-only modulation with binary modulation depth of $\pi/2$ and period 3.1 THz. (c) Periodic phase-only modulation with modulation depth of π. Source: With permission from [84].

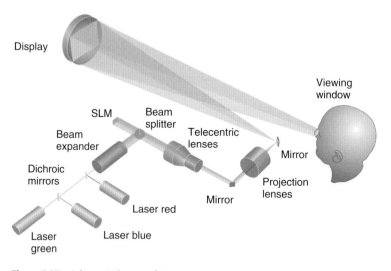

Figure 1.17 Schematic layout of 3D projection display. Source: With permission from [88].

Burckhardt or detour phase encoding [35]. A significant drawback of the system is its relatively low light efficiency, caused by the usage of an amplitude SLM. This is an unusual example of using amplitude modulation in digital holography, as 3D displays of this kind are thought to be realized with a phase modulator. However,

Figure 1.18 Photo of the display. (Source: With permission from [88]). (Please find a color version of this figure on the color plates.)

here the amplitude modulation provided good-quality reconstructions and helped to omit iterative calculations at the same time.

Light Blue Optics introduced full color 2D holographic projection unit with ferroelectric SLMs [90]. They demonstrated good quality of reconstructions, using high frame rate of ferroelectric SLMs to suppress perceivable speckle noise. Unfortunately, due to the nature of ferroelectric LCs, as it was already mentioned above, this approach shows relatively low light efficiency.

An implementation of the holographic visualization in the head mounted display (HMD) looks quite attractive as well. The features would be high brightness and capability of 3D representing information or objects overlapped with real objects (see-through). The reconstructed information can be adapted to the individual observer using the wavefront correction properties and thus will allow to compensate myopia or hyperopia, astigmatic errors of the eye, as well as other aberrations up to "supernormal vision" level as it was demonstrated by Liang *et al.* [91].

The basic layouts for the HMD can be different, e.g., projection of the SLM into the eye pupil, as it was already demonstrated for head-up display [92] or the projection of the Fourier transform of the SLM, as it was made in the projection holographic display of SeeReal Technologies (see earlier discussion). Main parameters of the HMD are the exit pupil and the FOV. Simple analysis shows that the product of the FOV and exit pupil in a digital holographic HMD is an invariant quantity that is proportional to the number of pixels at the SLM, regardless of the scheme used (as long as no pupil expander is used). This means that visual holographic systems are very critical to pixel count so that system becomes feasible only if the pixel amount exceeds 2–4 megapixels. Figures 1.19 and 1.20 show reconstructions from a prototype of a digital holographic see-through HMD, based on phase-only HOLOEYE SLM of the high-definition television (HDTV) resolution and a fiber-coupled red superluminescent diode. The numerical aperture of the camera was matched to that of the human eye. This ensures similarity of speckle characteristics between the captured pattern and the pattern perceived by the

(a)   (b)

Figure 1.19 Reconstruction of a hologram, captured at two reconstruction distances. The FOV is ≈6° in horizontal and 3.4° in vertical direction. The exit pupil is 12 mm. Note the difference in acuity of the background. (a) Fuel sign. (b) Stop sign.

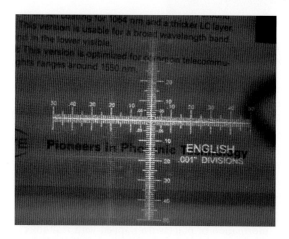

Figure 1.20 Reconstruction of a resolution test target. Field of view is ≈14°.

human eye [93]. The SLM is projected in the exit pupil plane with help of a telecentric system. The observer's eye makes a Fourier transform and gets an image at the retina. Figure 1.19a,b is the reconstruction from the same hologram. The hologram contains 3D information, which is reconstructed in two different planes. Another example shows approximately 14° field of view (Figure 1.20), where the eye can resolve a finer pattern.

The 3D display application, realized as an array of phase SLMs with Fresnel holograms, is given in Chapter 3. The readers are also referred to Chapter 2 for a general overview of the 3D display technologies.

In addition to the more established fields of applications such as microscopy, metrology, and holographic visualization, there is an apparently ever-growing number of other fields in which the potential benefits of using SLMs are expected. It can relate to optical processing of the information; for example, Marchese *et al.* [94] designed an SLM-based optronic processor for rapid processing of the synthetic-aperture radar (SAR) multilook images. As an adaptive optics example, Vellekoop and Mosk [95] focused light through a highly scattering medium. There are a number of scientific works, related to quantum optics. Becker *et al.* [96]

applied amplitude SLM to generate dark solitons in Bose–Einstein Condensate (BEC), whereas another group manipulated BEC with a ferroelectric phase SLM [97]. Bromberg et al. [98] implemented computational ghost imaging with only a single bucket detector, where the rotating diffuser was replaced with a phase SLM, thus allowing the computation of the propagating field and omitting the usage of the second high-resolution detector. Stütz et al. [99] generated, manipulated, and detected multidimensional entangled photons with a phase SLM. Following him Lima et al. [100] demonstrated manipulation with spatial qudit ("multilevel qubit") states.

1.4
Conclusion

LC-based SLMs have existed for the past 40 years and have been a basis for many studies in electro-optical effects. With the reflective LCOS microdisplay technology, one could realize components with parameters, unthought before (e.g., resolution, pixel size, fill factor, overall light efficiency, and driving electronics solutions).

Even though projection display and SLM target markets are quite different, trends in the LCOS microdisplay technology fit to some extent the requirements of SLM development. This fact, the accessibility of the technology, and the possibility of customization of the parameters also lead to considerably smaller investments. The background in consumer products also ensures achievement of stable, predictable, and high-performance commercial products with competitive pricing.

The availability of such SLMs has helped the scientific community to explore a wide range of potential applications. High-resolution devices were made possible, but for phase modulators, high diffraction efficiency along with high frame rate is still a challenge. Customization including SLM-specific backplane design together with mass-production-suitable production facilities opens the way for implementation in various commercial applications.

Current developments will bring 10 megapixel phase-only panels (e.g., 4160 × 2464 pixels) to the market. Pixel size will drop down further below 4 µm. The continuous progress in the development of driving electronics makes higher refresh rates available, together with a reduction of digital-specific noise. This will positively influence feasibility of SLMs for industrial applications.

Although LCOS SLMs have originally found their applications in scientific research, there is an increased interest in the commercial field and a transition from science to industry is expected in the near future.

References

1. Solgaard, O. (2009) *Photonic microsystems: micro and nanotechnology applied to optical devices and systems*, Springer.
2. Armitage, D., Underwood, I., and Wu, S.-T. (2006) *Introduction to Microdisplays*, John Wiley & Sons, Ltd, Chichester.
3. Knipe, R.L. (1996) Challenges of a digital micromirror device: modeling and design. SPIE Proceedings Micro-Optical

Technologies for Measurement, Sensors, and Microsystems, vol. 2783, pp. 135–145. doi: 10.1117/12.248483.
4. Amm, D.T. and Corrigan, R.W. (1998) Grating light valve technology: Update and novel applications. *SID Symp. Dig. Tech. Pap.*, **29** (1), 29–32. doi: 10.1889/1.1833752.
5. Schenk, H., Wolter, A., Dauderstaedt, U., Gehner, A., and Lakner, H. (2005) Micro-opto-electro-mechanical systems technology and its impact on photonic applications. *J. Microlith. Microfab. Microsyst.*, **4** (4), 041501. doi: 10.1117/1.2131824.
6. López, D., Aksyuk, V.A., Watson, G.P., Mansfield, W.M., Cirelli, R., Klemens, F., Pardo, F., Ferry, E., Miner, J., Sorsch, T.W., Peabody, M., Bower, J., Peabody, M., Pai, C.S., and Gates, J. (2007) Two-dimensional MEMS array for maskless lithography and wavefront modulation. *SPIE Proc.*, **6589**,. doi: 10.1117/12.724467.
7. Lapisa, M., Zimmer, F., Niklaus, F., Gehner, A., and Stemme, G. (2009) Cmos-integrable piston-type micro-mirror array for adaptive optics made of mono-crystalline silicon using 3-d integration. Proceedings of the IEEE International Conference on Micro Electro Mechanical Systems, pp. 1007–1010.
8. Doucet, M., Picard, F., Niall, K.K., and Jerominek, H. (2005) Operation modes for a linear array of optical flexible reflective analog modulators. SPIE Proceedings Cockpit and Future Displays for Defense and Security, vol. 5801, pp. 219–233. doi: 10.1117/12.604059.
9. Melcher, R.L. (2000) LCoS microdisplay technology and applications. *Inform. Display*, **16** (7), 20–23.
10. Fournier, F.R. and Rolland, J.P. (2008) Design methodology for high brightness projectors. *J. Display Technol.*, **4** (1), 86–91.
11. Sterling, R. (2008) JVC D-ILA high resolution, high contrast projectors and applications. *Proceedings of the 2008 Workshop on Immersive Projection Technologies/Emerging Display Technologies*, ACM, New York USA, pp. 10: 1–10:6.
12. Fujitsu Semiconductor Ltd (2007) Wafer foundry service. FIND Vol.25 No.1.
13. Neil Savage (2009) Digital spatial light modulators. *Nat. Photon.*, **3** (3), 170–172.
14. Cuypers, D., Van Doorselaer, G., Van Den Steen, J., De Smet, H., and Van Calster, A. (2002) Assembly of an XGA 0.9 LCOS display using inorganic alignment layers for VAN LC. Conf. Rec. Int. Disp. Res. Conf., Society for Information Display (SID), pp. 551–554.
15. Gandhi, J., Anderson, J.E., and Stefanov, M.E. (2001) Experimental comparison of contrast ratio vs. f/# for various reflective LCoS modes. *SID Symp. Dig. Tech. Pap.*, **32** (1), 326–329. doi: 10.1889/1.1831862.
16. Osten, S., Krüger, S., and Hermerschmidt, A. (2008) New HDTV (1920 × 1080) phase-only SLM, in *Adaptive Optics for Industry and Medicine* (ed. C. Dainty), SPIE, pp. 124–129.
17. Márquez, A., Iemmi, C., Moreno, I., Campos, J., and Yzuel, M. (2005) Anamorphic and spatial frequency dependent phase modulation on liquid crystal displays. optimization of the modulation diffraction efficiency. *Opt. Express*, **13** (6), 2111–2119. doi: 10.1364/OPEX.13.002111.
18. Karl Waterman, J. (2007) Display driver architecture for a liquid crystal display and method therefore. US Patent 7161570.
19. Worley, S.W. and Hong Chow, W. (2010) System and method for reducing inter-pixel distortion by dynamic redefenition of display segment boundaries. EP Patent 1093653 B1.
20. Spencer Worley, W., Lyle Hudson, E., and Hong Chow, W. (2008) Display with multiplexed pixels and driving methods. US Patent Application 2008/0225030.
21. Shurcliff, W.A. (1962) *Polarized Light, Production and Use*, Harvard University Press.

22. Robinson, M., Sharp, G., and Chen, J. (2005) *Polarization Engineering for LCD Projection*, John Wiley & Sons, Ltd, Chichester.
23. Yariv, A. and Yeh, P. (1984) *Optical Waves in Crystals*, John Wiley & Sons, Inc., New York.
24. Hermerschmidt, A., Quiram, S., Kallmeyer, F., and Eichler, H.J. (2007) Determination of the Jones matrix of an LC cell and derivation of the physical parameters of the LC molecules. *SPIE Proc*, **6587**,. doi: 10.1117/12.722895.
25. Duelli, M., Shemo, D.M., Hendrix, K.D., Ledeur, A., and Tan, K.L. (2005) High performance contrast enhancing films for VAN-mode LCOS panels. *SID Sympo. Dig. Tech. Pap.*, **36** (1), 892–895. doi: 10.1889/1.2036592.
26. Tan, K.L., Hendrix, K.D., Duelli, M., Shemo, D.M., Ledeur, A., Zieba, J., and Greenberg, M. (2005) Design and characterization of a compensator for high contrast LCoS projection systems. *SID Symp. Dig. Tech. Pap.*, **36** (1), 1810–1813. doi: 10.1889/1.2036370.
27. Davis, J.A., Moreno, I., and Tsai, P. (1998) Polarization eigenstates for twisted-nematic liquid-crystal displays. *Appl. Opt.*, **37** (5), 937–945.
28. Clemente, P., Durán, V., Martínez-León, Ll., Climent, V., Tajahuerce, E., and Lancis, J. (2008) Use of polar decomposition of Mueller matrices for optimizing the phase response of a liquid-crystal-on-silicon display. *Opt. Express*, **16** (3), 1965–1974.
29. Milewski, G., Engström, D., and Bengtsson, J. (2007) Diffractive optical elements designed for highly precise far-field generation in the presence of artifacts typical for pixelated spatial light modulators. *Appl. Opt.*, **46** (1), 95–105. doi: 10.1364/AO.46.000095.
30. Georgiou, A., Christmas, J., Collings, N., Moore, J., and Crossland, W.A. (2008) Aspects of hologram calculation for video frames. *J. Opt. A: Pure Appl. Op.*, **10**, 035302. doi: 10.1088/1464-4258/10/3/035302.
31. Hermerschmidt, A., Krüger, S., and Wernicke, G. (2007a) Binary diffractive beam splitters with arbitrary diffraction angles. *Opt. Lett.*, **32** (5), 448–450. doi: 10.1364/OL.32.000448.
32. Persson, M., Engström, D., Frank, A., Backsten, J., Bengtsson, J., and Goksör, M. (2010) Minimizing intensity fluctuations in dynamic holographic optical tweezers by restricted phase change. *Opt. Express*, **18** (11), 11250–11263. doi: 10.1364/OE.18.011250.
33. Birch, P.M., Young, R., Budgett, D., and Chatwin, C. (2000) Two-pixel computer-generated hologram with a zero-twist nematic liquid-crystal spatial light modulator. *Opt. Lett.*, **25** (14), 1013–1015. doi: 10.1364/OL.25.001013.
34. Fütterer, G., Leister, N., Häussler, R., and Lazarev, G. (2010) Three-dimensional light modulation arrangement for modulating a wave field having complex information. Patent Application WO 2010/149588 A1.
35. Burckhardt, C.B. (1970) A simplification of Lee's method of generating holograms by computer. *Appl. Opt.*, **9** (8), 1949–1949. doi: 10.1364/AO.9.001949.
36. van Putten, E.G., Vellekoop, I.M., and Mosk, A.P. (2008) Spatial amplitude and phase modulation using commercial twisted nematic LCDs. *Appl. Opt.*, **47** (12), 2076–2081. doi: 10.1364/AO.47.002076.
37. Rosen, J. and Brooker, G. (2007) Digital spatially incoherent Fresnel holography. *Opt. Lett.*, **32** (8), 912–914. doi: 10.1364/OL.32.000912.
38. Hermerschmidt, A., Osten, S., Krüger, S., and Blümel, T. (2007) Wave front generation using a phase-only modulating liquid-crystal-based micro-display with hdtv resolution. SPIE Proceedings, Adaptive Optics for Laser Systems and Other Applications, vol. 6584, p. 65840E. doi: 10.1117/12.722891.
39. Emery, Y., Cuche, E., Marquet, F., Aspert, N., Marquet, P., Kuhn, J., Colomb, T., Montfort, F., Charriere, F., Depeursinge, C., Debergh, P., and Conde, R. (2006) Digital holographic microscopy (dhm) for metrology and dynamic characterization of mems and

moems. *SPIE Proc.*, **6186**, 61860 N. doi: 10.1117/12.660029.

40. Kohler, C., Haist, T., Schwab, X., and Osten, W. (2008) Hologram optimization for slm-based reconstruction with regard to polarization effects. *Opt. Express*, **16** (19), 14853–14861. doi: 10.1364/OE.16.014853.

41. Kohler, C., Haist, T., and Osten, W. (2009) Model-free method for measuring the full Jones matrix of reflective liquid-crystal displays. *Opt. Eng.*, **48** (4), 044002. doi: 10.1117/1.3119309.

42. Meeser, T., von Kopylow, C., and Falldorf, C. (2010) Advanced digital lensless Fourier holography by means of a spatial light modulator. 3DTV-Conference: The True Vision - Capture, Transmission and Display of 3D Video (3DTV-CON), 2010, pp. 1–4, 79.

43. Falldorf, C., Klattenhoff, R., and Gesierich, A. (2009) Lateral shearing interferometer based on a spatial light modulator in the Fourier plane, in *Fringe 2009: 6th International Workshop on Advanced Optical Metrology* (eds W., Osten and M. Kujawinska), Springer Heidelberg, pp. 93–98.

44. Schaal, F., Warber, M., Rembe, C., Haist, T., and Osten, W. (2009) Dynamic multipoint vibrometry using spatial light modulators, in *Fringe 2009, 6th International Workshop on Advanced Optical Metrology* (eds M. Osten and W. Kujawinska), Springer, pp. 528–531.

45. Baumbach, T., Osten, W., von Kopylow, C., and Jüptner, W. (2006) Remote metrology by comparative digital holography. *Appl. Opt.*, **45** (5), 925–934. doi: 10.1364/AO.45.000925.

46. Jenness, N.J. (2009) Three-dimensional holographic lithography and manipulation using a spatial light modulator. PhD thesis. Duke University.

47. Jenness, N.J., Wulff, K.D., Johannes, M.S., Padgett, M.J., Cole, D.G., and Clark, R.L. (2008) Three-dimensional parallel holographic micropatterning using a spatial light modulator. *Opt. Express*, **16** (20), 15942–15948. doi: 10.1364/OE.16.015942.

48. Jenness, N.J., Hill, R.T., Hucknall, A., Chilkoti, A., and Clark, R.L. (2010) A versatile diffractive maskless lithography for single-shot and serial microfabrication. *Opt. Express*, **18** (11), 11754–11762. doi: 10.1364/OE.18.011754.

49. Li, J., Wen, C.-H., Gauza, S., Lu, R., and Wu, S.-T. (2005) Refractive indices of liquid crystals for display applications. *J. Display Technol.*, **1** (1), 51.

50. Gauza, S., Wen, C.-H., Tan, B., and Wu, S.-T. (2004) UV stable high birefringence liquid crystals. *Jpn. J. Appl. Phys.*, **43**, 7176–+. doi: 10.1143/JJAP.43.7176.

51. von Freymann, G., Ledermann, A., Thiel, M., Staude, I., Essig, S., Busch, K., and Wegener, M. (2010) Three-dimensional nanostructures for photonics. *Adv. Funct. Mater.*, **20** (7), 1038–1052. ISSN 1616-3028. doi: 10.1002/adfm.200901838.

52. Nikolenko, V., Watson, B.O., Araya, R., Woodruff, A., Peterka, D.S., and Yuste, R. (2008) SLM microscopy: scanless two-photon imaging and photostimulation with spatial light modulators. *Front Neural Circuits*, **2** (5), 1–14.

53. Fratz, M., Fischer, P., and Giel, D.M. (2009) Full phase and amplitude control in computer-generated holography. *Opt. Lett.*, **34** (23), 3659–3661. doi: 10.1364/OL.34.003659.

54. Kuang, Z., Perrie, W., Liu, D., Edwardson, S., Cheng, J., Dearden, G., and Watkins, K. (2009) Diffractive multi-beam surface microprocessing using 10 ps laser pulses. *Appl. Surf. Sci.*, **255** (22), 9040–9044. ISSN 0169-4332. doi: 10.1016/j.apsusc.2009.06.089.

55. Hermerschmidt, A., Krüger, S., Haist, T., Zwick, S., Warber, M., and Osten, W. (2007c) Holographic optical tweezers with real-time hologram calculation using a phase-only modulating LCOS-based SLM at 1064 nm. *SPIE Proc.*, **6905**,. doi: 10.1117/12.764649.

56. Hirvonen, L., Wicker, K., Mandula, O., and Heintzmann, R. (2009) Structured illumination microscopy of a living cell. *Eur. Biophys. J.*, **38**,

807–812. ISSN 0175-7571. doi: 10.1007/s00249-009-0501-6.

57. Chang, B.-J., Chou, L.-J., Chang, Y.-C., and Chiang, S.-Y. (2009) Isotropic image in structured illumination microscopy patterned with a spatial light modulator. *Opt. Express*, **17**, 14710–+. doi: 10.1364/OE.17.014710.

58. Heintzmann, R. (2003) Saturated patterned excitation microscopy with two-dimensional excitation patterns. *Micron*, **34** (9), 283–291. doi: 10.1016/S0968-4328(03)00053-2.

59. Hayasaki, Y., Sumi, S., Mutoh, K., Suzuki, S., Itoh, M., Yatagai, T., and Nishida, N. (1996) Optical manipulation of microparticles using diffractive optical elements, in *SPIE Proceedings*, vol. 2778, (eds J.-S. Chang, J.-H. Lee, S.-Y. Lee, and C.H. Nam), SPIE, pp. 229–+.

60. Dufresne, E.R. and Grier, D.G. (1998) Optical tweezer arrays and optical substrates created with diffractive optics. *Rev. Sci. Instrum.*, **69**, 1974–1977. doi: 10.1063/1.1148883.

61. Reicherter, M., Haist, T., Wagemann, E.U., and Tiziani, H.J. (1999) Optical particle trapping with computer-generated holograms written on a liquid-crystal display. *Opt. Lett.*, **24** (9), 608–610. doi: 10.1364/OL.24.000608.

62. Haist, T., Zwick, S., Warber, M., and Osten, W. (2006) Spatial light modulators - versatile tools for holography. *J. Hologr. Speckle*, **3** (12), 125–136. doi: doi:10.1166/jhs.2006.019.

63. Zwick, S., Warber, M., Gorski, W., Haist, T., and Osten, W. (2009) Flexible adaptive phase contrast methods using a spatial light modulator. Proceedings DGAO, vol. A3.

64. Situ, G., Warber, M., Pedrini, G., and Osten, W. (2010) Phase contrast enhancement in microscopy using spiral phase filtering. *Opt. Commun.*, **283** (7), 1273–1277. ISSN 0030-4018. doi: 10.1016/j.optcom.2009.11.084.

65. Warber, M., Zwick, S., Hasler, M., Haist, T., and Osten, W. (2009) SLM-based phase-contrast filtering for single and multiple image acquisition, in *SPIE Proceedings*, Optics and Photonics for Information Processing III, vol. **7442**, (K.M. Iftekharuddin and A.A.S. Awwal), SPIE. doi: 10.1117/12.825945.

66. Takeda, M., Wang, W., and Nai, D.N. (2009) Coherence holography: a thought on synthesis and analysis of optical coherence fields, in *Fringe 2009, 6th International Workshop on Advanced Optical Metrology* (eds M. Osten and W. Kujawinska), Springer, pp. 14–+.

67. Rosen, J. and Brooker, G. (2008) Non-scanning motionless fluorescence three-dimensional holographic microscopy. *Nat. Photonics*, **2**, 190–195. doi: 10.1038/nphoton.2007.300.

68. Brooker, G., Siegel, N., Wang, V., and Rosen, J. (2011) Optimal resolution in Fresnel incoherent correlation holographic fluorescence microscopy. *Opt. Express*, **19**, 5047–+. doi: 10.1364/OE.19.005047.

69. Katz, B. and Rosen, J. (2011) Could SAFE concept be applied for designing a new synthetic aperture telescope? *Opt. Express*, **19**, 4924–+. doi: 10.1364/OE.19.004924.

70. Maurer, C., Jesacher, A., Bernet, S., and Ritsch-Marte, M. (2011) What spatial light modulators can do for optical microscopy. *Laser & Photonics Rev.*, **5** (1), 81–101. ISSN 1863-8899. doi: 10.1002/lpor.200900047.

71. Auksorius, E. (2008) Multidimensional fluorescence imaging and super-resolution exploiting ultrafast laser and supercontinuum technology. PhD thesis. Imperial College London.

72. Auksorius, E., Boruah, B.R., Dunsby, C., Lanigan, P.M.P., Kennedy, G., Neil, M.A.A., and French, P.M.W. (2008) Stimulated emission depletion microscopy with a supercontinuum source and fluorescence lifetime imaging. *Opt. Lett.*, **33** (2), 113–115. doi: 10.1364/OL.33.000113.

73. Mudry, E., Le Moal, E., Ferrand, P., Chaumet, P.C., and Sentenac, A. (2010) Isotropic diffraction-limited focusing using a single objective lens. *Phys. Rev. Lett.*, **105** (20), 203903–+. doi: 10.1103/PhysRevLett.105.203903.

74. Beversluis, M.R., Novotny, L., and Stranick, S.J. (2006) Programmable vector point-spread function engineering. Opt. Express, **14** (7), 2650–2656. doi: 10.1364/OE.14.002650.
75. Chen, H., Zheng, Z., Zhang, B.-F., Ding, J., and Wang, H.-T. (2010) Polarization structuring of focused field through polarization-only modulation of incident beam. Opt. Lett., **35** (16), 2825–2827. doi: 10.1364/OL.35.002825.
76. Wang, X.-L., Ding, J., Ni, W.-J., Guo, C.-S., and Wang, H.-T. (2007) Generation of arbitrary vector beams with a spatial light modulator and a common path interferometric arrangement. Opt. Lett., **32** (24), 3549–3551. doi: 10.1364/OL.32.003549.
77. Chen, W. and Zhan, Q. (2010) Diffraction limited focusing with controllable arbitrary three-dimensional polarization. J. Opt., **12** (4), 045707.
78. Dorn, R., Quabis, S., and Leuchs, G. (2003) Sharper focus for a radially polarized light beam. Phys. Rev. Lett., **91** (23), 233901-+. doi: 10.1103/PhysRevLett.91.233901.
79. Chen, W. (2009) Focus engineering with spatially variant polarization for nanometer scale applications. PhD thesis. The School of Engineering of the University of Dayton.
80. Wang, H., Shi, L., Lukyanchuk, B., Sheppard, C., and Chong, C.T. (2008) Creation of a needle of longitudinally polarized light in vacuum using binary optics. Nat. Photonics, **2**, 501–505. doi: 10.1038/nphoton.2008.127.
81. Bashkansky, M., Park, D., and Fatemi, F.K. (2010) Azimuthally and radially polarized light with a nematic SLM. Opt. Express, **18** (1), 212–217. doi: 10.1364/OE.18.000212.
82. Beversluis, M.R. and Stranick, S.J. (2008) Enhanced contrast coherent anti-Stokes Raman scattering microscopy using annular phase masks. Appl. Phys. Lett., **93** (23), 231115-+. doi: 10.1063/1.3046719.
83. Zhan, Q. (2009) Cylindrical vector beams: from mathematical concepts to applications. Adv. Opt. Photon., **1** (1), 1–57. doi: 10.1364/AOP.1.000001.
84. Frumker, E. and Silberberg, Y. (2007) Phase and amplitude pulse shaping with two-dimensional phase-only spatial light modulators. J. Opt. Soc. Am. B, **24** (12), 2940–2947. doi: 10.1364/JOSAB.24.002940.
85. Kali, A., Tan, A., Gravey, P., Wolffer, N., Lelah, A., and Pincemin, E. (2003) Assessment of LCOS technology for the realization of scalable $n \times n$ optical switches. Proceedings of International Conference on Photonics in Switching (PS 2003).
86. Roelens, M.A., Bolger, J.A., Williams, D., and Eggleton, B.J. (2008) Multi-wavelength synchronous pulse burst generation with a wavelength selective switch. Opt. Express, **16** (14), 10152–10157. doi: 10.1364/OE.16.010152.
87. Frisken, S.J. (2009) Optical communications system. US Patent 7593608.
88. Leister, N., Schwerdtner, A., Fütterer, G., Buschbeck, S., Olaya, J.-C., and Flon, S. (2008) Full-color interactive holographic projection system for large 3D scene reconstruction. SPIE Proc., **6911**,. doi: 10.1117/12.761713.
89. Onural, L., Yaras, F., and Kang, H. (2010) Digital holographic three-dimensional video displays. Proc. IEEE, **99**,. doi: 10.1109/JPROC.2010.2098430.
90. Buckley, E. (2008) Holographic laser projection technology. SID Symp. Dig. Tech. Pap., **39** (1), 1074–1079. doi: 10.1889/1.3069321.
91. Liang, J., Williams, D.R., and Miller, D.T. (1997) Supernormal vision and high-resolution retinal imaging through adaptive optics. J. Opt. Soc. Am. A, **14** (11), 2884–2892. doi: 10.1364/JOSAA.14.002884.
92. Hendricks, U., Krüger, S., Brandt, P., Charle, H., and Sahlbom, D. (2008) Holographic information display. US Patent Application 2008/0192312 A1.
93. Goodman, J.W. (1976) Some fundamental properties of speckle. J. Opt. Soc. Am., **66** (11), 1145–1150. doi: 10.1364/JOSA.66.001145.
94. Marchese, L., Bourqui, P., Turgeon, S., Doucet, M., Vachon, C., Harnisch, B., Suess, M., Châteauneuf,

F., and Bergeron, A. (2010) A SAR multilook optronic processor for operational Earth monitoring applications, in *SPIE Proceedings*, SAR Image Analysis, Modeling, and Techniques X, vol. **7829**, (ed. C. Notarnicola), pp. 782904. doi: 10.1117/12.866986.
95. Vellekoop, I.M. and Mosk, A.P. (2007) Focusing coherent light through opaque strongly scattering media. *Opt. Lett.*, **32** (16), 2309–2311.
96. Becker, C., Stellmer, S., Soltan-Panahi, P., Dörscher, S., Baumert, M., Richter, E.-M., Kronjäger, J., Bongs, K., and Sengstock, K. (2008) Oscillations and interactions of dark and dark bright solitons in Bose Einstein condensates. *Nat. Phys.*, **4**, 496–501. doi: 10.1038/nphys962.
97. Boyer, V., Godun, R.M., Smirne, G., Cassettari, D., Chandrashekar, C.M., Deb, A.B., Laczik, Z.J., and Foot, C.J. (2006) Dynamic manipulation of Bose–Einstein condensates with a spatial light modulator. *Phys. Rev. A*, **73** (3), 031402–+. doi: 10.1103/PhysRevA.73.031402.
98. Bromberg, Y., Katz, O., and Silberberg, Y. (2009) Ghost imaging with a single detector. *Phys. Rev. A*, **79** (5), 053840. doi: 10.1103/PhysRevA.79.053840.
99. Stütz, M., Gröblacher, S., Jennewein, T., and Zeilinger, A. (2007) How to create and detect n-dimensional entangled photons with an active phase hologram. *Appl. Phys. Lett.*, **90** (26), 261114. doi: 10.1063/1.2752728.
100. Lima, G., Vargas, A., Neves, L., Guzmán, R., and Saavedra, C. (2009) Manipulating spatial qudit states with programmable optical devices. *Opt. Express*, **17** (13), 10688–10696. doi: 10.1364/OE.17.010688.

2
Three-Dimensional Display and Imaging: Status and Prospects

Byoungho Lee and Youngmin Kim

2.1
Introduction

Three-dimensional (3D) display and imaging are important topics in the display field because they are the only candidate for realizing natural views. Since Wheatstone's "mirror stereoscope" concept was revealed [1], the need for 3D display and imaging has been growing in importance and many different methods of 3D display and imaging have been announced, especially over the past few decades. Although there are some difficulties in capturing and displaying the tremendous amount of optical data required for high-quality 3D objects, 3D display and imaging have been researched with great enthusiasm. With the recent rapid advances in electronics and computer software technologies, 3D display and imaging can now be implemented with image pickup devices such as high-resolution cameras and high-resolution display devices [2]. Figure 2.1 shows the progress of the important technical developments related to 3D display and imaging.

The technologies and studies currently being pursued for 3D display and imaging can be broadly categorized into hardware systems, software systems, and human factors related to 3D display, although there are many different methods to realize 3D displays and some terminologies are not clear. The hardware systems can be categorized as follows: a stereoscopic display, an autostereoscopic display, a volumetric display, a holographic display, and so on. Table 2.1 shows 3D display technologies recently released by companies or institutes.

The software system, that is, content, is just as important as the hardware system. For the acquisition of 3D information to realize 3D contents, a method for extracting depth information for an object by detecting the disparity between elemental images is crucial. Although many 3D animations and movies are being produced, there are not yet enough for 3D broadcasting or wide use of 3D displays. It is possible to convert existing 2D movies to 3D contents [3, 4]. However, this requires adding intermediate-view information that was never there to begin with, so one of the candidate solutions is to generate intermediate-view images. This approach uses stereo or multiple cameras for depth extraction using stereo matching or uses a depth camera system such as ZCam [5].

Optical Imaging and Metrology: Advanced Technologies, First Edition.
Edited by Wolfgang Osten and Nadya Reingand.
© 2012 Wiley-VCH Verlag GmbH & Co. KGaA. Published 2012 by Wiley-VCH Verlag GmbH & Co. KGaA.

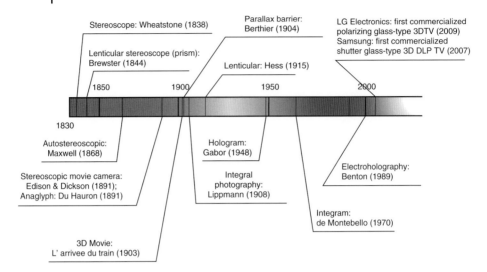

Figure 2.1 Progress of technical developments for 3D display and imaging. (Please find a color version of this figure on the color plates.)

Consumer acceptance and the response of the human visual system to 3D display deserve considerable thought. Many features of the 3D display itself influence 3D perception. Current commercial 3D display and imaging systems cannot provide all 3D perception cues. They cause visual fatigue such as eyestrain, eye discomfort, nausea, focusing difficulty, and headaches [6, 7]. The causes of visual fatigue in viewing 3D display are very diverse and complicated. Various studies and 3D display products have been investigated. This chapter explores recent 3D display techniques, content manipulation schemes, and visual fatigue related to human factors.

2.2
Present Status of 3D Displays

2.2.1
Hardware Systems

2.2.1.1 Stereoscopic Displays
Stereoscopic displays were recently introduced to the market by a few major display companies. These displays can be uncomfortable because of the glasses used, but they are easy to commercialize. In stereoscopic display, different perspective images are presented to the left and right eyes of the viewer. These images are combined to give 3D information to the brain; they only provide binocular disparity and convergence. Two representative methods have been used to carry the optical signals to the appropriate eyes in a stereoscopic display: polarizing glasses method (passive) and liquid crystal (LC) shutter glasses (active), as shown in Figure 2.2.

Table 2.1 Recently released 3D display techniques by companies or institutes.

Classification		Specifications	Companies or institutes
Stereoscopic display (requires glasses)	Polarizing glasses	84 in. ultrahigh-definition 3D theater (3840 × 2160)	LG Display
	LC shutter glasses	55 in. 240 Hz full HD Cross-talk-free 3D AMOLED TV (30 in. full HD)	Samsung Electronics
Autostereoscopic display (does not require glasses)	Parallax barrier	Time-multiplexing active parallax barrier	Samsung SDI
		Movable active parallax barrier	Seiko Japan
	Lenticular lens	Polarization-activated lens method	Ocuity
		Electric-field-driven lens (patterned electrode method)	LG Display
		Switchable LC lens (9 views, 640 × 320 3D resolution)	Samsung Electronics
		Liquid-crystal gradient index lens (9 views, 466 × 350 3D resolution)	Toshiba
	Integral imaging	Gradient index lens array	NHK
		Special color filter configuration (reduced moiré pattern)	Hitachi
		21 in. 9 parallax integral imaging (1280 × 720 3D resolution)	Toshiba
		Head tracking integral imaging/integral floating display	Seoul National University
	Volumetric display	Depth cube (20 LC projection screens with high-speed projector)	LightSpace Technologies
		Holografika (50.3 M pixels)	Holografika
		Interactive 360° light field display	University of Southern California
		Transpost (directional scenes from mirrors)	Hitachi
		Glass-free table-style 3D display (31 LCD projections)	NICT
		Rotating LED array (120M voxels)	Zhejiang University

(continued overleaf)

Table 2.1 (continued.)

Classification		Specifications	Companies or institutes
	Holographic display	Curved SLM arrays (9.4M pixels, field of view: 22.8°)	Seoul National University
		SeeReal (tracking)	SeeReal
		Refreshable holographic 3D display (45° viewing angle, 4 × 4 in. in size)	University of Arizona and *Nitto Denko Technical Corporation*
		Horizontal-parallax-only (HPO) hologram with high-speed SLM (field of view: 6.3°)	Tokyo University of Agriculture and Technology
	Other recent techniques	3D optical film (backlight multiplexing)	3M
		Depth-fused display	*Samsung*
		Super multiview projection (256 views)	Tokyo University of Agriculture and Technology

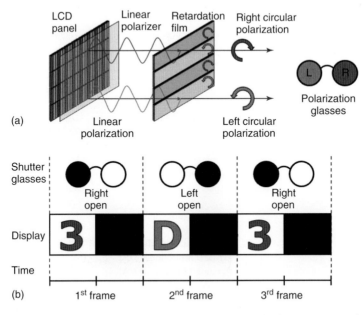

Figure 2.2 The principle of representative stereoscopic displays, using (a) polarizing glasses and (b) LC shutter glasses. (Please find a color version of this figure on the color plates.)

With the polarizing glasses method, two perspective images are displayed on the screen through orthogonal polarizing filters, and the viewer wears polarizing glasses. This method is very well suited for projection displays; most commercially available cinema systems use the polarization glasses method. To adapt the technique for monitors or flat panel display systems, a modification is made. Its basic principle is a spatial multiplexing of the display device, as shown in Figure 2.2a. This passive spatial multiplexing or polarization multiplexing method provides the viewer with only half of the display's vertical resolution because of the use of a single display panel and interleaved polarizers, as shown in Figure 2.2a. However, the cross talk between the perspective images is relatively low because of the good extinction ratio of polarizing glasses. This is a cost-effective way to realize a 3D display. Recently, a polarizing glasses system with high resolution and brightness, consisting of an active retarder synchronized with a display device, was demonstrated as shown in Figure 2.3 [8, 9]. It employs a time-multiplexing method for orthogonal polarization states.

The LC shutter glasses approach shown in Figure 2.2b is based on time multiplexing. LC glasses pass or block light in synchronization with the two images on the display device. Therefore, the frame rate of the glasses and the monitor or a projector must be doubled (more than 120 Hz) to allow alternation in blocking out the left and right eyes. The system provides the full resolution of the 2D display device and is simply implementable without modification of the conventional liquid crystal display (LCD) structure. However, the cross talk and ghost effect are relatively high because of the line-by-line driving architecture of LCD panels, which means that part of the left eye image may coexist with part of the right eye image, while the right eye is open and the left eye is shuttered.

The main issues for the stereoscopic displays are 3D resolution, brightness, and cross talk. A candidate solution to resolve these issues requires improved response times of each optical element, such as active retarders and LC shutters. In the near

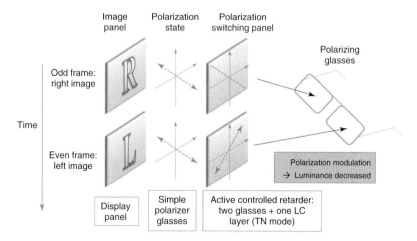

Figure 2.3 The principle of polarizing glasses displays using an active retarder.

Table 2.2 Comparison of the polarizing glasses method and the LC shutter glasses method for flat panel display.

Classification	3D resolution	Cross talk	Other factors	Candidate solution
Polarizing glasses	Half of 2D vertical resolution	Low	Low light efficiency	Fast polarization switching (120 Hz)
LC shutter glasses	Full 2D vertical resolution	High	High cost of LC shutter glasses	Fast LC switching (240 Hz)

future, the active matrix organic light-emitting diode (AMOLED) technology could be a candidate solution. Recently, a simultaneous emission driving scheme of 3D AMOLED TV for relieving cross talk was proposed [10]. Table 2.2 compares the main stereoscopic display methods for flat panel displays.

2.2.1.2 Autostereoscopic Displays

Current autostereoscopic display methods that are close to industrialization or have been commercialized can be classified into three methods: a parallax barrier method, a lenticular lens method, and integral imaging. The parallax barrier and lenticular lens methods have similar principles of implementing 3D imaging, as shown in Figure 2.4. The parallax barrier method was first proposed and

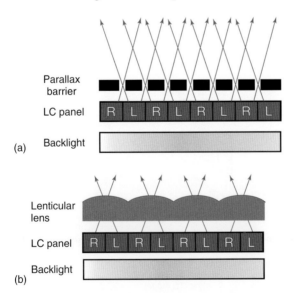

Figure 2.4 The principle and structure of (a) the parallax barrier method and (b) the lenticular lens method. (Please find a color version of this figure on the color plates.)

demonstrated by the French painter Bois-Clair in 1692. As a viewer walked by his paintings, they would appear to change from one picture to another. In 1904, Berthier came forward to claim credit for the first publication of the parallax barrier. The parallax barrier consists of black and white patterns. The white stripes are transparent, so light can be transmitted through them. In contrast, the black stripes block the light and act as optical barriers. Through the open windows, an observer at a specific position can recognize different perspective images. The parallax barrier system is relatively easy to fabricate and is low cost. Its key disadvantage is low 3D luminance. Therefore, in general, this method is used in mobile applications or 3D TVs having low or medium 3D image quality. A modified parallax barrier that is implemented by using a translucent LC parallax barrier is widely used for 2D/3D convertible displays. Here, the barrier is implemented with another LC layer on top of the display LC layer. Many display companies have developed prototypes of 2D/3D convertible displays based on the LC parallax barrier method [11–13]. Another approach is time multiplexing of the LC barrier, which is also called the *field-sequential method*. This method combines the time-multiplexing and the LC parallax barrier method. The resolution of the 3D image is doubled, while the system requires twice the driving speed of the conventional parallax barrier method. Table 2.3 shows recent products using the LC parallax barrier method by company.

In the lenticular lens method, the lenticular lens acts similarly to a parallax barrier. The lenticular lens method provides higher brightness than the parallax barrier method; however, it is not easy to manufacture large, uniform lenticular lenses. The lenticular lens of recent systems is commonly made of LC, using its birefringence characteristics. Moreover, it is possible to implement 2D/3D convertible displays with the LC active lenticular lens technique. There are three representative methods for achieving LC active lenticular lenses, as shown in Figure 2.5: a surface relief method, a polarization-activated lens method, and a patterned electrode method [14–16]. The surface relief method means that each LC lenticular lens cell can function as a lens or not by changing LC molecular

Table 2.3 Recent products using the LC parallax barrier method by company.

	LC parallax barrier method	Time-multiplexing method of the LC barrier
Company	Samsung SDI, LG Display, Sharp, Pavonine, Zalman, and others	Samsung Electronics and Samsung SDI
Properties	Resolution and brightness reduced by the number of views	Enhanced 3D resolution, local 3D display, high-speed switching, and moiré pattern decreased
Applications	Cellular phone, TV, and monitor	Monitor (prototype)

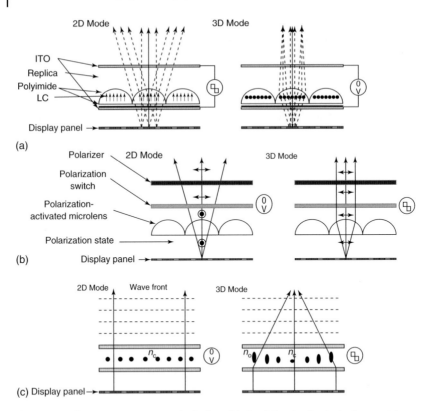

Figure 2.5 Three representative methods for achieving LC active lenticular lenses: (a) a surface relief method, (b) a polarization-activated lens method, and (c) a patterned electrode method. (Please find a color version of this figure on the color plates.)

directions electrically. Each LC active lens cell has the same refractive index as a replica outside the LC when a voltage is applied, so the LC lens cell does not perform a lens function. On the other hand, when no voltage is applied, there is an index difference between each LC lens cell and its replica, so the light (polarized along the LC molecular direction) from the display panel experiences different indexes when passing through the LC and the replica [14]. This method might have the disadvantages of complex fabrication and mismatch at the lens cell boundary. The polarization-activated lens method is based on a micro-optical lens array that gives different refractive indexes to light according to the polarization direction. This method might also suffer from complex fabrication and low light efficiency [15]. Recently, the patterned electrode method was proposed for a 2D/3D switching display [16]. The electric field at the lens edge is much stronger than that at the center of the lens cell. This nonuniform distribution of the electric field causes a nonuniform distribution of the tilt angle of the LC director, and the refractive index distribution changes spatially accordingly. This method requires a rather high operating voltage and causes high cross talk. Besides these methods, some other techniques have been suggested, such as the fast-switching Fresnel LC lens and LC graded-index lens [17, 18]. Fresnel lenses have been developed for wide viewing angle, low cost, and high resolution of 3D images; however, their system still remains as a prototype.

In 2D displays, the cross talk issue is not so severe because it matters only at the image boundaries with large gray scale differences. However, in 3D displays, much faster switching is required because pixel values are generally different for the left and right images. Therefore, key issues to commercialize autostereoscopic displays are high-definition display panels and high-speed driving circuits and optical elements. Companies are currently patenting many technologies for implementing 2D/3D switchable LC lenses.

Integral imaging, which was first introduced by Lippmann in 1908, is one of the most promising ways to realize 3D displays because it is possible to express both vertical and horizontal parallaxes with quasi-continuous viewpoints [19]. The structure and concept of the integral imaging system are shown in Figure 2.6. An integral imaging system is composed of a 2D elemental image set and a lens array. The lens array is a set of small lenses called *elemental lenses*. A 2D elemental image set is recorded on camera and displayed on either a film or a 2D display device [20].

The main trade-off for the full parallax is lower definition of the 3D image, and this is the main reason why developers in industry prefer horizontal-parallax-only (HPO) methods rather than integral imaging. Moreover, integral imaging suffers from inherent drawbacks such as limited viewing angle and limited image depth range. These limitations are related to the resolution of the display device, the gap between the display device and the 2D lens array, the number of lenses in the array, and the focal length of the lenses. Vigorous research has attempted to enhance the viewing quality of integral imaging [21–55]. Viewing angle enhancement is achieved by enlarging the area in the elemental image plane that corresponds to each elemental lens or by arranging it such that more elemental images can contribute to the integration of the 3D images [21–30]. Possible solutions for image depth range enhancement are to combine floating displays with the integral imaging

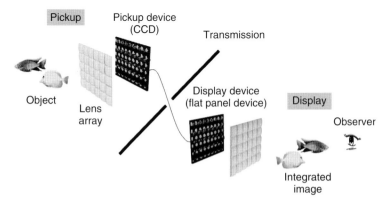

Figure 2.6 Concept of integral imaging.

and create multiple central depth planes (CDPs) [31–39]. The latter approach was accomplished by optical elements, such as a birefringent plate, multiple LCD panels, or multiple polymer-dispersed liquid crystal (PDLC) planes. Better viewing resolution of the 3D image was achieved by providing higher resolution display panels or using the spatial/temporal multiplexing method [40–46]. Recently, a 3D display system in which the 3D image casts its shadow or changes its body color differently according to the direction of illumination was developed [54]. Various 2D/3D convertible integral imaging methods have also been researched by using a pinhole array on a polarizer, light-emitting diode (LED) arrays, plastic fiber arrays, an organic light-emitting diode (OLED) panel, and an electroluminescence film [47–52]. The use of a boundary folding mirror scheme such as the Holografika system was proposed recently to enhance the uniformity of the angular resolution density by effectively using all elemental image regions without loss in integral imaging [53]. Table 2.4 summarizes recent technical methods for enhancing the viewing parameters of integral imaging.

2.2.1.3 Volumetric Displays

Volumetric displays and holographic displays, which are discussed in a later section, are fundamentally important technologies for displaying 3D objects because the object can be correctly reconstructed via the emission, scattering, or relaying of illumination from the 3D space with less eye fatigue. It can match accommodation and convergence cues and avoid the conflict generated by stereoscopic displays. Recently, with the growth of the display industry, LC spatial light modulators (SLMs) have been considered the most feasible devices for realizing volumetric and holographic displays [55].

However, it is often claimed that they possess some disadvantages, even though these volumetric displays match accommodation and convergence demands for objects. First, with some exceptions, volumetric displays are incapable of expressing occlusion and opacity because every point in a volumetric display is visible, even if a point on the rear side of an object should not be visible to the viewer. Second, their size is limited and they require bulky structures and sometimes coherent light sources. We also have limited content for volumetric displays, and, finally, they pose huge computational demands for data processing.

Various approaches have been proposed for realizing volumetric displays [56–63]. A three-color volumetric display based on two-step and two-frequency upconversion in rare-earth-doped heavy metal fluoride glass was proposed [63]. This method adopted infrared laser beams that intersect inside a transparent volume of optical material, but no further advances were achieved because of the use of laser sources and limited optical materials. Another method to implement volumetric displays was sweeping 2D slice images at high speed so that the observer can fuse each 2D slice image into a whole 3D image. The display surface can be diffusive or transmissive. The origin of this method was to use a vibrating mirror with a high-frame-rate 2D display in 1966. In 2004, the DepthCube 3D display for medical applications was introduced with the help of a high-speed digital light processing (DLP) projector and a stack of 20 polymer-stabilized cholesteric texture

Table 2.4 Recent technical methods for enhancing the viewing parameters of integral imaging.

Viewing parameters	Technologies	References
Viewing angle	Fresnel lens array with a small *f*-number	[21]
	Mechanical movement of lens array or barrier	[22, 23]
	Lens switching	[24]
	Orthogonal polarization switching	[25]
	Curved lens array and screen	[26]
	Embossed screen	[27]
	Multidirectional curved integral imaging with a large aperture lens	[28]
	Head tracking method	[29]
	A negative refractive index planoconcave lens array	[30]
Image depth range	Integral floating display	[31–35]
	Overlaid LCDs or active PDLC diffusers for multiple CDPs	[36, 37]
	Birefringent plate	[38]
	Dynamically moving elemental image plane	[39]
Viewing resolution	Spatial and/or temporal multiplexing	[40–43]
	Moving lenslet array	[22, 44]
	Rotating prism sheet	[45]
	Electrically controllable pinhole array	[46]
2D/3D convertible display	Electrically controllable diffuser (PDLC)	[47]
	Pinhole on a polarizer	[48]
	LED arrays, OLED panel, and electroluminescence film	[49–51]
	Plastic fiber arrays	[52]
Uniform angular resolution density of integral imaging	Boundary folding mirrors	[53]
6D integral imaging	Toward passive 6D reflectance field displays	[54]

screens [56]. This method has constantly been developed and has evolved toward several versions: an occlusion-capable volumetric 3D display using RGB image sources and a rotating diffuse projection screen (about 25 cm in diameter) [57], an interactive 360° light field display adopting a tracking technique for a 3D video teleconferencing system [58], a directional scene from multiple rotating mirrors [59], and a rotating LED array as a screen [62]. Another interesting approach to volumetric displays was to utilize multiple image sources and a directional screen [60, 61]. Two representative schemes of this approach are Holografika and

Table 2.5 Recent technical methods of volumetric displays.

Technology	Properties	References
DepthCube	A stack of 20 LC projection screens with a high-speed digital light processing projector Resolution: 1024 × 748 × 20 (without considering overlapped intensity modulation)	[56]
Perspecta	Occlusion-capable volumetric 3D display using RGB laser sources and a rotating diffuser screen Resolution: 768 × 768 × 198 slices	[57]
Interactive 360° light field display	High-speed projector, a rotating diffusing mirror, and head tracking Resolution: 768 × 768 Screen rotation frequency: 900–1200 rpm	[58]
Transpost: 360° viewable 3D display	Directional scenes from multiple rotating mirrors Frame rate: 60 Hz 24 views are provided by 24 mirrors	[59]
Holografika	Multiple SLMs and an asymmetric diffuser Resolution: 50.3 M pixels Viewing angle: 50–70°	[60]
Glass-free table-style 3D display	31 LCD microprojectors with a diffusing acrylic resin	[61]
Rotating LED arrays	Resolution: 320 × 256 color LED panel Rotating speed: 15 circles s^{-1}	[62]

a glass-free table-style 3D display. Both systems adopted multiple display modules, such as SLMs and an asymmetric diffuser. Table 2.5 describes the properties of each volumetric display.

2.2.1.4 Holographic Displays

Holographic displays allow both the amplitude and phase of the light scattered from an object to be recorded and the wave front from an object to be perfectly reconstructed. However, in general, they must use a coherent light source for recording in order to form interference patterns in the holographic materials. Temporal and spatial coherence is achieved when light waves have single frequency, and all parts of propagating waves of light are in phase. It is not easy to store and reconstruct images in real time because of bandwidth requirement and the laser safety requirements if coherent processing is made. The LC SLM is regarded as a feasible candidate device for embodying holographic displays. However, obstacles still remain in its implementation because the LC SLMs have limited bandwidth [64]. One idea to resolve this problem can be found in the holographic stereogram, which reduces the bandwidth of the hologram by defining the view scope by a viewing window.

An electroholography technique that can produce realistic 3D holographic images in real time was first demonstrated by several coresearchers [65–75]. A time-multiplexed scanned acousto-optics modulator (AOM) transferred the fringe pattern onto a beam of light [65]. Computer-generated holograms (CGHs) have been perceived as a promising technique for generating 3D images. However, more efficient and accurate algorithms for synthesizing CGHs are necessary because of the computational load. For example, a triangle-mesh model approach, which gives a reduced angular spectrum representation without interpolations, was derived [66]. Another approach for enhancing computational performance by harnessing the power of the graphics processing units (GPUs) uses a parallel computing architecture such as compute unified device architecture (CUDA) [67]. Creating and displaying a hologram for real-time reproduction can be achieved by merging the processing in a specialized one-unit system. A 3D image composed of 50 000 data points was displayed at a speed of 20 fps [68, 69]. Limiting the wave front information to the essential information, which was proposed by SeeReal, is also interesting [70–72]. They provide correct wave fronts only where they are actually required. To obtain a large number of data points and reduce the spatial bandwidth in SLMs, a dynamic holographic stereogram method using a curved array of SLMs was proposed [73]. Recently, to increase the image size and the viewing zone and angle of a hologram, an HPO hologram method using a high-speed SLM was also proposed [74].

Recording materials are critical for holographic displays. Photographic emulsions, photopolymers, and photorefractive polymers are representative materials for hologram recording [75]. Most of these holographic displays are write-only (or very slowly rewritable) systems. They have limited the use of holographic displays to static applications. To be used for updatable holography, the recording materials must satisfy certain conditions, such as high diffraction efficiency, high sensitivity, high spatial resolution, fast recording, and long storage time. Table 2.6 shows a comparison of holographic materials. Recently, a rewritable holographic stereogram that can generate 3D images in photorefractive polymers was introduced [76, 77]. It features a total horizontal viewing angle of $45°$ and a vertical view zone of $15°$ with uniform brightness 4×4 in. in size.

Table 2.6 Comparison between some holographic characteristics of representative recording materials.

Property	Bacteriorhodopsin	Azo dye	LC system	Photorefractive polymers
Efficiency	1–2%	<10%	1%	>90%
Response time	Seconds	Minutes	Hours	Seconds
Storage density (lines mm^{-1})	Thousands	Thousands	Hundreds	Thousands

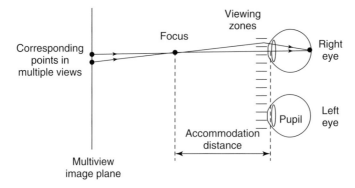

Figure 2.7 Super multiview display technique.

2.2.1.5 Recent 3D Display Techniques

Several techniques are receiving attention for novel 3D imaging. In one approach, a super multiview (SMV) display, as shown in Figure 2.7, has been developed as a glassless-type 3D display that is free from the visual fatigue caused by the accommodation–convergence conflict and provides smooth motion parallax. The SMV display generates dense viewing zones to make their pitch smaller than the pupil diameter of the human eye, and the retinal image changes smoothly with eye movement. However, a number of display sources are required for constructing an SMV display; an ultrahigh-definition display or tiled display techniques are sometimes necessary. Recently, to reduce the resolution requirement for the SMV display, a method was devised to generate viewing zones only around the left and right eyes [78].

An interesting pseudo-3D display technique, called depth-fused display (DFD), has been developed [79–82]. DFD perception occurs when two 2D images are displayed in a superimposed manner on two display devices, such as LCDs, transparent screens, or polarization-sensitive screens. Among the various versions of DFD, a technique using two fog screens and an optical tracking system provides a room-sized 3D display [82].

Unlike conventional display techniques that use a screen and display sources, a virtual retinal display projects an image directly onto the retina with an intensity-modulated light beam. By positioning the 3D image between the surface of a lens and its focal length, the 3D image can be magnified to occupy a virtual space. A challenge for the implementation of this method is to devise variable focal length lenses or deformable membrane mirrors with switching rates of the pixel. This approach is being studied by *OKO Technologies* [83].

2.2.2
Software Systems

For the acquisition of 3D information to realize 3D contents, a method for extracting depth information for an object by detecting the disparity between

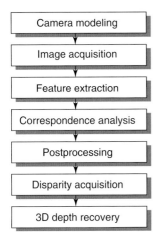

Figure 2.8 Steps for depth extraction from planner images.

elemental images is crucial. Although many 3D animations and movies have been produced, there are not yet enough for 3D broadcasting or wide use of 3D displays. Therefore, the 3D information about real objects should be acquired and manipulated. One issue is extracting spatial positions. When we know the spatial positions of the objects, we can manipulate them and combine other real objects with computer-generated 3D virtual objects. The 3D information can be extracted from two or more perspective view images by searching for the corresponding points in every perspective. However, the images seen by the two or more cameras must be calibrated, and a precise alignment is usually required [4]. Additional difficulties stem from the reconstruction of occluded regions of 3D objects. Optical flow with subpixel accuracy and point cloud representation may be one solution [84]. Figure 2.8 shows an example of the steps involved in depth extraction from planner images in integral imaging systems.

Another issue is generating an arbitrary 2D image. A representative method of 3D visualization is depth slice generation [85]. In an integral imaging system, it is also called computational integral imaging reconstruction (CIIR), as shown in Figure 2.9. CIIR computationally projects all elemental image points through the lens array to a given depth plane. The elemental image points that correspond to an object in the given depth plane are integrated at the same position, while the other elemental image points are dispersed in the given depth plane. Recently, arbitrary view image generation in perspective and orthographic geometry based on an integral imaging system with high resolution and a wide field of view was proposed [86].

The fundamental method for view image reconstruction is capturing 3D images using multiple cameras or a specialized depth camera for acquiring 3D depth information about the objects [87]. Each methodology employs either active or passive sensors such as a laser scanner, a time-of-flight (TOF) sensor, structured light, light field camera, or a depth camera (ZCam) [5]. With the acquisition of

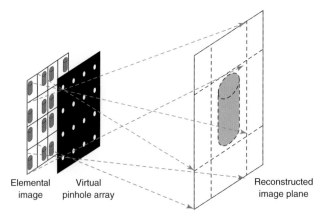

Figure 2.9 Concept of computational integral imaging reconstruction. (Please find a color version of this figure on the color plates.)

Figure 2.10 Demonstrations of *Microsoft*'s Kinect for the Xbox 360. (Please find a color version of this figure on the color plates.)

the 3D information for the objects, view images are generated by stereo cameras, multiple cameras, or a depth camera system. *Nintendo* Wii and Kinect for the Xbox by *Microsoft* are motion-sensing gadgets. In particular, Kinect includes a camera, an audio sensor, and motion-sensing capabilities that track 48 points of movement of the human body. Figure 2.10 shows Kinect's motion-sensing capabilities.

There is another possibility: converting existing 2D movies to 3D content [88]. With the recent rise of 3D display techniques, the desire for 3D displays has risen dramatically. However, because of the limited 3D content available at present, conversion of existing 2D content to 3D is necessary. However, this requires adding intermediate-view information that was never there to begin with, so one of the candidate solutions is an intermediate-view image generation method. For coordinating correct 3D information, defining a depth map of the objects and displacing each object according to the depth map are possible, and we hope this can be a real-time process. Recent 3D TV sets offer real-time conversion of 2D content to 3D with an integrated 2D/3D conversion chipset using motion accurate picture; examples are offered by *DDD* or *Trident Microsystems*. The technology requires further development to be commercialized.

2.3
The Human Visual System

With the progress of 3D displays, related visual fatigue has recently emerged as an important issue. Many stereoscopic films were produced in the early 1950s; however, public interest rapidly declined. One of the main reasons was associated with the visual fatigue caused by poor 3D effects. Since then, much effort has been devoted to clarifying the causes of visual fatigue related to 3D images.

Current commercial 3D display systems cannot provide all 3D perception cues. When a person observes an object, the observer uses both psychological and physiological cues to recognize depth information. The psychological cues include linear perspective, overlapping, shading, and texture gradient; these can be generated by computer graphics. The physiological cues can be classified into accommodation, binocular disparity, convergence, and motion parallax. Because a person's eyes are separated horizontally by about 6.5 cm on average, and each eye has its own perspective (variations in interpupillary distance for adult males range from 5.77 to 6.96 cm), slightly different images are projected onto the retina of each eye. The difference between a real object and a 3D image based on stereoscopic displays comes from artificial depth information. Although a 3D image using a binocular disparity is a plausible artifact, it does not provide all the depth cues that real objects do. Therefore, 3D image viewing may cause serious visual fatigue such as eyestrain, feeling of pressure in the eyes, eye ache, difficulty in focusing, and headaches [6].

Visual fatigue could be defined as a decline of the vision system. The causes of visual fatigue are diverse, and the issues are still highly controversial. We will not provide a comprehensive overview of all possible sources of viewer discomfort here, but some of the causes should be noted. These include keystone distortion, cross talk, cardboard effect, puppet theater effect, mismatch between accommodation and convergence, excessive binocular parallax, and individual differences in interpupillary distance. The literature mentions that the mismatch between accommodation and convergence is known to be the most significant factor for eye fatigue. In viewing real objects, both accommodation and convergence response are coupled reflexively [89]. On the other hand, in a binocular display system, the accommodative stimulus is fixed on the stereoscopic display, while the convergence stimulus varies depending on the screen disparity. This mismatch can induce visual fatigue; the situation is especially serious in near-range observation of 3D images. Depth perception at short distances may be affected by binocular disparity, accommodation, and convergence. The observation of 3D images can be mostly done within 2–3 m. Therefore, depth perception in stereoscopic displays could be strongly affected by mismatches inducing visual fatigue.

In spite of relatively vigorous studies of visual fatigue from stereoscopic displays, no definitive answers have been found regarding the causes leading to these symptoms. Two representative methods to measure visual fatigue are currently used: subjective and objective methods. A classic subjective assessment for visual fatigue is to use questionnaires as an evaluation indicator to determine the degree of visual fatigue. A simulator sickness questionnaire (SSQ), which was devised by Kennedy et al. [90], is a well-established tool for the investigation of the visual fatigue. Since then, many modified versions of the SSQ have been provided by researchers. However, these questionnaires covered a wide scope because they incorporated the fields of medical, mental, physiological, and social science. None ever became a standard protocol. Therefore, they are usually modified in accordance with experimental conditions and evaluation items. These subjective methods can be used in combination with objective assessment. Table 2.7 shows subjective methods using questionnaires to assess visual fatigue [91–96].

The most effective way to assess visual fatigue is to find biological indicators. Because visual fatigue is influenced by subjective experiences and is hard to assess, optometric devices or medical instruments are the potential candidates for assessing visual fatigue objectively. These devices are applied when measuring characteristic decline of the eye because of physical changes caused by visual fatigue. They could also be used in "before and after viewing stereoscopic displays" tests to determine the degree of visual fatigue. One effective way is to estimate the visual fatigue using analysis of comprehensive biological reactions [97–100]. However, these biological signals do not indicate visual fatigue directly and have complex patterns. Therefore, it is critical for better understanding of visual fatigue to analyze the captured biological reactions. Another proposal to assess visual fatigue is monitoring certain proteins in the body as markers for visual fatigue [101]. Another method is to assess direct indicators that appear in the fatigued eye. Optometric devices, such as refractometer and eye trackers, are typical examples of measurement tools [102–105]. However, experiments using these optometric devices are time consuming and have only been performed with a small number of subjects, as shown in Table 2.8.

2.4
Conclusion

It is very difficult for the current technology of 3D display and imaging to provide all the 3D depth perception cues. The computation of each display frame requires significantly more steps for a 3D display with wide viewing angle and high definition compared with a 2D display. Holography could be the final stage for natural 3D display. However, it is quite far from mass commercialization

Table 2.7 Recent subjective methods using questionnaires to assess the visual fatigue.

Experimental conditions	Features	Number of subjects	References
Head-mounted display (HMD)/TV	Questionnaire with 28 items	30	[91]
HMD	Visual reality symptom questionnaire (VRSQ)	16	[92]
HMD/video/projector	Questionnaire with 28 items based on SSQ Five factors for the assessment of visual fatigue	104	[91]
HDTV	Five subjective scores before and after reading stereoscopic test targets for about 1 h	6	[93]
Shaky video and stabilized video	Questionnaire with 28 items Five factors: eyestrain, general discomfort, nausea, focusing difficulty, and headaches	12	[91]
1 LCD projector/2 DLP projectors	Visual fatigue caused by color breakup Before and after 15 min of viewing movie	5	[94]
Visual display terminals (VDTs)	30 symptoms included in the questionnaire	30	[95]
Mobile stereoscopic game	SSQ and visual symptoms questionnaire (VSQ) Before and after playing a stereoscopic puzzle game for 40 min	20	[96]

yet because of the limitation in holographic display materials and demanding requirement in bandwidth. In addition, the barrier of significant computational time has prevented real-time natural 3D displays from becoming reality. No dominant technology in the commercial 3D display market has appeared. To process the high-density information of a 3D display in real time, the following issues must be considered: fast LC response time, fabrication cost, dynamic focus control of optical device, high-speed driving circuit for significant computational time, and high-definition display device. Three-dimensional displays are complex technologies that combine optical, display, and computer software technologies. Therefore, these technologies will require incorporated knowledge and fusion of the fields of optics, electronics, information processing, and psychological science.

Table 2.8 Recent objective methods to assess visual fatigue.

Measurement tools	Implementation tools	Number of subjects	References
Comprehensive biological signal analysis	EEG, MEG, EMG, ECG	3, 15, 52, 34	[97–100]
	Clinical demonstration using a protein (chromogranin A)	24	[101]
Direct indicator measurement	Refractometer	3, 9	[102, 103]
	Eye tracker	10, 47	[104, 105]

Acknowledgments

This work was supported by the National Research Foundation and the Ministry of Education, Science, and Technology of Korea through the Creative Research Initiative Program (2009-0063599).

References

1. Wheatstone, C. (1838) Contributions to the physiology of vision. Part the first. On some remarkable, and hitherto unobserved, phenomena of binocular vision. *Phil. Trans. R. Soc. Lond.*, **128**, 371–394.
2. Okano, F., Hoshino, H., Arai, J., and Yuyama, I. (1997) Real-time pickup method for a three-dimensional image based on integral photography. *Appl. Opt.*, **36** (7), 1598–1603.
3. Park, J.-H., Kim, Y., Kim, J., Min, S.-W., and Lee, B. (2004) Three-dimensional display scheme based on integral imaging with three-dimensional information processing. *Opt. Express*, **12** (24), 6020–6032.
4. Park, J.-H., Jung, S., Choi, H., Kim, Y., and Lee, B. (2004) Depth extraction by use of a rectangular lens array and one-dimensional elemental image modification. *Appl. Opt.*, **43** (25), 4882–4895.
5. Woods, A.J., Bolas, M.T., Merritt, J.O., and Benton, S.A. (eds) (2003) Depth keying. Proceedings of the SPIE: Stereoscopic Displays and Virtual Reality Systems X, May 30.
6. Howard, I.P. and Rogers, B.J. (1995) *Binocular Vision and Stereopsis*, Oxford University Press.
7. Lambooij, M., IJsselsteijn, W., Fortuin, M., and Heynderickx, I. (2009) Visual discomfort and visual fatigue of stereoscopic displays: a review. *J. Imaging Sci. Technol.*, **53** (3), 030201-14–030201-14.
8. Jung, S.-M., Park, J.-U., Lee, S.-C., Kim, W.-S., Yang, M.-S., Kang, I.-B., and Chung, I.-J. (2009) A novel polarizer glasses-type 3D display with an active retarder. Society for Information Display 2009 International Symposium (SID 2009) Digest of Technical Papers, San Antonio, May 31-June 5, 2009.
9. Kang, H., Roh, S.-D., Baik, I.-S., Jung, H.-J., Jeong, W.-N., Shin, J.-K., and Chung, I.-J. (2010) A novel polarizer glasses-type 3D displays with a patterned retarder. Society for Information Display 2010 International Symposium (SID 2010) Digest of Technical Papers, Seattle, May 23–28, 2010.
10. Lee, B.-W., Ji, I.-H., Han, S.-M., Sung, S.-D., Shin, K.-S., Lee, J.-D., Kim, B.H., Berkeley, B.H., and Kim, S.S. (2010) Novel simultaneous emission

driving scheme for crosstalk-free 3D AMOLED TV. Society for Information Display 2010 International Symposium (SID 2010) Digest of Technical Papers, Seattle, May 23–28, 2010.
11. Kang, H., Jang, M.K., Kim, K.J., Ahn, B.C., Yeo, S.D., Park, T.S., Jang, J.W., Lee, K.I., and Kim, S.T. (2006) The development of 42" 2D/3D switchable display. Proceedings of the 6th International Meeting on Information Display and The 5th International Display Manufacturing Conference (IMID/IDMC 2006), Daegu, August 22–25, 2006.
12. Lee, H.J., Nam, H., Lee, J.D., Jang, H.W., Song, M.S., Kim, B.S., Gu, J.S., Park, C.Y., and Choi, K.H. (2006) A high resolution autostereoscopic display employing a time division parallax barrier. Society for Information Display 2006 International Symposium (SID 2006) Digest of Technical Papers, San Francisco, June 4–9, 2006.
13. Hamagishi, G. (2009) Analysis and improvement of viewing conditions for two-view and multi-view 3D displays. Society for Information Display 2009 International Symposium (SID 2009) Digest of Technical Papers, San Antonio, May 31 – June 5, 2009.
14. de Zwart, S.T., IJzerman, W.L., Dekker, T., and Wolter, W.A.M. (2004) A 20-in. Switchable auto-stereoscopic 2D/3D display. Proceedings of the 11th International Display Workshops, Niigata, December 8–10, 2004.
15. Woodgate, G.J. and Harrold, J. (2005) A new architecture for high resolution autostereoscopic 2D/3D displays using free-standing liquid crystal microlenses. Society for Information Display 2005 International Symposium (SID 2005) Digest of Technical Papers, Boston, May 24–26, 2005.
16. Hong, H.-K., Jung, S.-M., Lee, B.-J., Im, H.-J., and Shin, H.-H. (2008) Autostereoscopic 2D/3D switching display using electric-field-driven LC lens (ELC lens). Society for Information Display 2008 International Symposium (SID 2008) Digest of Technical Papers, Los Angeles, May 18–23, 2008.
17. Chen, C.-W., Huang, Y.-C., and Huang, Y.-P. (2010) Fast switching Fresnel liquid crystal lens for autostereoscopic 2D/3D display. Society for Information Display 2010 International Symposium (SID 2010) Digest of Technical Papers, Seattle, May 23–28, 2010.
18. Takagi, A., Saishu, T., Kashiwagi, M., Taira, K., and Hirayama, Y. (2010) Autostereoscopic partial 2-D/3-D switchable display using liquid-crystal gradient index lens. Society for Information Display 2010 International Symposium (SID 2010) Digest of Technical Papers, Seattle, May 23–28, 2010.
19. Lippmann, G. (1908) La photographie integrale. *C. R. Acad., Sci.*, **146**, 446–451.
20. Lee, B., Park, J.-H., and Min, S.-W. (2006) in *Digital Holography and Three-Dimensional Display* (ed. T.-C. Poon), Springer, New York, pp. 333–378.
21. Woods, A.J., Bolas, M.T., Merritt, J.O., and Benton, S.A. (eds) (2001) Three-dimensional display system based on computer-generated integral photography. Proceedings of the SPIE, The 2001 Stereoscopic Displays and Applications Conference, Photonics West, San Jose, January 29, 2001.
22. Jang, J.-S. and Javidi, B. (2002) Improved viewing resolution of three-dimensional integral imaging by use of nonstationary micro-optics. *Opt. Lett.*, **27** (5), 324–326.
23. Choi, H., Min, S.-W., Jung, S., Park, J.-H., and Lee, B. (2003) Multiple-viewing-zone integral imaging using a dynamic barrier array for three-dimensional displays. *Opt. Express*, **11** (8), 927–932.
24. Lee, B., Jung, S., and Park, J.-H. (2002) Viewing-angle-enhanced integral imaging by lens switching. *Opt. Lett.*, **27** (10), 818–820.
25. Jung, S., Park, J.-H., Choi, H., and Lee, B. (2003) Wide-viewing integral three-dimensional imaging by use of orthogonal polarization switching. *Appl. Opt.*, **42** (14), 2513–2520.

26. Kim, Y., Park, J.-H., Min, S.-W., Jung, S., Choi, H., and Lee, B. (2005) Wide-viewing-angle integral three-dimensional imaging system by curving a screen and a lens array. *Appl. Opt.*, **44** (4), 546–552.
27. Min, S.-W., Kim, J., and Lee, B. (2004) Wide-viewing projection-type integral imaging system with an embossed screen. *Opt. Lett.*, **29** (20), 2420–2422.
28. Shin, D.-H., Lee, B., and Kim, E.-S. (2006) Multidirectional curved integral imaging with large depth by additional use of a large-aperture lens. *Appl. Opt.*, **45** (28), 7375–7381.
29. Park, G., Jung, J.-H., Hong, K., Kim, Y., Kim, Y.-H., Min, S.-W., and Lee, B. (2009) Multi-viewer tracking integral imaging system and its viewing zone analysis. *Opt. Express*, **17** (20), 17895–17908.
30. Kim, H., Hahn, J., and Lee, B. (2008) The use of a negative index planoconcave lens array for wide-viewing angle integral imaging. *Opt. Express*, **16** (26), 21865–21880.
31. Min, S.-W., Hahn, M., Kim, J., and Lee, B. (2005) Three-dimensional electro-floating display system using an integral imaging method. *Opt. Express*, **13** (12), 4358–4369.
32. Kim, J., Min, S.-W., Kim, Y., and Lee, B. (2008) Analysis on viewing characteristics of integral floating system. *Appl. Opt.*, **47** (19), D80–D86.
33. Kim, J., Min, S.-W., and Lee, B. (2009) Viewing window expansion of integral floating display. *Appl. Opt.*, **48** (5), 862–867.
34. Kim, J., Min, S.-W., and Lee, B. (2007) Viewing region maximization of an integral floating display through location adjustment of viewing window. *Opt. Express*, **15** (20), 13023–13034.
35. Kim, J., Min, S.-W., and Lee, B. (2008) Floated image mapping for integral floating display. *Opt. Express*, **16** (12), 8549–8556.
36. Kim, Y., Park, J.-H., Choi, H., Kim, J., Cho, S.-W., and Lee, B. (2006) Depth-enhanced three-dimensional integral imaging by use of multilayered display devices. *Appl. Opt.*, **45** (18), 4334–4343.
37. Kim, Y., Choi, H., Kim, J., Cho, S.-W., Kim, Y., Park, G., and Lee, B. (2007) Depth-enhanced integral imaging display system with electrically variable image planes using polymer-dispersed liquid-crystal layers. *Appl. Opt.*, **46** (18), 3766–3773.
38. Park, J.-H., Jung, S., Choi, H., and Lee, B. (2003) Integral imaging with multiple image planes using a uniaxial crystal plate. *Opt. Express*, **11** (16), 1862–1875.
39. Lee, B., Jung, S., Min, S.-W., and Park, J.-H. (2001) Three-dimensional display by use of integral photography with dynamically variable image planes. *Opt. Lett.*, **26** (19), 1481–1482.
40. Arai, J., Okui, M., Yamashita, T., and Okano, F. (2006) Integral three-dimensional television using a 2000-scanning-line video system. *Appl. Opt.*, **45** (8), 1704–1712.
41. Liao, H., Iwahara, M., Hata, N., and Dohi, T. (2004) Highquality integral videography using a multiprojector. *Opt. Express*, **12** (6), 1067–1076.
42. Liao, H., Iwahara, M., Koike, T., Hata, N., Sakuma, I., and Dohi, T. (2005) Scalable high-resolution integral videography autostereoscopic display with a seamless multiprojection system. *Appl. Opt.*, **44** (3), 305–315.
43. Kim, J., Kim, Y., Choi, H., Cho, S.-W., Kim, Y., Park, J., Park, G., Min, S.-W., and Lee, B. (2009) Implementation of polarization-multiplexed tiled projection integral imaging system. *J. Soc. Inform. Display*, **17** (5), 411–418.
44. Wang, X. and Hua, H. (2008) Theoretical analysis for integral imaging performance based on microscanning of a microlens array. *Opt. Lett.*, **33** (5), 449–451.
45. Liao, H., Dohi, T., and Iwahara, M. (2007) Improved viewing resolution of integral videography by use of rotated prism sheets. *Opt. Express*, **15** (8), 4814–4823.
46. Kim, Y., Kim, J., Kang, J.-M., Jung, J.-H., Choi, H., and Lee, B. (2007) Point light source integral imaging with improved resolution and viewing angle by the use of electrically movable

pinhole array. *Opt. Express*, **15** (26), 18253–18267.
47. Park, J.-H., Kim, H.-R., Kim, Y., Kim, J., Hong, J., Lee, S.-D., and Lee, B. (2004) Depth-enhanced three-dimensional-two-dimensional convertible display based on modified integral imaging. *Opt. Lett.*, **29** (23), 2734–2736.
48. Choi, H., Cho, S.-W., Kim, J., and Lee, B. (2006) A thin 3D-2D convertible integral imaging system using a pinhole array on a polarizer. *Opt. Express*, **14** (12), 5183–5190.
49. Cho, S.-W., Park, J.-H., Kim, Y., Choi, H., Kim, J., and Lee, B. (2006) Convertible two-dimensional-three-dimensional display using an LED array based on modified integral imaging. *Opt. Lett.*, **31** (19), 2852–2854.
50. Kim, Y., Kim, J., Kim, Y., Choi, H., Jung, J.-H., and Lee, B. (2008) Thin-type integral imaging method with an organic light emitting diode panel. *Appl. Opt.*, **47** (27), 4927–4934.
51. Jung, J.-H., Kim, Y., Kim, Y., Kim, J., Hong, K., and Lee, B. (2009) Integral imaging system using an electroluminescent film backlight for three-dimensional-two-dimensional convertibility and a curved structure. *Appl. Opt.*, **48** (5), 998–1007.
52. Kim, Y., Choi, H., Cho, S.-W., Kim, Y., Kim, J., Park, G., and Lee, B. (2007) Three-dimensional integral display using plastic optical fibers. *Appl. Opt.*, **46** (29), 7149–7154.
53. Hahn, J., Kim, Y., and Lee, B. (2009) Uniform angular resolution integral imaging display with boundary folding mirrors. *Appl. Opt.*, **48** (3), 504–511.
54. Fuchs, M., Raskar, R., Seidel, H.-P., and Lensch, H.P.A. (2008) Towards passive 6D reflectance field displays. *ACM Trans. Graph.*, **27** (3), 1–8.
55. Maeno, K., Fukaya, N., Nishikawa, O., Sato, K., and Honda, T. (1996) Electro-holographic display using 15mega pixels LCD. Proceedings of the SPIE, Practical Holography X, San Jose, January 29, 1996.
56. Sullivan, A. (2002) The DepthCube solid-state multi-planar volumetric display. Society for Information Display 2002 International Symposium (SID 2002) Digest of Technical Papers, Boston, May 19–24, 2009.
57. Cossairt, O.S., Napoli, J., Hill, S.L., dorval, R.K., and Favalora, G.E. (2007) Occlusion-capable multiview volumetric three-dimensional display. *Appl. Opt.*, **46** (8), 1244–1250.
58. Jones, A., McDowall, I., Yamada, H., Bolas, M., and Debevec, P. (2007) Rendering for an interactive 360° light field display. *ACM Trans. Graph.*, **26** (3), 1–8.
59. Otsuka, R., Hoshino, T., and Horry, Y. (2006) Transpost: 360°-viewable three dimensional display system. *IEEE Trans. Vis. Comput. Graph.*, **12** (2), 178–185.
60. Dingliana, J. and Ganovelli, F. (eds) (2005) A scalable hardware and software system for the holographic display of interactive graphics applications. Presented at the Eurographics 2005, Dublin, August 26 – September 2, 2005.
61. Yoshida, S., Yano, S., and Ando, H. (2010) Prototyping of glasses-free table-style 3D display for tabletop tasks. Society for Information Display 2010 International Symposium (SID 2010) Digest of Technical Papers, Seattle, May 23–28, 2010.
62. Wu, J., Yan, C., Xia, X., Hou, J., Li, H., and Liu, X. (2010) An analysis of image uniformity of three-dimensional image based on rotating LED array volumetric display system. Society for Information Display 2010 International Symposium (SID 2010) Digest of Technical Papers, Seattle, May 23–28, 2010.
63. Downing, E., Hesselink, L., Ralston, J., and Macfarlane, R. (1996) A three-color, solid-state, three-dimensional display. *Science*, **273** (5279), 1185–1189.
64. Jeong, T.H. and Bjelkhagen, H.I. (eds) (2004) Recent developments in computer-generated holography: toward a practical electroholography system for interactive 3D visualization. Proceedings of the SPIE, Practical Holography XVIII: Materials and Applications, San Jose, January 19, 2004.

65. Benton, S.A. (ed.) (1990) Electronic display system for computational holography. Proceedings of the SPIE, Practical Holography IV, Los Angeles, January 18, 1990.
66. Kim, H., Hahn, J., and Lee, B. (2008) Mathematical modeling of triangle-mesh-modeled three-dimensional surface objects for digital holography. *Appl. Opt.*, **47** (19), D117–D127.
67. Chen, R.H.-Y. and Wilkinson, T.D. (2009) Computer generated hologram from point cloud using graphics processor. *Appl. Opt.*, **48** (36), 6841–6850.
68. Ichihashi, Y., Masuda, N., Tsuge, M., Nakayama, H., Shiraki, A., Shimobaba, T., and Ito, T. (2009) One-unit system to reconstruct a 3-D movie at a video-rate via electroholography. *Opt. Express*, **17** (22), 19691–19697.
69. Ichihashi, Y., Nakayama, H., Ito, T., Masuda, N., Shimobaba, T., Shiraki, A., and Sugie, T. (2009) HORN-6 special purpose clustered computing system for electroholography. *Opt Express*, **17** (16), 13895–13903.
70. Schwerdtner, A., Leister, N., and Häussler, R. (2007) A new approach to electro-holography for TV and projection displays. Society for Information Display 2007 International Symposium (SID 2007) Digest of Technical Papers, California, May 20–25.
71. Bjelkhagen, H.I. and Kostuk, R.K. (eds) (2008) Large holographic displays for real-time applications. Proceedings of the SPIE, Practical Holography XXII: Materials and Applications, San Jose, January 20–23, 2008.
72. Javidi, B., Son, J.-Y., Thomas, J.T., and Desjardins, D.D. (eds) (2010) Generation, encoding and presentation of content on holographic displays in real time. Proceedings of the SPIE, Three-Dimensional Imaging, Visualization, and Display 2010 and Display Technologies and Applications for Defense, Security, and Avionics IV, Florida, April 6, 2010.
73. Hahn, J., Kim, H., Lim, Y., Park, G., and Lee, B. (2008) Wide viewing angle dynamic holographic stereogram with a curved array of spatial light modulators. *Opt. Express*, **16** (16), 12372–12386.
74. Takaki, Y. and Okada, N. (2009) Hologram generation by horizontal scanning of a high speed spatial light modulator. *Appl. Opt.*, **48** (17), 3255–3260.
75. Peyghambarian, N., Tay, S., Blanche, P.-A., Norwood, R., and Yamamoto, M. (2008) Rewritable holographic 3D displays. *Opt. Photonics News*, **19** (7), 22–27.
76. Tay, S., Blanche, P.-A., Voorakaranam, R., Tunc, A.V., Lin, W., Rokutanda, S., Gu, T., Flores, D., Wang, P., Li, G., St Hilaire, P., Thomas, J., Norwood, R.A., Yamamoto, M., and Peyghambarian, N. (2008) An updatable holographic three-dimensional display. *Nature*, **451**, 694–698.
77. Blanche, P.-A., Bablumian, A., Voorakaranam, R., Christenson, C., Lin, W., Gu, T., Flores, D., Wang, P., Hsieh, W.-Y., Kathaperumal, M., Rachwal, B., Siddiqui, O., Thomas, J., Norwood, R.A., Yamamoto, M., and Peyghambarian, N. (2010) Holographic three-dimensional telepresence using large-area photorefractive polymer. *Nature*, **468**, 80–83.
78. Takaki, Y., Tanaka, K., and Nakamura, J. (2011) Super multi-view display with a lower resolution flat-panel display. *Opt. Express*, **19** (5), 4129–4139.
79. Walton, E., Evans, A., Gay, G., Jacobs, A., Wynne-Powell, T., Bourhill, G., Gass, P., and Walton, H. (2009) Seeing depth from a single LCD. Society for Information Display 2009 International Symposium (SID 2009) Digest of Technical Papers, San Antonio, May 31–June 5, 2009.
80. Pham, D.-Q., Kim, N., Kwon, K.-C., Jung, J.-H., Hong, K., Lee, B., and Park, J.-H. (2010) Depth enhancement of integral imaging by using polymer-dispersed liquid-crystal films and dual-depth configuration. *Opt. Lett.*, **35** (18), 3135–3137.
81. Park, S.-G., Kim, J.-H., and Min, S.-W. (2011) Polarization distributed depth map for depth-fused

three-dimensional display. *Opt. Express*, **19** (5), 4316–4323.
82. Lee, C., DiVerdi, S., and Höllerer, T. (2009) Depth-fused 3D imagery on an immaterial display. *IEEE Trans. Vis. Comp. Graph.*, **15** (1), 20–33.
83. Schowengerdt, B.T. and Seibel, E.J. (2006) True 3D scanned voxel displays using single or multiple light sources. *J. Soc. Inform. Display*, **14** (2), 135–143.
84. Jung, J.-H., Hong, K., Park, G., Chung, I., Park, J.-H., and Lee, B. (2010) Reconstruction of three-dimensional occluded object using optical flow and triangular mesh reconstruction in integral imaging. *Opt. Express*, **18** (25), 26373–26387.
85. Hong, S., Jang, J., and Javidi, B. (2004) Three-dimensional volumetric object reconstruction using computational integral imaging. *Opt. Express*, **12** (3), 483–491.
86. Park, J.-H., Baasantseren, G., Kim, N., Park, G., Kang, J., and Lee, B. (2008) View image generation in perspective and orthographic projection geometry based on integral imaging. *Opt. Express*, **16** (12), 8800–8813.
87. Levoy, M. and Hanrahan, P. (1996) Light field rendering. Proceedings of SIGGRAPH '96 Proceedings of the 23rd Annual Conference on Computer Graphics and Interactive Technique, New Orleans, August 4–9, 1996.
88. Passalis, G., Sgouros, N., Athineos, S., and Theoharis, T. (2007) Enhanced reconstruction of three-dimensional shape and texture from integral photography images. *Appl. Opt.*, **46** (22), 5311–5320.
89. Irvin, I.M. (1970) *Clinical Refraction*, The Professional Press.
90. Kennedy, R.S., Lane, N.E., Berbaum, K.S., and Lilienthal, M.G. (1993) Simulator sickness questionnaire: an enhanced method for quantifying simulator sickness. *Int. J. Aviat. Psychol.*, **3** (3), 203–220.
91. Kuze, J. and Ukai, K. (2008) Subjective evaluation of visual fatigue caused by motion images. *Displays*, **29** (2), 159–166.
92. Ames, S.L., Wolffsohn, J.S., and Mcbrien, N.A. (2005) The development of a symptom questionnaire for assessing virtual reality viewing using a head-mounted display. *Optom. Vis. Sci.*, **82** (3), 168–176.
93. Yano, S., Ide, S., Mitsuhashi, T., and Thwaites, H. (2002) A study of visual fatigue and visual comfort for 3D HDTV/HDTV images. *Displays*, **23** (4), 191–201.
94. Ogata, M., Ukai, K., and Kawai, T. (2005) Visual fatigue in congenital nystagmus caused by viewing images of color sequential projectors. *J. Display Technol.*, **1** (2), 314–320.
95. Murata, K., Araki, S., Kawakami, N., Saito, Y., and Hino, E. (1991) Central nervous system effects and visual fatigue in VDT workers. *Int. Arch. Occup. Environ. Health*, **63** (2), 109–113.
96. Häkkinen, J., Pölönen, M., Takatalo, J., and Nyman, G. (2006) Simulator sickness in virtual display gaming: a comparison of stereoscopic and non-stereoscopic situations. Proceedings of the 8th Conference on Human-Computer Interaction with Mobile Devices and Services, Espoo, September 12–15, 2006.
97. Murata, K., Araki, S., Yokoyama, K., Yamashita, K., Okumatsu, T., and Sakou, S. (1991) Accumulation of VDT work-related visual fatigue assessed by visual evoked potential, near point distance and critical flicker fusion. *Ind. Health*, **34** (2), 61–69.
98. Yamamoto, S. and Matsuoka, S. (1990) Topographic EEG study of visual display terminal (VDT) performance with special reference to frontal midline theta waves. *Brain Topogr.*, **2** (4), 257–267.
99. Jap, B.T., Lal, S., Fischer, P., and Bekiaris, E. (2009) Using EEG spectral components to assess algorithms for detecting fatigue. *Expert Syst. Appl.*, **36** (2), 2352–2359.
100. Gao, F. and Zhu, X. (2009) Estimating VDT visual fatigue based on the features of ECG waveform. Proceedings of the 2009 International Workshop on Information Security and Application

(IWISA 2009), Qingdao, November 21–22, 2009.
101. Fujimoto, S., Nomura, M., Niki, M., Motoba, H., Ieishi, K., Mori, T., Ikefuji, H., and Ito, S. (2007) Evaluation of stress reactions during upper gastrointestinal endoscopy in elderly patients: assessment of mental stress using chromogranin A. *J. Med. Invest.*, **54**, 140–145.
102. Inoue, T. and Ohzu, H. (1997) Accommodation responses to stereoscopic three-dimensional display. *Appl. Opt.*, **36** (19), 4509–4515.
103. Shibata, T., Kawai, T., Ohta, K., Otsuki, M., Miyake, N., Yoshihara, Y., and Iwasaki, T. (2005) Stereoscopic 3-D display with optical correction for the reduction of the discrepancy between accommodation and convergence. *J. Soc. Inform. Display*, **13** (8), 665–671.
104. Miyao, M., Hacisalihzade, S.S., Allen, J.S., and Stark, L.W. (1989) Effects of VDT resolution on visual fatigue and readability: an eye movement approach. *Ergonomics*, **32** (6), 603–614.
105. Nguyen, H.T., Isaacowitz, D.M., and Rubin, P.A.D. (2009) Age- and fatigue-related markers of human faces: an eye-tracking study. *Ophthalmology*, **116** (2), 355–360.

3
Holographic Television: Status and Future
Małgorzata Kujawińska and Tomasz Kozacki

3.1
Introduction

The most futuristic vision of 3D television (3DTV) or video requires delivering of a ghostlike, high-quality optical replica of a moving object or scene, which is floating in space and can be viewed by an observer freely from different angles. This vision is often referred as True 3D [1]. However, it is very distinct from the recent developments in stereoscopic and autostereoscopic systems [2–4], which are based on a certain manipulation of two or more 2D images. These images, when converted by our brain, provide the depth information, allowing to some extent an immersive and presence feeling. Stereoscopic techniques, although relatively easy for commercial implementation, are far from the vision of true optical replicas and therefore have several drawbacks including a viewer's discomfort due to physical duplication of 2D light distribution that reaches his eyes. To describe this main disadvantage of stereoscopic systems the two depth cues that characterize the difference in spatial vision with eyes and on a stereoscopic display have to be discussed [5]. For normal viewing, an object is seen by both eyes. The eyes converge toward the object with a convergence angle α. The human vision system merges the two images seen by the eyes and deduces depth information. The convergence is one depth cue. The other depth cue is accommodation. The eye lens will focus on the object and thereby optimize the perceived contrast. Both depth cues provide the same depth information. Stereoscopic displays fail to provide a conclusive depth information. An autostereogram or a stereoscopic display provides two images with different perspective views. The eyes will have the correct convergence angle α as for the real object. However, the eye lenses will focus on the display plane as both images are displayed on the spatial light modulator (SLM) in the display plane. Therefore, there is a mismatch between the depth information from convergence and accommodation.

The most natural way to overcome this difficulty and to realize the postulate of True 3D video is application of holography [6, 7]. It is because the basic principle of holography ("holo" – whole, "graph" – record) involves recording of an entire optical field (complex amplitude) representing an object and later allows to recreate this optical field from the recording, in the absence of the original object. In this case, the situation of normal viewing also applies to holographic displays as they mimic a real existing object by reconstructing the light wave front that would be generated by an object.

The importance and new possibilities opened by holography in the creation of 3D videos and television had been recognized by Leith and Upatnieks already in 1962 [8]. However, because of several limitations connected with an analog but digital data capture and limited bandwidth of hologram transmission systems the real implementation of optical holography in holographic television and displays had not been realized. The information contained in the object beam is recorded in a fringe pattern (hologram), which is produced by interference between the object and reference beams, and therefore may be stored in the form of 2D images. Since the period of the fringes is inversely proportional to the angle formed by the object and reference beams, a process of recording and displaying wide viewing angle scenes requires materials or optoelectronic devices with a high spatial resolution (more than 2000 lines mm^{-1}). Also, the size (aperture) of the detector/display should be large enough to avoid the well-known "keyhole" problem, which limits the viewing zone of a reconstructed scene. This brings us to the most significant problem in 3D holographic TV or video: the required spatiotemporal bandwidth product of a typical 3D object and its hologram, which should be serviced by the entire chain of 3D image acquisition, encoding, transmission, processing, and finally, display, is extremely high and difficult to satisfy by the present technologies. The dormancy of the research works on holographic television and 3D holographic displays lasted until 1990 when high-resolution cameras and SLMs became available. The advances in matrix detectors, SLMs, as well as powerful computers and increasing bandwidth of telecommunication links have speeded up the works in both electroholographic videos and displays. These works had been enhanced by introducing both computer-generated holograms and digital holograms into multimedia applications [5]. At the moment, most of the electroholographic systems use computer-generated holograms created offline as the input data. Some of the systems provide a real-time computation of a hologram reconstructed in a display; however, in order to produce a realistic 3D scene the computational requirements are extremely high [9, 10].

In 1989, the first electroholographic video (Mark I) was developed at MIT Media Lab [11]. The display, based on an acousto-optic (AO) modulator and horizontal scanning by polygonal mirror, provided wide diffraction angle (15°), high diffraction efficiency, and spatial frequency (20 frames s^{-1}). The Mark system was in constant development over the past 20 years, and its last version (Mark III) uses as a modulator a surface acoustic wave (SAW) device and a holographic optical element instead of mirror-based scanning devices [12]. It provides reasonable viewing conditions: 24° view angle, 30 frames s^{-1} and 80 mm × 60 mm × 80 mm viewing volume [13]; however, AO devices are rarely used for 3D displays as they modulate the light in one direction and require scanning.

The most common 2D SLMs are liquid-crystal-based devices (liquid-crystal-based spatial light modulators (LCSLMs) [14] and digital-micromirror-based devices

(DMDs) [15]. DMDs are usually used for binary modulation, and they may result in strong speckle noise due to vibrations of micromirrors. LCSLMs are more suitable for electroholography because they provide multilevel phase or amplitude modulation, can be applied in transmission and reflection, and are easy to use. Therefore, even if the diffraction angle (connected with SLM pixel size) is not sufficient, many prototypes use various LCSLMs including electrically addressed LCSLMs working in transmission or reflection (liquid crystal on silicon, LCOS [16]) and optically addressed LCSLMs. Two main approaches to holographic displays using LCSLMs had been realized:

- Reconstruction of a small part of a wave front originating from an object (from a large volume of 3D scenes). Only two wave fronts from an object that actually hit the pupils of an observer (observer windows) are reconstructed. In this case, the requirements for LCSLM are much lower as the required diffraction angle is a fraction of a degree. Such approach had been introduced in SeeReal technologies [5]. Although the viewing angle for each eye is small, an eye tracking system provides the necessary flexibility. The observer window is shifted to the actual eye position by movement of a light source.
- Creating high-resolution, big aperture (with high pixel count) holographic display, which provides simultaneously total information about a displayed image within a wide viewing zone to multiple observers. The "active tiling system," developed by QuinetiQ [17], is a good example of such an approach. This system applies high-frame-rate electrically addressed SLMs and high-resolution optically addressed SLMs. It also uses spatial multiplexing method to obtain 3×8 billion pixels and provides full-parallax and full-color reconstruction. However, an enormous computational resource is required to generate a huge holographic pattern in real time [10].

This second approach is also realized by means of creating larger display through combining reflective type LCSLMs into a flat or curved panel. The most advanced systems were reported recently by the group from Korea [18], which proposed a kind of holographic stereogram with a curved array of LCoS SLMs. In the system, the viewing angle achieved is $22.8°$ and the loss of optical power is significantly reduced. Other advanced solutions using an array of LCoS SLM are reported by a group from Turkey, in which high-quality 3D images and videos are reconstructed from computer-generated holograms on holographic display with flat [19] and circular geometry [20, 21] with an increased viewing angle. All the above-mentioned works focus on holographic imaging of computer-generated holograms or stereograms, which have no restrictions and limitations connected with a data capture system. A big challenge still is to provide efficient digital holography approach from 3D object data capture to wide viewing angle display of this object and its viewing by an observer. Also, it is important to find a flexible method to couple a capture and display systems. In the following sections, after discussing the theoretical problems connected with space–bandwidth product in planar and circular holographic display configurations and its influence on visual perception, we present the full technology chain and main modules of the 3D holographic video system recently proposed in EU Real3D project [22]. On the basis of the results obtained in this system, we discuss the future prospects of 3D holographic video and TV.

3.2
The Concept of Holographic Television System

The full implementation of digital holography into 3D video or 3DTV systems requires development of a total chain of 3D acquisition, encoding, transmission, signal conversion, and display. Up to now, most works connected with holographic techniques have focused at the display end and the advanced systems had most often used computer-generated holograms as the input data [14, 23]. The better connection with the real world may be provided through digital holography, which, in this case, is the natural successor of digital 2D photography and video.

The first demonstration of a system fully based on digital holography was given nearly 20 years ago by Hashimoto and Morokawa [24]. The optical setup (Figure 3.1) consists of two modules: recording and reconstruction ones, both of them utilizing laser light. The collimated laser light is divided by a beam splitter. The reference beam strikes the CCD camera normal to the sensor surface. The object beam impinges at the object, from which a portion of the scattered beam passes through an imaging lens and creates at the CCD plane a real image of an object. The reference and object beams are collinear. They are combined on

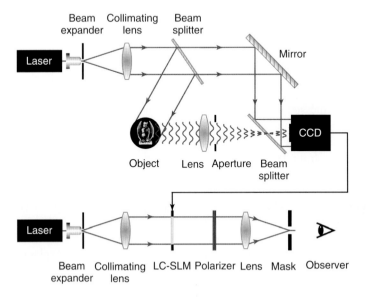

Figure 3.1 The optical scheme of electroholography including capture, transmission, and display based on a single CCD and transmissive LCSLM. (Please find a color version of this figure on the color plates.)

the CCD matrix and form an inline hologram (interference pattern), which is transmitted as a real-time electronic signal to the SLM. In the reconstruction module a special optical filtering at the plane of the mask is applied in order to remove a zero-order term from the reconstructed image. Even for the case of an inline hologram the system suffered strong restrictions of angular size and complexity of an object because of low number of relatively big pixels at both CCD camera and SLM.

Although the progress in matrix detectors and spatial modulators has been significant during the past two decades [25], we still lack an optoelectronic media with the parameters similar to holographic plates or film. The efficient way of overcoming this technological limitation is to form an extended digital holographic media by building a synthetic aperture hologram through using multiple cameras to capture optical field of an object and multiple SLMs to reconstruct its image. It is important to note that the available SLMs cannot be addressed with a complex amplitude, which fully represents optical wavefield, but they work in amplitude- or phase-only modes. In the discussion that follows, we assume implementation of phase SLMs as they provide much better diffraction efficiencies and allow to use directly the phase calculated at the hologram plane for an image reconstruction.

The general concept of such a system with multiple cameras around a 3D scene and the respective multiple SLMs display is shown in Figure 3.2 [21, 22, 26]. In the case of a static object the synthetic aperture hologram may be created through a sequential in-time capture of an optical field with a single camera, which is translated in space or alternatively by rotating or shifting an object in space.

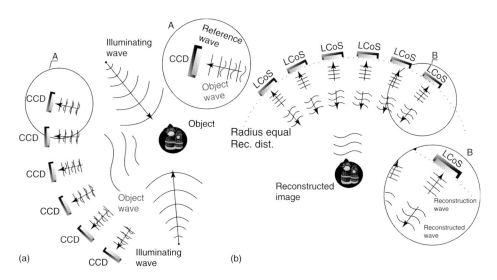

Figure 3.2 The general schemes of wide viewing angle capture and display systems based on (a) multiple CCDs and (b) multiple LCoS SLMs in circular configuration. (Please find a color version of this figure on the color plates.)

Before development of a 3D holographic TV or video system, there are several important questions to answer including

- how to arrange the multiple CCDs and SLMs;
- can a total optical field generated by a 3D scene be captured;
- what will be the effect of the gaps occurring in the captured or/and displayed optical field on the visual perception of a 3D scene,
- can we decouple the capture and display ends of 3DTV or 3D video.

The analysis of these problems is provided in Sections 3.3–3.6. The last question is strongly connected with the organization and capabilities of a holographic data processing module participating in the full technology chain of 3D holographic TV or 3D video.

An end-to-end 3DTV or video system has to include several functional modules starting from digital holograms capture, compression, transmission, and digital processing and ending with their optoelectronic reconstruction by means of a holographic display [22, 27–30]. Compression of a hologram involves transforming it so that a smaller sized description of that hologram can be found. The purpose of data compression (whether lossless or with loss) is to speed up the transfer of hologram data from the capture side to the display side. For transmission of holograms, a conventional transmission control protocol (TCP) socket connection (for example, FTP, HTTP) is envisaged at the moment; however, in future the broadband Internet connection will be required. The digital processing of holograms includes DC removal, twin removal/reduction, speckle reduction, resampling, tilting, defocusing, stitching, phase-only encoding, and other adaptations (size, bit depth) for the LCoS SLM display [16]. The compression needs to take place immediately before transmission, and the decompression, immediately after transmission. Figure 3.3 shows a general description of the full chain from capture to display [21] with different, possible options of the location of a digital processing module

1) after capture and before compression,
2) after decompression and before display, and
3) some before compression and some after decompression.

In option 1, the characteristics of the display (number of LCoSs, and position, orientation, and pixel properties of each) are known at the capture side. The holograms are processed to completion (to a set of phase-encoded holograms) before compression and transmission. This option allows simpler and more reliable hardware at the display side, which would just require copying the decompressed data directly to the appropriate LCoS. Also phase-only holograms at the correct bit

Figure 3.3 The chronological relationship between processing (P), compression (C), and decompression (D), and transmission; options (1), (2), (3) refer to the locations of the processing module.

depth for the LCoS pixels can be compressed and transmitted efficiently. However, this solution requires tight coupling between capture and display sides and the compression will be specific to this particular scenario.

In option 2, the raw data from the cameras is compressed and transmitted and all hologram processing takes place at the display side. This creates simpler and more reliable hardware on the capture side – no processing at all, just copying the data directly from the cameras to the compression/transmission routine, which works on real-valued nonnegative integers (as they are directly from a camera); digital holograms could be compressed quite efficiently. Also, the raw hologram data is available for alternative or complementary processing, and archival of raw data if required, after transmission. In this case, the chain as a whole, and the compression step itself, represents a more general solution with less coupling between capture and display, allowing simultaneous transmission to different hologram display technologies from the capture side. However, in the case of the techniques that require capturing of several holograms for one "display frame," this option is not efficient. This includes application of phase shifting digital holography (PSDH) [28] and multiframe speckle reduction techniques [29, 30]. Also, it cannot be applied in the case of the capture system that delivers complex amplitudes [26].

For option 3, some initial processing of holograms is performed before compression and transmission, and further processing is done at the display side after decompression. This option supports the concept of hologram capture camera in which the output is complex amplitude, not intensity. In such a case, the display-independent processing is done on the capture side (DC removal, twin removal, speckle reduction), and all display-specific processing is done on the display side. Some of the generalization properties from option 2 are retained because of the loose coupling with the display technology. However, the data will be complex valued (in particular, if the twin is removed) and the compression and transmission may be less efficient. It has the danger that even after compression, the data set will be larger than the original raw hologram data (for good visual quality data).

In conclusion, it seems that the option 2 is a most general solution for the end-to-end holographic real-time 3DTV or video and it enables to fulfill the requirement of decoupling the capture and display stages.

3.3
Holographic Display Configuration

3.3.1
Planar and Circular Configurations

SLMs available commercially have low resolution and small aperture. These features limit the viewing angle of holographic displays using SLMs. It mostly affects the user experience of observing optically reconstructed holographic images. To overcome this problem two major display configurations have been proposed: a flat geometry display [19, 21] and a circular one [20, 21]. The schemes of these configurations together with the important display features and their relation to

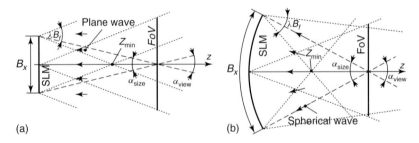

Figure 3.4 The general configuration of holographic displays: (a) planar and (b) circular; R – radius of curvature of display, the origin of z-axis is at the SLM plane.

SLM sampling parameters (sampling frequency $B_f = 1/\Delta$ (Δ – sampling pitch) and dimensions $B_x = N\Delta$ (N – number of pixels)) are presented in Figure 3.4. The circular configuration in Figure 3.4 is a purely theoretical one; SLM's pixels are distributed on a circle. In both configurations, the illumination waves are chosen to obtain maximum diffraction efficiency of SLM, that is, they impinge normal to the pixel area. For the planar configuration, it requires normal plane wave illumination, while for the spherical configuration, it applies a spherical wave with source at the modulator curvature. We mention here that the best reconstruction distance for circular configuration the curvature radius R. This will become apparent when we present Wigner distribution function (WDF) [31] analysis of both configurations (Section 3.2). The most important display futures are the minimum reconstruction distance and its field of view (FoV) and viewing angle.

Minimum reconstruction distance (z_{min}) for both configurations is a function of SLM Bx and B_f. For this distance, the entire SLM aperture can take part in generation of an axial point image. If we choose closer reconstruction distance the phase representation condition [32–34] will be weakened. The object wave will not be well approximated by the phase-only distribution. This will result in reduction of imaging quality. In the circular geometry this distance is additionally related to a modulator curvature. Therefore, the minimum reconstruction distance is closer for spherical configuration.

The distances are:

$$z_{min} = B_x \Delta \lambda^{-1} \quad \text{for planar configuration,}$$
$$z_{min} = B_x \left(\lambda B_f + B_x R^{-1}\right)^{-1} \quad \text{for circular configuration} \tag{3.1}$$

The size of a field of view (FoV – maximum image size) is defined as an area of the imaging space that it is free from aliasing images from neighboring diffraction orders. The situation is shown in Figure 3.5. When we consider zero-order image of letters CD-EFGH-IJ enclosed by the vertical lines, its minus-one and first-order diffraction images are just next to the zero-order one. In the areas of letters CD and IJ, there are losses of both intensity and resolution [35]. Applying this criteria the size of FoV for planar configuration is

$$\text{FoV} = \lambda z_r B_f \tag{3.2}$$

where z_r is a reconstruction distance.

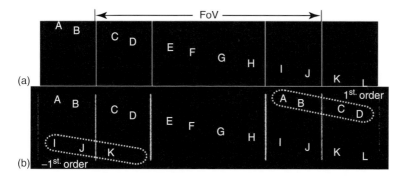

Figure 3.5 The image of a computer-generated object (a) and its optical reconstruction from computer-generated hologram performed in planar configuration of a holographic display (b); the vertical lines enclosing letters CD-EFGH-IJ define the size of FoV.

The size of FoV for circular display configuration is the same; the relation holds for paraxial optics approximation. On-axis viewing angle has its maximum value for z_{min} for both configurations. For planar configuration the maximum viewing angle equals the diffraction limiting angle $\alpha_{view} = \lambda B_f$. For display in circular configuration, the maximum viewing angle is the sum of angular size of the display at the center of curvature and the SLM diffractive angle $\alpha_{view} = \lambda B_f + \alpha_{size}$. However, when we consider optimal reconstruction distances for circular configuration (radius of curvature) and planar configuration ($z_r > z_{min}$), the viewing angle for both configuration is equals to α_{size}. At first glance the viewing angles are the same. To increase the viewing angle of planar configuration we increase its angular size by enlarging the modulator aperture. Unfortunately, there is a limit for a planar configuration. Let us take an exemplary reconstruction distance z_r. To increase the viewing angle we increase the SLM size, the minimum reconstruction distance grows linearly (Eq. (3.1)), and finally, we obtain the limit $z_r = z_{min}$ and $\alpha_{view} = \lambda B_f$. Further increase of the angular size of SLM will not extend the viewing angle. For circular configuration and optimal reconstruction distance R a circular shape gives rise to viewing angle only. Therefore, when we increase the angular size of the SLM, the viewing angle grows linearly: $\alpha_{view} = \alpha_{size}$.

3.3.2
Real Image Wigner Distribution Analysis of the Holographic Displays in Planar and Circular Configurations

In order to gain insight into the general display configurations presented in Figure 3.4, we use the phase space diagrams or "Wigner charts" [35, 36]. The Wigner distribution (WD) suites the analysis of holographic display since it is an analysis in both space and spatial frequency coordinates. There are both angular and spatial information describing a reconstructed image plane. This plane view representation of the WDF allows us to derive first-order approximations to the chosen important properties of the analyzed configurations: FoV, viewing angle, image resolution, and size of a perceived image. The analysis provides more general results; we can understand variation of display parameters such as viewing angle across FoV or size of a perceived image with an eye movement. We assume that the paraxial approximation can be used to describe propagation from the object plane to the plane of the SLM. Under this approximation, the WDF of the optical signal at two parallel planes is mapped as

$$W(x,f) \to W(x - \lambda z f, f) \tag{3.3}$$

3.3.2.1 Planar Configuration

Figure 3.6 illustrates WDF graphical representation of holographic signal bounds for three different planes $z = 0, z = z_{min}$, and $z = 3z_{min}$. SLM is limited in both frequency and spatial extents, WDF of light leaving SLM (zero diffraction order) is limited by the rectangle $B_x \times B_f$. In Figure 3.6b,c, the shaded areas represent energy of the signal that can contribute to a holographic image. The figures show utilization of image space–bandwidth product [32] in relation to a reconstruction distance. For the case presented in Figure 3.6b, 75% of the energy contributes to a holographic image, whereas for the next case (Figure 3.6c), it is 92%. Moreover, for the case in Figure 3.6b, within the entire FoV the reconstructed image intensity and resolution linearly decrease with an increase of $|x|$. For these reasons bigger reconstruction distances are preferable for planar configuration. Besides, it is very difficult to design digital holographic capture setup for distances corresponding to z_{min}, or close. For larger reconstruction distances (Figure 3.6c), in the central FoV region we have constant resolution $B_x/\lambda z$ with carrier frequency $-x/\lambda z$; viewing angle is then spatially dependent. Therefore, we calculate bounds of viewing angle: maximum α_{view}^+ and minimum α_{view}^-

$$\alpha_{view}^{\pm} = B_x z^{-1} \pm x z^{-1} \tag{3.4}$$

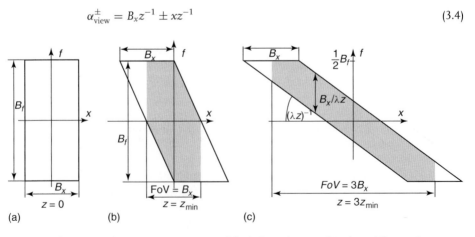

Figure 3.6 The WDF representation of the holographic signal in three different planes (a) $z = 0$, (b) $z = z_{min}$, and (c) $z = 3z_{min}$ for planar configuration.

where

$$\alpha_{\text{view}} = \alpha_{\text{view}}^+ - \alpha_{\text{view}}^- = B_x z^{-1}$$

3.3.2.2 Circular Configuration

We now turn our attention to display with spherical shape of SLM (Figure 3.4b). The bounds of WDF of optical signal reproduced with spherical SLM is presented in Figure 3.7. The charts visualize WDFs of holographic signal for the SLM plane and the plane of reconstruction. The reconstruction distance equals the radius of SLM curvature R. The WDF representation for the SLM plane ($z = 0$) is tilted according to inclination angle λR^{-1}. This is an expected feature given by a circular SLM. The tilted dashed line crossing in the middle of the WDF plot represents a spherical beam generated in the display replying constant phase signal. Locally, at any point of SLM we can change phase in the range of SLM resolution $(-\frac{1}{2}B_f, \frac{1}{2}B_f)$ relative to the local shape of SLM defined by a carrier frequency $-x(\lambda R)^{-1}$.

At the image plane ($z = R$) the effect of change of WDF shape can be illustrated by focusing by the SLM constant phase signal (dashed line). In the circular configuration, the entire SLM bandwidth is utilized (WDF shaded area), while (as shown in Figure 3.6) in the planar configuration there are some losses. For circular configuration WDF has a shape stretched in frequency (high resolution). Figure 3.7 illustrates the configuration in which most of the resolution is entirely defined by the SLM shape, and SLM bandwidth is used to change location of a point image only (Fourier configuration). When we apply a linear phase distribution onto spherical SLM we obtain a point image. When we replay a plane wave of maximum (minimum) SLM frequency we get marginal point images at $x = \pm \text{FoV}/2$. The resolution over entire image FoV is constant: $B_x/\lambda R$ with wave carrier $-x/\lambda R$. The viewing angle is constant within FoV as well

$$\alpha_{\text{view}} = \alpha_{\text{view}}^+ - \alpha_{\text{view}}^- = B_x R^{-1} \tag{3.5}$$

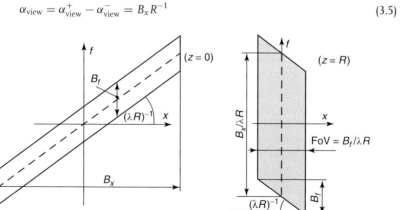

Figure 3.7 The WDF representation of the holographic signal in two different planes (a) $z = 0$ and (b) $z = R$ for circular configuration.

where the limiting angles are

$$\alpha_{\text{view}}^{\pm} = \pm \frac{B_x}{2R} - xR^{-1}$$

3.3.3
Visual Perception Analysis of Holographic Displays in Planar and Circular Configurations

In this section, we consider a holographic image observation in reconstruction geometry presented in Figure 3.8 using WDF representation [35, 36]. The real image is observed from location $[x_0, z_0]$. Once again the analysis is performed by means of WDF graphical representation of the holographic imaging process; this time we additionally introduce an eye aperture with pupil of size ϕ_0.

In Figure 3.9, the single eye off-axis observation of a real holographic image for planar and circular displays is represented in WDF chart. The single eye position is arbitrary, so extension to binocular observation is straightforward. The Wigner charts of an eye and a holographic image are shown for the reconstruction plane. For both displays we present WDF shapes representing an energy that can contribute to the generation of reconstructed image. The Wigner chart representing the effect of an eye in the reconstructed image plane is found by back propagating the WDF of a rectangular function of width ϕ_0 to the image plane. The overlap represents the energy from an image that is actually captured by an eye. The spatial size of this overlap defines the monocular field of view (MFoV).

For planar configuration according to the WDF representation we obtain

$$\text{MFoV} = \frac{B_x z_0}{z_0 + z_r} \tag{3.6}$$

In deriving Eq. (3.6) we have applied the criteria of 50% loss of observed energy and resolution. At marginal points of the observed MFoV region we view reconstructions of half the resolution and half the intensity. This is presented in Figure 3.9a, where size of MFoV is computed from projection of the central line of an eye slit in WD on x-axis. When we view the reconstruction in a real display, the size of MFoV appears to be larger than the computed one according to the

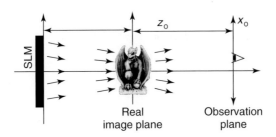

Figure 3.8 Monocular viewing of real image in holographic image.

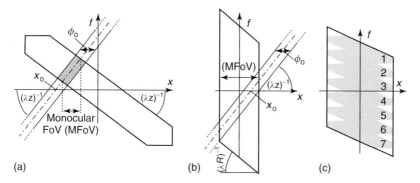

Figure 3.9 The WDF representation of visual perception of the holographic image: (a) planar configuration, (b) circular configuration, and (c) the WDF representation with planar SLMs arranged uniformly (no gaps) on a circle.

50% criteria. For planar configuration when an eye shifts (varying x_0) MFoV shifts linearly. The shift can be obtained from the WD chart

$$x_{\text{MFoV}} = \frac{z_2 x_0}{z_0 + z_2} \quad (3.7)$$

This is one of the problems of planar configurations. It is minimized in circular configuration in which we obtain constant MFoV (MFoV = FoV) with variation of an eye position. This results in constant size (position) of a viewed area for entire reconstructions from different perspectives, which is desired feature of holographic displays. Therefore, for circular configuration we characterize the observation area ($|x_0| < x_{0\,\text{lim}}^{(\text{MFoV}=\text{FoV})}$), where an eye can move and viewer sees an image of the same size and position (constant MFoV = FoV). This situation is illustrated in Figure 3.9b. If an eye moves beyond this range, the MFoV decreases linearly form maximum to zero width for $x_{0\,\text{lim}}^{(\text{MFoV}=\text{FoV})} > |x_0| > x_{0\,\text{lim}}^{(\text{MFoV}=0)}$. Using graphical representation from Figure 3.7b the relation for MFoV as a function of observer position x_0 is given as

$$\text{MFoV} = \begin{cases} \text{FoV} & |x_0| < x_{0\,\text{lim}}^{(\text{MFoV}=\text{FoV})}, \\ \dfrac{R}{z_0 + R}\left(\dfrac{B_x z_0}{2R} + \dfrac{\lambda B_f(z_0 + R)}{2} - x_0\right) & x_{0\,\text{lim}}^{(\text{MFoV}=\text{FoV})} > |x_0| > x_{0\,\text{lim}}^{(\text{MFoV}=0)}, \\ 0 & \text{otherwise}, \end{cases} \quad (3.8)$$

where

$$\left[x_{0\,\text{lim}}^{(\text{MFoV}=\text{FoV})} = \frac{B_x z_0}{2R} - \frac{\lambda B_f}{2}(z_0 + R)\right]$$

$$x_{0\,\text{lim}}^{(\text{MFoV}=0)} = \frac{B_x z_0}{2R} + \frac{\lambda B_f}{2}(z_0 + R)$$

3.3.4
Comparison of the Display Configurations for Commercially Available SLMs

The theoretical discussion presented in the above subsections applies paraxial approximation. Recently, it has been applied for the most popular reflective, phase-only type of LCoS SLM offered by Holoeye, model 1080P SLM [16, 37]. These SLMs are used in the experimental display configurations reported in Section 3.5. The 1080P SLM is characterized by 1920 × 1080 pixels and the pixel pitch is 8 μm. It diverges a beam in the range −1.9 to 1.9°. The commercial (but theoretical) SLMs applied to display are planar SLMs distributed on a circle. This affects parameters of holographic image and quality of visual perception. In Figure 3.9c, we illustrate WDF representation of bounds of image generated by holographic display with seven SLMs arranged on a circle; the dotted line shows WDF given by circular display with the same space–bandwidth product. The gray area represents WDF of display with planar SLMs. Every planar SLM has the same WDF shape but different frequency shift. The frequency shift is a function of SLM's angular position. Plot corresponds to the situation where active areas of SLMs are arranged next to each other (pixel to pixel). WDF representation for both display configurations are very similar. They have the same FoV. For $|x| < (\text{FoV} - B_x)/2$, there is no frequency gap in the WDF plot; for this region the viewing angles are the same. For larger $|x|$ there are frequency gaps and the viewing angle has corresponding gaps. These frequency gaps result in either smaller VFoV or modulation of visual perception (loss in perceived energy and resolution). Generally, the discussion of spherical display given in the above subsections corresponds to its practical implementation with planar SLMs arranged on a circle.

Let us now compare planar and circular configurations for multi-SLM display configurations. We consider the holographic display having 6 and 12 SLMs distributed on a plane and on a circle. The discussion here is related to the experimental configurations presented in Section 3.5.2.

For *planar configuration* there is a condition of good-quality imaging with no aliasing images; the minimum reconstruction distance for 6 and 12 SLMs arranged in the horizontal (X) direction is

$$z_{\min} = \begin{cases} 6B_x \Delta \lambda^{-1} = 1.39[m] & \text{for 6 SLMs} \\ 12B_x \Delta \lambda^{-1} = 2.77[m] & \text{for 12 SLMs} \end{cases} \quad (3.9)$$

For such long reconstruction distances there is no increase in viewing angle. We mention that in some limited area of *FoV* we can obtain good-quality images in planar configuration for smaller reconstruction distances. However, then we have to cope with aliasing effect [33] and SLM energy losses. Overall reconstructions will be of low quality; it is difficult to satisfy the condition of constant intensity at SLM plane. Therefore, we have selected the spherical configuration for a holographic display system, and the discussion in next sections refers to such geometry.

3.4
Capture Systems

3D holographic TV and video requires wide-angle holographic capture of varying-in-time real-world 3D objects and scenes. In order to fulfill this requirement the following two basic configurations are applied:

- A capture system with multiple cameras arranged in a circular configuration around an object (Figure 3.2a). This system allows to simultaneously capture digital holograms of a dynamic object or scene seen by the cameras from different perspectives. However, due to physical size of optical sensors it is not possible to capture total object optical field,
- A capture system with a single camera and a static object located at a rotational stage. Here, 360° object optical field is registered sequentially in time by capturing digital holograms of an object in different angular positions. Proper angular sampling of an object optical field allows to capture full information about an object.

3.4.1
The Theory of Wide-Angle Holographic Capture

In order to gain insight into the general capture setup shown in Figure 3.2a we again use "Wigner charts," which allows us to derive some important properties of the multicamera configuration. We apply our analysis only to two adjacent cameras, and in this case we can state that these two cameras are located in the same plane at a distance z_1 from the object and are separated by an angle α_1. We assume that the paraxial approximation can be used to describe propagation from the object plane to the plane of the two cameras [36]. The two camera capture setup is shown in Figure 3.10a. The cameras are positioned in a plane at distance z_1 from the object

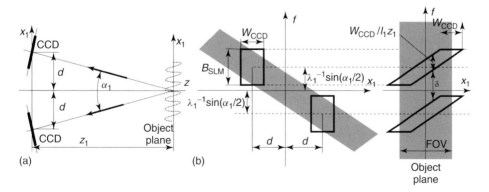

Figure 3.10 The multi-CCD capture setup: (a) the configuration for capturing two portions of the object wavefield, (b) the WDF of the signal in the capture plane, and (c) the corresponding WDF mapped back to the original object plane.

plane. The cameras are located at a distance D at either side of the optical axis and the angles they subtend to the optical axis are $\alpha_1/2$ and $-\alpha_1/2$. These two cameras are assumed to be identical in a width given by W_{CCD} and a bandwidth given by $B_{CCD} = 1/\Delta_1$ where Δ_1 is the pixel pitch of the camera. The WDF of the object signal is shown for the capture plane in Figure 3.10b and for the object plane in Figure 3.10c. In both cases, the lighter shaded areas represent this full continuous signal. The object will have a certain width W_{OBJ}, and this explains the finite support of the WDF shown in Figure 3.10c. The bandwidth of the signal extends over infinity, and hence the signals are shown not to be bounded in the f axis. The WDF of the propagated signal is illustrated in Figure 3.10b. The horizontal displacement results in the signal having a finite local bandwidth, and hence a CCD can be used to capture the complex wavefield using interferometry. The WDFs of the two cameras in the capture plane are shown in Figure 3.10b using thick black lines. These rectangular WDFs have a width and bandwidth given by W_{CCD} and B_{CCD}. Their centers are located at $\pm D$ in the x-direction and $\pm \lambda_1^{-1} \sin(\alpha_1/2)$ in y-direction. These two CCDs capture a portion of the energy of the object signal. This energy is shown as the darker shaded region in the figure. The corresponding region of energy in the original object plane and the WDF of the CCDs in this plane are shown in Figure 3.10c. Using basic geometry it can be proven that the local bandwidth of the regions is described by $W_{CCD}/\lambda_1 z_1$. Therefore the gap between the two shaded regions in Figure 3.10c is given by

$$\delta_1 = 2 \frac{\sin(\alpha_1/2)}{\lambda_1} - \frac{W_{CCD}}{\lambda_1 z_1} \qquad (3.10)$$

The equation gives an estimate of the frequency gap between two adjacent cameras. The same analysis can be applied to any two adjacent cameras in a wider assembly. In order to capture the full continuous field δ_1 should be zero. It will be achieved if

$$\sin(\alpha_1/2) = \frac{W_{CCD}}{2z_1} \qquad (3.11)$$

If this condition is fulfilled the capture fill factor equals unity ($FF_1 = 1$). In general, this case is not possible due to the external components of the camera, usually much bigger than the active areas of detector matrices.

3.4.2
The Capture System with Multiple Cameras

The physical realization of the theoretical multi-CCD holographic capture system (Figures 3.2a and 3.10) has been demonstrated recently by the Bremen Institute of Applied Beam Technology (BIAS) [22, 26]. The system in the partially circular six-sensor arrangement capturing digital holographic videos of dynamic scenes from six different views simultaneously in time is shown in Figure 3.11. The high pulse energy laser InnoLas YM-R 800 with a wavelength 532 nm, a pulse length of 6 ns, and a maximum pulse energy of 100 mJ had been chosen as the light source. Since the energy of the laser is too high to apply fibers the reference waves and the object illuminating wave are guided by beam splitters, plates, and mirrors, whereby

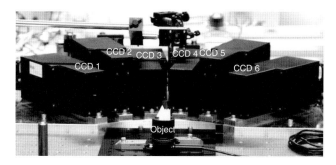

Figure 3.11 Photo of the setup used for capturing digital holographic videos – front view. (Source: Courtesy of BIAS [22].) (Please find a color version of this figure on the color plates.)

the reference light path length of one sensor can strongly differ from the reference light paths of the other sensors and the object light path. Thus, the coherence length of the laser has to be high enough. As verified by applying a Michelson interferometer the pulsed laser has a sufficient coherence length of ∼120 cm. In order to shape six plane reference waves illuminating each CCD entirely, a collimator is placed in front of the pulsed laser right before the reference wave is split into six reference waves. All six reference waves impinge perpendicular to the CCDs to provide an inline digital Fresnel holography capturing setup. A neutral density filter has been positioned in the reference wave arm (before it is separated into the six reference waves) and a second one in the object wave arm in order to control the relation between the object and the reference wave powers by varying those filters. The frame rate of the video is ∼10 Hz and the image size of each hologram is 1920 (width) × 1080 (height) pixels with 8 bit per pixel. The exposure time of each CCD has been set to 100 ms, but the real capturing time is just 6 ns because of the pulse length of the laser. For synchronizing the capturing time of the six sensors the trigger device has been connected to the six CCDs. The block diagram of the electronic components as well as the data and the control flow are shown in Figure 3.12. All sensors are controlled by one computer, the Main PC. This PC merges the captured digital holograms into one HDF5-file, compress the data, and transmits to the display side. The full process is automated. Owing to constraints specified by the capturing arrangement, the capture is subjected to the following restrictions:

- The object size is limited by the used wavelength λ, the pixel pitch $\Delta = 3.45$ μm of the CCDs, and the distance z between the axis of rotation of the partial circular arrangement and the capturing CCDs. This distance corresponds to the radius of the given partial circular arrangement of the CCDs, which is 260 mm in the final capturing arrangement. Applying the equation

$$h = \frac{z \cdot \lambda}{\Delta} \tag{3.12}$$

which is used for calculating the maximum size h of the object, a maximum object size of 40 mm is obtained for the given capturing arrangement.

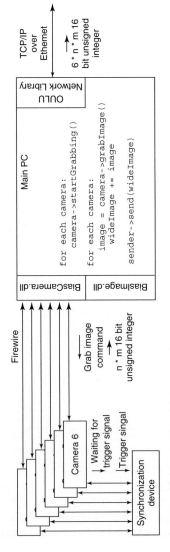

Figure 3.12 Block diagram of the electronic components and the data and control flow using one computer controlling all six sensors. (Source: Courtesy of BIAS [22].)

- The gap size: due to the dimensions of the sensor housings the six CCDs detect a smaller part of the wave field than the whole partial circular six-sensor arrangement covers. This leads to a gap between the captured digital holograms of the different sensors. A second factor defining the size of this gap is the distance between the axis of rotation of the partial circular arrangement and the CCDs. Regarding the given capturing setup, this distance is 260 mm and the width of the used CCD is 6.6 mm because only 1920 pixels are addressed. So each CCD detects an angular range of ~1.5° in horizontal direction, but the angle between the centers of two neighboring CCDs in this direction is about $15° \pm 1°$. This results in a minimum gap between two neighboring CCDs of at least 13.5° and the corresponding capture fill factor of $FF_1 = 0.1$,
- The capturing frame rate is limited by the maximum pulse rate of the pulsed laser used, which is 10 Hz. However, the actual maximum frame rate could be even lower depending on the used pixel number of the CCDs. The pixel number of the Stingray CCDs used is 2056 (width) × 2452 (height) pixels. Using all these pixels the capturing frame rate is limited to a maximum of 9 fps, but it can be increased by reducing the number of pixels used in terms of width. In order to adapt the number of pixels used the CCDs in the capturing arrangement to the pixel number of the SLMs in the display arrangement, the number of CCD pixels used has been reduced to the SLM pixel number of 1920 (width) × 1080 (height). This results in a maximum capturing frame rate of 9.8 fps.
- The transmission rate: the Fast Ethernet is used for live transmission of digital holographic videos to the display side. Owing to this connection the bandwidth is limited to a maximum of 12.5 Mbps. Using six CCDs with an image size of 1920 × 1080 pixels each leads without data compression to a theoretical frame rate of 1 frame s^{-1} at most. Practically, the real bandwidth achieved is about 8 Mbps, yielding a maximum frame rate of 0.6 fps.

As described above the capture system with multiple cameras provides holographic data of a real-word varying-in-time 3D object or scene. The data will be used as the input for the tests of the full technology chain of holographic 3D video. However, the main disadvantage of this capture system is the low value of capture fill factor. The reconstructed images will not provide a continuous object optical field, which will directly influence the comfort of observation.

3.4.3
The Capture System with a Single Camera and Rotated Static Object

As the system with multiple cameras is not able, at the moment, to capture a full object optical field, the alternative system based on a single camera and an object placed on a rotary table has been developed [30]. Here the required multiple digital holograms of 3D static objects are captured sequentially for different perspective views of an object. The angle α_1 between consecutive object positions provides an arbitrary capture fill factor. It is easy to reach $FF_1 = 1$ or a value matched to the display fill factor. Also, as we assume that the object is static it is possible to apply the PSDH [28], which requires capturing at least three phase-shifted holograms for calculation of a single-phase frame for the display. The main advantages of PSDH method are that we get the maximum space bandwidth product for the captured complex object wavefield, and this allows to use maximum object size and retrieve maximum angular perspective. The method is effectively noise free and provides the highest quality images [30].

The system for capture in 360° digital holograms of static 3D objects and scenes was built and optimized for a large variety of objects and the capture conditions adopted for reconstruction at a multi-SLM display. It is assumed that the reconstruction will be performed in two different modes.

- "naked eye" observation of small images seen in a SLM aperture;
- visualization of relatively big images at an asymmetrical diffuser with a big scattering angle in vertical direction (no vertical parallax) and small scattering in horizontal direction (reasonable horizontal parallax).

The capture system in a flexible geometrical configuration, which enables registration of holograms of static objects, is presented in Figure 3.13. The light source is a CW laser with the wavelength 785 nm and a coherence length of ∼1 m. The beam passes through a neutral density filter, linear polarizer, and quarter wave plate. By varying these elements, the relative power of the object and reference beams obtained behind the polarizing beam splitter is controlled. The microscopic objective and pinhole followed by a lens in the horizontal arm collimates the light and creates a plane wave reference beam that propagates directly toward the camera. The camera selected for the system is Basler CCD with 1920 × 1080 pixels and a pixel pitch of 7.45 μm. The relatively big pixel size and the available number of pixels are matched to the parameters of LCoS SLM used in the display.

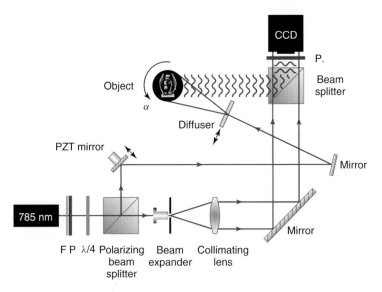

Figure 3.13 The holographic system for 360° capture of static 3D objects; P – polarizer, F – neutral density filter. (Please find a color version of this figure on the color plates.)

The second beam is reflected from a mirror attached to a piezoelectric motor, which can be moved by fractions of wavelength for PSDH. This beam is further directed by an additional mirror to a diffuser, which spreads the light out to illuminate the entire object. The light from an object is scattered directly toward the final beam splitter, which reflects the light toward the camera. The object is located on a stage that is used to rotate the object in a given static position so that we can simulate the presence of multiple cameras positioned in different angular positions. In many cases the objects are recorded using the PSI technique at 360 positions separated by a single degree. The diffuser used for object illumination is attached to a translation stage. For each object position, the diffuser is moved and a four-capture PSI hologram is recorded. This is repeated for a number of different diffuser positions. These holograms can then be superimposed by displaying them in sequence within the aperture time of the eye for effective speckle noise reduction.

This capture system has been optimized in order to capture the "golden master" of holographic data, which provides the best quality of phase for addressing LCoS SLMs and reconstruct the object optical wavefield with no spatial gaps. Also, the compression and decompression algorithms are avoided in order to maintain the highest quality. However, these data are not adapted for the requirements of full technology chain of holographic video. The holograms are sequentially written on a computer disk, and the data after processing are read from the disk and displayed in multi-SLMs display. This can be realized with the frequency 25 Hz and higher.

3.5
Display System

3.5.1
1LCoS SLM in Holographic Display

In this section, we present holographic display designs applying phase-only reflective LCoS SLMs. These SLMs have the best parameters for phase reproduction of holographic optical wave [16, 38]: relatively large (although limited) space bandwidth product, accurate phase reproduction, high diffraction efficiency, and high fill factor of the pixel matrix. The effect of limited space bandwidth product is discussed in Sections 3.3.2 and 3.3.3. Now, in the practical analysis of a planar LCoS SLM implemented in a display with circular configuration two important features should be considered: the shape of an illuminating beam and gaps in the system due to a local absence of pixels in a circular display (it is caused by the presence of a frame around an active SLM area).

In Section 3.3.2, we have shown that a display with SLMs of spherical shape focuses illumination wave into a single image point when a constant phase ("*0 state*") is applied. This is also possible in the case of a display with planar SLMs when it is illuminated with a converging beam focused at the display center. An optical layout buildup from planar SLMs that is entirely equivalent to the spherical SLM display (Figure 3.4b) is presented in Figure 3.14a. For presentation simplicity

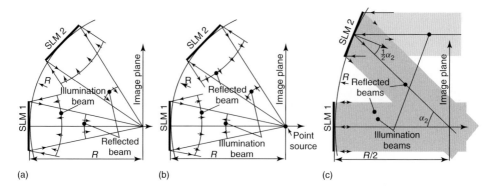

Figure 3.14 Illumination beam configurations of display with spherically configured planar SLMs with (a) convergent spherical illuminations, (b) divergent spherical illumination, and (c) plane wave illumination.

only two SLMs are considered; however, the discussion is general. The optical fields reflected from both displays at "0 state" are the same; they are assembles of a single spherical wave. Unfortunately, in the case of multi-SLMs display it is very difficult to generate such an illumination beam for all reflective SLMs, therefore alternative solutions are discussed below and later used in the systems reported in Section 3.5.2.

In the first practical configuration (Figure 3.14b), planar SLMs are aligned on the circle of radius R; SLM's normals point toward the center of alignment curvature. The divergent spherical illumination beam is generated from a point source at the curvature center. As the SLMs are flat the beam reflected from them diverges. This is a weakness of this configuration. A part of SLM space bandwidth product must be used to converge beam (to eliminate beam divergence). In the design discussed in Section 3.5.2.3, an astigmatic beam is applied; however, the discussion is equivalent for both spherical and astigmatic beams.

In the second practical configuration (Figure 3.14c) SLMs are aligned on a circle of radius R as well; however, an illumination beam is now a plane wave. For a circular SLM such a beam focuses at distance $R/2$ (image plane). This is a fundamental feature of this configuration. The SLMs are illuminated with tilted plane waves and the impinging angle could be high. This is the design weakness; the diffraction of light reproduced with SLM decreases with increasing illumination tilt. However, it is presented elsewhere that good-quality reconstructions are obtained for SLM tilts up to even 40° [39]. It means that with this design we can achieve a display with 160° viewing angle. The configuration with tilted plane wave illumination is demonstrated in Section 3.5.2.2.

The second issue to be discussed is the gap problem. Around an active area of LCoS SLM there is a mount, so it is not possible to bring the areas of neighboring SLMs close to each other (pixel to pixel). When SLMs are aligned on a circle there is a substantial gap in the display pixel arrangement (for Holoeye LCoS). In general, this problem can be solved in two ways: by optically rearranging SLMs in space or by means of temporal multiplexing.

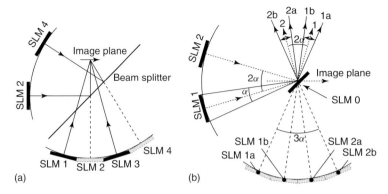

Figure 3.15 Elimination of gap problem in the holographic display: (a) with beam splitter and (b) with an additional SLM.

The spatial multiplexing method was proposed by the group in Bilkent University [20], and it is presented schematically in Figure 3.15a. There are four SLMs aligned on the circle in two groups: group 1 – SLMs 1 and 3 group 2 – SLMs 2 and 4. The groups are aligned on two arcs, separated angularly. The arc separation is chosen such that images of active areas of SLMs from group 2 given by beam splitter coincide with SLMs active area from group 1.

The second method proposed by our group [21, 36] realizes temporal multiplexing of SLMs by means of an additional SLM0 (Figure 3.15b). SLM0 is placed in the display image plane, where circular beam is focused. It is synchronized with SLMs arranged on the circle. SLM0 sequentially changes the direction of beams of all SLMs. For example, if SLM0 is switched off, the beam from SLM1 shall be reflected in direction (1). However, when we sequentially change the linear phase of SLM0, the reflected beam will be redirected in directions (1a) and (1b). SLM1 is synchronized with SLM0, so it can replay holograms as it would come from direction (1a) or at the next time step from (1b). This solution gives four virtual SLMs from two physically present, that is, instead of two holographic views there are four of them. The typical SLM refresh rate is 60 Hz, so it is possible to get from a single SLM four views without being recognized by a human eye and brain. This solution exchanges time bandwidth information into spatial ones.

3.5.2
Experimental Configurations of Holographic Displays

We now review three holographic display configurations using LCoS SLMs. The first display utilizes a divergent beam illumination [20]. Although at the moment the system is used for displaying holographic videos from synthesized world, in future it can be adapted to a full technology chain of holographic videos of real-world 3D scenes. The next two displays, developed at Warsaw University of Technology (WUT) [36, 40], use a plane beam illumination. They are specially designed to enable reconstructions of real-world objects and scenes, therefore the constrains coming from the capture setup are taken into account. All reported displays use LCoS SLMs in circular configuration, and the main differences come from the design of illumination module.

3.5.2.1 Display Basing on Spherical Illumination

The first display developed recently by Bilkent University [20, 22] is shown in Figure 3.16. The display consists of nine phase-only LCoS SLMs aligned on a circle. The SLMs are illuminated with a single astigmatic divergent beam; this illumination scheme is discussed in Section 3.5.1.

The circular shape of SLM alignment images astigatic beam (cone mirror and point source) into a reconstruction space. The beam splitter is used to tile the SLMs side by side without any gap between them, that is, at the circle there are both real SLMs and virtual images of SLMs. The side view of the illumination beam scheme is shown in Figure 3.16b. There is a point source and then cone mirror. Owing to the shape of the cone mirror and the point source the generated illumination beams are astigmatic. During computation of the holograms the shape of an illumination beam has to be taken into account. In addition, some part of the space–bandwidth product of SLM is used for eliminating the divergent nature of beam illumination. Anyhow, the system has full parallax; an observer can see the reconstruction floating in space if the reconstructed image is sufficiently small. The design aims at displaying holographic videos from the synthesized world. As the computer-generated holograms do not have the limitations and noises during the real holographic capture, the quality of the reconstructed images and videos is highly satisfactory.

3.5.2.2 Display Configuration Matching the Capture Geometry

In both single and multiple cameras capture systems, each CCD camera points toward the center of a 3D scene (Figures 3.2 and 3.12). Also, for simplicity of further reconstruction and alignment of reconstructed optical waves we had chosen plane reference beams. In order to match this capture geometry every SLM of the holographic display must be illuminated by a plane wave with a wave vector normal to the SLM and each SLM must point toward the center of rotation axis of the

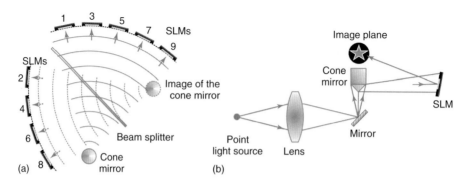

Figure 3.16 Holographic display design with a cone mirror: (a) top view and (b) side view.

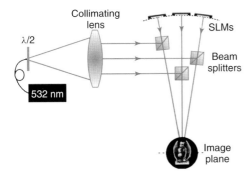

Figure 3.17 The scheme of a holographic display matching the capture geometry.

capture geometry. A practical realization of such display system with three SLMs in circular geometry is presented in Figure 3.17. The display is constructed so the captured wave in multi-CCD circular configuration can be directly reproduced in multi-SLM configuration. In the experimental setup this is realized by an accurate angular adjustment of beam splitting cubes and SLMs.

The big advantage of such system is the possibility to directly display the captured complex waves (their phase distributions). The drawback of this solution comes from the use of beam splitting cubes, each having different angular orientation. This limits the minimal angular separation of SLMs. More detailed technical information about the display and the number of obtained holographic images can be found in publication [21].

3.5.2.3 Display Based on Tilted Plane Wave Illumination

The much simpler display can be designed if an additional processing necessary to couple capture-display configurations is added [39]. Such holographic display system in a circular configuration with multiple LCoS SLMs illuminated by a single collimated beam is presented in Figure 3.18. The display consists of two modules: illumination and reconstruction. The task of the first one is to illuminate six SLMs with homogeneous and parallel beams of equal intensity. For this we use a single laser beam delivered by a single mode fiber and collimated by a well aberration corrected lens. Then the beam is divided into six parts by means of three beam splitter cubes. SLMs are illuminated with beams directed by a set of independent mirrors. It is necessary to separate incident and reflected beams vertically, so a small tilt between these beams, not larger than 1.5–2°, is introduced. Its negative effect on a phase can be neglected. The SLMs are aligned on a circle having radius twice the reconstruction distance. There are gaps between active SLM areas. Depending on the size of the reconstructed images the gaps are minimized (removed) using one of the viewing scheme: with an additional SLM or with an asymmetrical diffuser (Figure 3.18a). In order to couple capture–display systems (see Section 3.6.2) each SLM is tilted in horizontal direction with respect to the illumination beam. This requires additional processing of the phase with which the SLMs are addressed.

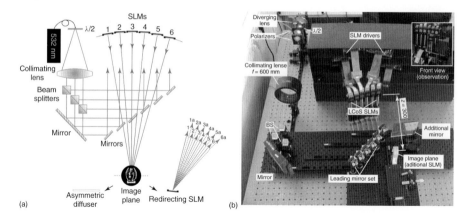

Figure 3.18 The general scheme of a holographic display (a) and the photo (b) of the display for "naked eye" observation mode. At the scheme there are two additional modules: asymmetric diffuser used for observation of images at large reconstruction distances (large FoV), redirecting SLM module for "naked eye" observation of close reconstruction (small FoV). (Please find a color version of this figure on the color plates.)

The display is capable of reconstructing holograms of real scenes as shown in Section 3.7. However, in this section we show the reconstructions of scenes generated in the computer graphic world. The holograms of computer models of a scene composed of three chairs (Figure 3.19) and Gargoyle statue (Figure 3.20) were computed for different perspective views. The synthetic holograms are calculated using iterative method adopted to the tilted plane geometry [39]. The algorithm is designed using propagation between tilted planes [41, 42] and parallel planes [43] followed by an error reduction [44] algorithm.

The reconstructions for two distances: 300 and 700 mm are presented. With the change of reconstruction distance the limits (spatial – FoV, frequency – viewing angle) exchange values. When one grows, another one decreases (Table 3.1); this surely affects the experience of observing optically reconstructed images.

Moreover, reconstructions for these two distances use different observation schemes illustrated in Figure 3.18a. Closer reconstruction is observed with

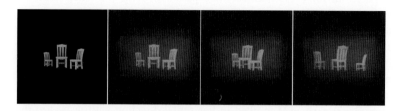

Figure 3.19 3D model of scene composed of chairs and the three views of holographic reconstruction. (Please find a color version of this figure on the color plates.)

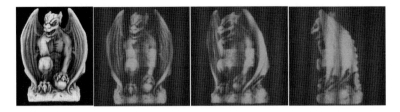

Figure 3.20 3D model of a Gargoyle statue and its reconstructed views (composed of six waves) as seen from three different perspectives. Reconstruction at an asymmetrical diffuser. (Please find a color version of this figure on the color plates.)

Table 3.1 Characterization of viewing scene of circular holographic display: FoV, viewing angle, limits of MFoV for constant MFoV = FoV and linearly decreasing to zero.

	6 SLMs		12 SLMs	
	$z_r = 300$ (mm)	$z_r = 700$ (mm)	$z_r = 300$ (mm)	$z_r = 700$ (mm)
FoV (mm)	19.9	46.5	19.9	46.5
$\alpha_{view}(°)$	17.5	7.5	35	15
($z_0 = 300$ mm)				
$x_{0\,lim}^{(MFoV=FoV)}$ (mm)	26	–	72	6
$x_{0\,lim}^{(MFoV=0)}$ (mm)	66	53	112	72
($z_0 = 600$ mm)				
$x_{0\,lim}^{(MFoV=FoV)}$ (mm)	62	–	154	35
$x_{0\,lim}^{(MFoV=0)}$ (mm)	122	83	214	122

additional SLM (redirecting SLM) and "naked eye" observation mode. The additional SLM is applied to eliminate the gap problem and to extend the viewing angle (Figure 3.15b) by virtual extension of the number of SLMs. By synchronizing display SLMs with SLM0 located at the image plane we double the number of SLMs. The display still works with 30 Hz (half of SLM frequency). For this system we have generated holograms for 12 SLM display and reconstruction distance of 300 mm. For this system we have received quite big viewing angle $\alpha_{view} = 35°$ and relatively small FoV: for horizontal direction MFoV = FoV = 20 mm and for vertical direction MFoV = 4.7 mm.

The reconstructions of the scene with chairs (Figure 3.19) are very small; they have to fit within the FoV. However, they can be viewed with a naked eye. This is an impressive feature, which is desired by holographic displays. On the other hand, a display should be capable of reconstructing images with much larger size. In Figure 3.20 we present the reconstruction of computer-generated holograms of Gargoyle statue with a reconstruction distance of 700 mm and FoV = 47 mm.

If we try to view it directly, the well-known keyhole problem appears. To extend the size of the MFoV in the y-direction in the system an asymmetric diffuser is placed at the center of the real image reconstruction space. The diffuser scatters the light in one direction only (y). This allows for a more convenient holographic image observation, that is, the eye may take a wide range of positions in the y-plane. We note that the perspective of the holographic image in the x-direction is unaffected.

Figure 3.20 presents three views of hologram reconstructions obtained with asymmetric diffuser. Four perpendicular stripes are seen. Each stripe represents the perspective given by different SLMs. It can be noted that there are no gaps (zero intensity) in the FoV, even that SLMs are distributed with substantial gaps (FF = 0.59). Analyzing Wigner chart of system with gap we can find that the gap size decreases as reconstruction distance increases [36].

While capturing images of the reconstructed scenes presented in this section and in Section 3.7 care was taken to present them as they are viewed by a human eye, that is, the views are captured with a digital camera adjusted to human eye observation conditions. The entrance pupil diameter of digital camera is set to 8.2 mm; this value is close to human observation condition in a dark room (eye aperture diameter 8 mm).

The two experiments for 300 and 700 mm reconstruction distances have mimicked the conditions for reconstructions of holograms captured by the system with a single camera and object at a rotary table. However, for the full technology chain of 3D holographic video the system matching multi-CCD camera systems is of highest importance. In this case the required image reconstruction distance is much bigger, that is, 1000–1400 mm, and the display fill factor is very low (FF = 0.1). We will not be able to see the continuous wide viewing angle image, but are able to get discrete views of an object from a different perspective. The display system configuration matched with multi-CCD capture system is presented in Figure 3.21. The observation of the image is performed by means of asymmetric diffuser; however, it is not convenient as due to large gaps in reconstructed optical fields it is not possible to merge the views (no proper monocular and binocular visual perception).

The image is observed by means of asymmetric diffuser; however, it is not convenient as due to large gaps in reconstructed optical fields the views cannot be merged (no proper monocular and binocular visual perception). This fact is well illustrated in Figure 3.22, which represents the visual field of view (VFoV) for on-axis binocular perception of an image reconstructed at 1000 (mm) in circular display with FF = 0.6 (Figure 3.22a) and FF = 0.3 (Figure 3.22b).

In the simulation shown in Figure 3.22, we use binocular distance $d_b = 65$(mm). With the blue and green colors, the VFOVs given by left and right eye, respectively, are shown. The overlapping regions show stereoscopic perception. For determination of VFOV we use the criterion of 50% drop of an image resolution and intensity. This is strong assumption for visual perception, so the gaps will be smaller during a hologram viewing. However, it is clearly seen that even for FF = 0.3 the conditions for continuous observation of an image from different perspectives is

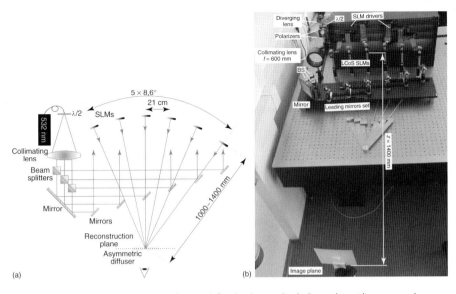

Figure 3.21 The scheme of the display used for displaying the holographic video captured by the multi-CCD holographic capture system shown in Figure 3.11. (Please find a color version of this figure on the color plates.)

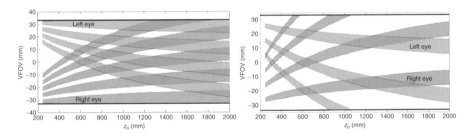

Figure 3.22 Representation of VFOV as a function of observation distance for reconstruction distance 1000 mm and on-axis binocular observation ($d_b = 65$ mm) for six SLM circular displays with (a) FF = 0.6 and (b) FF = 0.3. (Please find a color version of this figure on the color plates.)

impossible (in the current system FF = 0.1). This is the reason why in Section 3.7 the results from holographic video reconstructions are presented separately for each LCoS SLM.

3.6
Linking Capture and Display Systems

In order to reconstruct and display the captured wide-angle complex image in holographic multi-SLM display (Section 3.5.2.3) we have to consider two major problems. First, the capture and display devices are usually characterized by different sampling parameters, the capture and reconstruction wavelengths can differ as well. Second, display and capture systems may have decoupled configurations. It means that we cannot directly display captured complex fields. In the following subsections, we present a short analysis of these two issues.

3.6.1
Mismatch in Sampling and Wavelength

The captured optical field given by discrete samples is to be replicated by SLMs. A single CCD and SLM are characterized by the following sampling parameters: the number of pixels N and the size of a pixel Δ. We assume that the experimental capture–display devices CCDs and SLMs have comparable numerical space bandwidth products (number of pixels). Therefore, we can directly modulate SLM with a discrete captured field. Applying this we display holographic data with transverse magnification

$$M_t = \frac{\Delta_2}{\Delta_1} \tag{3.13}$$

and longitudinal (axial) magnification

$$M_l = \frac{\lambda_1}{\lambda_2} M_t^2 = \frac{\lambda_1 \Delta_2^2}{\lambda_2 \Delta_1^2} \tag{3.14}$$

where subscripts 1 and 2 relate to the capture and reconstruction processes, respectively.

The transverse and longitudinal magnifications are linked by the angular magnification

$$M_a = \frac{\sin \alpha_2}{\sin \alpha_1} = \frac{\lambda_2}{\lambda_1 M_t} \tag{3.15}$$

Concluding, the capture–display system mismatch gives the following:

- the transversely magnified (M_t) image
- the reconstruction distance and 3D depth are increased by a factor of the longitudinal magnification (M_l),
- the angular separation of SLMs is increased by a factor of the angular magnification M_a (when compared with the angular separation of CCDs).

3.6.2
Processing of Holograms to the Display Geometry

The display SLM pixels are addressed with a phase distribution of the object wave, which for real-world objects is delivered by means of the digital holographic capture system. During a hologram acquisition system the intensity fringes are captured. The capture and display system configurations differ (Section 3.4 – capture, Section 3.5.2.2 – display). To couple both systems special processing algorithms are developed, and they are schematically presented in Figure 3.23. Generally, algorithms process intensity fringes to the phase that can be directly replied to by

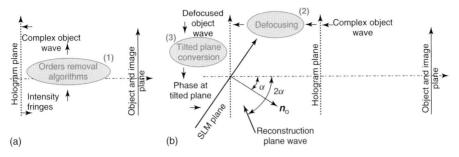

Figure 3.23 Visualization of the processing algorithms for coupling display and capture systems: (a) order removal algorithm and (b) defocusing and tilted plane conversion algorithms.

the display system producing holographic image. The processing algorithms shall accomplish the following three major tasks:

- Algorithm 1 – recovery of complex wave from intensity fringes.
- Algorithm 2 – propagation of complex wave to the defocused plane.
- Algorithm 3 – propagation of holographic signal to the tilted plane of SLM.

Algorithm 1 is visualized in Figure 3.23a. In holographic fringes zero order and twin image are present and they have to be filtered out, so a complex object wave can be recovered. The filtering algorithm construction depends on a holographic capture configuration. For Fourier digital holographic setup simple frequency-selective filtering techniques can be used. However, to recover complex object wave, reference beam has to be known exactly. In Fresnel digital holography, twin image and real image are not separated spatially within the same domain. Therefore, this configuration requires a more demanding algorithm. At the moment, the best method of twin image filtering is a segmentation algorithm [30, 45]. Within the algorithm, holographic signal is propagated from the object plane to the twin image plane sequentially. At the twin image plane where the twin is focused the mask is created, and then the twin can be filtered spatially and complex object wave can be recovered. To improve the algorithm a similar filtering technique can be additionally applied at the object plane.

Generally, a different reconstruction distance in display geometry than the one given by a capture setup may be required (Figure 3.23b). Therefore in Algorithm 2, the recovered complex wave is propagated to the parallel plane where the SLM is located. The task can be performed by one of the Fresnel propagation algorithms [33] that can be chosen according to the propagation distance [43]. During the propagation step, the optical field can be magnified using interpolation techniques [46] or frequency zero padding.

The last algorithm comes from the difference in configurations. The optical axis of the display for specific SLM is located at an angle α to SLM's normal (Figure 3.23b). The angle α violates Fresnel approximation. As a result, the reflected and refracted light from SLM generates off-axis field, which is paraxial around the optical axis. The field resulting from the first two processing algorithms gives object wave at the plane normal to the optical axis. Therefore, in order to display digital hologram in tilted configuration, we have to process it numerically to the tilted geometry of SLM [39]. The technique applies rigorous propagation of paraxial field between tilted planes [41].

It has to be noted that the hologram processing stage may be additionally modified by the tilting and defocusing procedures supporting the calibration procedure, allowing to match the real geometries of registration and display setups. An efficient approach to the capture–display system calibration has been proposed by a group from National University of Ireland [22]. In order to determine the parameters for processing, the calibration procedure that measures the camera positions and orientations in the recording set up has to be implemented [46]. This technique may be based on recording holograms of a known chessboard object with a well-defined size and a number of distinguishable features. By processing the numerical reconstructions from each camera in the capture setup the calibration procedures estimate the relative positions and orientations of the cameras. This information is then passed to the display side to aid in the alignment of LCoS SLMs, which should create the common reconstructed 3D scene [47].

3.7
Optical Reconstructions of Real Scenes in Multi-SLM Display System

The system presented in Section 3.5.2.3 allows to reconfigure the arrangement to match different capture setups and observation modes, which includes

- static object, "naked eye" observation mode with gap removal
- static object, "asymmetrical diffuser" observation mode
- dynamic objects, video and "asymmetrical diffuser" observation mode.

The display system in the "naked eye" observation mode has, due to physical dimensions of LCoS SLM, small FoV (a few millimeters). Therefore, we have produced (using 3D printing technology) a very small object: a model of molecule. The phase-shifted digital holograms of different views (0–360°) of the molecule were captured in the system with a single camera (Figure 3.12) and processed, and the calculated phases were displayed at LCoS SLMs. The final result of these reconstructions (combined image from 12 SLMs) for different perspective views of the molecule are shown in Figure 3.24. The image can be observed binocularly in a wide angular FoV as a floating-in-space high-quality reconstruction of a real-world object, so this would be the model configuration for the future holographic 3D video. Unfortunately, the size of reconstruction is very small (a few millimeters). Also, due to use of PSDH and the requirement for zero gap in reconstruction, the capture/transmission scheme cannot be at the moment integrated into the full chain of holographic video as described in Section 3.2.

The bigger real-world objects have been reconstructed at a distance of 750 mm by means of an asymmetrical diffuser as explained in Section 3.5.2.3. The reconstruction of printed models of a scene with two chairs captured (within 360°) in the

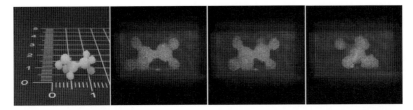

Figure 3.24 Printed model of ethanol molecule with 3D printer and views of optical reconstruction of digital holograms captured for this model in the system presented in Figure 3.17. (Please find a color version of this figure on the color plates.)

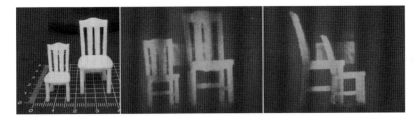

Figure 3.25 Printed model of chairs scene with 3D printer and views of optical reconstruction of digital holograms captured for this model in the system presented in Figure 3.17. (Please find a color version of this figure on the color plates.)

system with a single camera and an object on a rotary table is shown in Figure 3.25. The captured image is the result of simultaneous reconstruction by six SLMs, which provides a wide angular FoV and comfortable binocular observation. Although the display fill factor was not unity (FF = 0.6), the gaps in the reconstructed optical field can be seen by local lower intensity of the combined image.

The first proof of principle experiments for full technology chain of 3D holographic video was performed by BIAS (capture) and Warsaw University of Technology (display). Compression and transmission was supported by the Oulu University and processing by the National University of Ireland (all partners of the REAL 3D EU project [22]). The images had been reconstructed from six SLMs with the final frequency of ~1 Hz. Several objects and varying-in-time scenes had been transmitted. Two examples of reconstructions are shown in Figures 3.26 and 3.27. The first one is the running hand watch. Owing to large gaps in the optical field the reconstructions from sequential SLMs (the views of the watch with different perspectives) are shown separately in the first row (Figure 3.26). In the second row, the change in the position of the hand indicating seconds is clearly seen as these holograms represent another time frame selected from the transmitted video. The quality of reconstructed images are a bit worse that in the previous cases, but one has to keep in mind that they are based on the decompressed signal and less efficient (when compared with PSDH) procedures of zero and twin orders removal.

The second example of reconstructed video presents a more complex scene with significant depth of focus. The scene with three chicks (Figure 3.27a) was

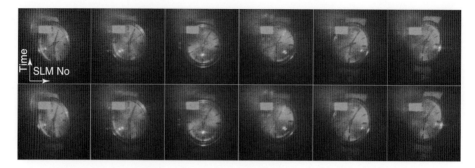

Figure 3.26 The exemplary reconstructions selected from a holographic video of the running watch. The photos in a row represent the images reconstructed by the sequential SLMs (1–6); the columns show two different states of the watch in time. (Please find a color version of this figure on the color plates.)

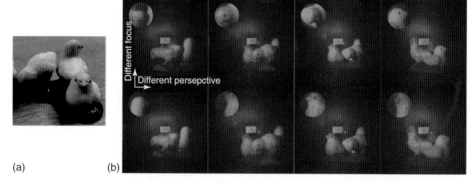

(a) (b)

Figure 3.27 The scene with three chicks at a table captured during rotation of the table: (a) the photo of the scene and (b) exemplary reconstructions selected from a holographic video of the scene. The photos represent reconstruction from the same SLM but captured (in a row) for different perspectives. In the columns, the photos are taken for two different focuses in order to show the depth of the scene. (Please find a color version of this figure on the color plates.)

located on a rotating table; holograms had been captured with frequency 10 Hz and were compressed and transferred (with losing some of the images due to lower compression and transmission speed) for processing and displaying at WUT. The images with different perspective of the scene and different focusing within the scene as reconstructed by each of the six SLMs are shown in Figure 3.27b. The reconstructed images were a few centimeters in size, had satisfactory quality, and had large depth of focus. The calibration of the system was correctly performed as there was no "jumping" of the reconstructed images when moving from one reconstruction to the next. The transmission sessions of different varying-in-time scenes last a few hours and have proved the sufficient functionality of the 3D holographic video system.

3.8
Conclusions and Future Perspectives

Looking at the development of home theater systems for 3D movies, it seems reasonable to assume that for the next 10 years the consumer market will focus on displays for autosteroscopic viewing. They allow for freely moving around while watching a movie and due to providing many views (typically five or nine) make the perceived 3D experience vivid. It can be expected that the number of views available will eventually be strong sales argument on the market of 3D home theater systems and 3DTV.

Holography inherently provides an almost "infinite" number of horizontal, vertical, and depth views. In that sense, holography appears to be the final stage of a development, which just begun in the 3D home theater and 3DTV market. However, development and widespread of utilization of holographic 3D videos and displays is still strongly hindered by current technological limits. It is less critical at the capture side where megapixel CCD detectors with 1 µm pixel size are already available than at the display side, where SLMs are still characterized by a low number of relatively large pixels. It causes either a very limited viewing angle or a small display size. To overcome this limitation there is a strong need of megapixel SLMs with small pixel and high frame rate. The hypothetical parameters for the ideal single phase-only SLM electroholographic display system for an observer free to move laterally and free to rotate (move in 0.2×0.2 m^2 area parallel to the hologram plane, rotation angle 15° both horizontally and vertically) include 1.2 µm pixel interval, 0.35×0.35 m^2 hologram size with its doubling by temporal multiplexing (60 Hz frame rate), and 300×300 K number of pixels. Holography is a very narrow market, so we depend on the product development driven by a wider consumer market. It can be predicted that the increasing number of views provided by a classic autostereoscopic display (while preserving the image resolution) will demand for increasingly higher number of pixels. Also, the SLMs frame rate is consequently increasing driven by the progress in microdisplays and picoprojectors. The SLMs with higher frame rate allow to introduce a temporal data multiplexing, which virtually extends the number of available pixels. Consequently, we believe that the main technological hurdle associated with the realization of holographic displays will be overcome in the next 5–15 years by both the technical progress in the consumer market and the research activities targeted to progress in digital holography.

The reported development of the full chain of 3D holographic video system composed of multi-CCD capture setup and multi-SLM circular display is a good example of the progress toward realization of futuristic 3D holographic video. Works have shown that it is possible to obtain a high-quality, wide viewing angle 3D image based on holographic data representing a real-world object. This image can be seen with the naked eye, floating in space; however, in such case the size of an image is limited to LCoS SLM aperture. If the aperture is increased (physically or by temporal multiplexing), the scheme can be applied for small format displays as in a smartphone. In this application, the system miniaturization will be the critical issue, especially if a color image is required. On the other hand, it was shown that the implementation of full technology chain from multihologram capture, through compression, transmission, decompression, sophisticated processing, and display, is possible and enables nearly real-time viewing of a dynamic 3D scene. Merging of the ideas developed for the static and dynamic object capture and display and combining them with the principles of visual perception brings us closer to holographic implementation of the True 3D vision systems.

References

1. Yaras, F., Kang, H., and Onural, L. (2010) State of the art in holographic displays: a survey. *J. Display Technol.*, **6** (10), 443–454.
2. Sexton, I. and Surman, Y. (1999) Stereoscopic and autostereoscopic display systems. *Signal Process. Mag., IEEE*, **16**, 85–99.
3. Son, J.-Y. and Javidi, B. (2005) 3-dimensional imaging systems based on multiview images. *J. Display Technol.*, **1**, 125–140.
4. Lee, B. and Kim, Y. (2012) Three-dimensional display and imaging: status and prospects, in *Optical Imaging and Metrology: Selected Topics*, Chapter 2 (eds W. Osten and N. Reingand), Wiley-VCH Verlag GmbH, pp. 31–54.
5. Häussler, R., Schwerdtner, A., and Leister, N. (2008) Large holographic displays as an alternative to stereoscopic displays. *Proc. SPIE*, **6803**, 68030M.
6. Poon, T.-C. (ed.) (2006) *Digital Holography and Three-Dimensional Display*, Springer, Berlin.
7. Sainov, V. and Stoykova E. (2012) Display holography – status and future, in *Optical Imaging and Metrology: Selected Topics*, Chapter 4 (eds W. Osten and N. Reingand), Wiley-VCH Verlag GmbH, pp. 93–118.
8. Leith, E.N., Upatnieks, J., Hildenbrand, B.P., and Haines, K. (1965) Requirements for the wavefront reconstruction television facsimile system. *J. SMPTE*, **74**, 893–896.
9. Tomoyoshi, I., Nobuyuki, M., Kotaro, Y., Atsushi, S., Tomoyoshi, S., and Takashige, S. (2005) Special-purpose computer HORN-5 for a real-time electro-holography. *Opt. Express*, **13** (6), 1923–1932.
10. Nakayama, H., Masuda, N., Shiraki, A., Ichihashi, Y., Ito, T., and Shimobaba, T. (2008) Development of special-purpose computer HORN-6 for holography. 3D Image Conference 2008.
11. Kollin, J.S., Benton, S.A., and Jepsen, M.L. (1989) Real-time display of 3-d computed holograms by scanning the image of an acousto-optic modulator. *Proc. SPIE*, **1136**, 178–185.
12. Smalley, D.E., Smithwick, Q.Y.J., and Bove, V.M. Jr. (2007) Holographic video display based on guided-wave acousto-optic devices. *Proc. SPIE*, **648**, 64880L.
13. MIT Multimedia Lab, http://en.wikipedia.org/wiki/Zebraimaging (accessed 2012).
14. Javidi, B., Okano F., and Son, J-Y. (eds) (2009) *Three-Dimensional Imaging, Visualization and Display*, Springer, Berlin.
15. Huebschman, M.L., Munlujun, B., and Garner, H.R. (2003) Dynamic holographic 3D image projection. *Opt. Express*, **11**, 437–445.
16. Lazarev, G., Hermerschmidt, A., Kruger, S., and Osten, S. (2012) LCoS spatial light modulators: trends and applications, in *Optical Imaging and Metrology: Selected Topics*, Chapter 4 (eds W. Osten and N. Reingand), Wiley-VCH Verlag GmbH, pp. 1–30.
17. Stanley, M. *et al.* (2003) 100-megapixel computer generated holographic images from active tiling: a dynamic and scalable electro-optic modulator system. *Proc. SPIE*, **5005**, 247–258.

18. Hahn, J., Kim, H., Lim, Y., Park, G., and Lee, B. (2008) Wide viewing angle dynamic holographic stereogram with a curved array of spatial light modulator. *Opt. Express*, **16**, 12372–12386.
19. Yaras, F., Kang, H., and Onural, L. (2010) Multi-SLM holographic display system with planar configuration. Proceedings of 3D TV Conference: The True Vision-Capture, Transmission and Display in 3D Video, IEEE.
20. Yaras, F., Kang, H., and Onural, L. (2011) Circular holographic video display system. *Opt. Express*, **19**, 9147–9156.
21. Finke, G., Kozacki, T., and Kujawinska, M. (2010) Wide viewing angle holographic display with multi spatial light modulator array. *Proc. SPIE*, **7723**, 77230A.
22. http://www.digitalholography.eu (accessed 2011).
23. Ozaktas, H.M. and Onural, L. (2008) *Three-Dimensional Television*, Springer, Berlin.
24. Hashimoto, N. and Morokawa, S. (1993) Real-time electroholographic system using liquid crystal television spatial light modulators. *J. Electron. Imaging*, **2**, 93–99.
25. http://www.photonics21.org/download/SRA 2010.pdf (accessed 2011).
26. Meeser, T., Falldorf, C., von Kopylow, C., and Bergmann, R.B. (2011) Reference wave adaptation in digital lensless Fourier holography by means of a spatial light modulator. *Proc. SPIE*, **8082**, 8082206.
27. Shortt, A., Naughton, T., and Javidi, B. (2006) Compression of digital holograms of three-dimensional objects using wavelets. *Opt. Express*, **14**, 2625–2630.
28. Yamaguchi, I. and Zhang, T. (1997) Phase-shifting digital holography. *Opt. Letters*, **22** (16), 1268–1270.
29. Monaghan, D., Kelly, D., Hennelly, B., and Javidi, B. (2010) Speckle reduction techniques in digital holography. *J. Phys.: Conf. Ser.*, **206**, 012026.
30. Kelly, D.P. et al. (2010) Digital holographic capture and optoelectronic reconstruction for 3D displays. *Int. J. Digital Multimedia Broadcasting*, **2010**, doi: 10.1155/2010/759323.
31. Bastiaans, M.J. (1980) Wigner distribution function and its application to first-order optics. *J. Opt. Soc. Am.*, **69**, 1710–1716.
32. Lohmann, A.W., Dorsch, R.G., Mendlovic, D., Zalevsky, Z., and Ferreira, C. (1996) Space-bandwidth product of optical signals and systems. *J. Opt. Soc. Am., A*, **13**, 470–473.
33. Goodman, J.W. (1996) *Introduction to Fourier Optics*, 2nd edn, McGraw-Hill, New York.
34. Oppenheim, A.V. (1981) The importance of phase in signals. *Proc. IEEE*, **69**, 529–550.
35. Kozacki, T. (2010) On resolution and viewing of holographic image generated by 3D holographic display. *Opt. Express*, **18**, 27118–27129.
36. Kozacki, T., Kujawinska, M., Finke, G., Zaperty, W., and Hennelly, B. (2012) Wide viewing angle holographic capture and display systems in circular configuration. *J. Display Technol.*, **8** (4), 225–231.
37. http://www.holoeye.com/spatial_light_modulators-technology.html (accessed 2011).
38. Michałkiewicz, A., Kujawinska, M., Kozacki, T., Wang, X., and Bos, P.J. (2004) Holographic three-dimensional displays with liquid crystal on silicon spatial light modulator. *Proc. SPIE*, **5531**, 85–94.
39. Kozacki, T. (2011) Holographic display with tilted spatial light modulator. *Appl. Opt.*, **50**, 3579–3588.
40. Zaperty, W., Kozacki, T., and Kujawinska, M. (2011) Multi SLMs holographic display with inclined plane wave illumination. *Proc. SPIE*, **8083**, 80830X.
41. Delen, N. and Hooker, B. (1998) Free-space beam propagation between arbitrarily oriented planes based on full diffraction theory: a fast fourier transform approach. *J. Opt. Soc. Am. A*, **15**, 857–867.
42. Kozacki, T., Krajewski, R., and Kujawinska, M. (2009) Reconstruction of refractive-index distribution in off-axis digital holography optical diffraction homographic system. *Opt. Express*, **17** (16), 13758–13767.

43. Kozacki, T. (2008) Numerical errors of diffraction computing using plane wave spectrum decomposition. *Opt. Commun.*, **281**, 4219–4223.
44. Fienup, J.R. (1982) Phase retrieval algorithms: a comparison. *Appl. Opt.*, **21**, 2758–2769.
45. Latychevskaia, T. and Fink, H.-W. (2007) Solution to the twin image problem in holography. *Phys. Rev. Lett.*, **98**, 23390–23393.
46. Heikkilä J. and Silvén O. (1997) *Proceedings of the IEEE Computer Society Conference on Computer Vision and Pattern Recognition (CVPR'97)*, IEEE Computer Society, pp. 1106–1112.
47. McElhinney, C.P., Hennelly, B.M., and Naughton, T.J. (2008) Extended focused imaging for digital holograms of macroscopic three-dimensional objects. *Appl. Opt.*, **47**, D71–D79.

4
Display Holography – Status and Future
Ventseslav Sainov and Elena Stoykova

4.1
Introduction

Display holography is one of the most attractive outcomes of the method for wave front reconstruction, proposed in 1948 by Dennis Gabor [1], a Nobel Prize winner in physics for 1971. Three-dimensional imaging involves two steps – recording of light reflected from the object and reconstruction of the object's wave front from the recorded data. In holography, such an inverse task is solved by recording the interference pattern produced by coherent summation of the object beam, scattered from the object, with the reference beam, both beams coming from the same coherent light source. Thus the phase distribution, which is related to 3D coordinates of the object, is encoded as an intensity of the recorded interference pattern. In such a way, the phase information in the wave front is preserved, and a 3D image of the object can be completely reconstructed by diffraction of light from the recorded pattern at illumination with a copy of the reference beam. The emerging at reconstruction phase-conjugated wave fronts correspond to real and virtual 3D images. Information about 3D coordinates is completely lost in usual photography, where only the intensity of light coming from the object is recorded.

At reconstruction of a Gabor's hologram, recorded by means of an in-line or on-axis recording scheme, the real and virtual images overlap. This inherent drawback became, for a while, a serious obstacle for implementation of display holography. However, two crucial steps forward have been made almost simultaneously: (i) Emmet Leith and Juris Upatnieks [2] discovered off-axis holography by the introduction of a spatial frequency carrier through angular separation between the reference and object beams and (ii) Yury Denisyuk [3] proposed recording with opposite beams (in reflection) by direct illumination of the object through a light-sensitive plate and made possible reconstruction with a "point" source of noncoherent white light. In both cases, only one – real or virtual – image is reconstructed in direction to the viewer. The history of display holography would be incomplete without the fact that Gabriel Lippmann [4] advanced for the first time the idea of making light diffract from the recorded interference pattern in his

Optical Imaging and Metrology: Advanced Technologies, First Edition.
Edited by Wolfgang Osten and Nadya Reingand.
© 2012 Wiley-VCH Verlag GmbH & Co. KGaA. Published 2012 by Wiley-VCH Verlag GmbH & Co. KGaA.

famous color photography, which was awarded the Nobel Prize in physics in 1908. In Lippmann photography, standing waves are formed in a layer of ultrafine-grain silver halide emulsion by interference of the plane photographic image of the object focused on the emulsion and the same image formed after reflection from behind. The reconstructed color image is 2D because of the complete loss of phase information. Nevertheless, Lippmann's silver halide emulsion continues its successful existence nowadays for recording of extremely high spatial frequency holograms.

Light coherence is a vital requirement for holographic recording in order to produce a stable high-contrast interference pattern. Invention of lasers as the best high-power coherent light sources solved this task. Lasers made possible not only high-quality reconstruction of 3D images of real objects but paved the ground for different novel applications of the wave front reconstruction in interferometry, diffractive optics, storage of information, and parallel signal processing. These applications exploit the flexibility of holographic techniques for high-density information recording and processing of signals.

Very soon the wave front reconstruction method was named *"holography"* on the basis of the Greek words "holos" – full, complete and "grafos" – write, record. The name, although quite pretentious, has been quickly accepted and widely used. Leading scholars, scientific groups, and research laboratories have been involved in profound studies for characterization of holograms, improvement of light-sensitive materials and laser sources, and practical and industrial applications of the method [5]. As a result, holography became a new and a very promising part of the modern optics with its own methodology and a large field of applications. Display holography is only one of its branches. As a whole, this branch comprises methods for creation of visually observed 3D images that exhibit a parallax. Display holography owes its success to the fact that wave front reconstruction is the only correct approach for 3D imaging [6]. In comparison with the other stereoscopic and autostereoscopic imaging techniques, only holographic reconstruction provides all depth cues such as perspective, binocular disparity, motion parallax, convergence, and accommodation [7]. Resolution and terabytes information capacity of optical holograms are by several orders higher in comparison with computer graphics and raster displays. Holograms are characterized by a continuous parallax. Natural multiview perception of the 3D images, which can replace the originals without the need for physical copies, makes display holography well accepted in museums, theme parks, and trade shows. Its potential is still to be exploited in modern design, art, engineering, training, and education.

The main shortcoming of the analog display holograms is lack of interactivity. That is why the research emphasis has been recently placed on development of the 4D holographic display technique on the basis of computer-generated holograms (CGHs). Holograms could be created by calculating interference patterns or by rendering 2D multiview images and combining them as a stereogram. Since vast amount of data need be numerically processed, transferred, and stored, optically displayed dynamic CGHs nowadays have a very small size and poor resolution and color characteristics.

4.2
Types of Holograms

Holograms substantially differ depending on the process of recording and reconstruction. There exists a great diversity of designations, although for all holograms imaging is accomplished through diffraction of light from the recorded interference patterns. A factor that defines the properties of holograms is the distance from the object to the recording medium. In accordance with the scalar diffraction theory [8], the holograms recorded in the near or far zones of the scattered light are called *Fresnel* or *Fraunhofer holograms*, respectively [9, 10]. Very often in display holography image formation is made with an objective for appropriate scaling of the recorded scene. A special case of this approach is image plane holography, when the image is focused in the plane of recording. If the object is in the front focal plane of the objective, and the image is recorded in the rear focal plane by a superposition of plane waves, the hologram is called a *Fourier hologram*. A similar recording is applied in modern holoprinters to form a synthesized autostereoscopic display from 2D intensity images of an object or a scene that have been recorded at different angles.

Holograms can be described as reflection or transmission holograms depending on the mutual orientation of the object and reference beams. In transmission holograms, the directions of both beams coincide, whereas in reflection holograms they are opposite. The type of diffraction determines two large classes of thin and thick (or volume) holograms. Thin holograms are characterized by Raman–Natt diffraction, which leads to multiple diffraction orders. Bragg's diffraction dominates in volume holograms producing a single diffraction order, which is essential for the display and visual perception. The type of modulation in the light-sensitive material divides holograms into amplitude holograms with modulation of absorption and phase holograms with modulation of refractive index [11]. Best visual perception of the reconstructed 3D images is ensured by the phase volume off-axis transmission holograms and Denisyuk's reflection holograms. Angular sensitivity of volume transmission holograms is better in comparison with the volume reflection holograms [5]; however, reconstruction of transmission holograms requires strictly monochromatic light sources, even lasers. For this reason, these holograms find limited application, mostly for producing master holograms for holographic reproduction and in commercial displays for projection of attractive 3D images in space in front of the hologram. Denisyuk's reflection holograms possess the highest possible spatial frequency of the recorded interference pattern (more than $6000\,\text{mm}^{-1}$ in the visible) due to recording in opposite beams. They work as narrowband interference mirrors and ensure higher resolution of the reconstructed image itself. Most essential for the display holography is their inherent feature for reconstruction in the natural white light from a "point" light source, which could be a usual projector or direct sunlight. Thanks to this feature, large holographic exhibitions to present monochrome, color, and pseudo-color 3D images of real and art objects have become possible [12–14].

A separate branch in display holography involves recording with pulse lasers to capture mechanically unstable or life objects. Pulse recording is mostly used for holographic portraitures [15]. The most attractive and acceptable visual perception portraitures involve a two-step process [16]. The first step implies recording of the so-called master hologram. For the second step, the reconstructed virtual image from the master is used as an object beam to produce the final image plane reflection hologram. The drawback of the first human holographic portraitures [17], made by a ruby laser (wavelength 694 nm), was penetration of radiation in the human skin. This effect, combined with the lack of movement, leads to unusual and nonesthetic appearance of the reconstructed images. Obviously, 30–50 ns exposures are good for research and interferometry, in particular, but they are questionable for holographic portraits. There are expectations that a solution to this task is to use holoprinters, holographic movies, and 3D dynamic holographic displays.

Another important branch in display holography is rainbow holography, proposed by Steven Benton in 1969, which yields high-quality images at reconstruction in white light from modified transmission holograms [18–21]. The method usually includes two steps: (i) holographic recording of a master transmission hologram of the object and (ii) holographic recording of the reconstructed image from the master virtual image, which acts as an object beam. The most essential is the use of a horizontal slit at recordings. The presence of the slit eliminates the vertical parallax in the reconstructed final image. Depending on the illumination and observation angles, horizontal-only parallax images appear in different narrow spectral bands in the visible resembling a rainbow modulated by the object, which gives the holograms their name. Rainbow holograms are suitable for creation of very attractive pseudo-color artistic scenes, even when the recording is performed at a single wavelength and especially when they are made using several master holograms. A similar two-stage technique is used for compensating light dispersion in achromatic holograms [6, 22] and for synthesis of holographic stereograms recorded from a sequence of 2D pictures.

Spatial frequency of rainbow and achromatic holograms is less than $1500\,\mathrm{mm}^{-1}$. This allows for surface relief modulation in the second step of the holographic process. Maximum diffraction efficiency is achieved at a depth of modulation less than or equal to a half of the wavelength used for reconstruction [23, 24]. The surface relief holograms emerge as a separate branch in display holography, paving the ground for mass production of low-price stamped holographic copies. Schematically, the technology for their production includes recording of a surface relief hologram from the master/s onto a positive or negative photoresist. After metallization and the so-called electroforming, the obtained relief is transferred to the metal, usually nickel matrix, for stamping the relief onto thermoplastic layers. Maximal diffraction efficiency of metallized surface relief holograms is 100%. This technique allows for mass production of identical low-price holographic replicas for holographic illustrations; security of documents, banknotes, and credit cards; and for packaging, labeling, decorations, jewelry, and so on. For the first time, such illustrations appeared in the cover pages of the *National Geographic* journal more than 20 years ago. Nowadays, production of stamped holograms is the most profitable holographic technology with more than 10% annual growth, achieving up to 10 billion US dollar annual turnover. However, one should have in mind that most of the above-mentioned industrial applications are not based strictly on the holographic principle. Instead of real holograms, simple diffractive optical elements are used for creation of 2D images in different colors, similar to rainbow holograms. Typical examples are the so-called "dot matrix holograms," which are CGHs composed from millions of tiny diffraction gratings ("holopixels") that have different orientations in a two-dimensional array. Illumination with white light leads to formation of 2D color images that represent variable views of an object or scene. This could also create an impression of a moving object. The main drawback of "holograms" composed from diffraction gratings when applied for security purposes is the risk of their reproduction since the period and orientations of the gratings can be easily determined. Volume reflection holograms ensure better protection against forgery. Image plane reflection holograms copied from a transmission or reflection master hologram provide 3D reconstruction with acceptable quality even at illumination with scattered light, but only for small depth of the objects. Mass production of such holograms onto photopolymers at flexible substrates for security purposes and illustrations has been already realized by the firm DuPont.

4.3
Basic Parameters and Techniques of Holographic Recording

Holographic recording and reconstruction of optical information comprise three successive steps: (i) formation of the interference pattern, (ii) its registration in a light-sensitive material, and (iii) reconstruction of the 3D object image. There are various parameters for quality assessment of the holographic process. Optical efficiency of the recorded hologram is characterized by diffraction efficiency. It gives the intensity of the diffracted light, which forms the useful image as a percentage of the intensity of the reconstructing light beam. Diffraction efficiency determines the brightness of the reconstructed image. Another essential parameter is the contrast, $V = (I_{max} - I_{min})/(I_{max} + I_{min})$, which shows the depth of modulation, where I_{max}, I_{min} are the maximum and minimum values of light intensity in the recorded fringe pattern.

In 1960, H. Kogelnik [25] proposed the so-called "coupled wave theory," which considers amplitude, phase, or amplitude-phase volume holograms as Bragg's phase gratings [8] and allows for evaluation of maximum values of diffraction efficiency at sinusoidal 100% modulation of the recorded fringe patterns for transmission and reflection holograms (Table 4.1). Diffraction efficiency depends on the modulation of the parameter that influences the reconstructed beam: absorption, refractive index, and thickness of the light-sensitive layer. The dependence of the modulated parameter on the exposure, defined as a product of light intensity and exposure time, $E = I \times t$, is called an *exposure characteristic*. For convenience,

Table 4.1 Theoretical maximum value of diffraction efficiency (%) [25].

Volume (thick) holograms				Plane (thin) holograms		
Transmission		Reflection		Transmission		Reflection
Amplitude	Phase	Amplitude	Phase	Amplitude	Phase	Phase
3.7	100	7.2	100	6.3	33.9	100

it is usually presented as a change of diffraction efficiency. The linear part of the exposure characteristic gives the dynamic range of recording, limited by the exposure values, which correspond to the minimum and maximum diffraction efficiency in the linear section of the exposure characteristic. The dynamic range determines the energy sensitivity and the working exposure in the holographic experiment. Exposure characteristics of real light-sensitive materials could be completely different from the ideal one, which exhibits linear increase starting from zero. Estimation of diffraction efficiency for multiexposure recording is a key issue for color, rainbow, and synthesized holograms. Since several interference patterns are registered successively on a single light-sensitive material, the limited dynamic range is divided between individual recordings, and each consequent recording exerts an impact on the previously made ones and decreases their efficiency. For multicolor recording, such a scenario could destroy completely the required color balance.

The next basic parameter of holograms is signal-to-noise ratio, which may have various definitions, for example, a ratio of the mean light intensity in the useful image to the averaged intensity of the noise in the observation plane. Roughly it gives the number of brightness gradations that can be perceived. Noise is any parasitic light that overlaps the reconstructed image. Noise of a different nature occurs both in recording and reconstruction. Light scattering in the bulk of the light-sensitive material, restricted dynamic range, and low resolution change the contrast of the interference pattern and cause nonlinear distortions and inaccurate recording of the wave front. Inevitable self-interference of the light waves coming from the object produces intermodulation noise. Its suppression is achieved by increasing the intensity of the reference beam at the expense of efficiency. Multiple noisy images appear in reconstruction with an extended light source and/or due to the background illumination. The consequence of all effects is parasitic diffraction and apparent images, which overlap in the direction of observation.

Spectral and angular selectivity of holograms describe the dependencies of diffraction efficiency on the wavelength and illumination angle at reconstruction. They give the acceptable misalignments in the geometry of the reconstruction scheme and determine the limits of multiexposure recording and reconstruction with white light. Similar to aberrations in classical optics, geometrical distortions may worsen the reconstructed image if the wavelength, direction, and the front of the wave that illuminates the hologram at reconstruction differ from the same parameters of the reference wave at recording. Careful analysis of the optical schemes for recording, copying, and reconstruction is mandatory in display holography.

Resolution of holograms is limited by diffraction. In accordance with the Rayleigh criterion, resolution in the object and image planes is given by $\delta \propto \lambda z/D$, where λ is the wavelength, z the distance between the hologram and the object or reconstructed image, and D the size of the hologram. The hologram angular resolution δ/z depends on its size, whereas its spatial resolution depends on its angular dimensions. These parameters are essential for the analysis of the relationship between holographic resolution and resolution of the light-sensitive material. The latter is given by the maximum number of fringes (lines) per unit length that are registered as separate. It should substantially exceed the spatial frequency of the recording. Resolution of the light-sensitive material is limited by the size of elementary regions or volumes, which create stable in space and time modulation. Increase in the linear dimensions of the hologram and resolution of the light-sensitive material do not always entail an increase in holographic resolution. For a plane reference beam, the resolution in the object plane is less than the resolution of the light-sensitive material; the latter could be reached only for a point object. This is valid for Fresnel and Fraunhofer holograms. Spatial resolution of Fourier holograms achieves values on the order of a wavelength, and therefore yields the highest values of information capacity for holographic recording. Theoretically, the maximum possible density of recording in the light-sensitive material in the 2D case is one bit in a region of size λ_C^2, whereas for the 3D case – one bit in a volume of size λ_C^3 where $\lambda_C = \lambda/n$, and n is the refractive index of the material. Information capacity for a unit square or a unit volume is respectively $C_{2D} = 1/\lambda_C^2$ and $C_{3D} = 1/\lambda_C^3$. At $\lambda = 0.5$ μm and $n = 1.5$ we have $C_{2D} = 4 \times 10^6$ bit mm^{-2} and $C_{3D} = 10^{10}$ bit mm^{-3}, respectively. The theoretical limit can be achieved if the resolution of holograms is equal to the wavelength. The limited aperture of optical systems and the related diffraction phenomena decrease the resolution. The resolution and signal-to-noise ratio of the light-sensitive material are also crucial. However, the information capacity of Fourier holograms is not limited by the resolution of the recording media as in the other cases and is an order of magnitude greater. Practically it could reach 10^9 bit mm^{-3} for the volume recording in the visible spectral region.

Technical equipment for display holography has been dramatically improved in recent years. Compact high-power continuous wave (CW) and pulse-generating diode and diode-pumped solid-state lasers with a proper coherence length in the visible, high-resolution digital cameras, spatial light modulators, computer techniques, and low-energy-consuming light sources with controllable intensities in blue, green, and red spectral regions for reconstruction, optical, optoelectronic, and mechanical elements, as well as different light-sensitive materials for holographic recording are available on the market.

The basic elements in the holographic scene are the light sources – laser, optical elements (beam splitters, beam expanders, spatial filters, collimators, and mirrors) and mechanical elements required for stable assembling on a holographic table.

The most appropriate for display holography are the CW lasers emitting in the visible: gaseous lasers as a He-Ne laser (632.8 nm), HeCd laser (442 nm), and ion lasers Ar^{2+} (456.5, 476.5, 488.0, and 514.5 nm), and Kr^{2+} (530.9, 647.1, and 676.4 nm); temperature-stabilized diode lasers (635–658 nm) and diode-pumped solid-state lasers on the second and third harmonic (440, 532, and 660 nm). The coherence length of the gas and especially ion lasers is limited, but it can be increased substantially by incorporation of Fabry–Perout etalons in their resonators. The primary task in the holographic experiment is stable recording with a maximum contrast of the interference pattern. The contrast of the interference pattern decreases when the difference between the optical paths of the two interfering beams is close to the coherence length of the laser. The contrast worsens at the nonparallel polarizations of the object and reference waves. Usually, linear polarization is used. Rotation of the polarization plane occurs at refraction, reflection, and scattering; for some diffusely reflecting surfaces, the object wave is totally depolarized. This makes necessary the usage of additional optical elements in the holographic scheme – polarizers, analyzers, and rotators.

Displacements of the elements in the holographic scheme, even at a fraction of a micrometer, lead to phase changes in the interfering beams, which decrease the contrast, causing partial or total vanishing of the pattern. Stability is especially crucial for holographic recording with CW lasers, for which exposure duration could be even minutes depending on the light power, scene dimensions, and sensitivity of the light-sensitive material. Mechanical stability is achieved by mounting the object and the holographic scheme on a vibration-insulated table. Mounting of lasers on the same table is not always justified, especially in the cases of their water or air cooling, which is a source of additional heat and vibration.

The setting of a holographic scene is a key element of display holography. The main art composition rules equally apply in this case. The desired artistic impact of the reconstructed image substantially depends on the spatial arrangement of the objects, background, shadows, and occlusion. This means that one should choose the geometry of recording for each object individually. The size and positioning of the light-sensitive material depends on the chosen recording scheme and the size and spatial position of the object. A successful solution is to use variable module structures with holders for the photoplate, object, and optical elements.

4.4
Light-Sensitive Materials for Holographic Recording in Display Holography

A light-sensitive material as a recording medium is a basic element in the holographic scheme. The media used in display holography are the media for permanent recording with high values of diffraction efficiency and signal-to-noise ratio. Materials with phase and amplitude-phase modulation satisfy the above requirements. Some of the frequently used light-sensitive materials in display holography are described below [26, 27].

4.4.1
Photoresists

Photoresists [28] are polymers that change their solubility under illumination. They are divided into negative or positive, depending on which part of the layer is removed by the solvent – non-illuminated or illuminated, respectively. After processing with the solvent, a surface relief, which is used to produce nickel matrices for stamping of rainbow or achromatized holograms on a thermoplastic layer, is formed. Although the resolution of photoresists in the visible is less than 3500 mm^{-1}, it exceeds the carrying frequency of the rainbow holograms. Photoresists are sensitive in the short wavelength region and require exposures of tens of millijoules per square centimeter. The stamped holograms are used in art holography, for illustrations and for the security of documents. Production of the metal matrix is a time-consuming and expensive process, so the price of the produced holograms becomes acceptable only in the case of mass production. A new development of surface relief holograms is based on diffraction of plasmons, propagating at the surface of a thin metal film, which covers a surface relief RGB hologram, recorded onto a photoresist with blue, green, and red lasers. The surface plasmons [29, 30], excited at different illumination angles with white light, diffract from the hologram and create full color 3D images whose color does not depend on the viewing angle and remains the same as that of the original object [31].

4.4.2
Dichromate Gelatin

The property of chromium ions for dubbing of gelatin, well known from polygraphy, is used for holographic recording in dichromated gelatin (DCG). In the presence of small quantities of ammonium, potassium, or other dichromate, transversal molecular cross-links build up in the gelatin matrix. At illumination in the short wavelengths range, this dubbing effect is increased by a photochemical valence transfer $Cr^{+6} \rightarrow Cr^{+3}$, which saturates new gelatin bonds. Local dubbing at the molecular level forms surface relief and very high modulation of the refractive index in the bulk of the layer. Holograms are produced in the following progression. Preliminary dubbed gelatin layers with thickness up to 20 μm are sensitized in a bath of 3–10% water solution of ammonium dichromate at room temperature for 5–10 min. During the bath the gelatin layer swells up and its initial thickness increases many times. After that the layer undergoes uniform and slow drying in darkness at room temperature. Thus sensitized, the photosensitive layer is comparatively stable for only several hours. However, industrial manufacture of stabilized DCG holographic plates with prolonged lifetime (up to 12 months) has been also reported [32]. After exposure the layer is washed up with water for removal of the dichromate remainder. The last step of the development process is fast dehydration in several successive baths of isopropyl alcohol with increasing concentration. Further, the hologram undergoes drying, thermal processing, or hermetical sealing, which makes impossible its moisturizing and hence deterioration of phase modulation. The hologram recording in the DCG layer has not yet been fully clarified. During development, the phase modulation is intensified through the dehydration stage; in addition, simultaneous shrinkage of the gelatin layer forms internal stresses the distribution of which corresponds to local dubbing induced by illumination. Depending on the degree of the total dubbing, the forces due to local stresses could evoke tearing of gelatin bonds and gradient splitting of the gelatin layer along the interference fringes in depth. In this case, peak values of diffraction efficiency for reflection holograms are observed in a widened spectral range. This creates impressive reconstruction of some black and white and metal objects from nickel, silver, metal alloys, and others with color tinctures at white light illumination. This processing is used successfully for holographic recording of numismatic collections. Modulation can be achieved also without splitting in the bulk of the layer because of the formed inner nonhomogeneities of the refractive index during the dubbing process. Formation of a complex compound from the isopropyl alcohol and the trivalent chromium ion at transversal stitching of gelatin molecules, which leads to a bulk phase modulation, has been achieved. In this case, reconstruction occurs in a narrow spectral range with the maximum at the recording wavelength. The main drawback of the DCG is comparatively low sensitivity, which entails high (20–100 mJ cm^{-2}) exposures in the short-wavelength spectral range. Improvement of sensitivity and sensitizing in the long-wavelength spectral range is achieved through different sensitizing inclusions. Another approach is to use silver-halide-sensitized gelatin [33]. Silver halide provides higher exposure sensitivity as well as local dubbing of gelatin during the chemical processing, being completely removed after that. In this case, pure phase modulation connected to the local dubbing of the gelatin layer is obtained. The research in this field, however, obviously ceases to be a pressing task because of recent advances of photopolymers as light-sensitive media in display holography. The resolution of DCG extends 6000 mm^{-1} lines in the visible spectral range. It has uniform response for all spatial frequencies and low absorption in a wide spectral range. The thickness of the DCG layer can be modified before or after exposure to obtain maximal efficiency at the desired wavelength. This is the main advantage in comparison with the other light-sensitive materials, which normally shrink after development. A self-developing property of DCG has been reported [34]; however, the obtained diffraction efficiencies are very low for display holography.

4.4.3
Photopolymers

Photoinduced changes in polymers are the result of various photochemical reactions such as cross-linking, photodestruction, and photoisomerization [35]. Cross-linking leads to polymerization of single molecules (monomers) in the material at light illumination. As a result, optical and mechanical parameters of the medium change, and phase modulation is formed both because of the creation of a surface relief and a change of refractive index in the bulk of the layer. Phase modulation in light-sensitive organic systems is also created by formation of transversal binding, photodestruction, used in photoresists, and trans–cis isomerization, which is typical for materials containing azo-groups – two phenyl rings – connected with a double nitrogen bond [28]. The two possible trans or cis states of azo-groups change at illumination, leading to phase and amplitude modulation. High value of photoinduced anisotropy could be obtained by introducing azo-dyes in the polymer matrix, used for polarization holographic recording [36]. Promising results have been obtained with asobenzene liquid crystalline and amorphous polymers in which the azo-groups are incorporated in the molecular chain, leading to surface relief and refractive index modulation. There are several known types of photopolymer systems. They contain acrylamide, acrylates and other monomers, organic dyes that make them sensitive in the visible, and sensitive in the UV region inclusion, which stops the process of polymerization when illuminated, and fixates the holograms. The whole processing is dry (waterless). There are also known *multicomponent systems* with different rates of photochemical reactions for different monomers. During the recording, the rapidly polymerizing components with changed viscosity are localized in the zones with maximum irradiance and displace the liquid monomer, which polymerizes afterwards. Owing to the change of the optical properties, phase modulation in the bulk of the layer occurs. Diffraction efficiency of the recorded volume holograms reaches almost 100% at exposures of the order of tens of millijoules per square centimeter, but resolution in this case is not higher than 6000 mm^{-1}. Deterioration of photopolymer resolution is observed because of growing of polymer chains away from their initial position.

The most promising recording media for holographic display are the novel composite and nanoparticles containing light-sensitive materials, which are also called *discrete carriers* for holographic recording. In the carrying polymer matrix of the light-sensitive layer there are inclusions of nanoparticles from SiO_2, TiO_2, diamond, zeolite, and others, as well as liquid-crystal droplets, known as *holographic-polymer-dispersed liquid crystals*, having a different refractive index from that of the carrying matrix [24, 28, 37–39]. For sensitizing in the whole visible spectral range, an appropriate combination of light absorbers (dyes) and photoinitiators has to be used. Light-excited organic molecules work as molecular engines for mass transfer and redistribution of the particles in the bulk of the photopolymer layer in accordance with the intensity of the recorded interference patterns. The result is high values of diffraction efficiency, resolution, and signal-to-noise ratio at substantially lower exposures. Very nice results have been already reported – several millijoules per square centimeter exposure energy in the visible for almost 100% diffraction efficiency and more than 6000 mm^{-1} resolution. Research efforts are focused on optimization of the processes for better reliability and repeatability [28].

4.4.4
Silver Halide Emulsions

Silver halide emulsions [27] are typical representatives of discrete carriers for permanent holographic recording, which occurs in separate silver halide crystals, isolated from each other, dispersed in a carrying matrix (gelatin). These media are still the main light-sensitive material in display holography. Their main advantages are the higher energy sensitivity and the wider dynamic range in comparison with the other materials for permanent holographic recording, as well as very high resolution, which at 10 nm average size of the crystals (ultrafine-grain emulsion) is much greater than 6000 mm^{-1}. The photoemulsion layer is usually very thin with thickness less than 10 μm. Since the natural spectral sensitivity of silver halides is in the short wavelengths range, they are sensitized in the visible using organic dyes (sensitizers). Energy sensitivity depends on the type and size of the silver halide crystals, efficiency of the sensitizer, and the developing composition used. High sensitivity is due to formation of the so-called hidden (latent) image, which is multiple times increased (more than 10^8 times) at development. During illumination of the silver-halide crystal the photons promote electrons to the conduction band, where the electrons migrate till the moment when they are trapped by a shallow electron trap (sensitivity speck), which may be a crystalline defect or a cluster of silver sulfide, gold, and other dopants. The electron recombines with a positively charged silver ion from the crystal lattice to form a silver atom, which is unstable and has a lifetime on the order of seconds. During this period, another capture of an electron and recombination of a silver ion are highly probable, which leads to formation of a stable metal speck from two or more silver atoms and hence to the hidden (latent) image. These centers of a latent image or developing centers are localized on the surface as well as inside the crystals. They undergo development, growing to silver grains, which contain millions of silver atoms. In practice, there are two developing processes: (i) chemical development at which silver is supplied due to chemical reconstruction of the silver-halide crystals and (ii) physical development at which silver is supplied from silver complexes in the developing solution. During the chemical development, the centers grow like fibers and form silver structures with a complex shape that cause strong absorption, typical for amplitude modulation. To obtain phase modulation one should use bleaching – transfer of the developed silver particles into transparent silver acid salts. At physical development, enlarging of compact silver particles (colloidal silver) occurs, resulting in low absorption in the complex refractive index, which ensures amplitude-phase modulation with high diffraction efficiency without bleaching. The main advantage of these holograms is their extremely high stability for a long time.

In photography, cinematography, and microelectronics the useful effect is absorption of the developed silver halide grains. The diffractive mechanism of image formation in holography sets new requirements on the developing process. Distribution of the developed grains should correspond to intensity distribution in the recorded interference pattern in the bulk of the light-sensitive layer. This entails higher resolution than that determined by the spatial frequency of the interference pattern, so in the case of Denisuyk holograms the resolution should be much greater than 6000 mm^{-1}. The development process should guarantee such size, absorption coefficient, reflection coefficient, and spatial distribution of the developed silver particles, as to achieve in-phase reflection. However, even at the optimum size and reflectivity, the migration of the developed silver particles from the position determined by the interference pattern in the bulk of the layer destroys the in-phase condition and the contrast and diffraction efficiency decrease.

The mentioned requirements are satisfied to a large extent at fixing developing, during which the initial silver halide crystals dissolve. As a first step, the surface centers increase at chemical developing, and after that – due to physical development. During fixation, the size of the silver halide crystals on the surface decreases and inner centers, which grow in a similar way, are revealed. As a result, the form of the developed grains is spherical, with a strongly reflecting surface. The growing centers are steadily trapped in the gelatin-carrying matrix and do not migrate. To ensure the required size, concentration, and distribution corresponding to irradiance, one should choose optimal rates of solving and developing processes. This is achieved by a proper dosage of solvents and developing and stabilizing agents in the developer composition. Fixation comprises solving and removal of the nondeveloped silver halide crystals from the emulsion layer. A typical fixing substance is sodium thiosulfate. In fixing developers, the unused silver halide dissolves completely during the development and no additional fixing is required. In separate cases, after the chemical and physical development, the holograms undergo no fixing, because the nondeveloped transparent silver halide crystals in the minima of the interference pattern have greater refractive index, leading to phase modulation, as in bleached holograms. The drawback of bleaching is a strong increase of noise, limitation of modulation-transfer characteristics and nonlinear distortions. Nevertheless, the procedure finds wide application in display holography because of technological simplicity and high brightness of the reconstructed images. After bleaching, the silver halide crystals darken in time (the so-called print-out effect) and diffraction efficiency decreases. Because of that, reducers are introduced for stabilization in order to make the probability of the silver-halide-to-pure-silver reconstruction less than the probability of the opposite process – binding of silver with halogen ions and formation of silver halide. For this purpose, different organic dyes, bismuth ions with desensitized action and ability to protect from short-wavelength radiation, are used. This partially solves the task and ensures stabilization for long-term storage and exploitation.

The dissolved and removed nonexposed silver halide crystals do not always compensate for the volume of the developed silver grains, which results in shrinkage of the layer after drying up. Shrinkage diminishes the distance between the interference fringes and changes the condition for Bragg diffraction. To compensate this effect, fillers are introduced to increase the thickness of the dried layer to its required value.

It is clear that the manifold influence of development, fixing, washing, swelling, and drying up processes on the gelatin layer makes the achievement of the desired correspondence between the spatial distributions of the developed grains and intensity at recording of the interference pattern difficult. Although the photographic silver halide emulsions have more than a century's history of development and application, currently there is a lack of commercially available suitable materials for multicolor holographic recording. Invasion of digital photosensors in the photographic industry reduced substantially the production of silver halide

Table 4.2 Characteristics of materials for permanent holographic recording in the visible.

Light-sensitive material	Thickness of the layer (μm)	Spectral sensitivity (nm)	Exposure energy (mJ cm^{-2})	Type of modulation	Maximal resolution (lines per mm)	Maximum diffraction efficiency (%)
Silver halide emulsion	5–10	400–700	0.05–2	Phase	1 000–10 000	80–98
DGG	5–30	400–700	10–300	Phase	10 000	99
Photopolymers	5–100	400–700	10–1 000	Phase	6 000	99
Photoresists	1	400–650	10–100	Phase	3 500	30–95

light-sensitive materials. Many of the former producers of such materials as Agfa, Kodak, and other firms have entirely stopped industrial production. Limited quantities of silver halide emulsion are produced in research laboratories but mainly for scientific applications [40]. Today, the materials produced by Slavich, Russia [32, 41] and Ultimate, France [42] are the only ones available on the market. Technologically, producing such materials is a rather complicated task because of the thermodynamic instability of the ultrafine grains and their photosensitizing. The thermodynamic instability leads to growth of the grains and deterioration of the holographic characteristics such as sensitivity, diffraction efficiency, and signal-to-noise ratio. It is well known that the temporal stability of the emulsion strongly correlates with its monodispersity. Another factor that substantially affects the lifetime of the silver halide holographic materials is the temporal stability of the used photosensitizers for recording in the green and red spectral regions. Development of a new high-sensitivity (0.1 mJ cm^{-2}) panchromatic silver halide emulsion as a commercial product will have many spheres of impact such as security, cultural heritage and modern art, education, advertising, and display systems, including future 3D dynamic holographic display.

The main characteristics of the materials for permanent holographic recording in the visible are summarized in Table 4.2.

4.5
Diffraction Efficiency of Discrete Carrier Holograms

Analysis of holographic characteristics of light-sensitive materials with dispersed nanosized particles in a carrying matrix can be made on the basis of the Gustav Mie theory for light scattering from spherical particles [43]. According to Mie's theory, the complex amplitude $\overline{S}(\theta, \overline{m}, x)$ of the light scattered from a single spherical particle depends on the scattering angle θ, complex refractive index $\overline{m}$, and effective size of the particle $x = kr$ ($k = 2\pi/\lambda$ and r is the radius of the particle). If the average distance between the particles is an order of magnitude higher than their size, multiple scattering of the light can be neglected and a complex, the so-called

4.5 Diffraction Efficiency of Discrete Carrier Holograms

"effective" refractive index, is introduced as follows:

$$\bar{n} = n + iNk^{-3}\bar{S}(\theta, \overline{m}, x)$$

where n is the refractive index of the carrying matrix (gelatin) and N is the concentration of the developed grains. At illumination with a linear polarized monochromatic light, we have

$$\bar{S}(\theta, \overline{m}, x) = \sum_{l=1}^{\infty} \frac{2l+1}{2(l+1)} \left(\bar{a}_l \pi_l + \bar{b}_l \tau_l \right)$$

where $\bar{a}_l, \bar{b}_l$ are the so-called Mie coefficients that depend on particle size and complex refractive index:

$$\bar{a}_l = \frac{A_l(y)\Psi_l(x) - \overline{m}\Psi'_l(x)}{A_l(y)\Psi_{\varsigma_l}(x) - \overline{m}\varsigma'_l(x)}, \quad \bar{b}_l = \frac{\overline{m}A_l(y)\Psi_l(x) - \Psi'_l(x)}{\overline{m}A_l(y)\Psi_{\varsigma_l}(x) - \varsigma'_l(x)}, \quad A_l(y) = \frac{\Psi_l(y)}{\Psi'_l(y)}$$

where $y = \overline{m}x$, $\Psi_l(x)$, $\Psi'_l(x)$, and $\varsigma_l(x)$, $\varsigma'_l(x)$ are the Ricatti–Bessel functions and their derivatives, respectively. The coefficients π_l, τ_l depend only on the scattering angle and can be expressed as

$$\pi_l(\theta) = \cos\theta \frac{2l-1}{l-1} \pi_{l-1}(\theta) - \frac{l}{l-1} \pi_{l-2}(\theta)$$

$$\tau_l(\theta) = \cos\theta \left[\pi_l(\theta) - \pi_{l-2}(\theta) \right] - (2l-1)\sin^2\theta \times \tau_{l-1}(\theta) + \tau_{l-2}(\theta)$$

with $\pi_0(\theta) = 0$, $\pi_1(\theta) = 1$, $\pi_2(\theta) = \cos\theta$, $\tau_0(\theta) = 0$, $\tau_1(\theta) = \cos\theta$, and $\tau_2(\theta) = 3\cos 2\theta$. The calculated values of the complex refractive index are used to obtain the diffraction efficiency for transmission and reflection holograms using the Kogelnik formulas [44, 45]. They give diffraction efficiencies as a function of the absorption coefficient, α, and the amplitude of its modulation, $\alpha_1/2$, amplitude of the refractive index modulation, n_1, thickness of the layer, d, and the Bragg's angle, θ_B. Maximal concentration of the particles is determined from the necessary requirements $n_1 \ll n$ and $\alpha_1 \ll 2\pi n/\lambda$ to apply the proposed approach. It can be applied to all nanoparticle-containing carriers of holographic recording.

In Tables 4.3 and 4.4, we present the results of calculations for $d = 8\,\mu m$ and $n = 1.55$. Table 4.3 gives the values of the diffraction efficiency for bleached holograms for diameters of the bleached grains 20 and 100 nm at $n_1 = 0.01$ and $\lambda = 632.8\,nm$. As expected, the diffraction efficiency reaches 95% in recording volume Bragg's holograms for particles with a diameter of 20 nm on increasing the concentration. Such high values of diffraction efficiency are achieved within a wide range of diffraction angles ($\theta_B \sim 10° - 85°$). The efficiency substantially decreases at a larger size (100 nm) of the developed bleached grains. Tables 4.4 and 4.5 give the results for the nonbleached volume holograms at the diameter of the developed silver grain 20 and 30 nm, respectively. In both cases, diffraction efficiency reaches values greater than 90%. However, at 30 nm the efficiency is higher, which could be explained with the so-called back scattering Mie' effect.

We should outline that the theoretical maximum diffraction efficiency is more than 90% for the small size of the developed grains at concentrations that are

Table 4.3 Diffraction efficiency for 20 and 100 nm size of the particles (%).

Concentration (μm^{-3})		Reflection		Transmission	
20 nm	100 nm	20 nm	100 nm	20 nm	100 nm
53 240	440	94.74	31.63	93.68	8.82
39 930	294	85.51	30.53	95.58	19.32
26 620	147	63.05	21.98	62.18	18.14
13 310	59	24.44	7.33	19.27	5.60
5 324	44	4.55	4.65	3.27	3.48
3 993	29	2.60	2.32	1.85	1.70
2 662	11	1.16	0.65	0.82	0.47
1 331	–	0.29	–	0.21	–

Table 4.4 Diffraction efficiency for 20 nm size of the particles (%).

Concentration (μm^{-3})	Reflection, λ (nm)				Transmission, λ (nm)			
	510 nm	550 nm	630 nm	650 nm	510 nm	550 nm	630 nm	650 nm
75 000	94.65	94.65	95.82	95.95	11.95	11.95	34.96	56.84
50 000	94.65	94.65	95.82	95.95	7.74	7.74	12.35	0.66
25 000	94.65	94.65	95.80	95.92	2.78	2.78	77.48	82.45
12 500	93.83	93.83	93.15	92.65	0.82	0.82	35.37	45.11
6 250	78.74	78.74	68.61	66.16	91.81	91.81	84.43	81.41
3 125	39.90	39.90	29.39	27.46	45.81	45.81	32.54	30.20
1 250	8.62	8.62	5.78	5.32	8.88	8.88	5.89	5.41
625	2.27	2.27	1.49	1.37	2.29	2.29	1.50	1.38

Table 4.5 Diffraction efficiency for 30 nm size of the particles (%).

Concentration (μm^{-3})	Reflection, λ (nm)				Transmission, λ (nm)			
	510 nm	550 nm	630 nm	650 nm	510 nm	550 nm	630 nm	650 nm
22 222	93.31	96.62	99.13	99.62	24.39	4.64	52.55	94.74
14 815	93.31	96.62	99.13	99.62	37.27	2.57	18.04	0.03
7 407	93.31	96.62	99.11	99.58	22.71	0.80	90.86	98.22
3 704	92.96	95.85	95.36	96.14	7.96	0.22	37.81	49.16
1 852	82.66	80.92	70.74	68.26	86.68	94.78	88.22	85.11
926	46.54	41.29	30.05	28.02	54.57	47.67	33.36	30.89
370	10.82	8.93	5.85	5.36	11.23	9.21	5.97	5.46
185	2.90	2.35	1.51	1.38	2.93	2.37	1.52	1.38

10 times smaller than the maximum admissible in accordance with the Mie and Kogelnik theories. However, the experimentally obtained values of diffraction efficiency are less than 60% for nonbleached and 80% for bleached silver halide reflection holograms. This is due to scattering of light at recording, nonuniform (gradient) development, growing up of residual stresses in the bulk of the layer; all these affect the developed grains distribution and decrease the contrast, efficiency, and signal-to-noise ratio. This hypothesis is confirmed by the experimentally observed rise of diffraction efficiency for nonbleached holograms after thermal processing and long-term storage, during which partial or total relaxation occurs within the layer. Such materials are successfully used for monochrome holographic recording of homogeneous objects – bronze, copper, gold, steel, stone, wood, and others. A proper processing, including swelling of the developed hologram, ensures Bragg's diffraction at wavelengths coinciding with the spectral characteristics of the object. The reconstructed images from nonbleached silver halide reflection holograms are presented in Figures 4.1–4.3.

4.6
Multicolor Holographic Recording

Manufacture of effective and reliable materials for multicolor (RGB) holographic recording is nowadays a mandatory task in view of the potential mass markets for holographic products. A lot of research has been dedicated to development of an ideal panchromatic medium that should exhibit high sensitivity, resolution, and signal-to-noise ratio across the visible region, and excellent performance in creation of stable permanent images. High reliability and reproducibility and easy usage are also obligatory requirements. The DCG, photopolymers, and silver halide emulsions are among the candidates for such media. Very low losses and high diffraction efficiency make the DCG an excellent choice for display holography; however, its

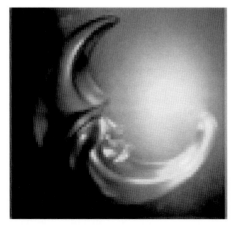

Figure 4.1 Art hologram, author Kalin Jivkov. (Please find a color version of this figure on the color plates.)

Figure 4.2 African mask, wood carving, 20th century A.D. (Please find a color version of this figure on the color plates.)

Figure 4.3 Thracian mask, bronze, fourth to third century B.C. (Please find a color version of this figure on the color plates.)

spectral response and sensitivity are far from the expected for multicolor holographic recording. Different types of photopolymers are efficient media and permit simple rapid processing, but the comparatively limited sensitivity, resolution, and dynamic range are still obstacles for multicolor recording. Currently, the most suitable are the silver halide emulsions that offer many advantageous features such as high sensitivity in a broad spectral range and long-term storage. Although they outperform the photopolymers in sensitivity, resolution, and dynamic range, they have lower diffraction efficiency and suffer from low signal-to-noise ratio in the blue spectral region because of increased light scattering. Increase in diffraction efficiency is achieved by bleaching of the silver halide holograms. Ultrafine-grain silver halide materials such as PFG-03C suitable for recording of volume reflection holograms are made available in the market by the firm Slavich, Russia. These materials suffer from low signal-to-noise ratio in the blue spectral region as the size of initial silver halide crystals are usually more than 10 nm. The scattering problem in this region is solved by the creation of a nanosized emulsion with the size of silver halide crystals about 8 nm, for example, the panchromatic holographic plates produced by Ultimate [42] or the plates HP-P [40, 46], which have been used for the results, presented below.

Preparation of the emulsion HP-P is based on the well-known "double jet" technique from Lippmann's time, not using "freezing and thawing proposed by Kirillov [47], as in the case of Slavich materials. The average size of the silver halide grains is less than 10 nm. The thickness of the layer coated onto the glass substrates is about 8 μm. The HP-P plates, having natural sensitivity in the blue spectral region (400–442 nm), are sensitized for the red and green with maximum absorptions at 630 and 530 nm. The transmission spectrum of the HP-P plate is shown in Figure 4.4a. To measure diffraction efficiency, test reflection holograms of two collimated beams are recorded at CW laser irradiation at different wavelengths – 442 nm (HeCd laser), 532 nm (frequency-doubled diode-pumped solid-state laser), and 632.8 nm (HeNe laser). Chemical processing comprises three steps: (i) amplitude development with SM-6 developer, (ii) bleaching with Slavich PBU-Amidol composition, and (iii) swelling in the bath of 5% water solution of collagen hydrolyzate for 5 min at 20 °C of holograms to compensate for the shrinkage of the layers after chemical processing and to satisfy the Bragg's condition for reconstruction at the wavelengths of recording. The typical exposure characteristics are shown in Figure 4.4b.

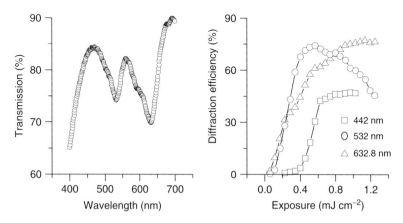

Figure 4.4 (a) Transmission spectrum of the HP-P holographic plate and (b) exposure characteristics for RGB recording.

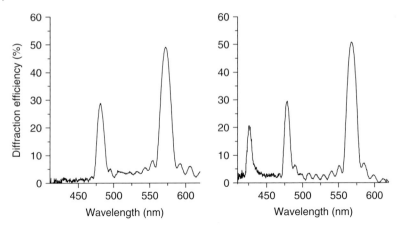

Figure 4.5 (a) Diffraction efficiency for successive recording at 532 and 632.8 nm and (b) diffraction efficiency for successive RGB holographic recording.

The maximal values of diffraction efficiency are almost 80, 75, and 50% for recording in the red, green, and blue spectral regions, respectively. The decrease in efficiency for diffuse objects is not more than 20% from the maximal values. This is due to the low light scattering of the initial silver halide crystals and weak migration of the developed grains, leading to the high signal-to-noise ratio (more than 100 : 1) that is most important for recording in the blue spectral region. The obtained results are promising for RGB recording of reflection holograms and for correct color balance in reconstruction with noncoherent white light. Exposure characteristics and swelling procedures are crucial for fulfillment of this task. As the dynamic range of the refractive index modulation for all bleached silver halide materials is limited, being typically not more than 0.08, successive recording onto a single plate diminishes the diffraction efficiency of the individual gratings by a factor equal to the number of recordings in power 1.5–2. This is illustrated in Figure 4.5. Implementation of the so-called sandwich structure [48, 49] with two holographic plates – one for blue recording and the other one for red and green – could be used for better efficiency in the reconstructed images. Optimization of simultaneous recording in red and green is shown in Figure 4.6. The proper choice of exposures is crucial for the quality of the RGB recording. Exact reproduction of the original color requires deviation in color coordinates to be less than 0.1%. This is not an easy task taking into account the different dynamic ranges and intensity profiles for the primary colors [6, 50, 51]. Uniform intensities of all recording beams, filtered with the same pinhole, have to be provided. Direct reconstruction from a color hologram could hardly ensure such a high accuracy without any additional correction. In order to achieve close to the original color perception, a specially designed light source with controllable color and intensity parameters, for example, a high-power light-emitting diode, can be used for a finer adjustment of the color balance in the reconstructed image. It is essential to point out that similar results, as presented above, were obtained also with 40 ns pulse lasers emitting at

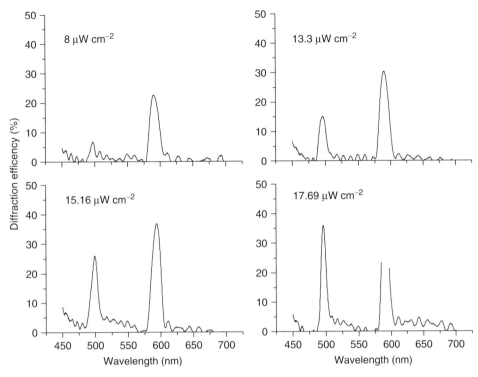

Figure 4.6 Diffraction efficiency spectra of a bleached nonswelled reflection hologram for simultaneous recording at 532 and 632.8 nm and exposure 1 mJ cm^{-2}; irradiance at 632.8 nm is 36 μW cm^{-2}; and the irradiance at 532 nm is given as a parameter.

440, 532, and 660 nm. This means that the well-known photography "reciprocity low" ($E = I \times t$), which is usually violated when diminishing the size of silver halide crystals, is completely valid only for the nanosized (~10 nm) silver halide emulsions. This result is important for high-quality reflection RGB holographic recording with pulse lasers in holoprinters, using coated onto glass or flexible substrate panchromatic silver halide emulsions [52].

4.7
Digital Holographic Display: Holoprinters

Digital holography [53] reveals new possibilities for implementation of 3D display by means of holography. Computed fringe patterns are fed to the medium for permanent or dynamic holographic recording; the latter can be realized with controllable spatial light modulators (SLMs) as liquid crystals or digital micromirror devices [54]. The goal is to project floating in space attractive 3D images of real or artistic objects. A comparatively low value of SLM resolution (less than

$1000\,\text{mm}^{-1}$) and apertures on the order of a few square centimeters are still limitations for high-quality visual perception. However, SLMs find application for generation of object beams for quasi-Fourier microholograms in modern high-resolution and high field-of-view digital holoprinters, capable of producing large size digital holograms of real world scenes [52]. Single and full parallax digital holograms, named i-Lumograms by Geola UAB, are recorded from a large number of parallax-related digital 2D images of a 3D scene. This technology is implemented in pulsed and CW laser-based holoprinters of Zebra Imaging Inc. in the United States [55], 3D Holoprint in France [56], and others. The technology includes three essential steps: (i) digital recording of a sequence of 2D color images at different viewing angles of static or moving objects; (ii) rearranging pixel by pixel the data from the recorded frames for creation of microholograms (named hogels in Zebra Imaging Inc. [55] and holopixels in Geola [32]), and (iii) RGB holographic recording of holopixels onto a single holographic plate or film. For example, the recording of i-Lumograms is performed in three parallel channels with solid-state lasers in pulse mode of generation (40 ns) emitting at 440, 532, and 660 nm. They are synchronized with mechanical translation of the carrier. Liquid crystal SLMs are used for formation of the object's beam for each red, green, and blue channel at successive recording of holopixels. Up to now, the only suitable registration medium for holoprinters with pulse holographic recording is the ultrafine-grain panchromatic silver halide material because of its wide dynamic range, spectral response, and sensitivity. The achieved impressive results at white light reconstruction are presented in Figure 4.7. Not only the volume and color characteristics but also the motion of objects and scenes with required scaling are reconstructed from a single holographic screen.

Figure 4.7 Three frames of the reconstructed 3D moving object from i-Lumogram; the total number of views is 640. (Source: With kind permission of Geola UAB, Vilnius, Lithuania.) (Please find a color version of this figure on the color plates.)

4.8 Conclusion

The different self-existing subdivisions of display holography could boast of almost half a century of development, which is a comparatively a long period for implementation of innovative products. Regardless of the attractive high-quality results that have been obtained, practical application of display holography is still limited. The subject of industrial production are only the security holograms. The reasons for this are not the price or quality of the reconstructed images. For example, at mass circulation of holographic illustrations the price will become comparable with the price of usual 2D pictures. A more substantial impediment is the need for a special reconstructing illumination. High-quality reconstruction in scattered light, which is usual for visual perception, is possible only for the plane objects in image plane holographic recording, which excludes the 3D effect. In addition, one could even argue that impressive illusion of volume can be achieved in 2D pictures through perspective, shadows, occlusion, gleams, and other effects and that precise reconstruction of 3D coordinates is not so crucial for 3D visual perception. Independent of such shortcomings, development of new approaches and digital methods in display holography continues. As in the early beginning of holography the reconstructed 3D images still attract public interest not only as a puzzling fancy but also as a serious base of knowledge with large information and esthetic capacity. A typical example is the so-called 4D display, realized by digitally arranged structures in holoprinters for reconstruction from a single screen not only of the volume and color characteristics but also of motion of the objects. Implementation of digital techniques as well as of newly developed light-sensitive materials could be a challenge that opens new horizons for future development of display holography.

References

1. Gabor, D. (1948) A new microscopic principle. *Nature*, **161**, 777–778.
2. Leith, E. and Upatnieks, J. (1962) Reconstructed wavefronts and communication theory. *J. Opt. Soc. Am. A*, **52**, 1123–1130.
3. Denisyuk, Y. (1962) On the reflection of optical properties of an object in the wave field of light scattered by it. *Dokl. Akad. Nauk SSSR*, **144** (6), 1275.
4. Lippmann, G. (1891) La photographie des couleurs. *CRAS (Paris)*, **112**, 274–275.
5. Collier, R., Burckhardt, C., and Lin, L. (1971) *Optical Holography*, Academic Press, New York.
6. Hariharan, P. (1996) *Optical Holography: Principles, Techniques and Applications*, 2nd edn, Cambridge University Press.
7. Lambooij, M., IJsselsteijn, W., Fortuin, M., and Heynderickx, I. (2009) Visual discomfort and visual fatigue of stereoscopic displays: a review. *J. Imaging Sci. Technol.*, **53**, 030201.
8. Goodman, J. (2005) *Introduction to Fourier Optics*, 3rd edn, Roberts and Company Publishers.
9. Cai, L.Z. and Guo, C.S. (1994) The fraunhofer-fresnel hologram and its applications. *Opt. Laser Technol.*, **26** (1), 55–58.
10. Ersoy, O. (2007) *Diffraction, Fourier Optics, and Imaging*, Wiley-Interscience, John Wiley & Sons, Inc.
11. Ackermann, G.K. and Eichler, J. (2008) *Holography: A Practical Approach*, Wiley online library, published Online.

12. Benton, S. (1980) Holographic displays: 1975–1980.. *Opt. Eng.*, **19**, 686.
13. Caulfield, J. (ed.) (2003) *The Art and Science of Holography: A Tribute to a Tribute to Emmett Leith and Yuri Denisyuk*, SPIE Press Monograph.
14. Johnston, S. (2006) *Holographic Visions: a History of New Science*, Oxford University press Inc., New York.
15. Shevtsov, M.K., Kornev, A.F., Pokrovskii, V.P., and Stupnikov, V.K. (2006) The GREEF portable holographic camera and its practical use. *J. Opt. Technol.*, **73**, 462–465.
16. Saxby, G. (2003) *Practical Holography*, 3rd edn, IOP Publishing.
17. Siebert, L. (1968) Large-scene front-lighted hologram of a human subject. *Proc. IEEE*, **56**, 1242.
18. Benton, S. (1969) Hologram reconstructions with extended incoherent sources. *J. Opt. Soc. Am.*, **59**, 1545A.
19. Leseberg, D. and Bryngdahl, O. (1984) Computer-generated rainbow holograms. *Appl. Opt.*, **23**, 2441–2447.
20. Odinokov, S.B. (2005) Researching quality parameters of rainbow holographic image by measuring modulated transfer function. *Opt. Mem. Neural Networks*, **V17** (2), 111–118.
21. Shi, S.Y., Wang, H., Li, Y., Jin, H., and Ma, L. (2009) Practical method for color computer generated rainbow holograms of real-existing object. *Appl. Opt.*, **48**, 4219–4226.
22. Halle, M., Benton, S., Klug, M., and Underkoffler, J. (1991) The ultragram: a generalized holographic stereogram. *Practical Holography V*, Proceedings of SPIE, Vol. 1461, 142–155.
23. Lih Yeh, S. (2006) A study of light scattered by surface-relief holographic diffusers. *Opt. Commun.*, **264** (1), 1–8.
24. Yatagai, T., Barada, D., Itoh, M., and Harada, K. (2006) Generation of surface relief hologram and nano structure on azobenzen polymer films and its numerical analysis by moving particle method. *Proc. SPIE*, **6252**, 0X - 5.
25. Kogelnik, H. (1969) Coupled wave theory of thick hologram gratings. *Bell Syst. Techn. J.*, **48** (9), 2909.
26. Denisyuk, Y., Sainov, V., and Stoykova, E. (eds) (2006) Materials and systems for optical data storage and processing of information. *Proc. SPIE*, **6252**, 625202–625215.
27. Bjelkhagen, H.I. (1995) *Silver-Halide Recording Materials for Holography and Their Processing*, Springer, Berlin.
28. Beev, K., Beeva, K., and Sainov, S. (2008) *Three-Dimensional Television: Capture, transmission, Display*, Springer, Berlin, pp. 557–598.
29. Nazarova, D., Mednikarov, B., and Sharlandjiev, P. (2007) Resonant optical transmission from a one-dimensional relief metalized subwavelength grating. *Appl. Opt.*, **46** (34), 8250–8255.
30. Sharlandjiev, P. and Nazarova, D. (2007) Nanosized particles in thin metal composite films and in gratings. *J. Opt. Adv. Mat.*, **9** (8), 2462–2467.
31. Ozaki, M., Kato, J., and Kawata, S. (2011) Surface-plasmon holography with white-light illumination. *Science*, **332**, (6026), 218–220.
32. Geola Digital UAB, Company (2011) URL http://www.geola.com (accessed 22 March 2012).
33. Kim, J., Choi, B., Kim, S., Kim, J., Bjelkhagen, H., and Phillips, N. (2001) Holographic optical elements recorded in silver halide sensitized gelatin emulsions-Part1. *Appl. Opt.*, **40**, 622–632.
34. Mehta, P.C. and Rampal, V.V. (1993) *Lasers and Holography*, World Scientific Publishing Co.Ptc.Ltd, Singapore.
35. Lawrence, J., O'Neill, F., and Sheridan, J. (2001) Photopolymer holographic recording material. *Optik*, **112** (10), 449–463.
36. Nikolova, L. and Ramanujam, P.S. (2009) *Polarization Holography*, Cambridge University Press, Cambridge.
37. Tomita, Y., Suzuki, N., and Chikama, K. (2005) Holographic manipulation of nanoparticle distribution morphology in nanoparticle-dispersed photopolymers. *Opt. Lett.*, **30** (8), 839–841.
38. Tomita, Y., Chikama, K., Nohara, Y., Suzuki, N., Furushima, K., and Endoh, Y. (2006) Two-dimensional imaging of atomic distribution morphology created

by holographically induced mass transfer of monomer molecules and nanoparticles in a silica-nanoparticle-dispersed photopolymer film. *Opt. Lett.*, **31** (10), 1402–1404.
39. Trainer, K., Wearen, K., Nazarova, D., Naydenova, I., and Toal, V. (2010) Optimization of an acrylamide-based photopolymer system for holographic inscription of surface patterns with sub-micron resolution. *J. Opt.*, **12** (12), 124012.
40. Sainov, V., Petrova, T., Ivanov, B., Zdravkov, K., Nazarova, D., Stoykova, E., and Minchev, G. (2009) Basic holographic characteristics of panchromatic light sensitive material for reflective auto stereoscopic 3D display. *EURASIP J. Adv. Sign. Process.*, Issue January, Capture, Transmission, and Display of 3D Video, Article No. 216341.
41. Integraf L.L.C., Company (2011) URL http://www.integraf.com (accessed 22 March 2012).
42. Gentet, Y. and Gentet, P. (2000) Ultimate emulsion and its applications: a laboratory made silver-halide emulsion of optimized quality for monochromatic pulsed and full color holography. *Proc. SPIE*, **4149**, 56–62.
43. Born, M. and Wolf, E. (2002) *Principles of Optics*, 7th edn., Cambridge University Press, Cambridge.
44. Kovachev, M., Sainov, V., and Mateeva, T. (1976) Diffraction efficiency of discrete carrier holograms, (in Russian). *Quantum Electron.*, **31** (113), 2399–2406.
45. Sainov, V., Mazakova, M., and Koleva, N. (1981) Characteristics of non-bleached reflection holograms. *C. R. Acad. Bulg. Sci.*, **34** (9), 1241–1244.
46. Sainov, V., Zdravkov, K., Stoykova, E., Ivanov, B., Nazarova, D., Markova, B., Sainov, S., and Petrova, T.S. (2009) Multi-color holographic recording. *JOAM*, **11**, 1448–1451.
47. Kirillov, N. (1979) *High Resolution Photo Materials for Holography and their Processing*, Nauka, Moscow.
48. Sainov, V., Spassov, G., and Sainov, S. (1981) Diffraction efficiency increases in color display holograms recording. *J. Sci. Appl. Phot.*, **6**, 413.
49. Kubota, T. (1986) Recording of high quality color holograms. *Appl. Opt.*, **25**, 4141–4145.
50. Ulibarrena, M., Mendez, M., Carretero, L., Madrigal, R., and Fimia, A. (2002) Comparison of direct, rehalogenating and solvent bleaching processes with BB640 plates. *Appl. Opt.*, **41**, 4120–4123.
51. Bjelkhagen, H. (2006) Color holography: its history, state-of-the-art, and future. *Proc. SPIE*, **6252**, 1U–11.
52. Brotherton-Ratcliffe, D., Zacharvas, S., Bakanas, R., Pileckas, J., Nikolskij, A., and Kuchin, J. (2011) Digital holographic printing using pulsed RGB lasers. *Opt. Eng.*, **50** (9), 091307.
53. Yaroslavsky, L. (2004) *Digital Holography and Digital Image Processing: Principles, Methods, Algorithms*, Kluwer Academic Publishers.
54. Onural, L. and Osaktas, H. (eds) (2008) *Three-Dimensional Television: Capture, Transmission, Display*, Springer, Berlin, pp. 557–598.
55. Zebra Imaging Inc., Company (2011) URL http://www.zebraimaging.com (accessed 22 March 2012).
56. 3D Holoprint, Company (2011) URL http://www.ultimate-holography.com (accessed 22 March 2012).

5
Incoherent Computer-Generated Holography for 3D Color Imaging and Display
Toyohiko Yatagai and Yusuke Sando

5.1
Introduction

Among many applications of holograms, computer-generated holography is one of the most important because it can reconstruct virtual objects and modulate wave fronts arbitrarily [1]. Various kinds of computer-generated holograms(CGHs) have been reported in order to reconstruct virtual three-dimensional objects [2–4]. Recently, research on new synthesizing techniques such as a polygon-based method [5] and a ray-sampling plane method [6] have also been proposed. However, the use of computer-generated holography is limited mainly to the display of virtual object synthesized in computers. In conventional holography, objects are recorded by a coherent interference effect. Imaging and reconstruction of real existing objects by CGHs have not been widely discussed, since the complex amplitude diffracted from an object is not obtained directly. In order to synthesize a hologram of incoherently illuminated real objects, new methods should be developed. Recently, Rosen *et al.* proposed some methods, including optical scanning holography [7], an incoherent correlation method [8], and a synthetic aperture Fresnel element method [9]. These methods require calculations of the projections for real or virtual objects [4]. Rosen *et al.* [10, 11] and Sando *et al.* [12–14] proposed new methods for creating real existing objects, as opposed to virtual objects, using image projection and computational processing. In these methods, CGHs are synthesized from several projection images recorded with white light. The absence of a coherent light source in these methods enables us to reconstruct real existing objects. This type of CGHs, however, requires a number of projection images to synthesize one hologram. Katz *et al.* [15] have succeeded in reducing the number of projections by interpolation of intermediate projection. On the other hand, Sando *et al.* have revealed the spectral relation between the 3D object and its diffracted wave fronts to reduce the number of projections imaged without any degradation. This relation also has the potential to be applied to optical 3D tomographic imaging. The superposition of three color CGHs corresponding to red, green, and blue components can reconstruct full-color 3D objects.

In this chapter, first, some CGH techniques for 3D imaging and display are reviewed and then, a method for the 3D reconstruction of full-color real existing objects is described. Angular ranges, angular increments, and projection numbers in CGHs are presented. Verification of this method by both numerical and optical reconstruction is also presented. Finally, application of this method to biological imaging is discussed.

5.2
Three-Dimensional Imaging and Display with CGHs

Even though there are some technologies for 3D display such as stereogram and integral photography, it is only holography that can display objects ideally and perfectly. The recent development of computers has brought enormous interests to researchers in the form of CGHs, digital holography, holographic encryption, and so on [1]. Of these, CGHs play an interesting role in 3D display because they can create virtual objects.

The display applications of CGHs have been proposed by many authors. Yatagai has proposed the use of CGHs in holographic stereoscope. A method to enlarge the viewing angle has been discussed [16], in which high-order diffraction of light by pixel structures of a display device are used for reconstructing objects. Hamano et al. [17] have discussed CGHs for color reconstruction where white light reconstruction is possible at the cost of vertical parallax. Methods that can encode both amplitude and phase information in a CGH have been reported [18]. Owing to complex encoding, undesired light, such as zero-order and minus-first-order light, is omitted.

Although CGHs can reconstruct virtual objects, it is difficult to make a CGH of real existing objects because of the necessity of 3D object information. Poon et al. [19] have discussed a method for obtaining holographic information of real existing objects by scanning two dimensionally, called *optical scanning holography*. However, it cannot be performed outdoors since it also records wave interference by coherent light.

As a solution to this problem, Li et al. [10] proposed another new method for obtaining holographic information of real existing objects. This method does not require coherent light because it is not the wave interferences but a series of projection images that are recorded. White light is sufficient. Moreover, compared with optical holography, this method is impervious to vibration. In this method, the vertical blur is inevitably caused in reconstructed images as projection images are recorded only along the horizontal direction. The type of CGHs are called *1D Fourier holograms* and need two cylindrical lenses in reconstruction. The peak corresponding to the bias component at the Fourier domain limits the dynamic range of the synthesized CGH.

Here, we present a new method for synthesizing Fresnel CGHs of real existing objects in which the objects are reconstructed without any blur in any direction [12–14]. In our method, which is based on the one proposed by Li et al., projection images are recorded by the 1D azimuth scanning with a white light and a color CCD. Three lasers (Ar-ion, YAG, and He-Ne) and three CGHs are used to generate full-color objects in reconstruction. The method to calculate a Fresnel CGH is also discussed. In the next section, the theory of the method proposed here is described in detail.

5.3
Theory of this Method

The method proposed here mainly consists of three steps. In the first step, projection images are recorded with a color CCD scanning azimuthally and parts of the 3D Fourier spectrum of objects are calculated in accordance with the principle of computed tomography (CT) [20]. In the second step, the Fourier components required to generate a CGH are extracted from the 3D Fourier spectrum. Finally, a Fresnel hologram is directly synthesized from it.

5.3.1
Relation between Object Waves and 3D Fourier Spectrum

A virtual optical system for recording 3D objects is shown in Figure 5.1. We assume that 3D object surfaces reflect waves emerging from external light sources isotropically. In general, the spatial reflectivity distribution of 3D objects is expressed as complex values. However, we have to treat the spatial phase distribution of the reflectivity as spatially unvaried because a CCD cannot detect the phase information without interferometer. The 3D spatial distribution with spatially unvaried phase is represented by $O(x, y, z)$, which corresponds to the square root of the intensity. The complex wave fronts reflected on the objects are observed in the Fourier plane in Figure 5.1. The distribution $g(x_0, y_0)$ in the Fourier plane is expressed as

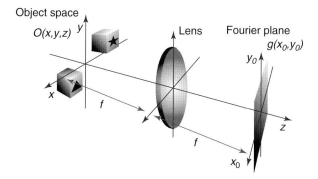

Figure 5.1 A virtual optical system for 3D objects.

follows [12]:

$$g(x_0, y_0) = \iiint O(x, y, z) \exp\left\{-\frac{i2\pi}{\lambda}\left[\frac{x_0 x + y_0 y}{f} - \frac{(x_0^2 + y_0^2)z}{2f^2}\right]\right\} dx\,dy\,dz \tag{5.1}$$

where λ and f are the virtual wavelength of incident light and the focal length of the lens introduced in Figure 5.1, respectively. Our principle is based on the relation between $g(x_0, y_0)$ and the 3D Fourier spectrum of the 3D distribution $O(x, y, z)$. The relation is revealed after substituting $u = x_0/\lambda f$ and $v = y_0/\lambda f$ in Eq. (5.1).

$$\begin{aligned} g(u, v) &= \iiint O(x, y, z) \exp\left\{-i2\pi\left[ux + vy - \frac{\lambda}{2}(u^2 + v^2)z\right]\right\} dx\,dy\,dz \\ &= \left\{\iiint O(x, y, z) \exp[-i2\pi(ux + vy + wz)] dx\,dy\,dz\right\}\Big|_{w=-\lambda(u^2+v^2)/2} \\ &= \mathcal{F}[O(x, y, z)]|_{w=-\lambda(u^2+v^2)/2}, \end{aligned} \tag{5.2}$$

where $\mathcal{F}[\cdot]$ denotes a 3D Fourier transform operator. Subscripts in Eq. (5.2) represent a paraboloid of revolution in 3D Fourier space (u, v, w). Consequently, we can determine that the wave front distribution at the Fourier plane in Figure 5.1 is completely identical to components on the paraboloid of revolution in the 3D Fourier space of $O(x, y, z)$ without any approximation, unlike other similar methods [11–13]. Moreover, if the wave front distribution at the Fourier plane is obtained, a Fresnel CGH can be readily calculated from it by the multiplication of the quadratic phase term corresponding to the diffraction distance and the inverse Fourier transform.

5.3.2
Extraction Method for Paraboloid of Revolution

Equation (5.2) implies that an indirect acquisition of the distribution $g(u, v)$ becomes possible by direct access to the 3D Fourier space of $O(x, y, z)$. To achieve this, the principle of the 3D central slice theorem (CST) is essential [20]. The principle of the 3D CST ensures that partial components of the 3D Fourier spectrum of a 3D object are obtained from an orthogonal projection image of the object. In this principle, at first, the 3D object is projected onto a plane whose normal vector is inclined by θ to the z-axis in the $z - x$ plane. Subsequently, the projection image is 2D Fourier-transformed. The 2D Fourier spectrum then corresponds with a sectional Fourier plane whose normal vector is inclined by θ to the w-axis in the $w - u$ plane in the 3D Fourier space of the object, as shown in Figure 5.2. Therefore, using this principle, it is possible to obtain partial components on the paraboloid of revolution represented by Eq. (5.2) from one projection image, as shown in Figure 5.3.

Figure 5.3a shows the components represented by Eq. (5.2). Only intersections between the sectional Fourier plane in Figure 5.2b and the paraboloid of revolution in Figure 5.3a can be extracted from the sectional Fourier plane. The intersections are calculated by solving the following simultaneous equations:

$$w \cos\theta + u \sin\theta = 0 \tag{5.3}$$

$$w = -\frac{\lambda}{2}(u^2 + v^2). \tag{5.4}$$

Equations (5.3) and (5.4) represent the planar equation shown in Figure 5.2b and the equation of the paraboloid of revolution shown in Figure 5.3a, respectively. These simultaneous equations give the following solution:

$$\left(u - \frac{\tan\theta}{\lambda}\right)^2 + v^2 = \left(\frac{\tan\theta}{\lambda}\right)^2, \quad w = -u\tan\theta \tag{5.5}$$

This solution shows that the intersections between the sectional Fourier plane and the paraboloid of revolution form an ellipse on the sectional Fourier plane, which is component extracted from one projection, as expressed with a red line in Figure 5.3b. Moreover, from this solution, it is found that the projection of the ellipse onto the $u - v$ plane in the 3D Fourier space becomes a circle with a radius $\tan\theta/\lambda$, as shown in Figure 5.4a. The position of the center and the radius depend on the direction of projection. On the other hand, obtaining all the components on the paraboloid of revolution is equal to filling the $u - v$ plane with red circles. In order to fill the 2D $u - v$ plane, 2D scanning of 3D objects is not necessarily needed. One-dimensional scanning is enough because the 1D components on the 2D $u - v$ plane are extracted from only one projection. Although there are some scanning methods to accomplish this, Figure 5.5 indicates the best scanning method in terms of the feasibility of a recording optical system. In this diagram, tangential θ and azimuth ϕ determine the radius of the extractive circle and the azimuth

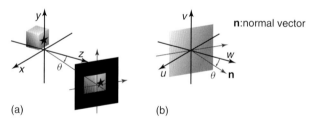

Figure 5.2 Schematics of the principle of the 3D central slice theorem (CST). (a) Orthogonal projection in the real space and (b) a sectional plane in the 3D Fourier space obtained from a projection image.

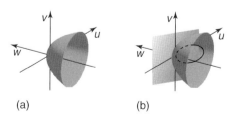

Figure 5.3 Paraboloid of revolution. (a) Components identical to objects waves and (b) intersections between the paraboloid of revolution and a sectional Fourier plane.

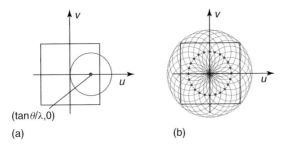

($\tan\theta/\lambda$,0)

(a)　　　　　　　　　　(b)

Figure 5.4 Extractive area on the $u - v$ plane from (a) one projection image and (b) a series of projection images. (Please find a color version of this figure on the color plates.)

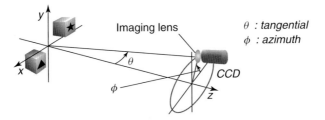

Figure 5.5 A recording optical system. (Please find a color version of this figure on the color plates.)

position of its center, respectively, ($\phi = 0$ in Figure 5.4a). Three-dimensional objects are imaged onto a CCD plane by an imaging lens. This projection is not exactly an orthogonal one; however, it can be approximated as an orthogonal projection in the case that the distance from the origin to the CCD camera is considerably longer than the depth of the 3D objects. The CCD camera records the intensity of such projection images by revolving around the z-axis. Since the locus of the CCD camera and the Fourier components shown in Figure 5.3a are in rotational symmetry with the z- and w- axes, respectively, the extractive components from a series of projection images recorded by this system can fill the $u - v$ plane with red circles, as shown in Figure 5.4b. Hence, it is found that this scanning method can provide all the components on the paraboloid of revolution and requires only 1D azimuth scanning of objects, unlike other similar methods [11–13].

5.3.3
Extension to Full-Color Reconstruction

It is easy to extend this method to full-color reconstruction if the projection images are recorded with a color CCD. The Fourier spectra for three colors are obtainable. The wavelength λ has not been related to the recording process or the CT principle. Wavelengths are defined freely in the virtual system (Figure. 5.1). Therefore, if

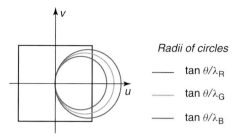

Figure 5.6 Adjustment of magnifications due to wavelength. (Please find a color version of this figure on the color plates.)

three wavelengths corresponding to R, G, and B components are introduced in the synthesizing process, it is easily possible to perform full-color reconstruction.

It should be noted that the magnification in the z-direction depends on the wavelength. In full-color reconstruction, it is very important that the magnifications are constant in spite of wavelength differences. There are a few ways to accomplish magnification adjustment. However, the object size W and the sampling number N cannot be changed because the difference of W gives the effects of the magnification of x and y-axes and N is constant.

So, in our method, the adjustment of the magnification of the z-axis is performed by changing the radius of an extractive circle for each wavelength as shown in Figure 5.6. This extraction enables us to obtain all the three color components by only a single azimuth scanning with a color CCD.

5.4
Imaging System and Resolution

5.4.1
Size of Object

Consider a point object located at (x, y, z). Using Eq. (5.2), its amplitude at the Fourier plane is given by

$$g(u, v) = \exp\left\{-i2\pi\left[ux + vy - \frac{\lambda}{2}(u^2 + v^2)z\right]\right\} \quad (5.6)$$

The local spatial frequency for u-direction is given by

$$\xi = \frac{1}{2\pi}\frac{\partial \arg[g(u, v)]}{\partial u} = -x + \lambda z u \quad (5.7)$$

where arg denotes argument of the complex number. Equation (5.7) means that the spatial frequency in the outer region is higher than that in the inner one. Consider that the size of a projection image is $W \times W$ and its sampling number is $N \times N$. The Fourier component ranges in $-N/2W \leq u \leq N/2W$ and $-N/2W \leq v \leq N/2W$. The maximum spatial frequency in the u-direction is given by

$$\xi_M = |x| + \frac{\lambda N}{2W}|z| \tag{5.8}$$

Since the sampling period is $1/W$ in Fourier space, the sampling condition for the diffraction wave from the point object should satisfy the following equation:

$$W \geq 2\xi_M = 2|x| + \frac{\lambda N}{W}|z| \tag{5.9}$$

Therefore, the size of the object $|x|$ and $|z|$ is limited as

$$W \geq |x| + \sqrt{|x|^2 + \lambda N|z|} \tag{5.10}$$

5.4.2
Spatial Resolution

The resolution is defined as the inverse of the bandwidth of the spatial frequency. The resolutions in x- and y-directions are given by W/N. On the other hand, according to Eq. (5.4), the maximum spatial frequency in the z-direction is given by

$$\zeta_M = \frac{\lambda}{2}(u_M^2 + v_M^2) \tag{5.11}$$

where u_M and v_M are the maximum spatial frequencies in the x- and y-directions, respectively. Since they are limited by the diameter of the extraction circle $R = 2\tan\theta/\lambda$ or the sampling period of the projection images, we have

$$u_M^2 + v_M^2 = \begin{cases} \frac{N^2}{2W^2} & : R \geq \frac{\sqrt{2}N}{2W} \\ \frac{4\tan^2\theta}{\lambda^2} & : R < \frac{\sqrt{2}N}{2W} \end{cases} \tag{5.12}$$

The spatial resolution δz in the z-direction is the inverse of Eq. (5.11)

$$\delta z = \frac{2}{\lambda(u_M^2 + v_M^2)} \tag{5.13}$$

5.4.3
Magnification along the z-direction

The magnification along the x-, y-, z- directions are 1, if the frequency components are extracted according to the theory. In practice, however, the diameter of the extracting circle $R = 2\tan\theta/\lambda$ is very large. From Eq. (5.12), the spatial resolution $\delta z/\lambda$ to the z-direction is given as

$$\frac{\delta z}{\lambda} = \frac{4W^2}{\lambda^2 N^2} \tag{5.14}$$

This means that the spatial recognition of the object in the z-direction is very difficult in the visible wavelength region.

If wavelength λ_r different from wavelength λ_s of the synthesizing process is used, we can optimize the spatial resolution in the z-direction. The minimum of $\delta z/\lambda$

is obtained in the case from Eq. (5.12). In this condition, the wavelength of the synthesizing process is given by

$$\lambda_s = \frac{4W \tan \theta}{\sqrt{2N}} \qquad (5.15)$$

The resolution along the z-direction is

$$\frac{\delta z}{\lambda_s} = \frac{1}{2 \tan^2 \theta} \qquad (5.16)$$

This means larger projection angle θ gives larger position recognition ability in the z-direction. In the case of optical reconstruction, the reconstruction wavelength λ_r differs from that of the synthesizing process λ_s, and so the magnification along the z-direction is given by

$$M_z = \frac{4W \tan \theta}{\sqrt{2N}\lambda_r} \qquad (5.17)$$

5.5
Experiments

5.5.1
Computer Simulation and Some Parameters

In order to verify the principle described above, we have demonstrated an experiment. A total of 90 projection images were recorded with a recording optical system proposed in Figure 5.5. Some typical examples of projection images are shown in Figure 5.7. A grape and a mushroom on square planes are located at $z \cong 0$ mm

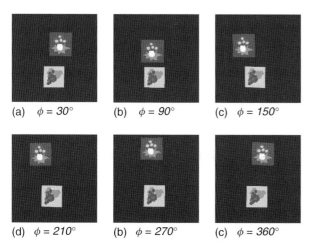

(a) $\phi = 30°$ (b) $\phi = 90°$ (c) $\phi = 150°$

(d) $\phi = 210°$ (b) $\phi = 270°$ (c) $\phi = 360°$

Figure 5.7 Color projection images at $\theta = 17°$. (Please find a color version of this figure on the color plates.)

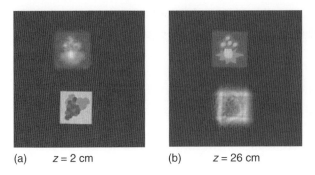

(a)   z = 2 cm   (b)   z = 26 cm

Figure 5.8 Numerical reconstructed images.

and $z \cong 4.9$ mm in the object space, respectively. A series of images shown in Figure 5.7 represent the parallax effect due to the difference of the locations. These images are divided into color components, and each component is binarized. The noise component is also removed. The size and pixel numbers of each image are 1×1 cm and 256×256 pixels, respectively. Three wavelengths, 632.8 nm, 514.5 nm, and 488.5 nm, corresponding to red, green, and blue, respectively, are used for full-color reconstruction. The angle between the optical axis of a CCD camera and the z-axis is $\theta \cong 17°$. Under these conditions, the magnification of the z-axis is ≈ 53 from Eq. (5.17). Therefore, the grape and the mushroom should be reconstructed at $z \cong 0$ cm and $z \cong 26$ cm, respectively. According to the principle proposed here, the distributions at the Fourier plane as described in Figure 5.1 are synthesized from the above projection images. Thus, the reconstructed images can be easily obtained by calculating back propagation from the distribution at the Fourier plane onto arbitrary sectional planes in the objects space. These procedures are performed for each color component, and the three color components are then superimposed in the final step. Such reconstructed images are shown in Figure 5.8. As can be observed from Figure 5.8, each of the two different objects can be clearly reconstructed without any blur at the corresponding positions. The adjustments of magnifications for each wavelength are also successful. Thus, it is verified that this method can reconstruct 3D full-color objects from substantially smaller number of projection images as compared to Ref. [13].

5.5.2
Optical Reconstruction

The optically reconstructed images from synthesized CGHs for another object are shown in Figure 5.9. The optical system for reconstruction is designed to match the distances from each CGH to the color CCD.

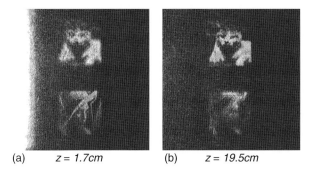

(a)   z = 1.7cm   (b)   z = 19.5cm

Figure 5.9 Experimental results of optical reconstruction. (Please find a color version of this figure on the color plates.)

5.6
Biological Specimen

For 3D observation and display of biological samples, such as pathological samples, a microscopic imaging system with rotation equipment is developed as shown in Figure 5.10. A telecentric imaging system is employed to perform orthogonal projection.

As one of the biological samples, breast cancer cells are prepared by a conventional formalin method. The volume of the sample is $20 \times 20 \times 2\,mm^2$. A total of 360 projected images are recorded by incrementing the viewing angle by $1°$. Figure 5.11 shows reconstructed 3D image of a human breast cancer sample.

Next, according to our method, a hologram is synthesized with the data accumulated and reconstructed optically. Images reconstructed are shown in Figure 5.12 with different focusing planes. White arrows show cancer cells which are calcificated.

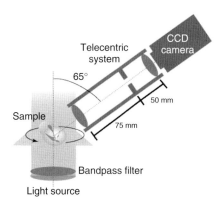

Figure 5.10 Telecentric microscope.

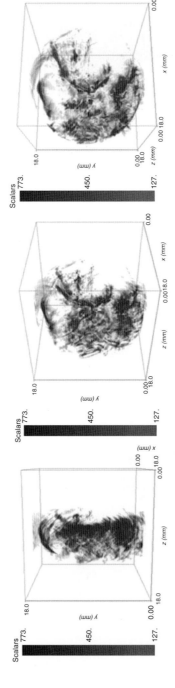

Figure 5.11 Reconstructed cancer images with different view angles. (Please find a color version of this figure on the color plates.)

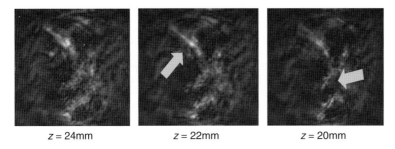

z = 24mm   z = 22mm   z = 20mm

Figure 5.12 Cancer cell images reconstructed optically. White arrows show calcificated cancer cells.

5.7 Conclusion

In conclusion, we have proposed a method for synthesizing Fresnel CGHs from a series of projection images recorded by the azimuth scanning with a color CCD. This method does not need a coherent light source and lenses in reconstruction and is impervious to vibration. This method essentially requires parts of 3D Fourier spectra. Therefore, it can be easily combined with methods that have connection with 3D Fourier spectra such as X-ray CT, ultrasonic CT, MRI imaging, and so on. The diagrams for the extraction of essential components are described. To verify our method, we have demonstrated the numerical and optical reconstruction including color process. The reconstructed images are clearly focused and have no blur in any direction, and the 3D reconstruction of objects is confirmed. Finally, as a biological application, our method has been applied for the 3D measurement and display of breast cancer cells. This means that the method proposed here has a high potential for application in wide fields such as holographic cameras, printers, security, and computer-graphics displays.

Acknowledgments

The authors express thanks to D. Barada and K. Miura for their fruitful discussions and helpful support in the subject.

References

1. Lohmann, A.W. and Paris, D.P. (1967) Binary Fraunhofer holograms, generated by computer. *Appl. Opt.*, **6**, 1739–1748.
2. Waters, J.P. (1968) Three-dimensional Fourier-transform method for synthesizing binary holograms. *J. Opt. Soc. Am.*, **58**, 1284–1287.
3. King, M.C., Noll, A.M., and Berry, D.H. (1970) A new approach to computer-generated holography. *Appl. Opt.*, **9**, 471–475.
4. Yatagai, T. (1976) Stereoscopic approach to 3-D display using computer-generated holograms. *Appl. Opt.*, **15**, 2722–2729.

5. Matsushima, M. and Nakamura, S. (2009) Extremely high-definition full-parallax computer-generated hologram created by the polygon-based method. *Appl. Opt.*, **48**, H54–H63.
6. Wakunami, K. and Yamaguchi, M. (2011) Calculation for computer generated hologram using ray-sampling plane. *Opt. Express*, **19**, 9086–9101.
7. Indebetouw, G., Tada, Y., Rosen, J., and Brooker, G. (2007) Scanning holographic microscopy with resolution exceeding the Rayleigh limit of the objective by superresolution of off-axis holograms. *Appl. Opt.*, **46**, 993–1000.
8. Shaked, N.T. and Rosen, J. (2008) Multiple-viewpoint projection holograms synthesized by spatially incoherent correlation with broadband functions. *J. Opt. Soc. Am.*, **A25**, 2129–2138.
9. Katz, B. and Rosen, J. (2010) Super-resolution in incoherent optical imaging using synthetic aperture with Fresnel elements. *Opt. Express*, **18**, 962–972.
10. Li, Y., Abookasis, D., and Rosen, J. (2001) Computer-generated holograms of three-dimensional realistic objects recorded without wave interference. *Appl. Opt.*, **40**, 2864–2870.
11. Abookasis, D. and Rosen, J. (2003) Computer-generated holograms of three-dimensional objects synthesized from their multiple angular viewpoints. *J. Opt. Soc. Am.*, **20**, 1537–1545.
12. Sando, Y., Itoh, M., and Yatagai, T. (2003) Holographic three-dimensional display synthesized from three-dimensional Fourier spectra of real existing objects. *Opt. Lett.*, **28**, 2518–2520.
13. Sando, Y., Itoh, M., and Yatagai, T. (2004) Color computer-generated holograms from projection images. *Opt. Express*, **12**, 2487–2493.
14. Sando, Y., Itoh, M., and Yatagai, T. (2004) Full-color computer-generated holograms using 3-D Fourier spectra. *Opt. Express*, **12**, 6246–6251.
15. Katz, B., Shaked, N.T., and Rosen, J. (2007) Synthesizing computer-generated holograms with reduced number of perspective projections. *Opt. Express*, **15**, 13250–13255.
16. Mishina, T., Okui, M., and Okano, F. (2002) Viewing-zone enlargement method for sampled hologram that uses high-order diffraction. *Appl. Opt.*, **41**, 1489–1499.
17. Hamano, T. and Kitamura, M. (2000) Computer-generated holograms for reconstructing multi 3-D images by space-division recording method. *Proc. SPIE*, **3956**, 23–32.
18. Yang, M. and Ding, J. (2002) Area encoding for design of phase-only computer-generated holograms. *Opt. Commun.*, **203**, 51–60.
19. Poon, T.C., Doh, K.B., Schilling, B.W., Wu, M.H., Shinoda, K., and Suzuki, Y. (1995) Three-dimensional microscopy by optical scanning holography. *Opt. Eng.*, **34**, 1338–1344.
20. Chiu, M.-Y., Barrett, H.H., and Simpson, R.G. (1980) Three-dimensional reconstruction from planar projections. *J. Opt. Soc. Am.*, **70**, 755–762.

6
Approaches to Overcome the Resolution Problem in Incoherent Digital Holography

Joseph Rosen, Natan T. Shaked, Barak Katz, and Gary Brooker

6.1
Introduction

Holography is an attractive imaging technique as it offers the ability to view a complete three-dimensional (3D) volume from a single two-dimensional (2D) image. However, holography is not widely applied to the regime of white-light imaging, because white-light is incoherent and creating holograms requires a coherent interferometer system. When practicing coherent light holography, we require both thermal and mechanical stability of the optical setup. All these factors often confine holography recording to the laboratory and have prevented hologram recorders from becoming as widely used for outdoor photography as conventional cameras. Thus, there is an ongoing effort to develop incoherent illumination holography recording process [1].

We review herein three new methods of generating digital incoherent holograms of 3D objects. These methods are known as digital incoherent protected correlation holography (DIPCH) [2], off-axis optical scanning holography (OSH) [3], and the holography method based on synthetic aperture Fresnel elements (SAFEs) [4, 5]. This review mainly focuses on one aspect of incoherent holographic imaging, namely, the imaging resolution problem. In each method, the imaging resolution of the generic configuration is discussed and a unique technique is implemented in order to improve the resolution.

The utilization of optical holography for improving the imaging resolution is usually restricted to coherent imaging [6–8]. Therefore, the use of holography for improving resolution is limited only to those applications in which the observed targets can be illuminated by a coherent laser. The holographic methods discussed in this review, on the other hand, are based on the recently invented techniques of incoherent holography [2–5]. Thus, holographic techniques of improving resolution can be implemented while the observed scene is illuminated by white light. In addition, the concept of the present methods can be applied to all practices of imaging, from microscopes to telescopes, as well as for both 2D and 3D imaging.

One way to record a white-light hologram is by specially processing multiple viewpoint projections (MVPs) of the observed scene. The next section deals with

Optical Imaging and Metrology: Advanced Technologies, First Edition.
Edited by Wolfgang Osten and Nadya Reingand.
© 2012 Wiley-VCH Verlag GmbH & Co. KGaA. Published 2012 by Wiley-VCH Verlag GmbH & Co. KGaA.

the technique of MVP hologram, in general, and with one member of the MVP family, the DIPCH, in particular. This type is a resolution-improved version of the original MVP [9]. Another way of recording white-light holograms is the OSH. In Section 6.3, we describe a special type of scanning hologram generated by a unique off-axis scanning system. We show that a composition of several off-axis scanning holograms can yield a hologram with higher resolution than the conventional scanning hologram. In Section 6.4, a review of the SAFE hologram is provided. This new technique of digital incoherent holography is extended to operate in a synthetic aperture mode in order to improve the imaging resolution performances of the original incoherent holography [10].

6.2
Digital Incoherent Protected Correlation Holograms

MVP holography [11] is a relatively new technique for generating digital and optical holograms of 3D real-existing objects under white-light illumination. To obtain these holograms, a conventional digital camera is used. The process does not require recording wave interference at all, and thus the twin image problem is avoided. MVP holograms are generated by first acquiring 2D projections (regular intensity images) of a 3D scene from various perspective points of view. Then, the acquired MVPs are digitally processed to yield the digital hologram of the scene. The MVP hologram is essentially equivalent to a conventional digital hologram of the scene recorded from the central point of view. It is acquired by white-light illumination, where neither an extreme stability of the optical system nor a powerful, highly coherent laser source is required as in conventional techniques of recording holograms.

Since the first work in this topic [12], several types of MVP holograms have been proposed. The list includes Fourier holograms [13, 14], digital incoherent modified Fresnel hologram (DIMFH) [9], and DIPCH [2]. The recording process of MVP holograms is performed by a regular digital camera and as mentioned above, under an incoherent white-light illumination. The recorded 3D scene can be reconstructed digitally by Fresnel back-propagator [9], in the case of DIMFH, or by other reconstruction techniques. Fresnel back-propagator means convolving the complex matrix of the MVP hologram with quadratic phase functions scaled according to the various axial reconstruction distances.

The DIPCH [2] is a special type of MVP hologram. This hologram has two advantages over the DIMFH. First, since a random-constrained point spread function (PSF) is used to generate the hologram, only authorized users who know this PSF can reconstruct the scene encoded into the hologram. Thus, the DIPCH can be used as a method of encrypting the observed scene. Second, the reconstruction obtained from the DIPCH, compared to DIMFH, has a significantly higher imaging resolution for the far objects in the 3D scene and that is the reason that in this chapter we review the DIPCH as a technique that can resolve observed objects better than the generic DIMFH method.

6.2 Digital Incoherent Protected Correlation Holograms

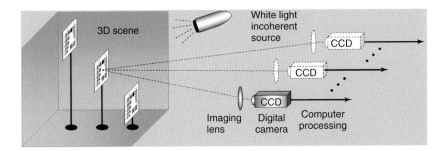

Figure 6.1 The optical system for acquiring MVPs of the 3D scene along the horizontal axis.

Both DIPCH and DIMFH belong to the same theoretical framework of generating and reconstructing MVP holograms synthesized by spatial correlation between the incoherently illuminated 3D scene and broadband spatial functions. The framework enables one to propose new types of digital holograms with certain advantages over the known holograms simply by choosing different PSF. Figure 6.1 illustrates a possible optical system for acquiring the MVPs of the 3D scene. In this scheme, the digital camera moves and acquires a different projection of the scene from each position. Instead of shifting the camera mechanically, the use of a microlens array for acquiring the entire viewpoint projections in a single camera shot [15] has also been proposed. Alternatively, one can acquire a small number of extreme projections simultaneously and predict the middle projections digitally by using the view synthesis algorithm [16]. Spatial multiplexing of several digital cameras is also possible. Following the acquisition stage, each projection is multiplied by the same 2D broadband complex function (which should have a phase-only Fourier transform (FT), as defined later), and the sum of the inner product is introduced into the corresponding pixel of the 2D hologram of the 3D scene. Alternatively, it is also possible to generate a one-dimensional (1D) hologram of the scene. In this case, each projection's row is multiplied by the same 1D broadband complex function and the column sum of all the inner products from each projection is introduced into the corresponding column in the hologram matrix. Since each projection is multiplied by the same spatial function, this process is actually a spatial correlation between the observed scene and PSF and the resulting matrix is obtained as an incoherent correlation hologram. However, in contrast to other correlation holograms [17, 18], the present holograms are produced from real-existing 3D objects illuminated by incoherent white light. The digital reconstruction of the incoherent correlation hologram is usually performed by convolving the hologram with the complex conjugate of the original generating PSF, scaled according to the reconstruction distance. Alternatively, the incoherent correlation hologram can be digitally converted to a known type of hologram (Fresnel, Fourier, etc.), so that the 3D scene can also be optically reconstructed by illuminating the hologram with coherent light.

In the DIMFH [9], the 3D scene is correlated with a PSF of a quadratic phase function. However, as mentioned above, in this chapter, we concentrate on a unique incoherent correlation hologram called *DIPCH*, which is characterized by a random broadband, space-limited, and secretive PSF. The DIPCH has a higher resolving power than the DIMFH, as well as the ability to encrypt the observed 3D information.

For every hologram type, 1D or 2D holograms can be synthesized depending on the nature of the MVP acquisition. In case the MVPs are acquired only horizontally (as illustrated in Figure 6.1) or along a different transverse axis, a 1D incoherent correlation hologram is generated. When the MVPs are acquired on a 2D grid of positions on the transverse plane, a 2D incoherent correlation hologram is generated. The 1D holograms are easier to produce because the projections are acquired along a single axis only. However, the 2D holograms have the advantage of encoding the 3D information into both axes.

Next, we describe the theory of generating and reconstructing 2D incoherent correlation holograms, where the theory of 1D incoherent correlation holograms can be straightforwardly derived from the general 2D case. The 2D incoherent correlation hologram is synthesized from $2K+1$ horizontal by $2K+1$ vertical projections of the 3D scene. We number the projections by m and n, so that the middle projection is denoted by $(m,n)=(0,0)$, the upper-right projection by $(m,n)=(K,K)$, and the lower-left projection by $(m,n)=(-K,-K)$. The (m,n)th projection $P_{m,n}(x_p, y_p)$ is multiplied by a PSF and the product is summed to the (m,n)th pixel in the following complex matrix:

$$H_2(m,n) = \iint P_{m,n}(x_p, y_p) E_2(x_p, y_p) dx_p dy_p \tag{6.1}$$

where $E_2(x_p,y_p)$ represents the generating PSF of the 2D hologram and (x_p, y_p) are the coordinates of the camera plane. This PSF is given by

$$E_2(x_p, y_p) = A_2(bx_p, by_p) \exp[ig_2(bx_p, by_p)] \tag{6.2}$$

where A_2 and g_2 are functions depending on (x_p, y_p) and may be chosen differently for each type of incoherent correlation hologram and b is an adjustable parameter (with units that preserve the arguments of A_2 and g_2 as unitless quantities). In addition, the function $E_2(x_p, y_p)$ has the property that its FT is a phase-only function. As is shown in Ref. [2], this condition is necessary to guarantee that the generated hologram can be reconstructed properly. The process manifested by Eq. (6.1) is repeated for all the projections, so that at the end of this digital process the resulting 2D complex matrix $H_2(m,n)$ represents the 2D incoherent correlation hologram of the 3D scene.

The reconstructed planar image $s_2(m,n;z_r)$ located at an axial distance z_r from the 2D hologram is obtained by digitally convolving the hologram with the reconstructing PSF as follows:

$$s_2(m,n;z_r) = |H_2(m,n) * R_2(m,n;z_r)| \tag{6.3}$$

where * denotes a 2D convolution and $R_2(m,n;z_r)$ is the reconstructing PSF of the 2D hologram. This PSF is given by

$$R_2(m,n;z_r) = A_2\left(\frac{m\Delta p}{z_r}, \frac{n\Delta p}{z_r}\right) \exp\left[-ig_2\left(\frac{m\Delta p}{z_r}, \frac{n\Delta p}{z_r}\right)\right] \tag{6.4}$$

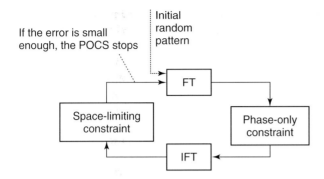

Figure 6.2 Schematic of the POCS algorithm for finding the PSF of the DIPCH.

where A_2 and g_2 are the same functions used for generating the PSF of Eq. (6.2) and Δp is the pixel size of the digital camera. A similar theory can also be applied for generating and reconstructing 1D incoherent correlation holograms [9]. However, in this case, each row in the mth projection $P_m(x_p, y_p)$ is multiplied by the same 1D PSF and the result is summed to the mth column in the 1D hologram. The reconstructed planar image, located at an axial distance z_r from the 1D hologram, is obtained by convolving the rows of the hologram matrix with the 1D reconstructing PSF.

In case of DIPCH, the generating PSF is a space-limited random function that fulfills the constraint that its FT is a phase-only function. In order to find this PSF, we use the projection onto the constraint sets (POCS) algorithm [19, 20]. The POCS algorithm used to find this PSF is illustrated in Figure 6.2. The POCS is an iterative algorithm that bounces from the PSF domain to its spatial frequency spectrum domain and back, using an FT and inverse Fourier transform (IFT). In each domain, the function is projected onto the constraint set. The two constraints of the POCS express the two properties required for the PSF of the DIPCH. First, the FT of the PSF should be a phase-only function. This requirement enables to reconstruct the DIPCH properly. So, the constraint of the POCS in the spectral domain is the set of all phase-only functions, and each Fourier transformed PSF is projected onto this constraint by setting its magnitude distribution to the constant 1. The other property of the PSF is that it should be space limited into a relatively narrow region close to but outside the origin. This condition reduces the reconstruction noise from the out-of-focus objects because the overlap, occurring during the cross-correlation between the resampled space-limited reconstructing PSF and the hologram in out-of-focus regions, is lower than the overlap in case of using a widespread PSF. Of course, the narrower the existence region of the PSF, the lower is the noise. However, narrowing the existence region makes it difficult for the POCS algorithm to converge to a PSF that satisfies both constraints within an acceptable error.

The constraint set in the PSF domain is the entire complex function that is identically equal to zero in any pixel outside the predefined narrow existence region. The POCS in the PSF domain is performed by multiplying the PSF by a function that is equal to 1 inside the narrow existence region of the PSF and equals 0 elsewhere. In the case of the 1D DIPCH, the random-constrained PSF is limited to a narrow strip of columns, whereas in the case of the 2D DIPCH, this PSF is limited to a narrow ring. At the end of the process, the POCS algorithm yields a suitable random-constrained PSF that is used in the hologram generation process.

Let us compare the reconstruction resolutions of the DIMFH and the DIPCH. Far objects captured by the DIMFH are reconstructed with a reduced resolution because of two reasons: (i) Owing to the parallax effect, farther objects "move" slower throughout the projections, and therefore they sample a magnified version of the generating PSF. This magnified version has narrower bandwidth, and thus the reconstruction resolution of farther objects decreases. (ii) The quadratic phase function used in the DIMFH has lower frequencies as one approaches its origin. Since far objects are correlated with the central part of the quadratic phase function along a range that becomes shorter as the object is farther, the bandwidth of the DIMFH of far objects becomes even narrower beyond the bandwidth reduction mentioned in (i). In contrast to the DIMFH, the spatial frequencies of the DIPCH's PSF are distributed uniformly all over its area. Therefore, the DIPCH sustain resolution reduction of far objects only due to reason (i). Hence, the images of far objects reconstructed from the DIPCH, besides being protected by the random-constrained PSF, also have higher transverse resolution.

The performances of 2D DIMFH and DIPCH were compared in Ref. [2]. To generate the 2D DIMFH, each projection is multiplied by a 2D quadratic phase function, and then each of the inner product is summed to the corresponding pixel in the hologram. Figure 6.3a presents the magnitude and phase of the 2D DIMFH generated by this process. Reconstructing the 3D scene from this hologram digitally is carried out by convolving the hologram with 2D quadratic phase functions with a phase sign opposite to the generating PSF. The three 2D quadratic phase functions yielding the best in-focus reconstructed planes are shown in Figure 6.3b. The corresponding reconstructed planes are shown in Figure 6.3c. In each plane, only a single United States Air Force (USAF) resolution chart is in focus, whereas the other two charts are out of focus. This holographic phenomenon validates that the volume information is indeed encoded into the hologram. Figure 6.3d presents zoomed-in images of the best in-focus USAF resolution charts shown in Figure 6.3c. Evidently from these figures, the resolution of the reconstructed charts decreases as the distance of the objects from the acquisition plane increases.

The same computer-generated projections were used for generating the 2D DIPCH. For this purpose, each projection was multiplied by the PSF computed by the POCS algorithm. The inner product between each projection and the PSF was summed to a single complex value, which was introduced into the corresponding pixel in the 2D DIPCH. The magnitude and the phase of the resulting 2D hologram are shown in Figure 6.3e. The reconstruction from this hologram was obtained by convolving it with the conjugate function of the generating PSF scaled differently in order to reconstruct a different transverse plane with each scaled PSF. The phases of the three PSFs that yield the best in-focus reconstructed planes are shown in

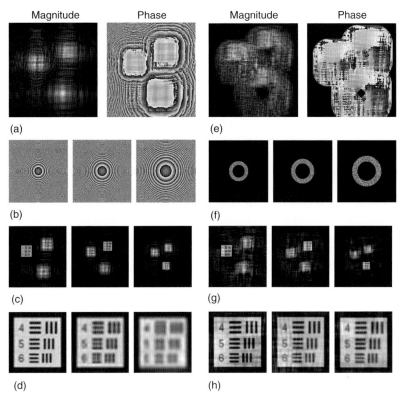

Figure 6.3 (a) Magnitude and phase of the 2D DIMFH; (b) the phase distributions of the reconstructing PSFs used for obtaining the three best in-focus reconstructed planes from the hologram in (a); (c) the corresponding three best in-focus reconstructed planes along the optical axis; (d) zoomed-in images of the corresponding best in-focus reconstructed objects; (e) magnitude and phase of the hologram the 2D DIPCH; (f) the phase distributions of the reconstructing random-constrained PSFs used for obtaining the three best in-focus reconstructed planes from the hologram in (e); (g) the corresponding three best in-focus reconstructed planes along the optical axis; and (h) zoomed-in images of the corresponding best in-focus reconstructed objects.

Figure 6.3f. The best in-focus reconstructed planes are shown in Figure 6.3g. Here again, the fact that in each of the reconstructed planes one USAF resolution chart is in focus, whereas the other two charts are out of focus demonstrates that the 3D information is encoded properly into this hologram. Figure 6.3h shows zoomed-in images of the best in-focus USAF resolution chart in each of these resampled reconstructed planes. Comparing between the best in-focus reconstructed charts obtained from the DIPCH (Figure 6.3h) and from the DIMFH (Figure 6.3d), one realizes that the far objects reconstructed from the DIPCH have a significantly better resolution than those reconstructed from the DIMFH, although the former are a bit noisier because of the nonuniformity of the gray levels of the random-constrained PSF generated by the POCS algorithm.

6.3
Off-Axis Optical Scanning Holography

Scanning holography [21–23] has demonstrated the ability to produce a Fresnel hologram of the incoherent light emission distributed in a 3D structure. The required correlation between an object function and a quadratic phase function is performed by a mechanical movement. More specifically, a certain pattern, which is the Fresnel zone plate (FZP), is projected on the observed object, whereas the FZP moves at a constant velocity relative to the object (or the object moves relative to the FZP). During the movement, the product of the FZP and the object is summed by a detector in discrete times. In other words, the pattern of the FZP scans the object and at each and every scanning position, the light intensity is integrated by the detector. The resulting electric signal is a sampled version of the 2D correlation between the object and the FZP. The dependence of the FZP in the axial position of object points is achieved by interfering two mutually coherent, monochromatic, spherical waves on the object surface. The number of cycles per radial distance in each of the spherical waves is dependent on their axial interference location. Since these beams interfere on the object, the axial distance of each object point is stored in the hologram due to the fact that each point is correlated with an FZP, the cycle density of which is dependent on the point's 3D location.

In conventional scanning holography [21–23], two pupils are combined by a beam splitter and superposed in the entrance pupil of the objective. One pupil is a spherical wave filling the objective's pupil disk and having a curvature chosen to produce, in the back focal plane of the objective (which is also the object's space), a spherical wave with some radius of curvature. The other is a point at the center of the entrance pupil of the objective, producing a plane wave in the object's space. The interference of these two waves results in a 3D FZP that is scanned over the specimen in a 2D raster. The scattered or fluorescent light is collected by a non-imaging detector (photodiode, or photomultiplier). A single-sideband hologram can be obtained from this data in two ways: either by heterodyne detection using a frequency difference between the two pupils [21–23] or by homodyne detection [24] using three scans with three fixed phase differences between the two pupils. The hologram can be reconstructed numerically by Fresnel back propagation or by correlation with an experimental PSF [3]. The spatial resolution of the reconstructed image is determined by the numerical aperture (NA) of the FZP scanned over the object. With an online hologram, the spatial frequency spectrum of the reconstruction is confined to the disk of the objective's pupil, which has a cutoff frequency $\rho_{MAX} = NA_{OBJ}/\lambda$, where NA_{OBJ} is the numerical aperture of the objective and λ is the wavelength of the radiation. In Ref. [3], it has been demonstrated experimentally that synthesizing a pupil exceeding the objective's pupil disk is easily implemented in scanning holographic microscopy and leads to images of 3D fluorescent specimens reconstructed with a spatial resolution exceeding the Rayleigh limit of the objective. In the following discussion, we summarize the research work of Indebetouw et al. [3].

To extend the spatial spectrum beyond the objective's cutoff in a particular direction specified by a unit vector $\hat{n}$, one can record an off-axis hologram by scanning an off-axis FZP on the specimen. If the spatial frequency offset of the FZP is $\rho_0 \hat{n}$, the spatial frequency spectrum of the reconstruction is extended up to a value $\rho_{MAX} + \rho_0$ in the direction of $\hat{n}$. By combining several holograms with shifts in different angular directions, it is thus possible to tile an object's spatial frequency spectrum that extends, in principle, far beyond the objective's pupil disk.

The two pupils needed to create an off-axis FZP are

$$\tilde{P}_1(\vec{\rho}) = \exp\left(i\pi \lambda z_0 \rho^2\right) \text{Disc}\left(\rho/\rho_{MAX}\right)$$
$$\tilde{P}_{2j}(\vec{\rho}) = \delta\left(\vec{\rho} - \rho_0 \hat{n}_j\right) \quad (6.5)$$

$\text{Disc}(x) = 1$ for $x < 1$ and $\text{Disc}(x) = 0$ for $x > 1$. In these expressions, $\vec{\rho}$ is the transverse spatial frequency vector in the pupil plane, proportional to the transverse spatial coordinate vector $\mathbf{r}_P$ in the pupil, that is, $\vec{\rho} = \mathbf{r}_P / \lambda f_{OBJ}$, where f_{OBJ} is the focal length of the objective [25]. In the object's space, $P_1(\mathbf{r})$ is a converging spherical wave with radius of curvature z_0, and $P_{2j}(\mathbf{r})$ is a plane wave with a transverse spatial frequency $\rho_0 \hat{n}$. After heterodyne detection, or phase-shifted homodyne detection, each object point is encoded as an off-axis spherical wave $S_j(\mathbf{r}, z)$, where z is the axial coordinate in the object space measured from the focal plane of the objective. In Fourier space, we have

$$\tilde{S}_j(\rho, z) = \tilde{P}_1(\vec{\rho}, z) \oplus \tilde{P}_{2j}(\vec{\rho}, z) = \exp\left\{i\pi \lambda \left[z_0 \rho_0^2 + (z_0 - z)\left(\rho^2 - 2\vec{\rho} \cdot \rho_0 \hat{n}_j\right)\right]\right\}$$
$$\times \text{Disc}\left(\left|\vec{\rho} - \rho_0 \hat{n}_j\right|/\rho_{MAX}\right) \quad (6.6)$$

where $\oplus$ symbolizes a correlation integral, and

$$\tilde{P}_{1,2}(\vec{\rho}, z) = \tilde{P}_{1,2}(\vec{\rho}) \exp\left(i 2\pi z \sqrt{\lambda^{-2} - \rho^2}\right) \quad (6.7a)$$

or in paraxial approximation

$$\tilde{P}_{1,2}(\vec{\rho}, z) = \exp\left(i 2\pi z/\lambda\right) \exp\left[i\pi \lambda (z_0 - z) \rho^2\right] \text{Disc}(\rho/\rho_{MAX}) \quad (6.7b)$$

The FT of the specimen hologram is given by

$$\tilde{H}_{Oj} = \int dz \tilde{I}(\vec{\rho}, z) \tilde{S}_j(\vec{\rho}, z), \quad (6.8)$$

where $\tilde{I}(\vec{\rho}, z)$ is the 2D transverse FT of the 3D object intensity distribution $I(\mathbf{r}, z)$. The reconstruction of the hologram in a chosen axial plane of focus $z = z_R$ can be obtained by digital Fresnel back propagation from the hologram for a distance $z_0 + z_R$. In the experiment discussed below, the hologram is reconstructed by correlation with the experimental hologram of a subresolution point source. As has been shown previously [26], this method offers a way of reducing the phase aberrations of the objective. With high-NA objectives, these aberrations are attributed to the fact that the objective is not (and cannot be) used in the geometry for which it was designed, and can be severe. In the experiment, this scheme is implemented by recording a reference hologram $H_{Rj}(\mathbf{r})$ of a point source $\delta(\mathbf{r}, z)$ and propagating it to the desired reconstruction plane by using a propagation factor $P_j(\mathbf{r}, z_R)$. In Fourier space, we have

$$\tilde{H}_{Rj}(\vec{\rho}) = \exp\left[i\pi\lambda z_0 \left(|\vec{\rho} - \rho_0\hat{n}_j|^2\right)\right] \text{Disc}\left(|\vec{\rho} - \rho_0\hat{n}_j|/\rho_{\text{MAX}}\right) \quad (6.9)$$

$$\tilde{P}_j(\vec{\rho}, z_R) = \exp\left[-i\pi\lambda z_R \left(\rho^2 - 2\vec{\rho}\cdot\rho_0\hat{n}_j\right)\right] \quad (6.10)$$

The FT of the reconstructed image is then given by

$$\tilde{R}_j(\vec{\rho}, z_R) = \tilde{H}_{Oj}(\vec{\rho})\left[\tilde{H}_{Rj}(\vec{\rho})\tilde{P}_j(\vec{\rho}, z_R)\right]^*$$

$$= \int dz \tilde{I}(\vec{\rho}, z) \exp\left[i\pi\lambda\,(z_R - z)\left(\rho^2 - 2\vec{\rho}\cdot\rho_0\hat{n}_j\right)\right]\text{Disc}\left(|\vec{\rho} - \rho_0\hat{n}_j|/\rho_{\text{MAX}}\right)$$

$$(6.11)$$

where the asterisk represents a complex conjugation. In the plane of focus $z = z_R$, the reconstruction has a spatial frequency spectrum consisting of the object's spatial frequencies located within a disk of radius ρ_{MAX}, centered at $\rho_0\hat{n}_j$. By adding coherently the reconstruction amplitudes of a number of holograms with different spatial frequency shifts, one can, in principle, tile a synthetic pupil with an area far exceeding the pupil disk of the objective.

It must be emphasized that although the pupil representation in Eq. (6.5) is, strictly speaking, valid only in the paraxial regime, the method of scanning holographic microscopy is not limited to small NAs. In fact, since we actually measure the PSF, and reconstruct the holograms by this experimental PSF, the method is valid regardless of the system's NA, and it is independent of the theoretical representation of the PSF.

The experimental setup of a scanning holographic microscope has been described in detail in previous publications [3, 26]. For completeness, a sketch is given in Figure 6.4. The only addition to the arrangement of Ref. [26] is the introduction of a wedge prism in a conjugate object plane to shift the spatial frequency of the plane wave creating the FZP by a chosen amount in a chosen direction. The wedge is then rotated in discrete angular steps to cover a desired area of the object's spatial frequency spectrum. For experimental simplicity we chose to introduce both the spherical wave and the off-axis plane wave forming the FZP through the pupil of the objective. With this method, the largest spatial frequency shift achievable is the objective's frequency cutoff, namely, $\rho_0 \leq \rho_{\text{MAX}}$. Opting for this geometry is not a necessary requirement. In principle, it is possible to introduce the off-axis plane waves externally at any angle, although this practice may be difficult to implement at high NA. Nevertheless, the transfer function of such system could, in principle, be extended to its theoretical frequency limit of $2/\lambda$ and cover the entire spectral region inside this limit.

To assess the gain in the resolution of the system, the reconstruction of a subresolution pinhole (0.5 μm diameter) was used to estimate the size of the PSF [3]. With NA of 0.42, the theoretical Rayleigh resolution limit of the objective is $1.22\lambda_{\text{EM}}/2\text{NA} \approx 0.9$ μm, where $\lambda_{\text{EM}} = 600$ nm is the emission wavelength of the beads. The wrapped phase of the online hologram of the 0.5 μm diameter pinhole is shown in Figure 6.5a, and its reconstruction is shown in Figure 6.5b. The radius

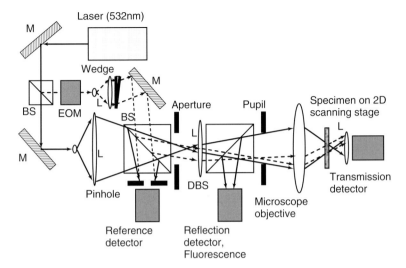

Figure 6.4 Schematic of an off-axis scanning holographic microscope. M: mirror; BS: beam splitter; DBS: dichroic beam splitter; and EOM: electro-optic phase modulator. $L_{1,2}$ are achromat doublet lenses having 16 cm focal length, L_3 is a collecting lens with 1 cm focal length. The wedge on a rotating stage is used to create off-axis Fresnel patterns on the specimen.

of curvature of the hologram is found to be $z_0 \approx 50$ μm, and its equivalent NA is $a/z_0 \approx 0.35$. This leads to an expected theoretical Rayleigh limit ≈ 0.9 μm, which is close to the Rayleigh resolution of the objective. The observation that the smaller equivalent NA of the hologram leads to the same resolution limit as that of the objective is attributable to the fact that the objective forms an image at the emission wavelength while the hologram is formed at the shorter excitation wavelength. The reconstruction shown in Figure 6.5b has a FWHM equal to ≈ 1.0 μm. Since this is the convolution of the actual PSF with the 0.5 μm pinhole, the size of the experimental PSF is estimated to be ≈ 0.9 μm, or smaller. Three off-axis holograms of the pinhole were recorded with a spatial frequency offset $\rho_0 \approx \rho_{MAX} \approx 0.66$ μm^{-1} in directions 120° apart. The wrapped phases of the three holograms are combined in Figure 6.5c to illustrate the wider spatial frequency coverage of the composite hologram. The reconstruction of the pinhole from this hologram is shown in Figure 6.5d. Its FWHM is ≈ 0.7 μm. The actual size of the PSF of the composite hologram is thus estimated to be smaller than ≈ 0.6 μm, which corresponds to an equivalent NA larger than ≈ 0.54. The reduction of the resolution limit by a factor ~ 0.6 or smaller could be further improved by combining more than three holograms. If the off-axis FZP is introduced on the specimen through the objective, as done in the experiment, we can expect a resolution limit down to a factor ~ 0.5 that of the objective.

To demonstrate the reality of the achieved resolution improvement, we chose a sample consisting of fluorescent beads with a diameter slightly larger than the ≈ 0.9 μm resolution limit of the objective at the emission wavelength and slightly

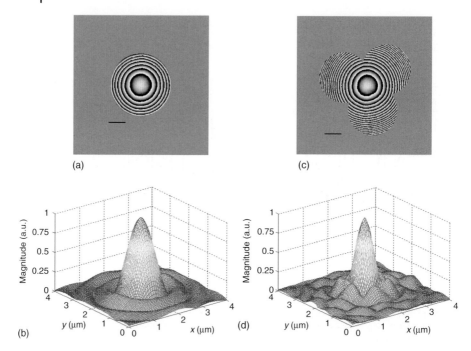

Figure 6.5 (a) Wrapped phase of the on-axis hologram of a 0.5 μm diameter pinhole. The scale bar is 10 μm. The phase distribution has a radius ~18 μm, a Fresnel number ~12, and a radius of curvature ~50 μm. (b) Amplitude of the reconstruction of the 0.5 μm pinhole using the online hologram. FWHM ~1.0 μm. (c) Wrapped phase of three off-axis holograms of the 0.5 μm pinhole illustrating the idea of pupil synthesis. The scale bar is 10 μm. (d) Amplitude of the reconstruction of the 0.5 μm pinhole using the composite off-axis holograms. FWHM ~0.7 μm. (Please find a color version of this figure on the color plates.)

larger than the ≈ 0.9 μm resolution limit of the on-axis hologram at the excitation wavelength. The expectation is that the reconstruction of the on-axis hologram will show, at best, barely unresolved beads, while the composite reconstruction should reveal beads that are resolved. The three holograms of the fluorescent beads were recorded in reflection and reconstructed individually and the amplitudes of their reconstructions were added coherently. Figure 6.6a,b shows the reconstructions of the on-axis hologram in two different planes. As expected, individual beads are detected in the best plane of focus ($z = 47.5$ μm in Figure 6.6a), but the clusters are not resolved. Shifting the focal plane to $z = 49$ μm does not reveal anything different because the two planes with an axial separation of 1.5 μm are well within the Rayleigh focal distance of ≈ 3.5 μm. The composite reconstructions of the three off-axis holograms in the same two focal planes are shown in Figure 6.6c,d. The clusters of beads are now well resolved. Furthermore, Figure 6.6c,d reveals that different clusters are in best focus in different planes. This is to be expected since the two planes are 1.5 μm apart, while the axial Rayleigh distance of the composite hologram is ≈ 1.8 μm.

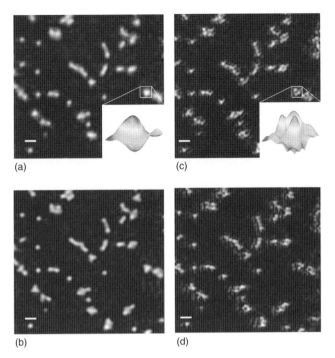

Figure 6.6 (a) Reconstruction of the on-axis hologram of a collection of ∼ 1.0 μm fluorescent beads at the "best focus" distance of 47.5 μm from the focal plane of the objective. The scale bar is 5 μm. Bead clusters are just barely resolved. (b) Same reconstruction at a focus distance of 49 μm. The two planes are within the Rayleigh range of the on-axis scanning FZP. (c) "Best focus" at 47.5 μm from the focal plane of the objective, where coherent sum of the complex amplitudes of the reconstructions of three off-axis holograms was recorded with off sets 120° apart. (d) Same reconstruction at a focus distance of 49 μm. The distance between the two planes is close to the Rayleigh range of the synthesized FZP, and different beads clusters are focused in different planes. (Please find a color version of this figure on the color plates.)

6.4
Synthetic Aperture with Fresnel Elements

SAFE is based on Fresnel incoherent correlation holography (FINCH) [10, 27], the recently invented method of a single-channel incoherent interferometer employed for generating digital Fresnel holograms [1]. In this nonscanning holographic technique, incoherent light is reflected or emitted from a 3D object, then propagates through a spatial light modulator (SLM), and is finally recorded by a digital camera. The SLM is used as a diffractive beam splitter of the incoherent interferometer, so that each spherical beam, originating from each object point, is split into two spherical beams with two different curve radii. Accumulation of the entire interferences within all of the couples of spherical beams creates the Fresnel hologram of the observed object. Three holograms are recorded sequentially, each for a different phase factor of the SLM. The three holograms are superposed in the computer so that the result is a complex-valued Fresnel hologram that does not contain the twin image and the bias term.

SAFE is actually a FINCH with an extended synthetic aperture aimed to improve the transverse and axial resolutions beyond the classical limitations. The synthetic aperture is implemented by shifting the system, located across the field of view, among several viewpoints. At each viewpoint a different mask is displayed on the SLM, and a single element of the Fresnel hologram is recorded. The various elements, each of which is recorded by the real aperture system during the capturing time, are tiled together so that the final mosaic hologram is effectively considered as captured from a single synthetic aperture that is much wider than the actual aperture.

An example of a system with the synthetic aperture, which is three times wider than the actual aperture, can be seen in Figure 6.7. To simplify the demonstration, the synthetic aperture was implemented only along the horizontal axis. In principle, this concept can be generalized for both axes and for any ratio of synthetic to actual apertures. Imaging with the synthetic aperture is necessary for cases in which the angular spectrum of the light emitted from the observed object is wider than the NA of a given imaging system. In SAFE shown in Figure 6.7, the SLM and the digital camera move in front of the object. The complete Fresnel hologram of the object, located at some distance from the SLM, is a mosaic of three holographic elements, each of which is recorded from a different point of view by the system with the real aperture, which is $A_x \times A_y$ in size. In this specific example, the complete hologram tiled from the three holographic Fresnel elements has the synthetic

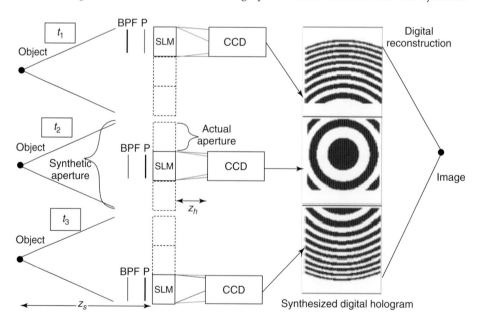

Figure 6.7 Scheme of SAFE operating as synthetic aperture radar to achieve super-resolution. P, polarizer and BPF, band-pass filter.

aperture, which is $3 \times A_x \times A_y$ in size, and it is three times larger than the real aperture at the horizontal axis. An object point located at the point (x_s, y_s, z_s), at a distance z_s from the SLM, induces a tilted diverging spherical wave of the form of $C_1(\bar{r}_s) Q[1/z_s] L[-\bar{r}_s/z_s]$ on the SLM plane (x,y), where for the sake of simplification, the quadratic phase function is designated by the function Q, such that $Q[s] = \exp[i\pi s\lambda^{-1}(x^2+y^2)]$, and the linear phase function is designated by the function L, such that $L[\bar{s}] = \exp[i2\pi\lambda^{-1}(s_x x + s_y y)]$, $\bar{r}_s \equiv (x_s, y_s)$, and $C_1(\bar{r}_s)$ is a complex constant dependent on the source point's location.

Each tilted diverging spherical wave in the (m,n)th exposure is split into two waves by the SLM mask, which is a sum of two aperture-limited quadratic phase functions of the form, $(C_2 Q[-1/f_1] + C_3 Q[-1/f_2]) \mathrm{rect}[(x - A_x \cdot m)/A_x, (y - A_y \cdot n)/A_y]$ where f_1 and f_2 are real constants indicating the focal distances of the two diffractive lenses, $C_{2,3}$ are complex constants, and

$$\mathrm{rect}\left(\frac{x}{\alpha}, \frac{y}{\beta}\right) \equiv \begin{cases} 1 & (|x|, |y|) \leq (\alpha/2, \beta/2) \\ 0 & \text{otherwise} \end{cases}$$

From the SLM plane, the two waves propagate a distance z_h till they are recorded by the digital camera. The complex amplitude on the camera plane (x_0, y_0) is computed as a free-space propagation under Fresnel approximation or, in other words, as a convolution between the complex amplitude on the SLM and the function $Q[1/z_h]$. A complete Fresnel hologram of the object point located at (x_s, y_s, z_s) is a sum of $M \times N$ holographic elements, each of which is the intensity recorded from the (m,n)th location by the digital camera as the following:

$$I_h(x_0, y_0; x_s, y_s, z_s) = \sum_{n=\frac{1-N}{2}}^{\frac{N-1}{2}} \sum_{m=\frac{1-M}{2}}^{\frac{M-1}{2}} \left| C_1(\bar{r}_s) Q\left[\frac{1}{z_s}\right] L\left[\frac{-\bar{r}_s}{z_s}\right] \right.$$
$$\times \left(C_2 Q\left[\frac{-1}{f_1}\right] + C_3 Q\left[\frac{-1}{f_2}\right] \right)$$
$$\left. \times \mathrm{rect}\left(\frac{x - A_x \times m}{A_x}, \frac{y - A_y \times n}{A_y}\right) \times Q\left[\frac{1}{z_h}\right] \right|^2 \quad (6.12)$$

where without the loss of generality, we assume that M and N are odd numbers. Following straightforward calculations detailed in Ref. [4], the intensity distribution recorded by the digital camera is expressed as the following:

$$I_h(x_0, y_0; x_s, y_s, z_s) = \left(C_4 + C_5(\bar{r}_s) Q\left[\frac{-1}{z_r}\right] \times L\left[\frac{-\bar{r}_r}{z_r}\right] + C_5^*(\bar{r}_s) Q\left[\frac{1}{z_r}\right] \times L\left[\frac{\bar{r}_r}{z_r}\right] \right)$$
$$\times \sum_{n=\frac{1-N}{2}}^{\frac{N-1}{2}} \sum_{m=\frac{1-M}{2}}^{\frac{M-1}{2}} \mathrm{rect}\left(\frac{x_0 - A_x \times m}{A_x}, \frac{y_0 - A_y \times n}{A_y}\right) \quad (6.13)$$

where

$$z_r = \frac{(f_1 z_s - z_h z_s + f_1 z_h)(f_2 z_s - z_h z_s + f_2 z_h)}{z_s^2 (f_1 - f_2)} \underset{f_2 \to \infty}{=} -\frac{(z_s + z_h)(f_1 z_s - z_h z_s + f_1 z_h)}{z_s^2}$$

$$\bar{r}_r = (x_r, y_r) = \frac{\bar{r}_s z_h}{z_s}$$

and $C_{4,5}$ are complex constants. z_r is the reconstruction distance of the point image from an equivalent optical hologram, although in the present case, the hologram is digital and the reconstruction is done by the computer. Note that z_r is obtained specifically in the case that one of the phase masks on the SLM is constant ($f_2 \to \infty$). This choice is usually made because practically the fill factor of the SLM is <100%, and therefore, the constant phase modulation inherently exists in the SLM. Consequently, choosing $f_2 < \infty$ could cause unwanted three, instead of two, waves mixing on the hologram plane, one wave due to the constant phase and the other two from the two different diffractive lenses.

Equation (6.13) is the expression of the transparency function of a hologram created by an object point and recorded by a conventional lensless FINCH with a synthetic aperture, which is $M \cdot A_x \times N \cdot A_y$ in size. This hologram has several unique properties. The transverse magnification M_T is expressed as $M_T = \partial x_r / \partial x_s = z_h / z_s$ (In contrast to a conventional Fresnel hologram [25], where $M_T = z_r / z'_s$, and z'_s represents the distance between the object and the recording medium). The axial magnification is $M_A = \partial z_r / \partial z_s = z_h \left(2 f_1 z_s + 2 f_1 z_h - z_h z_s\right) / z_s^3$ (In contrast to a conventional hologram [25], where $M_A = M_T^2$). Based on these properties, and assuming the system is diffraction limited, the resolution limitations of the FINCH as an imaging system is that the minimum resolved object size is given by

$$\Delta_{\min} = \max \{\lambda / NA_{in}, \lambda / (M_T NA_{out})\} = \max \{2\lambda z_s / D_{SLM}, 2\lambda z_r / (M_T D_{CCD})\}$$

(6.14)

where D_{SLM} and D_{CCD} are the diameters of the SLM and the digital camera, respectively. NA_{in} and NA_{out} are the numerical apertures at the input and output of the complete holographic system, respectively. This result is in contrast to the resolution limit of a conventional Fresnel hologram in which the transverse magnification is given by $M_T = NA_{in} / NA_{out}$, and therefore, the minimum resolved object size in case of the Fresnel coherent hologram is $\Delta_{\min} = 2\lambda z_s / D_H = 2\lambda z_r / M_T D_H$, where D_H is the diameter of a hologram. In other words, in FINCH, the resolution limitation can be dictated by either the input or the output apertures. In any event, however, the synthetic aperture system is an extension of both apertures. Substituting the various parameters in Eq. (6.14) indicates that for $D_{SLM} = D_{CCD} = D$ and for $f_1 < 0$, the resolution is always determined by the output aperture as proved by the following inequality:

$$\left| \frac{2\lambda z_s}{D_{SLM}} \right| = \left| \frac{2\lambda z_s}{D} \right| < \left| \frac{2\lambda z_r}{M_T D_{CCD}} \right| = \left| \frac{2\lambda z_r z_s}{z_h D} \right| = \left| \frac{2\lambda}{D} \times \frac{(z_s + z_h)(f_1 z_s - z_h z_s + f_1 z_h)}{z_h z_s} \right|$$

Using negative diffractive lens with $f_1 < 0$ is preferred in the lensless FINCH setup because the diverging lens guarantees a high visibility of the holographic interference fringes on the camera plane for any z_h distance.

Because the sum of the all of the rect functions in Eq. (6.13) becomes one rect function of the form $\text{rect}[x/(M \times A_x), y/(N \times A_y)]$, it is evident from Eq. (6.13) that the complete hologram tiled from $M \times N$ elements is a Fresnel hologram of a point with a synthetic aperture, which is $M \cdot A_x \times N \cdot A_y$ in size, and it is $M \times N$ times larger than the real aperture. Therefore, for $N = M$, $f_1 < 0$, and $f_2 \to \infty$, the transverse resolution power of SAFE is higher than that of the real aperture system by a number that is the inverse ratio of the minimal resolved sizes in these two cases as given by

$$\frac{\Delta_{\min}^{RA}}{\Delta_{\min}^{SA}} = \frac{2\lambda z_r^{RA}/M_T D_{CCD}^{RA}}{2\lambda z_r^{SA}/M_T D_{CCD}^{SA}} = \frac{N(|f_1|z_s + z_h z_s + |f_1|z_h)}{N|f_1|z_s + z_h z_s + N|f_1|z_h} \quad (6.15)$$

where the superscript RA and SA stand for real and synthetic apertures, respectively. It is assumed in Eq. (6.15) that because of the finite size of the SLM pixels, increasing the SLM aperture (even synthetically) by N times compels us to increase the focal distance $|f_1|$ by the same rate. Equation (6.15) indicates that using synthetic aperture always improves the resolution performance of lensless FINCH, but the improvement is less than the ratio of the synthetic to real apertures.

Equation (6.13) describes the Fresnel hologram obtained from a single object point, and therefore $I_h(x_0, y_0; x_s, y_s, z_s)$ is the PSF of the recording system in the synthetic aperture mode. The complete Fresnel hologram of a general incoherently illuminated object $I_s(x_s, y_s, z_s)$ is an integral of the entire PSFs given by Eq. (6.13) over all object intensity distribution and is defined by:

$$H(x_0, y_0) = \iiint I_s(x_s, y_s, z_s) I_h(x_0, y_0; x_s, y_s, z_s) dx_s dy_s dz_s \quad (6.16)$$

Since the overall PSF given in Eq. (6.13) is identical to the PSF of the previous works [10, 27], the mosaic hologram given in Eq. (6.16) is a Fresnel incoherent hologram of the object but with the property that this hologram has been recorded with the effective aperture, which is $M \cdot A_x \times N \cdot A_y$ in size.

The method to eliminate the twin image and the bias term (two terms of the three presented in Eq. (6.13)) is the same as has been used before with the FINCH [10, 27]; three elemental holograms of the same object and for each point of view are recorded, each of the holograms has a different phase constant of the SLM's phase mask. The final holographic element is a specific superposition of the three recorded elements. The digital reconstruction of the final complex-valued mosaic hologram is conventionally computed by Fresnel back propagation.

SAFE indeed improves the resolution performance of lensless FINCH; however, the resolution gain with this method has been less than the ratio of the synthetic to real apertures [4]. Next we describe a different scheme of SAFE, originally proposed in Ref. [5], and discuss its potential in the future to become a telescope for imaging nearby objects. Owing to its potential task, and in order to distinguish it from the earlier system [4], this configuration is dubbed "telescopic synthetic aperture Fresnel element" (T-SAFE). In T-SAFE, due to a modification of the original SAFE setup, the resolution gain is indeed equal to the ratio of synthetic to real apertures.

The T-SAFE system contains a chromatic band-pass filter (BPF) if the source is polychromatic rather than quasimonochromatic. In case T-SAFE is implemented using polarization-sensitive SLM, a polarizer is introduced in the system. However, in the following discussion, we show an alternative configuration of T-SAFE, and for such systems neither SLMs nor polarizers are needed. As in SAFE, the subholograms are recorded by a digital camera, then they are arranged in the computer, and the mosaic hologram is digitally reconstructed.

As mentioned above, Eq. (6.13) is the expression of the transparency function of a Fresnel hologram created by an object point and recorded by a conventional lensless FINCH [4] with a synthetic aperture, which is $M \cdot A_x \times N \cdot A_y$ in size. The transverse magnification M_T is expressed as $M_T = \partial x_r/\partial x_s = z_h/z_s$ As expressed by Eq. (6.14), in FINCH, the resolution limitation can be dictated by either input or output apertures. The NA_{in} is determined by the SLM size and cannot be changed by the system's free parameters. However, the product $NA_{out}M_T$ in Eq. (6.14) is dependent on the system parameters and our goal should be to keep this product equal or larger than NA_{in}, in order not to reduce the resolution dictated by the input aperture. Therefore, referring to Eq. (6.14), an optimal FINCH satisfies the inequality as follows:

$$\frac{z_s}{D_{SLM}} \geq \frac{|z_r|}{M_T D_H} \tag{6.17}$$

In this inequality all the parameters are well defined, except the diameter of the hologram. The hologram size is dependent on the overall size of the reconstructed image. Assuming the observed object is much smaller than the overall size of the field of view, as is frequently the case in telescopic observations, its image is much smaller than the SLM size. Figure 6.8a,b presents the two possible configurations of FINCH for the case of pointlike image, (a) is for $f_d > z_h$ and (b) is for $f_d < z_h$. From Figure 6.8, it is easy to see that the diameter of the hologram is determined by

$$D_H = D_{SLM} \frac{|f_d - z_h|}{f_d} \tag{6.18}$$

Substituting $M_T = z_h/z_s$, $|z_r| = |f_d - z_h|$ and Eq. (6.18) into Eq. (6.17), the following upper limit for the length of f_d is obtained:

$$f_d \leq z_h \tag{6.19}$$

In other words, the optimal configuration, in terms of resolution for T-SAFE, is achieved when the focal length of the positive diffractive lens displayed on the SLM is smaller than the distance between the SLM and CCD. This condition is fulfilled by the setup of Figure 6.8b. In this case, the resolution is determined by the input aperture and therefore, synthetically increasing the input aperture by N times improves the overall resolution by N times. It should be noted that, as has been demonstrated in Refs. [28, 29], in case that the observed object occupies most of the field of view, the image has the size of order of the SLM dimensions. In that case, it can be assumed that $D_H \cong D_{SLM}$, and the system remains limited by the input aperture if $f_d \leq 2z_h$. The lower limit of f_d for D_H in both the above-mentioned cases is determined by the SLM's pixel size Δ and the number of pixels K along the SLM's diameter, according to the well-known inequality $f_d \geq K\Delta^2/\lambda$.

6.4 Synthetic Aperture with Fresnel Elements

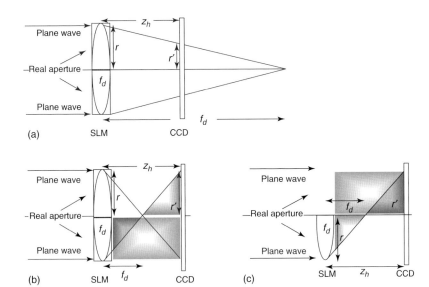

Figure 6.8 Possible configurations of recording holograms in the case of point-like object: (a) for FINCH where $f_d > z_h$. In this configuration, a hologram can be recorded, but, as indicated in the text, this hologram is suboptimal. (b) For FINCH where $f_d < z_h$. In this configuration, a hologram cannot be recorded because there is no interference between the plane and the spherical waves arriving from the same part of the SLM. (c) For T-SAFE, where $f_d < z_h$. In this configuration, the recorded hologram is optimal. The red and green areas indicate the spherical and plane waves, respectively. The rectangles in (a) and (b) symbolize the diffractive element of constant phase, where the lens symbol in all of the figures stands for the quadratic phase element, both the constant phase and quadratic phase elements are displayed on the same SLM. (Please find a color version of this figure on the color plates.)

Under the condition of Eq. (6.19), it is obvious from Figure 6.8b that if each submask of T-SAFE contains both quadratic and constant phase patterns, the split waves are detected in two separate areas on the CCD and therefore, there is no interference between them. In particular, in order to get interference between the spherical and the plane waves for every subhologram of T-SAFE and for every object point, we should use two separate submasks located on opposite sides of the optical axis, each one for different type of waves: the spherical and the plane waves. This situation is depicted in Figure 6.8c. We conclude that for the case of T-SAFE, contrary to SAFE of Ref. [4], it is not enough to move the observation system as one unit, like it is shown in Figure 6.7. In order to record each subhologram, two optical elements have to be moved in opposite directions as shown in Figure 6.9. On the one hand, this configuration has the disadvantages of shifting two elements in opposite directions and of interfering the two waves from a distance that grows with the size of the synthetic aperture. However, on the other hand, the important advantage of T-SAFE over SAFE and FINCH is that both phase functions, the quadratic and constant phases, are displayed individually on a different optical element. Consequently, we can display them directly on the phase-only SLM

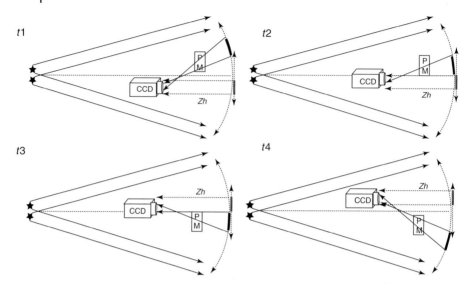

Figure 6.9 Proposed design of T-SAFE, which is based on spherical (black line) and flat (red line) mirrors rather than on SLMs. Four interfering steps, needed to obtain the synthetic aperture hologram, are shown. PM, phase modulator. (Please find a color version of this figure on the color plates.)

without any manipulations of spatial multiplexing. The analysis described above is no longer an approximation but an accurate description of the system. Moreover, since most large telescopes in the optical regime are equipped with mirrors, we suggest implementing T-SAFE with set of two types of mirrors, planar for the constant phase and spherical (or paraboloidal in case of non-paraxial imaging [30]) for the quadratic phase, as is shown in Figure 6.9. This figure presents four interfering steps needed to obtain the synthetic aperture hologram, which is four times larger than the available real aperture. In this process, at each step, a different part of the complete quadratic function is carried out by the spherical mirror (or displayed on the SLM in our experiment). The converging spherical wave interferes with the wave arriving from the planar mirror located on the opposite side of the optical axis.

The required mnth subpattern $P_{mn}(x, y)$ for the mnth subhologram of T-SAFE is given by the following expression:

$$P_{mn}(x, y) = C_2 Q\left[\frac{-1}{f_d}\right] \text{rect}\left[\frac{x - A_x(m + 1/2)}{A_x}, \frac{y - A_y(n + 1/2)}{A_y}\right]$$
$$+ C_3 \text{rect}\left[\frac{x + A'_x(m + 1/2)}{A_x}, \frac{y + A'_y(n + 1/2)}{A_y}\right] \quad (6.20)$$

It is clear from Eq. (6.20) that the subpattern $P_{mn}(x, y)$ comprises two separate parts, one with the quadratic phase mask, which is located at (mA_x, nA_y), and the other with the constant phase mask which is located at $(-mA'_x, -nA'_y)$, where the ratio of the steps of two parts is given by Eq. (6.21) as follows:

$$\frac{A'_x}{A_x} = \frac{A'_y}{A_y} = \frac{z_h - f_d}{f_d} \tag{6.21}$$

The new PSF of the overall recording system of the entire $M \times N$ subhologram is given by

$$I_h(x_0, y_0; x_s, y_s, z_s) = \sum_{n=\frac{N}{2}}^{\frac{N}{2}-1} \sum_{m=\frac{M}{2}}^{\frac{M}{2}-1} \left| C_1(x_s, y_s) Q\left[\frac{1}{z_s}\right] L\left[\frac{-x_s}{z_s}, \frac{-y_s}{z_s}\right] \right.$$

$$\left. \times P_{mn}(x, y) * Q\left[\frac{1}{z_h}\right] \right|^2 \tag{6.22}$$

where without the loss of generality we assume that M and N are even.

Following straightforward calculations, the intensity distribution recorded by the digital camera is again expressed by Eq. (6.13). Since the overall PSF given in Eq. (6.13) is identical to the PSF of the previous works [10, 27–29], the mosaic hologram given in Eq. (6.13) is a Fresnel incoherent hologram of the object, but with the property that this hologram has been recorded with the effective aperture, which is $M \cdot A_x \times N \cdot A_y$ in size.

In the specific example where $I_s(x_s, y_s, z_s)$ contains two separate points, at $(x_{s1}, y_{s1}, -z_s)$ and $(x_{s2}, y_{s2}, -z_s)$, the complete Fresnel hologram, obtained from Eq. (6.13) after eliminating the twin image and the bias term, is as follows:

$$H(x_0, y_0) = C_5(x_{s1}, y_{s1}) Q\left[\frac{-1}{z_r}\right] L\left[\frac{(x_{r1}, y_{r1})}{-z_r}\right]$$

$$+ C_5(x_{s2}, y_{s2}) Q\left[\frac{-1}{z_r}\right] L\left[\frac{(x_{r2}, y_{r2})}{-z_r}\right]$$

$$= 2C_5 Q\left[\frac{-1}{z_r}\right] L\left[\frac{(x_{r1}+x_{r2}),(y_{r1}+y_{r2})}{-2z_r}\right]$$

$$\times \cos\left\{\frac{\pi}{\lambda z_r}\left[(x_{r2}-x_{r1})x_0 + (y_{r2}-y_{r1})y_0\right]\right\} \tag{6.23}$$

where

$$x_{rk} = \frac{x_{sk} z_h}{z_s}, \quad y_{rk} = \frac{y_{sk} z_h}{z_s}, \quad k = 1, 2$$

It is assumed in Eq. (6.23) that both object points are close enough to the origin so that we can assume $C_5(x_{s1}, y_{s1}) \cong C_5(x_{s2}, y_{s2}) = C_5$. According to Eq. (6.23), a feasible separation of two points at the object plane is subject to the ability to record at least one cycle of the cosine term by the effective aperture of the system.

T-SAFE has been tested in the laboratory by the system shown in Figure 6.10. Note that in present indoor experiments, the phase-only SLM is still used rather than the mirrors configuration, because it is easier to implement the phase shifting procedure by the SLM, and there is no need for an additional phase modulator (PM), as it is in case of the mirrors configuration. The tested object in all the

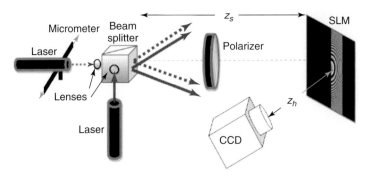

Figure 6.10 Experimental setup. The two uncorrelated object points are created by two He-Ne lasers and imaged by the T-SAFE.

experiments has been represented by two uncorrelated point sources created by two independent He-Ne lasers, each with 632 nm wavelength. The two laser beams have been focused by two lenses at focal distance of 6 cm, creating two diverging spherical waves with a transverse cross section on the SLM plane that is much wider than the SLM size. One of the lasers has been placed on top of a horizontal micrometer shifter to allow modification of the distances between the two point source objects. The distance z_s between the point source objects, that is, from the back focal point of two lenses, and the SLM is 8 m. The distance z_h between the phase-only SLM and the digital camera is 88 cm. The real aperture in the experiments has been chosen to be one-fourth of the complete physical aperture of the SLM, that is, 1080 × 480 pixels, and the synthetic aperture has been chosen to be as large as the complete physical aperture of the SLM, that is, 1080 × 1920 pixels.

In the first experiment, a comparison has been made between the complete aperture and the real aperture holograms recorded according to the setup of a regular FINCH. In the first experiment, the gap between the points has been set at 1 mm. Figure 6.11a–d shows one of the three masks displayed on the SLM and the corresponding recorded (b) and computed (c) magnitude and (d) phase holograms for the complete aperture. Figure 6.11e–h shows the same, but this time, the aperture is four times narrower. The focal distance of the diffractive lens on the SLM for both apertures is 55 cm. This focal distance guarantees optimal resolution, according to Eq. (6.19), and also assures that the complete hologram diameter obtained with the SLM's full aperture is equal to the detector size, according to Eq. (6.18). Each of the three masks on the SLM has sequentially one of the three different phase factors: 0, 120, or 240°. As mentioned above, these three phase masks with different phase constants are required in order to eliminate the bias term and the twin image from the holographic reconstruction.

The three recorded holograms are superposed according to the same superposition equation given in Ref. [10]. Figure 6.11c,d and Figure 6.11g,h show the magnitude and phase of the superposed holograms for the complete and narrow apertures, respectively. Resolved and unresolved best in-focus reconstructed

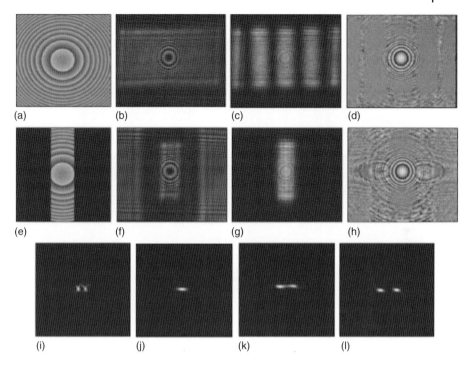

Figure 6.11 Experimental results of a regular FINCH obtained for the complete and narrow apertures: (a) is one of the three masks displayed on the SLM for the complete wide aperture; (b) the corresponding recorded hologram; (c) the computed absolute; (d) phase of the complex-valued hologram of 1 mm gap between the source points; (e–h) the same as (a–d) but for the narrow aperture; (i) the best in-focus reconstructed plane in case of the complete aperture, for a gap between the source points of 1 mm; (j–l) the same as (i) but in case of narrow apertures for a gap between the source points of 1, 1.5, and 2 mm, respectively.

planes, for 1 mm gap between the source points, are shown in Figure 6.11i for the complete, and in Figure 6.11j for the narrow apertures. It can be seen that the resolution along the horizontal axis of the reconstructed image, computed by Fresnel back propagation, is worse for the narrow aperture than for the complete aperture because the hologram of the former contains less than one cycle of the cosine term presented in Eq. (6.23). In terms of the Rayleigh criterion, our narrow aperture system can resolve a point gap larger than $1.22\lambda z_s/D = 1.6$ mm, where $z_s = 8$ m and D is the width of the aperture ($D = 480 \cdot 8$ μm $= 3840$ μm) is the width of the aperture. For comparison purposes, Figure 6.11k and Figure 6.11l present the best in-focus reconstructed planes for a gap between the points set to 1.5 and 2 mm, respectively. It can be seen that the narrow aperture system can hardly resolve a point gap of 1.5 mm as indicated by the Rayleigh criterion; however, for the 2 mm gap, a considerable separation is clearly seen.

In the second experiment, the synthetic aperture mechanism of T-SAFE has been tested and verified. In this case, the distance z_h between the SLM and CCD

158 | *6 Approaches to Overcome the Resolution Problem in Incoherent Digital Holography*

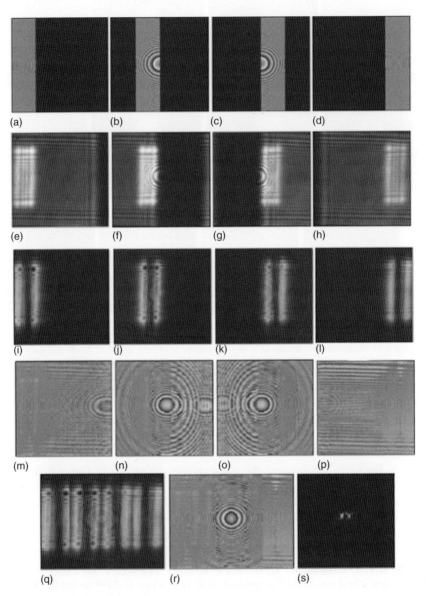

Figure 6.12 Experimental results obtained for the case of synthetic aperture: (a–d) present 4 of 12 masks displayed on the SLM during the recording process of the synthetic aperture hologram, and the corresponding recorded holograms are shown in (e–h); (i–l) computed magnitude and (m–p) phase of the complex holograms; (q) magnitude and (r) phase of the computed synthetic aperture hologram; and (s) best in-focus reconstructed plane for the synthetic aperture for the gap between the points of 1 mm.

and the focal distance of the diffractive lens displayed on the SLM have remained the same, 88 and 55 cm, respectively. A total of 12 different phase masks have been displayed on the SLM, three for each subaperture; leftmost, left, right, and rightmost. Each of the masks has real aperture of 1080 × 480 pixels. In each of the four locations, masks at every location sequentially have one of the three different phase factors: 0, 120, or 240°. The parts of the SLM, on which there is no quadratic function, use flat mirrors with the same constant phase value for the entire holograms.

Figure 6.12a–d shows the phase distribution of the four masks out of 12, each of which has been displayed at different times and different locations along the horizontal axis of the SLM. Note that unlike the masks of Figure 6.11a,e, in this T-SAFE experiment, the masks of Figure 6.12a–d are purely quadratic phase on each of the four areas. Figure 6.12e–h shows the recorded holograms obtained after displaying the masks of Figure 6.12a–d on the SLM, respectively. The object in this experiment is again two source points separated by 1 mm gap. The superposed complex-valued holographic element from each system's viewpoint is stored in the computer. On completing the scanning procedure along the entire synthetic aperture, all four holographic elements are arranged as a single mosaic hologram. Figure 6.12i–l and Figure 6.12m–p are, respectively, the magnitude and the phase of the subholograms from which the synthetic aperture mosaic hologram is assembled. Figure 6.12q and Figure 6.12r are the magnitude and the phase of the mosaic hologram, respectively. The best in-focus reconstruction result of the mosaic hologram, computed by the Fresnel back propagation, is depicted in Figure 6.12s. The two observed point source objects with a 1 mm gap between them are seen well in the reconstructed image, indicating that the synthetic aperture is wide enough to acquire the significant part of the horizontal spectral information of the objects.

6.5
Summary

We have reviewed three different methods of improving the imaging resolution by techniques of incoherent digital holography. In the DIPCH, synthesizing the PSF wisely improves the overall resolution of the 3D image, while in off-axis OSH and SAFE, the resolution power is increased because of use of synthetic apertures.

The DIPCH enables one to encode the 3D scene in a secure way and to image the scene with a significantly improved resolution. More important, the proposed framework that has yielded DIPCH might be used to find new types of incoherent correlation holograms for other purposes and with other advantages over the existing types of holograms.

Off-axis OSH shows that it is possible to overcome the spatial resolution limit of the objective in holographic microscopy by combining a number of off-axis scanning holograms to synthesize a pupil area larger than that of the objective. The principle of holographic synthetic aperture has been demonstrated using incoherent holography, in general, and scanning holographic fluorescent microscopy, in particular.

The main feature of T-SAFE is that its synthetic aperture increases the transverse resolving power by the ratio of the real to synthetic apertures. The captured hologram is of Fresnel type rather than Fourier type, which means that the information distribution over the hologram can be controlled by modifying the system parameters. Drawing nearer to the image plane also increases the similarity between the hologram and the observed object, whereas recording the hologram far from the image plane distributes the recorded information more globally over the hologram plane. The nature of the observed objects determines which of the above options is preferred, but using Fresnel-type holographic system enables us to choose between local and global distribution of the information.

Acknowledgments

Part of the work described herein was done in collaboration with G. Indebetouw and Y. Tada. This research was supported by Israel Ministry of Science and Technology (MOST) Grant GR-2234 (3-6807).

References

1. Rosen, J., Brooker, G., Indebetouw, G., and Shaked, N.T. (2009) A review of incoherent digital Fresnel holography. *J. Holography Speckle*, **5** (2), 124–140.
2. Shaked, N.T. and Rosen, J. (2008) Multiple-viewpoint projection holograms synthesized by spatially incoherent correlation with broadband functions. *J. Opt. Soc. Am., A*, **25** (8), 2129–2138.
3. Indebetouw, G., Tada, Y., Rosen, J., and Brooker, G. (2007) Scanning holographic microscopy with resolution exceeding the Rayleigh limit of the objective by superposition of off-axis holograms. *Appl. Opt.*, **46** (6), 993–1000.
4. Katz, B. and Rosen, J. (2010) Super-resolution in incoherent optical imaging using synthetic aperture with Fresnel elements. *Opt. Express*, **18** (2), 962–972.
5. Katz, B. and Rosen, J. (2011) Could SAFE concept be applied for designing a new synthetic aperture telescope?. *Opt. Express*, **19** (6), 4924–4936.
6. Beck, S.M., Buck, J.R., Buell, W.F., Dickinson, R.P., Kozlowski, D.A., Marechal, N.J., and Wright, T.J. (2005) Synthetic-aperture imaging laser radar: laboratory demonstration and signal processing. *Appl. Opt.*, **44** (35), 7621–7629.
7. Mico, V., Zalevsky, Z., García-Martínez, P., and García, J. (2006) Synthetic aperture superresolution with multiple off-axis holograms. *J. Opt. Soc. Am., A*, **23** (12), 3162–3170.
8. Martínez-León, L. and Javidi, B. (2008) Synthetic aperture single-exposure on-axis digital holography. *Opt. Express*, **16** (1), 161–169.
9. Shaked, N.T. and Rosen, J. (2008) Modified Fresnel computer-generated hologram directly recorded by multiple-viewpoint projections. *Appl. Opt.*, **49** (19), D21–D27.
10. Rosen, J. and Brooker, G. (2007) Digital spatially incoherent Fresnel holography. *Opt. Lett.*, **32** (8), 912–914.
11. Shaked, N.T., Katz, B., and Rosen, J. (2009) Review of three-dimensional holographic imaging by multiple-viewpoint-projection based methods. *Appl. Opt.*, **48** (34), H120–H136.
12. Li, Y., Abookasis, D., and Rosen, J. (2001) Computer-generated holograms

of three-dimensional realistic objects recorded without wave interference. *Appl. Opt.*, **40** (17), 2864–2870.
13. Sando, Y., Itoh, M., and Yatagai, T. (2003) Holographic three-dimensional display synthesized from three-dimensional Fourier spectra of real-existing objects. *Opt. Lett.*, **28** (24), 2518–2520.
14. Park, J.-H., Kim, M.-S., Baasantseren, G., and Kim, N. (2009) Fresnel and Fourier hologram generation using orthographic projection images. *Opt. Express*, **17** (8), 6320–6334.
15. Shaked, N.T., Rosen, J., and Stern, A. (2007) Integral holography: white-light single-shot hologram acquisition. *Opt. Express*, **15** (9), 5754–5760.
16. Katz, B., Shaked, N.T., and Rosen, J. (2007) Synthesizing computer generated holograms with reduced number of perspective projections. *Opt. Express*, **15** (20), 13250–13255.
17. Javidi, B. and Sergent, A. (1997) Fully phase encoded key and biometrics for security verification. *Opt. Eng.*, **36** (3), 935–942.
18. Aboukasis, D., Batikoff, A., Famini, H., and Rosen, J. (2006) Performance comparison of iterative algorithms for generating digital correlation holograms used in optical security systems. *Appl. Opt.*, **45** (19), 4617–4624.
19. Fienup, J.R. (1982) Phase retrieval algorithms: a comparison. *Appl. Opt.*, **21** (15), 2758–2769.
20. Stark, H. (ed.) (1987) *Image Recovery: Theory and Application*, Academic Press, pp. 29–78 and 277–320.
21. Poon, T.-C., Doh, K.B., Schilling, B.W., Wu, M.H., Shinoda, K., and Suzuki, Y. (1995) Three-dimensional microscopy by optical scanning holography. *Opt. Eng.*, **34** (5), 1338–1344.
22. Schilling, B.W., Poon, T.-C., Indebetouw, G., Storrie, B., Shinoda, K., Suzuki, Y., and Wu, M.H. (1997) Three-dimensional holographic fluorescence microscopy. *Opt. Lett.*, **22** (19), 1506–1508.
23. Indebetouw, G., Klysubun, P., Kim, T., and Poon, T.-C. (2000) Imaging properties of scanning holographic microscopy. *J. Opt. Soc. Am. A*, **17** (3), 380–390.
24. Rosen, J., Indebetouw, G., and Brooker, G. (2006) Homodyne scanning holography. *Opt. Express*, **14** (10), 4280–4285.
25. Goodman, J.W. (1996) *Introduction to Fourier Optics*, 2nd edn, McGraw-Hill.
26. Indebetouw, G. and Zhong, W. (2006) Scanning holographic microscopy of three-dimensional fluorescent specimens. *J. Opt. Soc. Am., A*, **23** (7), 1699–1707.
27. Rosen, J. and Brooker, G. (2008) Non-scanning motionless fluorescence three-dimensional holographic microscopy. *Nat. Photonics*, **2**, 190–195.
28. Katz, B., Wulich, D., and Rosen, J. (2010) Optimal noise suppression in Fresnel incoherent correlation holography (FINCH) configured for maximum imaging resolution. *Appl. Opt.*, **49** (30), 5757–5763.
29. Brooker, G., Siegel, N., Wang, V., and Rosen, J. (2011) Optimal resolution in Fresnel incoherent correlation holographic fluorescence microscopy. *Opt. Express*, **19** (6), 5047–5062.
30. Saleh, B.E.A. and Teich, M.C. (2007) *Fundamentals of Photonics*, 2nd edn, John Wiley & Sons, Ltd, p. 7.

7
Managing Digital Holograms and the Numerical Reconstruction Process for Focus Flexibility

Melania Paturzo and Pietro Ferraro

7.1
Introduction

Reduced depth of field is a severe limitation in microscopy since high magnification optics, because of their high numerical aperture, squeeze the depth of field and, consequently, only a thin slice of the imaged volume can be in good focus at the same time [1]. In fact, the depth of field of a microscope, depending on the different conditions of use, is not sufficient to obtain a single image in which all details of the object, even if they are located at different planes along the longitudinal direction, are still in focus [2]. Currently, the solution to this problem is to extend the depth of focus. Essentially, up to now in classical microscopy, only two methods have been proposed to obtain an extended depth of focus.

One solution provides a so-called extended focus image (EFI) in which all parts of the object, even in areas at different depths, are in good focus. The EFI is composed by selecting different portions in sharp focus in each image from a stack of numerous images recorded at different distances between the microscope objective (MO) and the object. Modern microscopes are equipped with micrometric mechanical translators, actuated by means of piezoelectric elements, which move the stage along the optical axis with a desired and opportune number of steps between the highest and the lowest point of the object. Essentially, what is performed is a mechanical scanning to image the object by means of a discrete number of planes across all the volume that is occupied by it [3, 4]. For each longitudinal step, an image is recorded and stored in a computer and linked with information about the depth at which it has been taken. The in-focus portion of each image is identified through contrast analysis. In fact, from each image, by means of numerical algorithms, the portion of the object that appears to be, or is numerically recognized as in good focus, is extracted. Then, the different portions are composed together to give single images in which all details are in focus, the EFI. In the EFI, all points of an object are in focus independently from the height they are in the topography of the object. Of course, the smaller the stepping increments performed in the mechanical scanning, the more accurate the result of the EFI. On the contrary, with more steps, the time taken for the

Optical Imaging and Metrology: Advanced Technologies, First Edition.
Edited by Wolfgang Osten and Nadya Reingand.
© 2012 Wiley-VCH Verlag GmbH & Co. KGaA. Published 2012 by Wiley-VCH Verlag GmbH & Co. KGaA.

acquisition increases and more computation is needed to obtain the EFI. Even if the computing time is not a serious problem, the length of the acquisition process (i.e., the time required to record all images at different depths) poses a severe limitation in obtaining an EFI for dynamic objects or process. Indeed, in the case of objects that experience dynamic evolution, for example, live biological samples, it is impossible to obtain an EFI by the above-described method.

An alternative solution that has been investigated is based on the use of a specially designed phase plate in the optical path of the microscope that allows an extension of the depth of focus of the images observable by the microscope. The phase plate introduces aberrations on the incoming optical rays at the expense of some distortion and a blurring effect, which are capable of having an effect of extending the depth of the focus. This method is called the *wave-front-coding approach* and has the drawback that a phase plate must be specifically designed and fabricated as a function of the specific object and the specific kind of adopted optical system [5–7].

Recently, the possibility of creating an EFI by digital holography (DH) has been demonstrated. In fact, holography has the unique feature of recording and reconstructing the amplitude and phase of a coherent wave front that has been scattered by an object. The reconstruction process allows the entire volume to be imaged. In classical film or plate holography, the reconstruction process is performed optically by illuminating the recorded hologram by the same reference beam.

In DH, the reconstruction process is accomplished by processing the digital hologram. The use of the Rayleigh–Sommerfeld diffraction formula allows the whole wave field in amplitude and phase to be reconstructed backward from the CCD array at any single image plane in the imaged volume [8].

Reconstructions at different image planes can be performed along the longitudinal axis (z-axis) by changing the distance of back propagation in the numerical diffraction process. Therefore, holographic microscope configurations have 3D imaging capability but suffer at the same time from a limited depth of focus. In fact, at a fixed reconstruction distance d, only some portion of the object will be in focus for 3D objects [9].

Of course, it is possible to obtain the entire volume by reconstructing a number of image planes in the volume of interest along the z-axis. In this way, a stack of images of the entire volume can be easily obtained in analogy with the classical microscopes.

The phase map of the object, which can be easily obtained by a single digital hologram, contains the information for a correct selection of the in-focus portions that have to be selected from each image stack. It will result in the correct construction of the final EFI. It is important to note that a significant advantage of this method is the possibility of obtaining an EFI of a microscopic object without a mechanical scanning operation. All the 3D information intrinsically contained in a single digital hologram is usefully employed to construct a single image with all portions of a 3D object in good focus. In this case, dynamic events can be studied. It means the EFI of a dynamic process can be obtained using a number of holograms recorded sequentially [10]. The limitation of the cited DH method is the complexity of the numerical computations. In fact, the EFI is built by using numerically retrieved phase maps; they could also contain phase unwrapping errors that would disturb the EFI formation. In the next two paragraphs, a new method to manage the depth of focus in a simple but effective way is shown. It consists in applying different kinds of geometrical deformations to the acquired holograms. Firstly, we apply a linear deformation to Fresnel holograms, that is, obtained using a plane wave as the reference beam, in order to change the real reconstruction distance. Then, we describe how it is possible to obtain the EFI for tilted objects by means of quadratic deformation of Fourier holograms, that is, acquired with a reference wave having the same curvature of the beam diffracted by the object. The numerical reconstruction of the object wavefield is achieved through a simple Fourier transform algorithm for the Fourier holograms, while, for the Fresnel holograms, by the numerical simulation of the Rayleigh–Sommerfeld diffraction integral.

The Rayleigh–Sommerfeld diffraction integral can be written as [11]

$$b'(x',y') = \frac{1}{i\lambda} \iint h(\xi,\eta) r(\xi,\eta) \frac{e^{ik\rho}}{\rho} \cos\Omega \, d\xi \, d\eta \tag{7.1}$$

where the integration is carried out over the hologram surface and $\rho = \sqrt{d'^2 + (x'-\xi)^2 + (y'-\eta)^2}$ is the distance from a given point in the hologram plane to a point of observation. The quantity $\cos\Omega$ is an *obliquity factor* [12] normally set to 1 because of small angles. If the reference is a collimated and normally impinging wave front, as in this case, one can set $r(\xi,\eta) = |r| = 1$, thus simplifying the numerical implementation. There are two ways of implementing Eq. (7.1) in a computer: the Fresnel and the convolution methods. Both approaches convert the Rayleigh–Sommerfeld diffraction integral into one or more *Fourier transforms*, which make the numerical implementation easy [13]. The Fresnel method is based on approximations of the ρ by applying the binomial expansion of the square root, which represents a *parabolic* approximation of spherical waves, and it is possible only for reconstruction distances $z \geq NP^2/\lambda$.

7.2
Fresnel Holograms: Linear Deformation

In Ref. [14], a method to change the depth of focus in DH is reported. Various geometrical deformations are applied to Fresnel digital holograms, that is, obtained when the distance between the object and the hologram plane is such that we are in the Fresnel region of diffraction. If a linear deformation is applied to the original hologram, the reconstruction distance changes. In fact, we consider a digital Fresnel hologram recorded with the object at distance d from the CCD, such that the numerical reconstruction of the object wavefield can be obtained through the Fresnel approximation of the Rayleigh–Sommerfeld diffraction integral, given by the formula

$$b(x,y,d) = \frac{1}{i\lambda d}\iint h(\xi,\eta)r(\xi,\eta)e^{ikd\left[1+\frac{(x-\xi)^2}{2d^2}+\frac{(y-\eta)^2}{2d^2}\right]}d\xi d\eta \qquad (7.2)$$

where $h(\xi,\eta)$ and $r(\xi,\eta)$ are the recorded hologram and the reference wave that in this case is $r(\xi,\eta) = 1$. The recorded hologram can be stretched, applying an affine geometric transformation, that is, $[\xi'\eta'] = [\begin{array}{ccc} \xi & \eta & 1 \end{array}]T$ where the operator T is

$$T = \begin{bmatrix} a & 0 \\ 0 & a \\ 0 & 0 \end{bmatrix}$$

obtaining the transformed hologram $h(\xi',\eta') = h(a\xi, a\eta)$ where α represents the scaling factor. Therefore, it results that the integral in Eq. (7.2) changes to

$$\begin{aligned} b(x,y,d) &= \frac{1}{i\lambda d}e^{ikd}\iint h(\alpha\xi,\alpha\eta)e^{\frac{ik\alpha^2(x-\xi)^2}{2d\alpha^2}}e^{\frac{ik\alpha^2(y-\eta)^2}{2d\alpha^2}}d\xi d\eta \\ &= \frac{1}{i\alpha^2\lambda d}e^{ikd}\iint h(\xi',\eta')e^{\frac{ik(x'-\xi')^2}{2D}}e^{\frac{ik(y'-\eta')^2}{2D}}d\xi'd\eta' \\ &= \frac{1}{\alpha^2}b(x',y',D) \end{aligned} \qquad (7.3)$$

with $D = d/\alpha^2$ and $x' = x/\alpha, y' = y/\alpha$. From Eq. (7.3), it is clear that even if the stretched hologram is reconstructed at a distance d, we observe the plane at a distance $D = d/\alpha^2$ from the CCD. Therefore, through successive reconstructions of the same hologram, stretched with a variable scaling factor α, it is possible to observe different object planes.

For example, we have employed an object made of three different wires located at three different distances from the CCD sensor: a horizontal wire, a twisted wire, and a vertical one set at distances of 100, 125, and 150 mm, respectively. The numerical reconstructions at those distances produce an image in which each wire is in good focus, as reported in Figure 7.1a–c.

When the hologram $h(x,y)$ is uniformly deformed with an elongation factor along both dimensions (x,y) of $\alpha = 1.1$, the horizontal wire results to be in focus at a distance $d = 150$ mm (Figure 7.1d) instead of 100 mm. If $\alpha = 1.22$, the twisted wire with the eyelet is in focus at $d = 150$ mm (Figure 7.1e) instead of $d = 125$ mm. Anyway, the results shown in Figure 7.1f demonstrate that the depth of focus can be managed by adopting this sort of uniform stretching of the digital holograms as expressed in Eq. (7.3).

A further test has been performed in order to demonstrate that, for speckle holograms, the linear deformation can also be applied to the hologram Fourier transform. In fact, two different holograms, corresponding to two different objects, are acquired with diverse distances between the object and the CCD camera. In this case, the Pinocchio recording distance is 720 mm, while the matryoshka one is 950 mm. Then, the Fourier transform of both holograms are performed; in the Fourier space, one of the twin images is filtered while the other one is

7.2 Fresnel Holograms: Linear Deformation

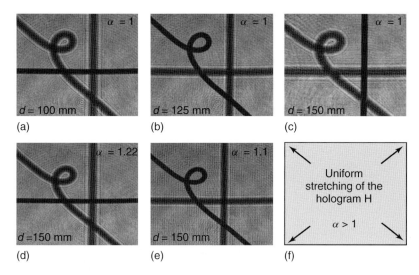

Figure 7.1 Reconstruction of a digital hologram as recorded (not deformed) at different distances: (a) 100, (b) 125, and (c) 150 mm. For $\alpha = 1.22$, the horizontal wire is in focus at (d) 150 mm, and for $\alpha = 1.1$, the curved wire is in focus at (e) 150 mm. (f) Conceptual drawing of the stretching.

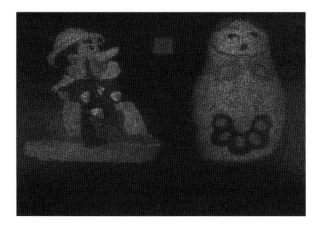

Figure 7.2 Numerical reconstruction of a multiplexed speckle hologram, obtained by adding two different holograms, which Fourier transform has been opportunely stretched in order to get both images in focus using a single reconstruction distance.

stretched according to the desired elongation parameter. Finally, the two resulting matrices are added to each other and a Fourier transform is applied to come back in the hologram space. Reconstructing the obtained hologram, using a unique reconstruction distance, we get the image shown in Figure 7.2, where both the images are in good focus if correct elongation factors are used.

7.3
Fresnel Holograms: Quadratic and Polynomial Deformation

In addition, a polynomial deformation of the form $[\xi'\eta'] = \begin{bmatrix} 1 & \xi & \eta & \xi*\eta & \xi^2 & \eta^2 \end{bmatrix} T$ can be taken into consideration to deform the digital holograms with different scopes. We adopted firstly a quadratic deformation such that the operator T is now expressed by

$$T = \begin{bmatrix} 0 & 0 \\ 1 & 0 \\ 0 & 1 \\ 0 & 0 \\ \beta & 0 \\ 0 & \gamma \end{bmatrix} \qquad (7.4)$$

The quadratic deformation can be used to create an EFI for a tilted object in a DH microscope (Figure 7.3). In the past years, different methods for digital holography on tilted (DHT) plane have been presented. De Nicola et al. [15] makes use of the angular spectrum of plane waves for fast calculation of optical diffraction. Fast Fourier transformation is used twice, and a coordinate rotation of the spectrum enables reconstruction of the hologram on the tilted plane. Interpolation of the spectral data is shown to be effective for correcting the anamorphism of the reconstructed image. Moreover, Jeong et al. [16] present an effective method for the pixel-size-maintained reconstruction of images on arbitrarily tilted planes in DH. The method is based on the plane wave expansion of the diffraction wave fields and the three-axis rotation of the wave vectors. The images on the tilted planes are reconstructed without loss of the frequency contents of the hologram and have the same pixel sizes. In a similar way, macroscopic flat objects have also been reconstructed in a single reconstruction step by Matsushima et al. [17].

Here, we show a different approach applied to a real case. We used, as object, a silicon wafer with the letters "MEMS" (microelectromechanical systems) written on it, tilted at an angle of 45° in respect to the optical axis. The deformation was applied only along the x-axis with $\beta = 0.00005$. Figure 7.4a shows the reconstruction of the hologram at distance of $d = 265$ mm as it has been recorded. It is clear from the picture that the portion of the object with the letter "S" is in good focus, while

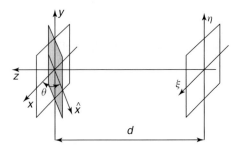

Figure 7.3 Schematic representation for reconstructing digital holograms on tilted planes.

7.3 Fresnel Holograms: Quadratic and Polynomial Deformation | 169

Figure 7.4 Quadratic deformation applied along the x-axis to a Fresnel hologram of a tilted object: amplitude of numerical reconstructions of the hologram as acquired by the CCD (a), after the quadratic deformation (EFI image) (b), and phase difference between the two reconstructions (c).

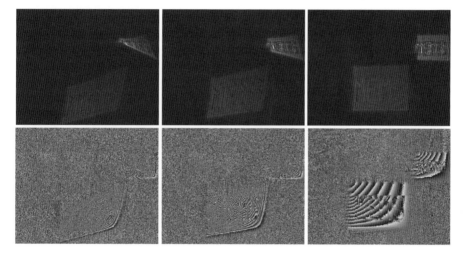

Figure 7.5 Three amplitudes and the corresponding phase reconstructions obtained varying the value $T(4,1)$ in Eq. (7.3).

the other letters are gradually out of focus. Figure 7.4b shows the reconstruction obtained by the digital hologram that has been quadratically deformed, where all the letters "MEMS" are in good focus. Since the quadratic deformation is very small, we can approximate it to linear deformation for each small portion of the Fresnel hologram. Basically, we can say that, for each portion of the digital hologram, the focus was changed in a different way, as shown by the parabolic shape of the phase map difference (Figure 7.4c) calculated by subtracting the phase of the deformed hologram from the undeformed one. The phase distribution demonstrates that the defocus tilt has been mainly removed, even if some residual errors remains there. In fact, in order to correct exactly the tilt term, the wrapped phase map should have a saw-toothed shape.

A further example is shown in Figure 7.5, where three amplitudes and the corresponding phase reconstructions obtained by varying another parameter in

Eq. (7.3) (i.e., the value T(4,1)) are exposed. By an opportune choice of the elements of the T matrix, it is possible to obtain whatever deformation of the hologram. Therefore, in principle, this technique could be used to reconstruct holograms of a generic not-plane object, if the relation between each kind of object surface deformation and the corresponding T matrix was known.

7.4
Fourier Holograms: Quadratic Deformation

For digital holograms recorded in Fourier configuration, that is, with the spherical reference wave diverging from a point in the object plane, the simple linear deformation furnishes different results in respect to the Fresnel holograms. In fact, applying a linear stretching to a Fourier hologram, no changes in the imaged plane can be observed after the reconstruction process. In the Fourier configuration, the numerical reconstruction of the hologram $f(\xi, \eta)$ is obtained by simply calculating its spatial FT $\hat{f}(x, y)$.

When the hologram is linearly deformed, it becomes

$$h(\xi, \eta) = f(a\xi, a\eta) \tag{7.5}$$

Because of the Fourier transform properties, we have

$$\hat{h}(x, y) = \frac{1}{a} \hat{f}\left(\frac{x}{a}, \frac{y}{a}\right) \tag{7.6}$$

where $\hat{h}(x, y)$ corresponds to the new numerical reconstruction and it results to be equal to the reconstruction of the original hologram with scaled dimensions and a scaled intensity. Therefore, as a result of Eq. (7.6), the reconstruction distance, that is, the distance at which the object results to be in focus, is not affected.

On the contrary, if a quadratic deformation is applied to the digital hologram, the in-focus reconstruction distance changes according to a linear law expressed by the formula

$$D = L(1 + 2\alpha l') \tag{7.7}$$

where l' is the horizontal coordinate in the reconstruction plane, α the deformation parameter along the l' coordinate, and L the reconstruction distance of the not deformed hologram (as recorded). Therefore, by this type of hologram deformation, it is possible to retrieve the EFI of a tilted object, putting all parts of the object in focus in a single reconstructed image [18]. To demonstrate how the in-focus distance depends on the quadratic deformation, we need to recollect some Fourier transform properties for a composite function $h(x) = g(f(x))$.

If we consider the following composite function

$$h(x) = g(f(x)) = \int G(l) e^{i2\pi l \cdot f(x)} dl \tag{7.8}$$

where $G(l)$ is the Fourier transform of $g(y)$, the FT of $h(x)$ results to be

$$\hat{h}(k) = \int e^{-i2\pi k \cdot x} \int G(l) e^{i2\pi l \cdot f(x)} dl \, dx = \int G(l) P(k, l) \, dl \tag{7.9}$$

where $P(k,l) = \int e^{-i2\pi k \cdot x} e^{i2\pi l \cdot f(x)} dx$. For a quadratic coordinate transformation of the hologram $g(x)$, it results that $f(x) = x + \alpha x^2$

Then, we obtain $P(k \cdot l) = \sqrt{\frac{\pi}{2\pi\alpha l}} e^{i\pi^2 \frac{(k-l)^2}{2\pi\alpha l}} e^{i\frac{\pi}{4}}$

Therefore,

$$\hat{h}(k) = \int G(l) P(k,l) \, dl = e^{i\frac{\pi}{4}} \int G(l) \sqrt{\frac{1}{2\alpha l}} e^{i2\pi \frac{(k-l)^2}{4\alpha l}} dl \qquad (7.10)$$

where $\hat{h}(k)$ is the reconstruction of the deformed hologram, while $G(l)$ is the reconstruction of the original hologram. If we set $k = \frac{k'}{\lambda L}$ and $l = \frac{l'}{\lambda L}$ in order to change the coordinates from the spatial frequency domain to the space domain, then, in Eq. (7.10), the exponential term $e^{i2\pi \frac{(k-l)^2}{4\alpha l}}$ becomes $e^{\frac{i2\pi}{\lambda} \frac{(k'-l')^2}{2d'}}$ where $d' = 2L\alpha l'$.

Consequently, the reconstruction of the quadratically deformed hologram can be read as the reconstruction of the original hologram at distance L and further propagated to a distance d' that depends linearly on the coordinate l' (x-axis in the reconstruction plane), stretching parameter α, and distance L (Figure 7.6). To demonstrate experimentally the presented thesis, we used as object a USAF target that was tilted at an angle $\varphi = 55°$ in respect to the laser light illumination direction along the x-direction. To recover the EFI image, we deform the hologram only

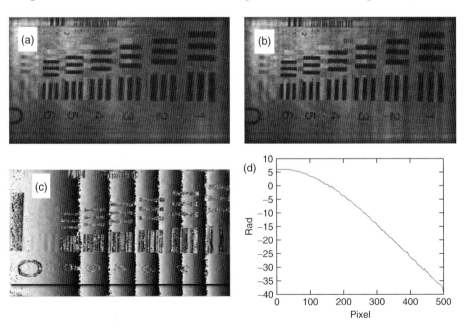

Figure 7.6 Numerical reconstruction of the holograms for an object tilted with an angle of 55° as acquired by the CCD (a) and after the quadratic deformation (b). Phase map difference between the reconstructions of the deformed and undeformed holograms (c) and the unwrapped profile corresponding to the black line of figure 7.6c (d).

along the direction of the tilt. Looking at Figure 7.6a that shows the numerical reconstruction of the acquired hologram, the region where the number is "0," that is, the part of the target closer to the CCD (at a distance L) results to be in focus, while the right side of the target (number "1") results to be clearly out of focus. The focusing gradually worsens going from left to right. Applying the quadratic deformation before the numerical reconstruction process, we obtained the amplitude image as shown in Figure 7.6b. In this case, parameter α was fixed at 2.1×10^{-5}. It is clear that in this case all parts of the target are clearly in focus. The phase map difference calculated by subtracting the phases of the two reconstructions, obtained by the deformed and undeformed holograms, respectively, is shown in Figure 7.6c, while Figure 7.6d shows the unwrapped profile corresponding to the black line of Figure 7.6c. Its shape indicates that the defocus tilt has been mainly removed. The deviation from the linear profile is due to the square root term in Eq. (7.10). Summarizing, if we apply a quadratic deformation to the hologram of a tilted object along the tilting direction, it is possible to recover the EFI of the tilted object. In fact, the reconstruction distance changes pixel by pixel along the tilting direction l' according to Eq. (7.7).

7.5
Simultaneous Multiplane Imaging in DH

A different approach to overcome the problem of the reduced depth of field in microscopy is presented in Ref. [19], where it is demonstrated that, adopting a numerical, quadratically deformed, diffraction grating into the reconstruction process of digital holograms, it is possible to image three planes at different depths simultaneously.

In fact, in many fields of science, such as imaging particles field, *in vivo* microscopy, or medical imaging, it is desirable to visualize simultaneously several layers within the imaged volume [20–24]. To image simultaneously more than one plane at different depths, various approaches can be followed. A first way makes use of lens array in which each lens can image a single plane in focus. Nevertheless, this approach is quite impractical and has different limitations [25]. A different method to bring different objects in focus at the same time in DH has been suggested using independent component analysis and image fusion [26].

A novel and smart approach has been proposed in the work of Blanchard and Greenaway in which a diffraction grating has been adopted in the optical setup to split the propagating optical field in three diffraction orders (i.e., -1, 0, and $+1$). The grating was distorted with an opportune quadratic deformation; consequently, the wavefield resulting from each diffraction order can form an image of a different object plane [27]. By this method, in Ref. [28], a quadratic deformed grating is used for multiplane imaging of biological samples to demonstrate nanoparticle tracking with nanometer resolution along the optical axis. In a recent work, a different approach named "depth-of-field multiplexing" is reported [29]. A high-resolution spatial light modulator was adopted in a standard microscope to generate a set of superposed multifocal off-axis Fresnel lenses, which sharply image different focal planes. This approach provides simultaneous imaging of different focal planes in a sample by a single camera exposure.

In this approach, a synthetic diffraction grating is easily included into the numerical diffraction propagation reconstruction of digital holograms with the aim of imaging simultaneously and in the same field of view multiple object planes. The advantage is dispensing with the use of real optical components and the related complex fabrication process as occurs in the methods of Refs [27, 28] that require precise design and fabrication of nonlinear grating. Even if the numerical reconstruction at any plane is already possible in DH, the advantage of the proposed method lies in the fact that having three image planes for each single reconstruction gives the possibility of reducing the computation time and judging the best focus of an object by visualizing simultaneously three image planes at a time. A better evaluation of the good focus is useful especially for pure-phase objects as *in vitro* biological cells. The flexibility of the proposed method is very wide thanks to the numerical nature of the diffraction grating. For example, depending on the grating period and the amount of deformation, the distance of the multiple planes can be easily changed and adapted to the need of the observer.

The holograms were always reconstructed through a Fresnel–Kirchhoff integral. In the numerical reconstruction algorithm, the hologram is multiplied for the transmission function of a quadratically distorted grating

$$T = a + b\cos\left(A\left(X^2 + Y^2\right) + C\left(X + Y\right) + D\right) \tag{7.11}$$

The insertion of such a digital grating in the numerical reconstruction process allows to image, in the same field of view, three object planes at different depths, simultaneously. In fact, the digital deformed grating, as happens for a physical one [27], has a focusing power in the nonzero orders and, therefore, acts as a set of three lens of positive, neutral, and negative power.

In the reconstruction plane, three replicas of the image appear, each of which is associated with a diffraction order and has a different level of defocus. The distance from the object plane, corresponding to the mth order, to that in the zero order is given by

$$\Delta d(m) = \frac{-2md^2 W}{N^2 Px^2 + 2mdW} \tag{7.12}$$

where d is the reconstruction distance, N is the number of pixel of size Px, while $W = \frac{AN^2\lambda}{2\pi}$ is the coefficient of defocus, with λ the laser wavelength. The use of a numerical grating, instead of a physical one, has the great advantage of increasing the flexibility of the system. In fact, varying the amount of quadratic deformation (i.e., the value of A) and the grating period (C value), we can change the distance between the multiple planes and the lateral separation between the three images, respectively. Moreover, by changing the value of parameters a and b in Eq. (7.11), it is possible to modify the relative contrast between the images corresponding to the orders ± 1 and the central one (the zero-order image).

In the first example (Figure 7.7), we use the same hologram shown in the first paragraph, where three different wires were positioned at different distances

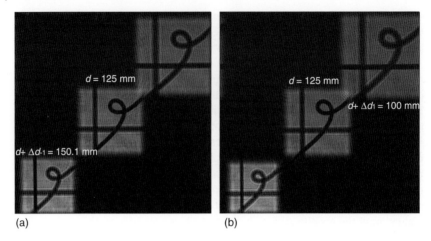

Figure 7.7 Numerical reconstructions of the "three wires" hologram at 125 mm, the in-focus distance of the twisted wire, with two different value of the quadratic deformation of the numerical grating in order to obtain (a) the vertical wire in focus in the −1 order image and (b) the horizontal wire in focus in the +1 order image.

from the CCD array of 100, 125, and 150 mm, respectively. We performed two numerical reconstructions of the corresponding hologram at 125 mm, the in-focus distance of the twisted wire, but with two different quadratic deformations of the numerical grating, that is, two different values of the parameter A. Figure 7.7 shows the amplitudes of the obtained reconstructions. In both cases, in the central (zero-order) image, the twisted wire is in good focus. In Figure 7.7a, the A value has been chosen in such a way to obtain also the vertical wire in focus in the −1 order image. On the contrary, in Figure 7.7b, we managed the reconstruction with the aim of obtaining in focus the horizontal wire in the +1 order image.

Obviously, it is possible to exploit all the display in Figure 7.7, inserting four numerical grating, as shown in Figure 7.8. We demonstrate that up to nine images can be obtained. The computational time is one-ninth in respect to the case of separated reconstructions, thus increasing the computational efficiency.

As a further experiment, we applied the method to holograms of a biological sample. The specimen is formed by three *in vitro* mouse preadipocyte 3T3-F442A cells that are at different depths. Figure 7.9 shows the amplitude reconstruction at a distance $d = 105$ mm at which the cell indicated by the blue arrow is in focus (see the zero-order image). The +1 order corresponds to a distance $d = 92.7$ mm at which the cell indicated by the yellow arrow is in good focus, while the −1 order corresponds to a depth of $d = 121$ mm where the filaments are well visible (highlighted by the red ellipse in Figure 7.9).Therefore, by this technique, we were able to image in a single reconstruction three different object planes allowing us to see cells lying on different planes all in focus simultaneously.

Figure 7.8 Amplitude reconstructions of the "three wires" hologram using four numerical grating in the algorithm.

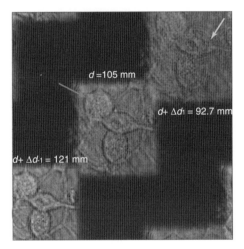

Figure 7.9 Amplitude reconstruction of a "cell" hologram at a distance $d = 105$ mm at which the cell indicated by the blue arrow is in focus. The +1 order correspond to a distance $d = 92.7$ mm at which the cell indicated by the yellow arrow is in good focus, while the −1 order correspond to a depth of $d = 121$ mm where the filaments are well visible, highlighted by the red ellipse. (Please find a color version of this figure on the color plates.)

7.6
Summary

We described some geometric transformations for Fresnel and Fourier holograms that allow us to manage the depth of focus of the reconstructed object. Exploiting this technique, which allows manipulation of the object's position and size in

3D, it is possible to synthesize a 3D scene combining multiple digital holograms of different objects [30]. The proposed method has been easily applied to digital holograms where the coordinate deformation is performed numerically. However, we think that for holograms recorded on a specific material (physically recorded on a photosensitive media) it is possible to foresee and apply an adaptive mechanical deformation to extend the depth of focus even in that case.

Finally, we show how, inserting a numerical quadratic deformed diffraction grating into the reconstruction process of digital holograms, it is possible to image three planes at different depths simultaneously.

References

1. Fraser, S.E. (2003) Crystal gazing in optical microscopy. *Nat. Bio.*, **21**, 1272–1273.
2. Mertz, L. (1965) *Transformation in Optics, 101*, John Wiley & Sons, Inc., New York.
3. Hausler, G. (1972) A method to increase the depth of focus by two step image processing. *Opt. Commun.*, **6**, 38.
4. Pieper, R.J. and Korpel, A. (1983) Image processing for extended depth of field. *Appl. Opt.*, **22**, 1449–1453.
5. Dowski, E.R. Jr. and Cathey, W.T. (1995) Extended depth of field through wavefront coding. *Appl. Opt.*, **34**, 1859–1866.
6. Barton, D.L. et al. (2002) Wavefront coded imaging system for MEMS analysis. Presented at International Society for Testing and Failure Analysis Meeting, Phoeneics, AZ, November 2002.
7. Marks, D.L., Stack, D.L., Brady, D.J., and Van Der Gracht, J. (1999) Three-dimensional tomography using a cubicphase plate extended depth-of-field system. *Opt. Lett.*, **24**, 253–255.
8. Ferraro, P., DeNicola, S., Finizio, A., Coppola, G., Grilli, S., Magro, C., and Pierattini, G. (2003) Compensation of the inherent wave front curvature in digital holographic coherent microscopy for quantitative phase-contrast imaging. *Appl. Opt.*, **42**, 1938–1946.
9. Ferraro, P., Coppola, G., Alfieri, D., DeNicola, S., Finizio, A., and Pierattini, G. (2004) Controlling image size as a function of distance and wavelength in fresnel transform reconstruction of digital holograms. *Opt. Lett.*, **29**, 854–856.
10. Ferraro, P., Grilli, S., Alfieri, D., De Nicola, S., Finizio, A., Pierattini, G., Javidi, B., Coppola, G., and Striano, V. (2005) Extended focused image in microscopy by digital Holography. *Opt. Express*, **13**, 6738–6749.
11. Kreis, T. and Jüptner, W. (1997) Principles of digital holography. *Proc. Fringe '97*, Akademie Verlag Series in Optical Metrology, Vol. 3, p. 353.
12. Goodman, J.W. (1996) *Introduction to Fourier Optics*, 2nd edn, McGraw-Hill Companies Inc.
13. Kreis, T. (2004) *Handbook of Holographic Interferometry: Optical and Digital Methods*, Wiley-VCH Verlag GmbH, Berlin.
14. Ferraro, P., Paturzo, M., Memmolo, P., and Finizio, A. (2009) Controlling depth of focus in 3D image reconstructions by flexible and adaptive deformation of digital holograms. *Opt. Lett.*, **34**, 2787–2789.
15. De Nicola, S., Finizio, A., Pierattini, G., Ferraro, P., and Alfieri, D. (2005) Angular spectrum method with correction of anamorphism for numerical reconstruction of digital holograms on tilted planes. *Opt. Express*, **13** (24), 9935–9940.
16. Jeong, S.J. and Hong, C.K. (2008) Pixel-size-maintained image reconstruction of digital holograms on arbitrarily tilted planes by the angular spectrum method. *Appl. Opt.*, **47**, 3064–3071.
17. Matsushima, K., Schimmel, H., and Wyrowski, F. (2003) Fast calculation method for optical diffraction on tilted

planes by use of the angular spectrum of plane waves. *J. Opt. Soc. A*, **20**, 1755–1762.

18. Paturzo, M. and Ferraro, P. (2009) Creating an extended focus image of a tilted object in fourier digital holography. *Opt. Express*, **17**, 20546–20552.

19. Paturzo, M., Finizio, A., and Ferraro, P. (2011) Simultaneous multiplane imaging in digital holographic microscopy. *J. Display Technol.*, **7**, 24–28.

20. Ortyn, W.E., Perry, D.J., Venkatachalam, V., Liang, L., Hall, B.E., Frost, K., and Basiji, D.A. (2007) Extended depth of field imaging for high speed cell analysis. *Cytometry A*, **71A** (4), 215–231.

21. Westphal, V., Rizzoli, S.O., Lauterbach, M.A., Kamin, D., Jahn, R., and Hell, S.W. (2008) Video-rate far-field optical nanoscopy dissects synaptic vesicle movement. *Science*, **320** (5873), 246–249.

22. Rosen, J. and Brooker, G. (2008) Non-scanning motionless fluorescence three-dimensional holographic microscopy. *Nat. Photonics*, **2** (3), 190–195.

23. Xu, W., Jericho, M.H., Kreuzer, H.J., and Meinertzhagen, I.A. (2003) Tracking particles in four dimensions with inline holographic microscopy. *Opt. Lett.*, **28** (3), 164–166.

24. Holtzer, L., Meckel, T., and Schmidt, T. (2007) Nanometric three-dimensional tracking of individual quantum dots in cells. *Appl. Phys. Lett.*, **90** (5), 053902–053904.

25. Verveer, P.J., Swoger, J., Pampaloni, F., Greger, K., Marcello, M., and Stelzer, E.H.K. (2007) High-resolution three-dimensional imaging of large specimens with light sheet-based microscopy. *Nat. Methods*, **4** (4), 311–313.

26. Do, C. and Javidi, B. (2007) Multifocus holographic 3D image fusion using independent component analysis. *IEEE J. Display Technol.*, **3** (3), 326–332.

27. Blanchard, P.M. and Greenaway, A.H. (1999) Simultaneous multiplane imaging with a distorted diffraction grating. *Appl. Opt.*, **38**, 6692–6699.

28. Dalgarno, P.A., Dalgarno, H.I.C., Putoud, A., Lambert, R., Paterson, L., Logan, D.C., Towers, D.P., Warburton, R.J., and Greenaway, A.H. (2010) Multiplane imaging and three dimensional nanoscale particle tracking in biological microscopy. *Opt. Express*, **18**, 877–884.

29. Maurer, C., Khan, S., Fassl, S., Bernet, S., and Ritsch-Marte, M. (2010) Depth of field multiplexing in microscopy. *Opt. Express*, **18**, 3023–3034.

30. Paturzo, M., Memmolo, P., Finizio, A., Näsänen, R., Naughton, T.J., and Ferraro, P. (2010) Synthesis and display of dynamic holographic 3D scenes with real-world objects. *Opt. Express*, **18**, 8806–8815.

8
Three-Dimensional Particle Control by Holographic Optical Tweezers

Mike Woerdemann, Christina Alpmann, and Cornelia Denz

8.1
Introduction

Light that is reflected, refracted, or absorbed by small particles, in general, underlies a change in momentum. In turn, the particles experience an analogous change in momentum, that is, a resulting force. It was demonstrated more than 40 years ago that radiation pressure from a (laser) light source can accelerate microscopic particles [1]. The historically most important insight, however, was that microscopic particles can not only be pushed by the radiation pressure but also at will be confined in all three dimensions, leading to the powerful concept of optical tweezers [2].

The basic concept of optical tweezers can be significantly extended by the use of computer-generated holograms. These holograms can structure the light field in order to, for example, produce a multitude of optical tweezers simultaneously. This approach, commonly known as holographic optical tweezers (HOT), has found a vast number of applications owing to its unrivaled versatility [3]. Each individual pair of optical tweezers can be freely positioned in the vicinity of the specimen, and with adequate driving hardware and software, interactive real-time control is possible.

After a short overview on the basic physical principles of optical tweezers and HOT, representative experimental implementation of HOT is briefly described. Three examples of the state-of-the-art applications of HOT are presented. These examples are selected to represent the wide range of possible applications and demonstrate applications in colloidal sciences, single cell biology, and supramolecular chemistry.

Despite the versatility of HOT there are strong demands for alternative and complementary concepts for optical manipulation and organization of microscopic matter. Optical micromanipulation with nondiffracting and advanced self-similar beams is introduced, which are expected to offer entirely new perspectives for optical molding and organization of matter.

8.2
Controlling Matter at the Smallest Scales

Understanding how smallest particles can be controlled and manipulated solely by optical means requires knowledge of the involved forces. In order to calculate the force on a particle, we need to calculate the difference between the (vectorial) momentum of the incident light field and the momentum of the light field after being altered by the particle. This leads to two tasks that need to be solved. First, an appropriate model of the incident light field is required – a nontrivial task if we want to include all effects that result from the nonparaxial nature of a strongly focused light beam. Second, the interaction with the particle needs to be described. The problem increases significantly in complexity if not only isotropic, spherical particles but also arbitrary shapes are considered. Despite the relatively complex situation, the system can be described accurately by a rigorous theory of light scattering based on the Maxwell equations, for example, by the generalized Lorenz–Mie theory [4]. This approach is very useful if quantitative values are required, but as the problem can be usually solved only numerically, it does not yield an intuitive understanding [5].

If one is interested in a more intuitive, analytical description, a limiting regime of very small or very large particles – always compared to the laser wavelength – can be considered. Sufficiently small spherical dielectric particles in an electromagnetic field can be treated as induced point dipoles [6]. This regime is called Rayleigh or dipole approximation. The radiation force exerted on the induced dipole has two components: a scattering force originating from light scattered at the sphere and a gradient force originating from the Lorentz force acting on the induced dipole. The scattering force is proportional to the local laser light intensity and always points in the direction of the beam, while the gradient force is proportional to the gradient of the light intensity. For a laser beam with a Gaussian TEM_{00} mode profile, this results in a gradient force that always points to the beam axis. If the beam is focused, the gradient force has an additional component along the beam axis in the direction of the focal point. Obviously, the axial intensity gradient increases when a laser beam is focused stronger, that is, by a lens with a shorter focal length or higher numerical aperture. If the beam is focused very tightly, the axial gradient force in the direction of the focus can exceed the scattering force in the direction of the laser beam, as depicted in Figure 8.1. In other words, this means that the particle experiences a resulting force always pointing to one point in space – its equilibrium position – close to the focus of the laser beam, but slightly shifted in the direction of beam propagation. This is the basic concept of a single-beam-gradient trap, or optical tweezers.

A qualitatively similar result can be derived when rays from the incident light field are traced and reflection, refraction, and absorption at a particle are considered. This ray optics approximation neglects diffraction effects and interference phenomena and thus implies particles that are large compared to the laser wavelength in order to yield quantitatively valid results. Again, for lenses with very high numerical

8.2 Controlling Matter at the Smallest Scales

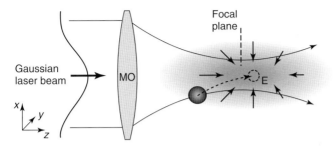

Figure 8.1 Basic principle of optical tweezers in the Rayleigh regime. A fundamental Gaussian laser beam is focused tightly through a microscope objective (MO) and accelerates a colloidal particle along the dashed trajectory toward the equilibrium position (E). The force field due to the optical gradient and scattering forces is indicated by the black arrows.

aperture, there exists an equilibrium position for the particle in the vicinity of the focal point [7].

8.2.1
Applications of Optical Tweezers

Optical tweezers have found a huge number of applications since their first demonstration by Ashkin and colleagues 25 years ago [2]. In particular, biological questions on a single cell or single-molecule [8, 9] level can be well addressed with optical tweezers for two reasons. First, there is no other tool available that enables handling of single cells, organelles, and macromolecules with such flexibility and precision at the same time without any physical contact that could possibly contaminate a sample. Second, optical tweezers can be used to exert defined forces and more importantly measure extremely small forces with an unrivaled precision [10–14].

Further applications of optical tweezers and closely related methods can be found in such diverse fields as colloidal sciences [15], microfluidics [16, 17], microscopic alignment [18, 19], particle separation [20] and sorting [17, 21, 22], or molecular motor dynamics [23, 24]. Optical tweezer experiments can answer fundamental physical questions, including the direct transfer of optical angular momentum [25, 26], hydrodynamic interactions [27, 28], and – of course – light-matter interaction [29]. This list is by no means exhaustive or complete but represents a small selection of interesting applications; an excellent overview can be found, for example, in Ref. [30].

8.2.2
Dynamic Optical Tweezers

Although single optical tweezers at a fixed position already enable many applications, it often is desirable to have a trap that can be moved in the sample chamber. In Figure 8.2a, a simplified picture of optical tweezers is depicted. A collimated

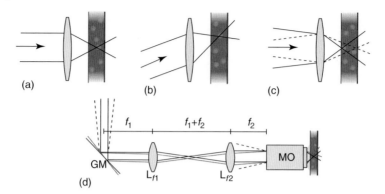

Figure 8.2 Basic principle of position control in optical tweezers. (a–c) The position of the laser focus and hence the optical trap is translated three dimensionally by variation of the incidence angle and divergence of the laser beam. (d) Technical realization with a telescope (L_{f1}, L_{f2}), and beam manipulation in a conjugate plane of the back aperture of a microscope objective (MO).

laser beam is focused through a lens with short focal length, which usually is a microscope objective, into a sample chamber that contains a fluid with dispersed particles. In order to move the focal spot and thus the optical trap to a different position in the plane orthogonal to the beam axis, the incident laser beam needs to have an angle with respect to the beam axis as shown in Figure 8.2b. A diverging or converging beam on the other hand would shift the focal plane along the beam axis (Figure 8.2c).

However, it is not sufficient to set the angle or divergence of the incident beam just somewhere in the optical path. It is important that the beam hits the back aperture of the microscope objective always with the same diameter and at the same centred position in order to keep the optical trap operating and its properties unchanged [7, 31]. One possibility is to use an afocal telescope of two lenses to create an optically conjugated plane (GM) of the back aperture of the microscope objective (cf. Figure 8.2d). Any angle introduced at this plane, for example, by a gimbal mounted mirror [31], will result in a corresponding angle at the back aperture of the microscope objective without a shift in position. Similarly, any divergence introduced with a constant beam diameter at this plane will be reproduced with a constant beam diameter at the back aperture of the microscope objective.

Position control can be automated if computer-controlled scanning mirrors are used [32–34]. A similar approach uses acousto-optic deflectors (AODs) at the conjugate plane [35]. AODs can introduce an angle by utilizing a dynamic Bragg grating inside a piezoelectric material, and this function principle allows for an extremely high rate of different deflection angles to be set. One powerful application is time-shared optical tweezers, where the laser beam is directed to one position, held there for a short time, and then directed to the next position. If this is done iteratively and the stopover at each position is long enough to pull back a particle to the center, and also the absence of the laser beam is short enough to prevent the particles escaping due to Brownian motion, many particles can be trapped quasi simultaneously [32, 34, 36, 37].

8.3
Holographic Optical Tweezers

One ingenious way to realize control of beam angle and beam divergence in one particular plane without mechanical manipulation is diffraction at computer-generated holograms, also known as diffractive optical elements (DOEs) in this context. The hologram can be imprinted statically in optical materials [38], for example, by lithographic methods, or alternatively displayed by a computer-controlled spatial light modulator (SLM) [39, 40]. The latter implementation enables versatile spatiotemporal structuring of the light field, leading to dynamic HOT [41]. The classical use of HOT is the generation of multiple optical traps simultaneously. More complex geometries with light fields that are tailored to a specific application are introduced in Section 8.5.

8.3.1
Diffractive Optical Elements

The problem to be solved in order to displace an optical trap can be reduced to the control of angle and divergence of the constituting laser beam by means of suitable DOE. A general grating is known to diffract an illuminating wave into several diffraction orders. The blazed diffraction grating depicted in Figure 8.3a in contrast deflects light into only one diffraction order with an angle Θ. With the

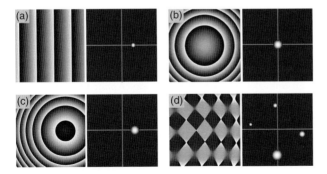

Figure 8.3 Basic function of DOEs in HOT. Each subfigure shows the phase distribution of a DOE in gray values (left) and the resulting position of the optical trap in the specimen plane (right). The position of the trap along the beam axis is indicated by the diameter of the point marking the position in the specimen plane.

grating periods Λ_x, Λ_y, the phase distribution of the grating is given by

$$\phi(x,y) = \left(\frac{2\pi}{\Lambda_x}x + \frac{2\pi}{\Lambda_y}y\right) \mod 2\pi \tag{8.1}$$

The diffraction angle introduced by this simple DOE is directly transferred into a lateral repositioning of the optical trap from the center. Furthermore, one can use a quadratic phase distribution, the holographic equivalent of a refractive lens, of the kind $\phi(x,y) = \Gamma(x^2 + y^2) \mod 2\pi$ in order to influence the divergence of the diffracted light (cf. Figure 8.3b). The combination of both finally gives full 3D control of the position of a single trap (cf. Figure 8.3c):

$$\phi(x,y) = \left(\frac{2\pi}{\Lambda_x}x + \frac{2\pi}{\Lambda_y}y + \Gamma(x^2 + y^2)\right) \mod 2\pi \tag{8.2}$$

The full potential of DOEs, however, can be realized when considering the complex superposition of multiple gratings and holographic lenses [40]. Figure 8.3d shows an example where four traps at different three-dimensional positions are generated by a DOE that is calculated as the argument of the complex sum of four phase distributions calculated from Eq. (8.2). Effectively, this means it is possible to control the position of one single trap and that the laser beam can be split in a multitude of traps, each of which can be controlled independently from the others.

The analytical gratings and lenses approach to generate a DOE for multiple beams in HOT grants an intuitive understanding of the basic physics involved in the diffractive element. The most important advantage, however, is the relatively low computational expense, which enables fast calculation of holograms in real time.

8.3.2
Iterative Algorithms

Besides the obvious advantage of low computational expense, the analytic calculation of holograms using the gratings and lenses approach has several practical and conceptional drawbacks. The most prominent limitations are strong possible ghost traps [42, 43], inhomogeneity of the individual optical traps [44], and the limitation to discrete spots in contrast to extended, continuous optical landscapes.

An alternative approach for calculating holograms is the use of iterative Fourier transformation algorithms (IFTAs) [45]. Therefore, we recall that the focal planes of a lens are related by an optical Fourier transformation. Applied to HOT, that means the light field distribution in the focal plane of the microscope objective can be described as the Fourier transform of the light field at the back focal plane of the objective and thus at the position of the DOE. If we start with an arbitrary intensity distribution in the focal plane and calculate the Fourier transform, however, the result in general is a complex-valued light field, structured in amplitude and phase. This is not feasible in most situations for two reasons. First, all available SLMs can only modulate either amplitude or phase and not both [46], and second, the modulation of the amplitude means an unavoidable loss of light intensity at the SLM and thus a poor efficiency.

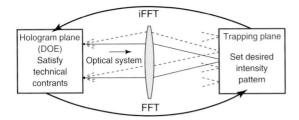

Figure 8.4 Sketch of the basic function of IFTAs. The hologram and trapping planes of the optical tweezers are connected by an optical Fourier transformation through the microscope objective. Numerically, fast Fourier transformations are used to propagate the light field iteratively between both planes and include physical constraints.

The task to solve is finding a phase-only hologram or a DOE that generates the desired intensity distribution and thus configuration of optical traps in the focal plane of the microscope objective. This problem cannot be solved analytically, but an iterative algorithm as sketched in Figure 8.4 is usually applied [47]. The basic idea is to iterate numerically between the hologram (DOE) and trapping planes. In the hologram plane, the technical constraints are included such as the requirement for pure phase modulation, plane wave or TEM_{00} illumination, discretization of phase values, or pixilation of the SLM. In the trapping plane, the intensity is always set to or approached to the desired intensity distribution. The algorithm can converge to a hologram only if there are enough degrees of freedom that can be chosen as required in every iteration step. Usually, the phase in the trapping plane is not of interest and can serve as a free parameter. Furthermore, often the intensity in the periphery of the field of view is of minor interest and can be an additional free parameter [46]. The algorithm can be terminated when a suitable convergence criterion is fulfilled, like a certain number of iterations or when the deviation of the calculated pattern from the desired pattern is below a threshold.

The clear advantage of IFTAs is their flexibility and the quality of the resulting holograms. Furthermore, it is straightforward to extend the algorithms to 3D intensity patterns in the trapping plane [48]. The versatility comes at the cost of a relatively high computational expense, limiting possible real-time applications.

8.3.3
Experimental Implementation

HOT require a laser that is modulated by a DOE, relayed appropriately, and focused through a high-numerical-aperture lens. As it is usually desired to observe what is being manipulated with the optical tweezers, they are commonly integrated into an optical microscope.

A typical experimental realization of HOT in our laboratories is depicted in Figure 8.5. The microscope can be easily identified. It consists of an illumination, a condenser, a translation stage that holds the sample, the microscope objective (MO), a tube lens (L_T), and a video camera. An inverted layout of the

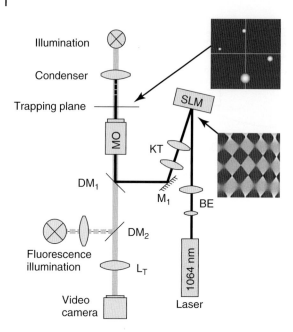

Figure 8.5 Experimental implementation of holographic optical tweezers. MO, microscope objective; DM1 and DM2, dichroic mirrors, M1, relay mirror; KT, telescope (not to scale); BE, beam expansion.

microscope – as in this implementation – is favorable for many biological samples, but not required for HOT in general. The depicted microscope also includes optional fluorescence illumination, which is integrated by means of a dichroic mirror DM_2. Conveniently, the MO used for microscopic investigations at the optical resolution limit is required to have a high numerical aperture. It is therefore possible to reuse the objective from the microscope in order to focus the laser light field with the required high numerical aperture.

The HOT part of the experimental setup basically consists of a laser, suitable relay optics, a phase-only SLM, and the microscope objective. For the majority of applications, a continuous wave (CW) laser is utilized. In the described setup, a $P_{max} = 2.5$ W CW Nd:YVO$_4$ laser emitting at $\lambda = 1064$ nm is used. Using near-infrared light minimizes photodamage to biological samples compared to wavelengths in the visible regime [49]. The required laser power depends strongly on the application scenario. Although stable optical trapping of 1 µm particles with less than 200 µW is possible with a perfectly corrected system [50], a more practical value – especially if a particle is to be moved fast – is a few milliwatts per particle. The laser beam is expanded and collimated in order to illuminate the active area of the SLM, which acts as the DOE. After the light field is structured by the SLM, it is relayed and imaged onto the back focal plane of the microscope objective. The illumination beam path and the laser beam path are separated by means of a dichroic mirror (DM_1).

From the multitude of available (dynamic) SLM technologies, including various liquid crystal display (LCD) implementations and microelectromechanical systems (MEMS) such as digital micromirror devices (DMD), LCDs have become most widespread in HOT applications. The reason for the popularity of LCDs is a set of favorable features, most importantly, their capability of pure phase modulation with a high number of different phase levels together with their potentially very high spatial resolution. An overview of the state of the art in the LCD technology and, in particular, in liquid crystal on silicon (LCOS) displays can be found in Chapter 1.

8.4
Applications of Holographic Optical Tweezers

One obvious though powerful application of HOT is the parallel execution of many experiments, each of which could already be realized with single optical tweezers. An example could be the parallel measurement of forces on multiple cells in a biological assay. This enables processing studies that require a large number of statistically uncorrelated measurements rapidly and efficiently. Of particular interest, however, are scenarios that cannot be realized with single optical tweezers. For example, in order to measure the binding force of two microscopic objects, it might be necessary to approach both objects and then separate them in a defined way. With nonspherical objects, it furthermore is possible to use multiple traps – tailored to the object shape – in order to rotate the object into a defined orientation and align it at will.

8.4.1
Colloidal Sciences

Colloidal suspensions, that is, particles suspended in a fluid, are a perfect field of activity for optical tweezers [15]. Single or double optical tweezers have been used to measure electrostatic [51] or hydrodynamic [27, 28, 52] interactions between two spherical particles. Optical tweezers can aid here in two ways. First, the absolute position of the particles under test with respect to other particles and surfaces can be defined and their relative position can be controlled precisely [28, 51]. Second, the optical trapping potential can be used directly to measure interaction forces [27, 52]. As soon as interactions between more than a very limited number of particles are involved, however, it is not reasonable to use a discrete combination of single optical tweezers.

Already the early works with HOT demonstrate that multiple spherical particles can be trapped simultaneously at different three-dimensional positions [40]. Soon it was shown that even large arrays of colloidal spheres can be created with the holographic technique [53]. These demonstrations were restricted to static configurations as they employed prefabricated, static DOEs.

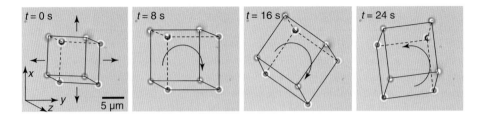

Figure 8.6 Eight 1 μm polysterene beads arranged in a simple cubic unit cell and resized and rotated as a whole. The particle positions are connected with lines that visualize the geometric configuration.

The utilization of computer-controlled SLMs enabled the creation of complex, three-dimensional structures of multiple particles that can be dynamically rearranged [41]. Although there are competing concepts for dynamic three-dimensional control of multiple particles, such as a combination of time-shared traps for transversal positioning and an SLM for axial positioning [54], HOT have established as the first choice for many applications.

One important field is the assembly of three-dimensional crystalline [55] and quasicrystalline [56] structures with HOT. In Figure 8.6, a basic example of a three-dimensional crystalline structure is shown that demonstrates the most important aspects. For this purpose, a suspension of polystyrene beads with a diameter of 1 μm was prepared in demineralized water. Eight arbitrarily selected beads were trapped with HOT and arranged in the three-dimensional configuration in Figure 8.6a. The three-dimensional positioning can be done interactively as we use the analytic approach to calculate the holograms, which is fast enough for real-time control [57]. The structure in Figure 8.6a corresponds to a simple cubic unit cell with roughly equal edge lengths. While the position of each bead can be chosen rather freely, the range of maximum possible transversal and axial displacements is limited to several tens of micrometers, depending on the resolution of the SLM used [58]. In the further sequence of Figure 8.6, the unit cell is resized dynamically and then rotated as a whole. The high degree of control is basically attributed to the software that keeps track of the relative particle positions and calculates the appropriate holograms following the user input.

This way, even highly demanding structures of tens and hundreds of individual particles can be arranged and dynamically rearranged as required. This does not only allows for a deeper understanding of particle–particle interactions beyond the experimental nearest neighbor approximation but also enables the creation of structures dedicated for a particular purpose. One exciting example is the creation of photonic band gap materials with HOT [59]. The main advantage over competing techniques such as photolithography, two-photon polymerization, holography, or self-assembly of colloids is that the material is built with control on every single unit. By this means, it is straightforward to introduce defects, including other materials, which is essential for tailoring the band gap structure [60].

8.4.2
Full Three-Dimensional Control over Rod-Shaped Bacteria

Many exciting applications require a high degree of control on micro- and nanoparticles that are not spherical but feature arbitrary shapes. Two important examples are discussed. The first is the class of bacteria, which, from the physical point of view, can be seen as dielectric and often nonspherical microscopic objects with a broad range of applications from the synthesis of substances [61], the use as active components of biohybrid systems [62], to microfluidic lab-on-a-chip applications such as mixing processes [63]. In particular, rod-shaped, self-propelled bacteria that achieve motility by rotational motion of their helical flagella filaments are of topical interest because they feature one of the smallest known rotational motors, a "nanotechnological marvel" [64], and because their motion is well adapted to constraints of movements at low Reynolds numbers [65]. The latter gives rise to a multitude of possible microfluidic applications, and curiosity about the fundamental principles is a strong motivating force toward a deeper understanding of bacterial interaction [66], interaction with the microscopic environment [67], or the formation behavior of biofilms [68].

Optical tweezers are ideal tools to confine bacterial cells [69, 70] and thus can be the starting point for a thorough analysis of hydrodynamic parameters, mutual interaction, or cooperative effects. However, like all elongated objects, rod-shaped bacteria always align their long axis with the axis of the focused laser beam (cf. Figure 8.7a), resulting in a strong limitation of possible (lateral) observation and interaction scenarios. Defined control of orientation would allow, for example, detailed studies on bacterial motility [71]. The strong demand for three-dimensional

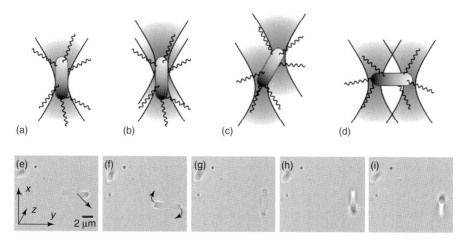

Figure 8.7 (a–d) Basic principle of full three-dimensional control over bacteria. (e–i) Demonstration of control over orientation and position of a duplicating bacterial cell (length of approximately 6 μm). (e–g) Orientation of the cell in the observation plane, (h) angle α between observation plane and bacterial cell is $(35 \pm 5)°$ (i) angle $\alpha = (27 \pm 5)°$.

orientation control has led to a number of methods that extend the basic concept of optical tweezers [72, 73], including support by a surface [74, 75], Hermite-Gaussian modes [76], linetraps [19, 77], oscillating traps [78], Mathieu beams [79], or multiple beam traps [71, 80, 81]. For complete and utter control over the three-dimensional position and at the same time over two or even all three rotational degrees of freedom, usually two or three single traps that can be steered individually to some extent are utilized [82–84]. Most of these approaches for complete control of rod-shaped bacteria, however, are restricted to one single bacterium or a very low quantity because they have strong requirements with respect to the mechanical or optical system, including timing of mechanically operating components [83, 84], or a direct correlation between quantity of desired traps and complexity of the setup [71, 83].

Here, the ability of HOT to create an almost arbitrary number of individual optical traps is of crucial value. Figure 8.7a–d shows the basic principle of orientation control over rod-shaped bacteria with a diameter and length of roughly 1 μm and 2–3 μm, respectively. First, the bacterial cell is confined in a single trap, aligned with the beam axis. A second trap is generated with HOT and placed in the vicinity of the first trap. The second trap is translated along an appropriately chosen trajectory, such that the geometric distance between both laser foci is kept constant and approximately equals the length of the cell. By this means, both traps exert force on the poles of the cell and induce the amount of torque required for defined alignment. Any arbitrary angle between the bacterial long axis and the beam axis can be achieved, including horizontal positioning of the cell (Figure 8.7d). During the whole process, the bacterial cell is trapped three-dimensionally, that is, not supported by any surface. With the available software, both traps can be moved while keeping their relative positions and thus the orientation of the cell [72, 82].

Although the alignment of a single cell itself is of highest relevance [71], the full strengths of HOT are revealed with two advanced examples. First, it is possible to have full orientation and translation control not only over one single bacterial cell but also over a multitude of bacteria simultaneously with independent control on any individual cell. By this means, tens of bacteria can be positioned and arbitrarily aligned, providing ideal initial situations for complex interactions scenarios or massively parallel investigations [72]. Furthermore, strongly elongated cells that cannot be handled with the simple two-trap approach can also be controlled with HOT [80]. Figure 8.7e–i shows a long bacterial cell or, more precisely, a bacterial chain just before cell division, with a length of $\approx$ 6 μm. In order to achieve full orientation control, four individual traps in a linear configuration were used [72].

8.4.3
Managing Hierarchical Supramolecular Organization

HOT are well suited to confine, manipulate, and arrange microscopic and nanoscopic container particles. One particularly promising application is the optical assembly and organization of nanocontainers that are loaded with specific guest molecules which themselves are highly ordered inside the containers

[85, 86]. The synthesis of molecular structures and materials that are held together by noncovalent interactions, commonly named supramolecular organization, can be one way toward the design of novel, functional materials with tailored properties that exploit the strong relationship between molecular arrangements and resulting macroscopic properties [87]. In particular, the hierarchical organization of pre-ordered structures is a promising approach to bridge different ordering scales – from the molecular up to the macroscopic level [86].

Microporous molecular sieves such as zeolite crystals are well suited host materials as they feature pores or cavities whose high geometric order directly transfers to a high degree of order of the guest molecules [88]. This first level of organization is relatively well accessible by chemical means [89]. The further organization of the host particles is usually accessed by means of self-organization. While this approach is very efficient for large-scale arrangements, it provides only very limited control over the individual hosts [90].

HOT, on the other hand, are perfectly adapted to the precise control of a finite number of particles. In the following discussion, zeolite L crystals are chosen to demonstrate hierarchical supramolecular organization induced by HOT. Zeolite L are crystalline aluminosilicates with a cylindrical shape that feature strictly parallel nanochannels arranged in a hexagonal structure and running parallel to the cylinder axis [88]. Zeolite L crystals are versatile hosts and can be loaded with a wide range of inorganic and organic guest molecules including many dyes [91].

Figure 8.8a shows an example of 4 × 4 zeolite L crystals arranged in a rectangular Bravais lattice configuration. All crystals are trapped simultaneously with HOT and their relative and absolute three-dimensional position is exactly defined by the laser light field. The whole structure can be easily translated in x-, y-, or z-direction without being disturbed. It is obvious that this high degree of control is not possible with classical manipulation methods or with chemical means. Furthermore, it is possible to increase the degree of organization dynamically, for example by ordering the crystals in the array by their size (Figure 8.8b). Finally, Figure 8.8c shows that the lattice can be reconfigured into a centered rectangular lattice. This is done

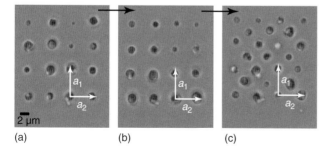

Figure 8.8 Dynamic patterning of cylindrical zeolite L crystals with nominal diameter and length of 1 μm. The crystals are organized in a rectangular Bravais lattice configuration (a), ordered by their size (b), and reconfigured dynamically into a centered rectangular lattice (c).

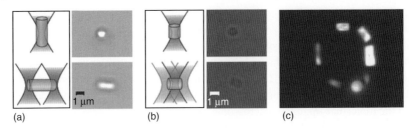

Figure 8.9 Full three-dimensional translational and orientation control over zeolite L crystals. Clearly elongated crystals (a) and crystals without clear elongation (b) are aligned with suitable optical trapping configurations, enabling elaborate two- or three-dimensional assemblies of arbitrarily aligned crystals (c) that can contain various molecular guest species.

interactively and in real-time, and the configuration is given by the user, not by any constraints of the method.

For the most versatile degree of control, it is desirable to additionally have full orientation control over each individual particle. With the longer zeolite L crystals, it is relatively easy to achieve orientation control with two or more traps in a linear configuration as shown in Figure 8.9a. Crystals with a cylinder length close to their diameter, however, require a more sophisticated approach because of their small asymmetry. With optimized viscosity of the solvent and a tailored trapping configuration consisting of a strong central trap, which defines the position of the crystal, and two weaker side traps, which induce the rotation, it is possible to rotate these crystals into horizontal orientation [85] (cf. Figure 8.9b).

A simple but conclusive example of organization of loaded zeolite L nanocontainers is shown in Figure 8.9c. Here, crystals of different sizes and shapes, loaded with DXP (N,N'-bis(2,6-dimethyl phenyl)-3,4:9,10-perylentetracarboxylic diimide) dye, are organized in a ringlike structure. Except for one crystal that is intentionally aligned in vertical orientation, all crystals are horizontally aligned. Again, the whole structure is solely confined by optical means, not supported by any surface. The fluorescence signal of the DXP dye is monitored, and it is clearly seen that the loading does not prevent the optical control over the nanocontainers.

The high degree of control over individual container particles that HOT provide in tandem with their ability to control tens or even hundreds of particles is a promising approach for designing novel materials with exciting properties that cannot be achieved by conventional means.

8.5
Tailored Optical Landscapes

Recent developments in optical micromanipulation emphasize the demand for versatile three-dimensional optical landscapes to build, for example, a wide range of configurations of elaborate particle structures with individual specifications [92]. Especially in nano- and material sciences, recent progress increases the need for novel tools and techniques enabling individual manipulation and organization of geometrically complex or functionalized materials [85, 88, 93]. At present, nanostructuring of surfaces and materials with chemical techniques or self-organization in the form of collective arrangements of molecules is already established [90, 94]. In addition to this collective organization, however, the individual manipulation and selection gets more and more important for the functionalization and three-dimensional structuring of matter, so that new approaches are needed.

HOT already provide almost arbitrary configurations in discrete layers that are calculated either iteratively or as a superposition of diffractive lenses and gratings, as described in Section 8.3. With three-dimensional generalization of the iterative algorithms, arbitrary three-dimensional intensity distributions can be approximated within the physically realizable limits [95], but the task remains computationally expensive. More importantly, however, the propagation properties of the resulting light fields are undefined in a sense that the intensity distribution is only defined in a volume considered by the algorithm and the optical phase serves as the free parameter.

Therefore, in recent years, complementary approaches for the modulation of light fields have emerged, occasionally termed tailored light fields. Instead of iteratively shaping a light beam, special analytical light modes are calculated, which can be experimentally realized by adequate phase holograms. Such tailored light fields can be used for the optical manipulation of extended three-dimensional, highly ordered structures owing to their diversity in transverse intensity distributions as well as their defined propagation properties.

After a short survey of previous work, in the following discussion, we focus on light fields in elliptical geometries. Owing to an additional degree of freedom, the ellipticity ϵ, they offer both a wide range of beam shapes and the inclusion of polar and Cartesian geometries. Two different beam classes are discussed: non-diffracting and self-similar beams. Both provide a diversity of transverse modes, but show significantly different, characteristic propagation properties. Figure 8.10 provides an overview of the relevant coordinate systems and the corresponding beam classes of nondiffracting and self-similar beams, respectively.

8.5.1
Nondiffracting Beams

It is well known that all propagating electromagnetic wavefields are subject to the phenomenon of diffraction and the resulting spreading during propagation. However, there are beam classes for which this law appears not to be valid because they maintain their transversal shape on propagation. These beams became famous as nondiffracting beams [96]. Their transverse intensity profile is a result of interference of plane waves whose wave vectors have the same projection to the optical axis [97], which means that both their structure and their spatial extent is maintained during propagation [98].

These properties can be best understood with one of the simplest nondiffracting beams, depicted in Figure 8.11a. It is produced by interference of two plane waves, leading to a transverse cosine lattice. The restriction on possible wave vectors

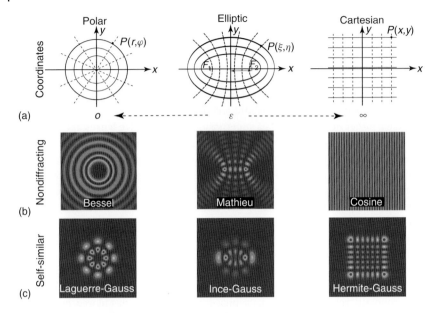

Figure 8.10 Nondiffracting and self-similar beams in polar coordinates (a), Cartesian coordinates (c), and elliptic coordinates (b), which include polar and Cartesian coordinates with an appropriately chosen ellipticity parameter ϵ. (Please find a color version of this figure on the color plates.)

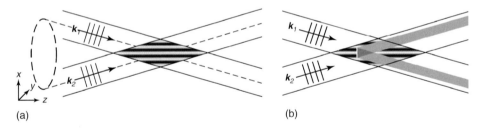

Figure 8.11 Nondiffracting cosine lattice. (a) Interference of two plane waves. (b) Self-healing of nondiffracting beams after disturbance of a small obstacle. (Please find a color version of this figure on the color plates.)

requires all additional plane waves, which might be used to produce higher order transverse lattices, to have a fixed angle with the axis of propagation. This constraint is equivalent with all constituting waves lying in an infinitesimally narrow ring in Fourier space, which obviously is impossible to realize in an experiment. Furthermore, the ideal, endless expansion of the nondiffracting area of nondiffracting beams would only be achievable with infinite energy. In consequence, all experimentally produced beams are approximations of nondiffracting beams, which always have a finite volume of interference and a finite nondiffracting region [99]. Another important feature of nondiffracting beams is their self-healing property, which means the reconstruction of their transverse intensity distribution after disturbance by small obstacles, as shown in Figure 8.11b.

Both features unique to nondiffracting beams, that is invariance on propagation and their self-healing property, make them ideally suited for the optical stacking of several particles in propagation direction, and extended three-dimensional structures become possible. In the context of optical micromanipulation, nondiffracting beams have been studied in different configurations. First experiments were realized with Bessel beams [100, 101], solutions of the Helmholtz equation in polar coordinates, which provide radially symmetric trapping potentials and allow for simultaneous micromanipulation in multiple planes [102]. In addition, they are able to transfer optical orbital angular momentum because of their rotating phase distribution caused by a phase singularity on the optical axis [103, 104].

In recent years, Airy beams have also attracted interest in different fields of optics and photonics [105–108]. They belong to the class of accelerating nondiffracting beams which, in contrast to the hitherto discussed straight nondiffracting beams [98], propagate in curved trajectories. In optical micromanipulation, they are proposed as "optical snow blower" because of their transverse accelerating property [109].

The circumstance that, for (straight) nondiffracting beams, the theoretical field distribution in the Fourier plane is represented by an infinitesimally narrow ring has practical implications. If, analogous to HOT, the nondiffracting beam is modulated by means of an SLM in the Fourier plane of an optical tweezers' trapping plane, the diffraction efficiency would normally be quite low since only a small fraction of the SLM's active area is utilized. To increase diffraction efficiency, preshaping of the hologram illumination to a ring is a promising approach to using all available power and reducing diffraction losses [79]. An axicon (conical lens) and a positive lens can be used to concentrate the light on a ring; the resulting beam path is shown in Figure 8.12a. The corresponding modifications to a generic HOT setup are illustrated in Figure 8.12b. A Gaussian TEM_{00} beam is converted to a ring-shaped hologram illumination in the modulation plane of the SLM. For practical purposes, in this case it is advisable to unwrap the hologram field distributions in a way that only the angular dependence is maintained, as illustrated in Figure 8.12c. The unwrapping reduces alignment demands and increases flexibility for different beam diameters. The combination of preshaping and unwrapping results in a powerful technique for efficient generation of straight nondiffracting beams.

The generation of complex transverse modes of nondiffracting beams can be achieved by imprinting the amplitude as well as the phase information of the distinct beam. One option is using two different SLMs, one modulating the phase and the other modulating the amplitude of the light field. An elegant alternative is encoding the amplitude information on a phase-only SLM together with the phase information [110]. For this purpose, an additional blazed grating is weighted with the amplitude [111] to encode the complete amplitude and phase information in one hologram. In combination with the preshaped hologram illumination, this provides a general technique for holographic generation of nondiffracting beams in the Fourier plane.

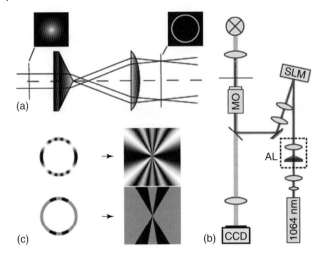

Figure 8.12 (a) Preshaping of the hologram illumination by a combination of axicon and lens results in a narrow ring. (b) Adaptation (AL) to the HOT setup from Figure 8.5. (c) Schematic sketch of amplitude (top, gray values indicate normalized amplitude) and phase (bottom, gray values indicate 0 and π radians) unwrapping.

8.5.1.1 Mathieu Beams

Nondiffracting beams exist in several geometries. Among them, elliptic symmetries are of higher interest because they coevally include Cartesian and polar geometries. The nondiffracting solutions of the Helmholtz equation in elliptical coordinates are given by the class of Mathieu beams [112], which provides a variety of transverse field distributions. Mathematically, their transverse field distributions are given by [113]

$$M_m^e(\eta, \xi, q) = C_m Je_m(\xi, q)\, ce_m(\eta, q), \qquad m = 0, 1, 2, 3\ldots \tag{8.3}$$

$$M_m^o(\eta, \xi, q) = S_m Jo_m(\xi, q)\, se_m(\eta, q), \qquad m = 1, 2, 3\ldots, \tag{8.4}$$

where η and ξ are elliptical coordinates, m gives the order of the beam, $q = f^2 k^2/4$ is a parameter of the ellipticity, f denotes the eccentricity of elliptical coordinates, and k is the transverse wave vector. C_m and S_m are normalizing constants and Je_m and Jo_m and ce_m and se_m are the radial and angular Mathieu functions, respectively. Beside even (M_m^e) and odd (M_m^o) Mathieu beams, the combination of both, referred to as *helical Mathieu beams*, is also of interest in optical micromanipulation because of its ability to transfer orbital angular momentum [103, 114]. A selection of intensity distributions of even, odd, and helical Mathieu beams of different orders is shown in Figure 8.13, illustrating characteristic properties such as their radial and angular nodal lines or the change from vertical to elliptical structures with increasing order m.

For three-dimensional organization of matter, even and odd Mathieu beams are of special interest because of their angularly substructured, highly ordered field distributions. They enable, for example, stacking micrometer spheres in their three-dimensional optical potential wells. The cartoon in Figure 8.13a illustrates five

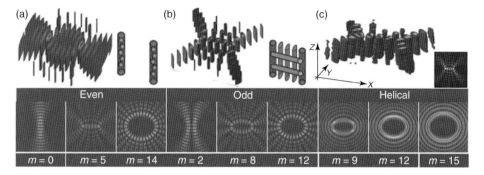

Figure 8.13 Nondiffracting Mathieu beams for three-dimensional particle manipulation. Cartoon and three-dimensional intensity distribution of (a) even Mathieu beam of seventh order enabling organization of microspheres and (b) even Mathieu beam of fourth order enabling orientation and organization of elongated particles. (c) Experimental intensity distributions of an even Mathieu beam of fourth order in the focal plane of the optical tweezers setup (left: three-dimensional measurement and right: transverse cut). Bottom row: selection of intensity distributions of even, odd, and helical Mathieu beams of different orders. (Please find a color version of this figure on the color plates.)

silica spheres stacked in each of the two main intensity maxima of a Mathieu beam of seventh order. The resulting three-dimensional structure has been visualized with a stereoscopic microscope [79]. Alternatively, Mathieu beams can be used to align elongated particles perpendicular to the direction of propagation, as depicted in the cartoon in Figure 8.13b. As well as for spheres, this configuration has been realized in a stereoscopic microscope. It is possible to stack several elongated particles with the same transverse orientation in the z-direction [79] and rotate the resulting, three-dimensional structure.

Summing up, nondiffracting beams, and especially Mathieu beams, facilitate a variety of extended three-dimensional trapping scenarios with future perspectives in many fields of organization and assembly.

8.5.2
Self-Similar Beams

Self-similar beams enable a completely different approach to optical micromanipulation compared to nondiffracting beams. Their transverse field distributions are preserved during propagation and merely scale due to divergence. For the Fourier transform of a self-similar wave field, this means that it only distinguishes by a scaling factor from the original field. The most prominent representative of the class of self-similar beams is the lowest order Gaussian beam, also known as TEM_{00} mode, which plays a dominant role in optical tweezers and already shows many desirable properties of the whole class. For example, parameters such as beam waist, Rayleigh range, and divergence have equivalents for all self-similar beams. The important role of the lowest order Gaussian beam is reflected by the nomenclature

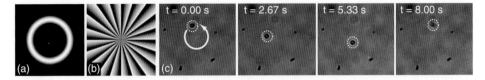

Figure 8.14 Transfer of optical orbital angular momentum with an $l = 20$ LG beam. (a) Transverse intensity distribution. (b) Transverse phase distribution. (c) Image sequence of a 1 μm polystyrene particle rotating in a Laguerre-Gaussian beam in the focal plane of optical tweezers.

of the individual beam classes: Hermite-Gaussian (HG), Laguerre-Gaussian (LG), and Ince-Gaussian (IG) beams, representing solutions of the paraxial Helmholtz equation in Cartesian, polar, and elliptical coordinates, respectively. Self-similar beams are paraxial wave fields, as their normal to the wavefronts encloses small angles with the optical axis. This is also the reason why self-similar beams are solutions of the resonator problem and hence often available directly from a laser source [115, 116].

Higher orders of HG beams and LG beams are well established in optical micromanipulation, for example, to align nonspherical particles [76] or transfer orbital angular momentum. In the latter case, LG beams were of significant importance for the experimental proof that orbital angular momentum can be transferred from light to matter [26]. For illustration, Figure 8.14 shows an example of a 1 μm polystyrene particle, continuously rotated by an LG beam with a topological charge of $l = 20$. From this feature, many applications emerged, including three-dimensional particle control [117] or microfluidic applications where artificial micromachines are driven [73, 118].

All favorable properties of HG and LG beams can also be exhibited by the more general class of IG beams. For example, they can be generated in any arbitrary plane of an optical setup, including the Fourier plane or transfer orbital angular momentum. This unified approach, and more importantly, their high variety of transverse field distributions makes them highly attractive for applications in optical micromanipulation.

8.5.2.1 Ince-Gaussian Beams

A combination of Ince functions [119], solving the paraxial Helmholtz equation in elliptical coordinates, gives a complete set of solutions. The whole beam class of IG beams is abbreviated by $IG_{p,m}$ and splits into even and odd IG beams, which are mathematically described by

$$IG^e_{p,m}(\eta, \xi, z, \epsilon) = C \cdot C^m_p(i\xi, \epsilon) C^m_p(\eta, \epsilon) \cdot e^{ip \cdot \psi_G(z)} \cdot \Psi_{\text{Gauss}} \qquad (8.5)$$

$$IG^o_{p,m}(\eta, \xi, z, \epsilon) = S \cdot S^m_p(i\xi, \epsilon) S^m_p(\eta, \epsilon) \cdot e^{ip \cdot \psi_G(z)} \cdot \Psi_{\text{Gauss}} \qquad (8.6)$$

where $\psi_G(z) = \arctan(z/z_R)$ is the Gouy phase, Ψ_{Gauss} gives the wave function of a fundamental Gaussian beam and C and S are normalizing constants [115, 120].

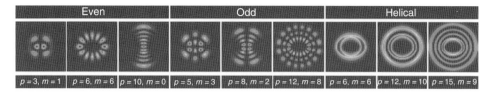

Figure 8.15 Selection of intensity distributions of even, odd, and helical self-similar Ince-Gaussian beams of different orders. (Please find a color version of this figure on the color plates.)

As tailored optical landscapes, IG beams offer a huge variety of field distributions for organizing and structuring of dielectric objects. A selection of intensity distributions of even, odd, and helical IG beams of different orders p and degrees m is shown in Figure 8.15. Characteristic properties as, for example, their radial and angular nodal lines or the change from vertical to elliptical structures with increasing degree, are illustrated. Experimental images of both, a conjugate and a Fourier plane of the optical trapping plane, are shown in Figure 8.16a. It can be seen that their transverse intensity distribution only changes by a scaling factor between both planes. Recently, optical micromanipulation with IG beams could be demonstrated [121]. In Figure 8.16b,c the organization of silica particles in different transverse field distributions is shown. From left to right, the theoretical amplitude and phase distributions, the combined phase hologram and an experimental image of optically ordered 1.5 μm particles are depicted. Furthermore, the interesting phenomenon of optical binding has been observed in optical micromanipulation experiments with IG beams, as shown in Figure 8.16b, declaring them as a promising platform for investigation of particle–particle and particle–light interactions. All these features, in combination with the variety of field distributions, offer new possibilities in optical micromanipulation with many fields of applications.

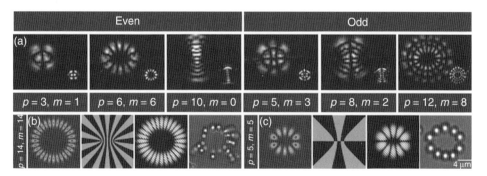

Figure 8.16 Experimental images of self-similar Ince-Gaussian beams. (a) Simultaneous imaging of both a conjugate and a Fourier plane of the trapping plane; the scaling between both planes can be seen clearly. (b,c) Intensity, phase, hologram, and particle configurations. In addition, in (b) the phenomenon of optical binding can be observed. (Please find a color version of this figure on the color plates.)

8.6
Summary

We have briefly reviewed the basic principles of optical tweezers and shown how computer-generated holograms can be utilized to extend the versatility of this instrument dramatically. Appropriate holograms can multiplex a multitude of single optical tweezers and position each freely in three dimensions, leading to the idea of HOT. With computer-addressed SLMs, the hologram can be updated with video frame rate and all trapped particles or only those selected can be dynamically rearranged. This enables rather exciting applications such as colloidal studies, alignment of nonspherical particles, or organization of matter. With computer-generated holograms, however, not only can a light field be split into many beams but also almost arbitrary light modes can be generated. Well-known examples are LG beams that can transfer orbital angular momentum or nondiffracting Bessel beams. We have presented a concise survey of applications of these advanced light modes in optical trapping and have highlighted two important examples: nondiffracting Mathieu beams for three-dimensional molding of matter and self-similar IG beams for the organization of microparticles. These tailored light fields open up promising perspectives for two- and three-dimensional assemblies and manipulation as an approach complementary to classical HOT. The strongest impact of tailored light fields can be assumed in applications in which optically biased organization of medium and larger numbers of colloidal particles is desired rather than utter control over individual particles.

References

1. Ashkin, A. (1970) Acceleration and trapping of particles by radiation pressure. *Phys. Rev. Lett.*, **24**, 156–159.
2. Ashkin, A., Dziedzic, J.M., Bjorkholm, J.E., and Chu, S. (1986) Observation of a single-beam gradient force optical trap for dielectric particles. *Opt. Lett.*, **11**, 288–290.
3. Grier, D.G. (2003) A revolution in optical manipulation. *Nature*, **424**, 810–816.
4. Gouesbet, G. (2009) Generalized Lorenz-Mie theories, the third decade: A perspective. *J. Quant. Spectrosc. Radiat. Transf.*, **110** (14-16), 1223–1238.
5. Nieminen, T.A., Loke, V.L.Y., Stilgoe, A.B., Knoner, G., Branczyk, A.M., Heckenberg, N.R., and Rubinsztein-Dunlop, H. (2007) Optical tweezers computational toolbox. *J. Opt. A: Pure Appl. Opt.*, **9** (8), S196–S203.
6. Harada, Y. and Asakura, T. (1996) Radiation forces on a dielectric sphere in the Rayleigh scattering regime. *Opt. Commun.*, **124** (5-6), 529–541.
7. Ashkin, A. (1992) Forces of a single-beam gradient laser trap on a dielectric sphere in the ray optics regime. *Biophys. J.*, **61**, 569–582.
8. Svoboda, K. and Block, S.M. (1994) Biological applications of optical forces. *Annu. Rev. Biophys. Biomol. Struct.*, **23**, 247–285.
9. Stevenson, D.J., Gunn-Moore, F., and Dholakia, K. (2010) Light forces the pace: optical manipulation for biophotonics. *J. Biomed. Opt.*, **15** (4), 041503.
10. Ghislain, L.P. and Webb, W.W. (1993) Scanning-force microscope based on an optical trap. *Opt. Lett.*, **18** (19), 1678–1680.

11. Florin, E.L., Pralle, A., Stelzer, E.H.K., and Horber, J.K.H. (1998) Photonic force microscope calibration by thermal noise analysis. *Appl. Phys. A*, **66**, S75–S78.
12. Neuman, K.C. and Block, S.M. (2004) Optical trapping. *Rev. Sci. Instrum.*, **75**, 2787–2809.
13. Berg-Sørensen, K. and Flyvbjerg, H. (2004) Power spectrum analysis for optical tweezers. *Rev. Sci. Instrum.*, **75**, 594–612. doi: 10.1063/1.1645654.
14. Jahnel, M., Behrndt, M., Jannasch, A., Schäffer, E., and Grill, S.W. (2011) Measuring the complete force field of an optical trap. *Opt. Lett.*, **36** (7), 1260–1262.
15. Grier, D.G. (1997) Optical tweezers in colloid and interface science. *Curr. Opin. Colloid Interface Sci.*, **2** (3), 264–270.
16. Leach, J. (2006) An optically driven pump for microfluidics. *Lab Chip*, **6**, 735–739.
17. MacDonald, M.P., Spalding, G.C., and Dholakia, K. (2003) Microfluidic sorting in an optical lattice. *Nature*, **426** (6965), 421–424.
18. Friese, M.E.J., Nieminen, T.A., Heckenberg, N.R., and Rubinsztein-Dunlop, H. (1998) Optical alignment and spinning of laser-trapped microscopic particles. *Nature*, **394** (6691), 348–350.
19. O'Neil, A.T. and Padgett, M.J. (2002) Rotational control within optical tweezers by use of a rotating aperture. *Opt. Lett.*, **27**, 743–745.
20. Imasaka, T., Kawabata, Y., Kaneta, T., and Ishidzu, Y. (1995) Optical chromatography. *Anal. Chem.*, **67** (11), 1763–1765.
21. Perch-Nielsen, I., Palima, D., Dam, J.S., and Glckstad, J. (2009) Parallel particle identification and separation for active optical sorting. *J. Opt. A: Pure Appl. Opt.*, **11**, 034013.
22. Jonas, A. and Zemanek, P. (2008) Light at work: the use of optical forces for particle manipulation, sorting, and analysis. *Electrophoresis*, **29** (24), 4813–4851.
23. Asbury, C.L., Fehr, A.N., and Block, S.M. (2003) Kinesin moves by an asymmetric hand-over-hand mechanism. *Science*, **302** (5653), 2130–2134.
24. Maier, B. (2005) Using laser tweezers to measure twitching motility in neisseria. *Curr. Opin. Microbiol.*, **8** (3), 344–349.
25. O'Neil, A.T., MacVicar, I., Allen, L., and Padgett, M.J. (2002) Intrinsic and extrinsic nature of the orbital angular momentum of a light beam. *Phys. Rev. Lett.*, **88** (5), 053601.
26. He, H., Friese, M.E.J., Heckenberg, N.R., and Rubinsztein-Dunlop, H. (1995) Direct observation of transfer of angular momentum to absorptive particles from a laser beam with a phase singularity. *Phys. Rev. Lett.*, **75** (5), 826–829.
27. Meiners, J.C. and Quake, S.R. (1999) Direct measurement of hydrodynamic cross correlations between two particles in an external potential. *Phys. Rev. Lett.*, **82** (10), 2211–2214.
28. Crocker, J.C. (1997) Measurement of the hydrodynamic corrections to the brownian motion of two colloidal spheres. *J. Chem. Phys.*, **106** (7), 2837–2840.
29. Dholakia, K. and Zemanek, P. (2010) Colloquium: Gripped by light: Optical binding. *Rev. Mod. Phys.*, **82** (2), 1767.
30. Padgett, M.J., Molloy, J.E., and McGloin, D. (2010) *Optical Tweezers: Methods and Applications (Series in Optics and Optoelectronics)*, Thaylor and Francis Group.
31. Fällman, E. and Axner, O. (1997) Design for fully steerable dual-trap optical tweezers. *Appl. Opt.*, **36**, 2107–2113.
32. Sasaki, K., Koshioka, M., Misawa, H., Kitamura, N., and Masuhara, H. (1991) Pattern-formation and flow-control of fine particles by laser-scanning micromanipulation. *Opt. Lett.*, **16** (19), 1463–1465.
33. Misawa, H., Sasaki, K., Koshioka, M., Kitamura, N., and Masuhara, H. (1992) Multibeam laser manipulation and fixation of microparticles. *Appl. Phys. Lett.*, **60** (3), 310–312.
34. Visscher, K., Brakenhoff, G.J., and Krol, J.J. (1993) Micromanipulation

by multiple optical traps created by a single fast scanning trap integrated with the bilateral confocal scanning laser microscope. *Cytometry*, **14** (2), 105–114.

35. Simmons, R.M., Finer, J.T., Chu, S., and Spudich, J.A. (1996) Quantitative measurements of force and displacement using an optical trap. *Biophys. J.*, **70** (4), 1813–1822.

36. Mio, C., Gong, T., Terray, A., and Marr, D.W.M. (2000) Design of a scanning laser optical trap for multiparticle manipulation. *Rev. Sci. Instrum.*, **71** (5), 2196–2200.

37. Mirsaidov, U., Scrimgeour, J., Timp, W., Beck, K., Mir, M., Matsudaira, P., and Timp, G. (2008) Live cell lithography: using optical tweezers to create synthetic tissue. *Lab Chip*, **8** (12), 2174–2181.

38. Dufresne, E.R. and Grier, D.G. (1998) Optical tweezer arrays and optical substrates created with diffractive optics. *Rev. Sci. Instrum.*, **69**, 1974–1977.

39. Reicherter, M., Haist, T., Wagemann, E.U., and Tiziani, H.J. (1999) Optical particle trapping with computer-generated holograms written on a liquid-crystal display. *Opt. Lett.*, **24**, 608–610.

40. Liesener, J., Reicherter, M., Haist, T., and Tiziani, H.J. (2000) Multi-functional optical tweezers using computer-generated holograms. *Opt. Commun.*, **185**, 77–82.

41. Curtis, J.E., Koss, B.A., and Grier, D.G. (2002) Dynamic holographic optical tweezers. *Opt. Commun.*, **207**, 169–175.

42. Hesseling, C., Woerdemann, M., Hermerschmidt, A., and Denz, C. (2011) Controlling ghost traps in holographic optical tweezers. *Opt. Lett.*, **36**, 3657–3659.

43. Polin, M., Ladavac, K., Lee, S.H., Roichman, Y., and Grier, D.G. (2005) Optimized holographic optical traps. *Opt. Express*, **13**, 5831–5845.

44. Curtis, J.E., Schmitz, C.H.J., and Spatz, J.P. (2005) Symmetry dependence of holograms for optical trapping. *Opt. Lett.*, **30** (16), 2086–2088.

45. Fienup, J.R. (1982) Phase retrieval algorithms - a comparison. *Appl. Opt.*, **21** (15), 2758–2769.

46. Zwick, S., Haist, T., Warber, M., and Osten, W. (2010) Dynamic holography using pixelated light modulators. *Appl. Opt.*, **49** (25), F47–F58.

47. Gerchberg, R. and Saxton, W. (1972) A practical algorithm for determination of phase from image and diffraction plane pictures. *Optik*, **35** (2), 237–246.

48. Haist, T., Schönleber, M., and Tiziani, H.J. (1997) Computer-generated holograms from 3D-objects written on twisted-nematic liquid crystal displays. *Opt. Commun.*, **140**, 299–308.

49. Neuman, K.C., Chadd, E.H., Liou, G.F., Bergman, K., and Block, S.M. (1999) Characterization of photodamage to Escherichia coli in optical traps. *Biophys. J.*, **77** (5), 2856–2863.

50. Cizmar, T., Mazilu, M., and Dholakia, K. (2010) In situ wavefront correction and its application to micromanipulation. *Nat. Photonics*, **4** (6), 388–394.

51. Crocker, J.C. and Grier, D.G. (1994) Microscopic measurement of the pair interaction potential of charge-stabilized colloid. *Phys. Rev. Lett.*, **73** (2), 352–355.

52. Reichert, M. and Stark, H. (2004) Hydrodynamic coupling of two rotating spheres trapped in harmonic potentials. *Phys. Rev. E*, **69** (3), 031407.

53. Korda, P., Spalding, G.C., Dufresne, E.R., and Grier, D.G. (2002) Nanofabrication with holographic optical tweezers. *Rev. Sci. Instrum.*, **73**, 1956–1957.

54. Melville, H., Milne, G.F., Spalding, G.C., Sibbett, W., Dholakia, K., and McGloin, D. (2003) Optical trapping of three-dimensional structures using dynamic holograms. *Opt. Express*, **11**, 3562–3567.

55. Sinclair, G., Jordan, P., Courtial, J., Padgett, M., Cooper, J., and Laczik, Z.J. (2004) Assembly of 3-dimensional structures using programmable holographic optical tweezers. *Opt. Express*, **12**, 5475–5480.

56. Roichman, Y. and Grier, D.G. (2005) Holographic assembly of quasicrystalline photonic heterostructures. *Opt. Express*, **13** (14), 5434–5439.
57. Leach, J., Wulff, K., Sinclair, G., Jordan, P., Courtial, J., Thomson, L., Gibson, G., Karunwi, K., Cooper, J., Laczik, Z.J., and Padgett, M. (2006) Interactive approach to optical tweezers control. *Appl. Opt.*, **45** (5), 897–903.
58. Sinclair, G., Jordan, P., Leach, J., Padgett, M.J., and Cooper, J. (2004) Defining the trapping limits of holographical optical tweezers. *J. Mod. Opt.*, **51**, 409–414.
59. Benito, D.C., Carberry, D.M., Simpson, S.H., Gibson, G.M., Padgett, M.J., Rarity, J.G., Miles, M.J., and Hanna, S. (2008) Constructing 3 d crystal templates for photonic band gap materials using holographic optical tweezers. *Opt. Express*, **16** (17), 13 005–13 015.
60. Braun, P.V., Rinne, S.A., and Garcia-Santamaria, F. (2006) Introducing defects in 3 d photonic crystals: state of the art. *Adv. Mater.*, **18** (20), 2665–2678.
61. Rosenberg, E. and Ron, E.Z. (1999) High- and low-molecular-mass microbial surfactants. *Appl. Microbiol. Biot.*, **52** (2), 154–162.
62. Sokolov, A., Apodaca, M.M., Grzybowski, B.A., and Aranson, I.S. (2010) Swimming bacteria power microscopic gears. *Proc. Natl. Acad. Sci. U.S.A.*, **107** (3), 969–974.
63. Kim, M.J. and Breuer, K.S. (2007) Controlled mixing in microfluidic systems using bacterial chemotaxis. *Anal. Chem.*, **79** (3), 955–959.
64. Berg, H.C. (2003) The rotary motor of bacterial flagella. *Annu. Rev. Biochem.*, **72**, 19–54.
65. Purcell, E.M. and Am, J. (1977) Life at low reynolds-number. *Am. J. Phys.*, **45** (1), 3–11.
66. Cisneros, L.H., Cortez, R., Dombrowski, C., Goldstein, R.E., and Kessler, J.O. (2007) Fluid dynamics of self-propelled microorganisms, from individuals to concentrated populations. *Exp. Fluids*, **43** (5), 737–753.
67. Darnton, N., Turner, L., Breuer, K., and Berg, H.C. (2004) Moving fluid with bacterial carpets. *Biophys. J.*, **86** (3), 1863–1870.
68. Kolter, R. and Greenberg, E.P. (2006) Microbial sciences - the superficial life of microbes. *Nature*, **441** (7091), 300–302.
69. Ashkin, A., Dziedzic, J.M., and Yamane, T. (1987) Optical trapping and manipulation of single cells using infrared-laser beams. *Nature*, **330** (6150), 769–771.
70. Ashkin, A. and Dziedzic, J.M. (1987) Optical trapping and manipulation of viruses and bacteria. *Science*, **235** (4795), 1517–1520.
71. Min, T.L., Mears, P.J., Chubiz, L.M., Rao, C.V., Golding, I., and Chemla, Y.R. (2009) High-resolution, long-term characterization of bacterial motility using optical tweezers. *Nat. Methods*, **6** (11), 831–835.
72. Hörner, F., Woerdemann, M., Müller, S., Maier, B., and Denz, C. (2010) Full 3 d translational and rotational optical control of multiple rod-shaped bacteria. *J. Biophoton.*, **3** (7), 468–475.
73. Padgett, M. and Bowman, R. (2011) Tweezers with a twist. *Nat. Photonics*, **5** (6), 343–348. doi: 10.1038/NPHOTON.2011.81.
74. Paterson, L., MacDonald, M.P., Arlt, J., Sibbett, W., Bryant, P.E., and Dholakia, K. (2001) Controlled rotation of optically trapped microscopic particles. *Science*, **292** (5518), 912–914.
75. Moh, K.J., Lee, W.M., Cheong, W.C., and Yuan, X.C. (2005) Multiple optical line traps using a single phase-only rectangular ridge. *Appl. Phys. B*, **80**, 973–976.
76. Sato, S., Ishigure, M., and Inaba, H. (1991) Optical trapping and rotational manipulation of microscopic particles and biological cells using higher-order mode ND-YAG laser-beams. *Electron. Lett.*, **27** (20), 1831–1832.
77. Dasgupta, R., Mohanty, S.K., and Gupta, P.K. (2003) Controlled rotation of biological microscopic objects using optical line tweezers. *Biotechnol. Lett.*, **25** (19), 1625–1628.

78. Carmon, G. and Feingold, M. (2011) Rotation of single bacterial cells relative to the optical axis using optical tweezers. *Opt. Lett.*, **36** (1), 40–42.
79. Alpmann, C., Bowman, R., Woerdemann, M., Padgett, M., and Denz, C. (2010) Mathieu beams as versatile light moulds for 3 d micro particle assemblies. *Opt. Express*, **18** (25), 26084–26091.
80. Agarwal, R., Ladavac, K., Roichman, Y., Yu, G., Lieber, C.M., and Grier, D.G. (2005) Manipulation and assembly of nanowires with holographic optical traps. *Opt. Express*, **13**, 8906–8912.
81. Gibson, G., Carberry, D.M., Whyte, G., Leach, J., Courtial, J., Jackson, J.C., Robert, D., Miles, M., and Padgett, M. (2008) Holographic assembly workstation for optical manipulation. *Appl. Phys. B-Lasers O*, **10** (4), 044 009.
82. Bingelyte, V., Leach, J., Courtial, J., and Padgett, M.J. (2003) Optically controlled three-dimensional rotation of microscopic objects. *Appl. Phys. Lett.*, **82**, 829–831.
83. Tanaka, Y., Hirano, K., Nagata, H., and Ishikawa, M. (2007) Real-time three-dimensional orientation control of non-spherical micro-objects using laser trapping. *Electron. Lett.*, **43** (7), 412–414.
84. Tanaka, Y., Kawada, H., Hirano, K., Ishikawa, M., and Kitajima, H. (2008) Automated manipulation of non-spherical micro-objects using optical tweezers combined with image processing techniques. *Opt. Express*, **16** (19), 15115–15122.
85. Woerdemann, M., Glsener, S., Hrner, F., Devaux, A., Cola, L.D., and Denz, C. (2010) Dynamic and reversible organization of zeolite l crystals induced by holographic optical tweezers. *Adv. Mater.*, **22** (37), 4176–4179.
86. Woerdemann, M., Devaux, A., De Cola, L., and Denz, C. (2010) Managing hierarchical supramolecular organization with holographic optical tweezers. *OPN (Optics in 2010)*, **21**, 40.
87. Elemans, J.A.A.W., Rowan, A.E., and Nolte, R.J.M. (2003) Mastering molecular matter. supramolecular architectures by hierarchical self-assembly. *J. Mater. Chem.*, **13** (11), 2661–2670.
88. Bruhwiler, D. and Calzaferri, G. (2004) Molecular sieves as host materials for supramolecular organization. *Micropor. Mesopor. Mater.*, **72** (1-3), 1–23.
89. Calzaferri, G., Meallet-Renault, R., Bruhwiler, D., Pansu, R., Dolamic, I., Dienel, T., Adler, P., Li, H.R., and Kunzmann, A. (2011) Designing dye-nanochannel antenna hybrid materials for light harvesting, transport and trapping. *Chemphyschem*, **12** (3), 580–594.
90. Ruiz, A.Z., Li, H.R., and Calzaferri, G. (2006) Organizing supramolecular functional dye-zeolite crystals. *Angew. Chem. Int. Ed.*, **45** (32), 5282–5287.
91. Megelski, S. and Calzaferri, G. (2001) Tuning the size and shape of zeolite l-based inorganic-organic host-guest composites for optical antenna systems. *Adv. Funct. Mater.*, **11**, 277–286.
92. Dholakia, K. and Cizmar, T. (2011) Shaping the future of manipulation. *Nat. Photonics*, **5** (6), 335–342.
93. Calzaferri, G. and Lutkouskaya, K. (2008) Mimicking the antenna system of green plants. *Photochem. Photobiol. Sci.*, **7** (8), 879–910.
94. Leiggener, C. and Calzaferri, G. (2004) Monolayers of zeolite a containing luminescent silver sulfide clusters. *Chemphyschem*, **5** (10), 1593–1596.
95. Whyte, G. and Courtial, J. (2005) Experimental demonstration of holographic three-dimensional light shaping using a Gerchberg Saxton algorithm. *New J. Phys.*, **7**, 117–128.
96. Durnin, J. (1987) Exact solutions for nondiffracting beams. I. The scalar theory. *J. Opt. Soc. Am. A*, **4**, 651–654.
97. Bouchal, Z. (2005) Physical principle of experiments with pseudo-nondiffracting fields. *Czech. J. Phys.*, **55** (10), 1223–1236.
98. Mazilu, M., Stevenson, D.J., Gunn-Moore, F., and Dholakia, K. Light beats the spread: "non-diffracting" beams. (2009) *Laser & Photon. Rev.*, **4**, 529–547.

99. Gutierrez-Vega, J. and Bandres, M. (2005) Helmholtz-gauss waves. *J. Opt. Soc. Am. A*, **22** (2), 289–298.
100. Vasara, A., Turunen, J., and Friberg, A.T. (1989) Realization of general nondiffracting beams with computer-generated holograms. *J. Opt. Soc. Am. A*, **6** (11), 1748–1754.
101. Arlt, J., Garces-Chavez, V., Sibbett, W., and Dholakia, K. (2001) Optical micromanipulation using a Bessel light beam. *Opt. Commun.*, **197**, 239–245.
102. Garcés-Chávez, V., McGloin, D., Melville, H., Sibbett, W., and Dholakia, K. (2002) Simultaneous micromanipulation in multiple planes using a self-reconstructing light beam. *Nature*, **419**, 145–147.
103. Chavez-Cerda, S., Padgett, M.J., Allison, I., New, G.H.C., Gutierrez-Vega, J.C., O'Neil, A.T., MacVicar, I., and Courtial, J. (2002) Holographic generation and orbital angular momentum of high-order mathieu beams. *J. Opt. B: Quantum Semiclass. Opt.*, **4** (2), S52–S57.
104. Volke-Sepulveda, K., Chavez-Cerda, S., Garces-Chavez, V., and Dholakia, K. (2004) Three-dimensional optical forces and transfer of orbital angular momentum from multiringed light beams to spherical microparticles. *J. Opt. Soc. Am. B*, **21** (10), 1749–1757.
105. Siviloglou, G.A., Broky, J., Dogariu, A., and Christodoulides, D.N. (2007) Observation of accelerating airy beams. *Phys. Rev. Lett.*, **99** (21),
106. Broky, J., Siviloglou, G.A., Dogariu, A., and Christodoulides, D.N. (2008) Self-healing properties of optical airy beams. *Opt. Express*, **16** (17), 12880–12891.
107. Ellenbogen, T., Voloch-Bloch, N., Ganany-Padowicz, A., and Arie, A. (2009) Nonlinear generation and manipulation of airy beams. *Nat. Photonics*, **3** (7), 395–398.
108. Chong, A., Renninger, W.H., Christodoulides, D.N., and Wise, F.W. (2010) Airy-bessel wave packets as versatile linear light bullets. *Nat. Photonics*, **4** (2), 103–106.
109. Baumgartl, J., Mazilu, M., and Dholakia, K. (2008) Optically mediated particle clearing using airy wavepackets. *Nat. Photonics*, **2** (11), 675–678.
110. Davis, J.A., Cottrell, D.M., Campos, J., Yzuel, M.J., and Moreno, I. (1999) Encoding amplitude information onto phase-only filters. *Appl. Opt.*, **38** (23), 5004–5013.
111. Bentley, J., Davis, J., Bandres, M., and Gutierrez-Vega, J. (2006) Generation of helical Ince-Gaussian beams with a liquid-crystal display. *Opt. Lett.*, **31** (5), 649–651.
112. Gutierrez-Vega, J., Iturbe-Castillo, M., and Chavez-Cerda, S. (2000) Alternative formulation for invariant optical fields: Mathieu beams. *Opt. Lett.*, **25** (20), 1493–1495.
113. Gutierrez-Vega, J., Rodriguez-Dagnino, R., Meneses-Nava, M., and Chavez-Cerda, S. (2003) Mathieu functions, a visual approach. *Am. J. Phys.*, **71** (3), 233–242.
114. Lopez-Mariscal, C., Gutierrez-Vega, J.C., Milne, G., and Dholakia, K. (2006) Orbital angular momentum transfer in helical mathieu beams. *Opt. Express*, **14** (9), 4182–4187.
115. Bandres, M. and Gutierrez-Vega, J. (2004) Ince-Gaussian modes of the paraxial wave equation and stable resonators. *J. Opt. Soc. Am. A*, **21** (5), 873–880.
116. Schwarz, U.T., Bandres, M.A., and Gutierrez-Vega, J.C. (2004) Observation of Ince-Gaussian modes in stable resonators. *Opt. Lett.*, **29** (16), 1870–1872.
117. Rodrigo, J.A., Caravaca-Aguirre, A.M., Alieva, T., Cristóbal, G., and Calvo, M.L. (2011) Microparticle movements in optical funnels and pods. *Opt. Express*, **19** (6), 5232–5243.
118. Asavei, T., Nieminen, T.A., Heckenberg, N.R., and Rubinsztein-Dunlop, H. (2009) Fabrication of microstructures for optically driven micromachines using two-photon photopolymerization of uv curing resins. *J. Opt. A: Pure Appl. Opt.*, **11**, 034001.

119. Ince, E. (1923) A linear differential equation with periodic coefficients. *Proc. London Math. Soc.*, **23**, 56–74.
120. Bandres, M.A. and Gutierrez-Vega, J.C. (2004) Ince Gaussian beams. *Opt. Lett.*, **29** (2), 144–146.
121. Woerdemann, M., Alpmann, C., and Denz, C. (2011) Optical assembly of micro particles into highly ordered structures using Ince-Gaussian beams. *Appl. Phys. Lett.*, **98**, 111101.

9
The Role of Intellectual Property Protection in Creating Business in Optical Metrology
Nadya Reingand

> The progressive development of man is vitally dependent on invention. It is the most important product of his creative brain. Its ultimate purpose is the complete mastery of mind over the material world, the harnessing of the forces of nature to human needs.
>
> *Nikola Tesla*

9.1
Introduction

In this chapter, we analyze the role of intellectual property in creating business on examples from optical metrology. In general, creating business includes a variety of tasks: business plan, strategy, financial arrangement, facilities managements, and so on. A scientist, who decides to commercialize his invention, meets a number of new challenges and has to develop new administrative skills, negotiation acumen, and executive expertise. Before starting this thorny path of high-tech business, a number of precautions should be taken. Nowadays, it takes only a few days or weeks to copy and counterfeit manufacture goods. After years of perseverance in building your product, making it profitable, you may face the situation that your market niche is slim because of multiple competitors, who took advantage of your research. The only option is to protect your intellectual property, to keep out others from the competition thus achieving exclusivity for about 20 years in the market. According to *The Economist*, the company value is determined up to 75% by its IP. Therefore, in any scenario, if you plan to either organize profitable mass production or to sell the company, protection of its intellectual property is one of the major parameters of success.

Although the basic principles of IP protection are similar in all countries, there are certain differences in legislation, and the readers should keep in mind that this chapter is written based on US rules.

9.2
Types of Intellectual Property Relevant to Optical Metrology

It may be worthwhile to illustrate this view of classification . . .

Charles Darwin

First of all, we should distinguish between the main types of IP, such as patents, trademarks, copyrights, and trade secrets and their relevance to optical metrology [1]. To grasp the differences between these IP types, we give some examples of their application.

Patent is a document granted by a government to the authors of an invention, which confirms their priority in the discovery and asserts that certain rights are granted to the inventors. These rights are given only over a limited time period and in exchange for that the authors agree to a public disclosure of their invention. Thus, if you happen to be an inventor, getting these rights might be worth all the trouble because once the patent is granted, everyone but you is excluded from making, using, selling, importing, or offering your invention for sale. Having said so, one must realize that the power of patent is not unlimited as it does not preclude other scientists from work on the same problem and improving existing solutions. This chapter is mostly devoted to patent practice since it is the most interesting aspect of IP in a majority of technical fields, including optical metrology.

There exist three types of patents: utility patents, design patents, and plant patents, each having innate merits and caveats. First and foremost, there are utility patents. This is the prevalent type of patents, which covers the inventions that produce utilitarian results. Practically anything made by humans can be patented as long as it is novel. Novelty may include just about anything: new materials and devices, their methods of usage and the processes involved in their manufacturing, new technological procedures, internet techniques, some novel genes sequence, new ways to facilitate business, innovative software and hardware, pharmaceutical products, and many other man-made items and actions limited only by one's imagination. Figure 9.1 shows an example of such a patent with the capacious title "optical metrology."

Not to be confused with utility patents, the other patent species is the design patent. A design patent covers the unique appearance of an object. It should

United States Patent [19]   [11] Patent Number: 4,725,884
Gurnell et al.   [45] Date of Patent: Feb. 16, 1988

[54] **OPTICAL METROLOGY**   e.g. the width of a line on a semiconductor wafer, hav-
[75] Inventors: **Andrew W. Gurnell,** Clifton; **Keith**   ing a light microscope, a video system for receiving an
                  Horner, Haxby; **Richard R. Jackson,**   optical image from the microscope and for displaying
                  York, all of England   the image on a display surface having a reference da-
[73] Assignee: **Vickers PLC,** London, England   tum, and an optical system for transmitting the optical
                                                               image from the light microscope to the video system. In

Figure 9.1 Utility patent for optical metrology.

9.2 Types of Intellectual Property Relevant to Optical Metrology

(12) **United States Design Patent**	(10) Patent No.:	US D514,531 S
Takagi et al.	(45) Date of Patent: **	Feb. 7, 2006

(54) OPTICAL MODULE

(75) Inventors: Shinichi Takagi, Tokyo (JP); Hiroshi Aruga, Tokyo (JP); Kiyohide Sakai, Tokyo (JP)

(73) Assignee: Mitsubishi Denki Kabushiki Kaisha, Tokyo (JP)

OTHER PUBLICATIONS

NEC Compound Semiconductor Devices, "Data Sheet, laser Diode NX7313UA NEC", 10 pages, 2002.
Product Overview, wysiwyg://mainframe.1101/http://www.shinko.co.jp/e.product/e.glass/e.glass.4.htm, "Glass−to−Metal Seals", 2 pages, 2001−2002.
Patent Abstracts of Japan, JP 10−284640, Oct. 23, 1998.
Patent Abstracts of Japan, JP 08−031970, Feb. 2, 1996.

Figure 9.2 Design patent for optical metrology.

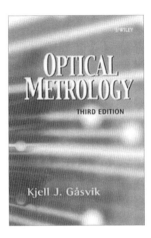

Figure 9.3 An example of copyrighted material in optical metrology.

be purely aesthetic, not functional, in design. If the subject of a design patent in addition to being pleasant to one's eye also happens to have some useful functionality, then the design must be covered by a utility patent. Figure 9.2 shows an example of a design patent applicable to optical metrology. Note that design patent numbers start with letter D.

And now, we come to the third sibling of the patent family – plant patents that protect asexually reproduced novel plants, such as flowers, trees, shrubs, and also algae, and macro fungi. This type of patent is obviously irrelevant to optical metrology.

Copyright protects the creative or artistic expression of an idea. Copyrights, identified by the symbol ©, do not cover ideas and information themselves, only the form or manner in which they are expressed. "Original works of authorship" are protected by copyright law once the author writes them on paper or places them on the drive of the computer (Figure 9.3). Software, which is an essential part of modern optical metrology systems, may be protected by copyright. The law today does not require attaching a notice of copyright to the work or registration; the author is the copyright owner even without these formalities.

Unlike copyright, the valid patent does not protect the expression of the idea but its underlying substance. For example, a utility patent application for a microchip addresses not the mask itself or the particular integrated circuit layout, but the idea that the given circuit can be organized and operated in a particular manner (Figure 9.4). A mask work as a two- or three-dimensional layout or topography of an integrated circuit, that is, transistors and passive electronic components such as

(12) **United States Patent**
Chen et al.

(10) **Patent No.: US 7,014,956 B2**
(45) **Date of Patent: Mar. 21, 2006**

(54) **ACTIVE SECONDARY EXPOSURE MASK TO MANUFACTURE INTEGRATED CIRCUITS**

(75) Inventors: **Fred Chen**, Cupertino, CA (US); **Jeff Farnsworth**, Los Gatos, CA (US); **Wen-hao Cheng**, Fremont, CA (US)

(73) Assignee: **Intel Corporation**, Santa Clara, CA (US)

6,541,165 B1 * 4/2003 Pierrat 430/5
6,544,695 B1 * 4/2003 Wang et al. 430/5
6,632,576 B1 10/2003 Tejnil
6,821,690 B1 * 11/2004 Bae 430/5

OTHER PUBLICATIONS

Warren Grobman and Yao-Ting Wang. Practical IC Design in the Subwavelength Regime. Web Page [online]. Inte-

Figure 9.4 Mask work protection by utility patent.

Word Mark OMISTRAIN

Goods and Services IC 009. US 021 023 026 036 038. G & S: Optical metrology apparatus and instrumentation, namely metrology tools for 2D and 3D visualization and measurement of the deformation motion and strain field in advanced electronic packages, system-on-chip and MIEMS/MOEMS (micro-electro-mechanical systems I micro-opto-electromechanical systems) product engineering; computer software used for image acquisition, stressing control, image processing.

Figure 9.5 Trademark for optical metrology business.

resistors and interconnections, may be protected by exclusive rights "mask work" Ⓜ, which are similar to copyright. This right provides time-limited exclusivity to reproduction of this particular layout; it is identified by a sign Ⓜ with the owner name written aside.

Trademark (or servicemark) defends distinctive terms, marks, and names that are used in relation to products (or services) as indicators of origin. A trademark and servicemark, identified by the symbols ™ and ^{SM} (not yet registered) and ® (registered), is a distinctive sign used to discern the products and/or services to consumers. An example of a trademark in optical metrology business is shown in Figure 9.5.

Trade secret is some confidential information that is kept secret, which provides advantage over the competitors. Obviously, trade secret is the best way to protect your innovation as long as you can keep it secret. Key employees' departure and joining a competitor company put trade secrets at great risk. For example, Nortel Networks (Canada) sued Optical Networks, a small US fiber optics company, for violating its IP. Optical Networks lured away dozens of key technical workers to learn Nortel's trade secrets and then built a similar product. Since Optical Networks does not do business in Canada, the charges were fairly narrow. This example shows fragility of trade secrets; a special care is required to keep them really secret.

9.3
What Kind of Business Does Not Need IP Protection?

> If you love somebody, set them free
>
> *Sting*

1) Classified topics.
   Optical metrology as any other high-tech discipline may be a part of classified project; for example, it may be used in missiles construction. In case it is applicable only to the specified goal, it cannot and need not be protected by IP.

2) Government contractors, where the market is the government only.
   Certain industrial areas, such as nuclear plant building or collider construction, belongs exclusively to the government. If optical metrology devices are designed purposely to solve specific problems for those government endeavors, they do not need IP protection.
3) If the technology is too complex, it requires advantageous technology, such as super power atomic force microscope or unique machinery; therefore, it is hard to copy those technologies, and they do not require IP protection. However, one should not underestimate the potential of one's competitors.
4) Technologies that can be protected by a trade secret, as with Coca-Cola, do not need patenting, given that the secret is kept.

9.4
Does IP Protect Your Product from Counterfeiting?

> Marconi is a good fellow. Let him continue.
> He is using seventeen of my patents.
>
> *Nikola Tesla*

To a certain extent, IP definitely protects your rights. However, according to the International Chamber of Commerce, ~8% of world trade every year is in counterfeit goods. Most commonly affected by IP theft are industries related to manufacturing consumer goods, high tech technology, software, and biotechnology, including pharmaceuticals.

IP enforcement is extremely expensive, but sometimes, it is the only way to defend important innovation. The most famous example of a trial about optical device: Kodak lost a patent battle with Polaroid Corporation on the instant camera case of 1986. At that time, everyone thought that instant camera would replace film cameras that were photographic standard equipment at that time. Experts considered that soon everyone would switch to instant cameras. Those cameras included photomaterial, and the image processing of that material was performed inside the camera itself. After releasing half-processed photo, one needed to expose it to light to finalize the development procedure. Polaroid was awarded damages in the patent trial in the amount of US $909 457 567.00, a record at the time [2].

IP enforcement has certain limitations: it only works for the country where IP has been registered, for example, patent filed. Therefore, there are no restrictions to reproduction of your equipment or software, say, in China, if you did not file your patent there. Another problem that may arise even if you register your patent rights: it may be exceptionally hard to impose those rights because of a lack of history of the IP enforcement in those countries.

Here, we touched an issue of national character of IP rights, and we discuss it in more detail in the following section.

9.5
Where to Protect Your Business?

> ... They have a world to gain
>
> *Karl Marx*

A brief answer to this question will be on those territories where you have business with your product. Patent gives the right to exclude others in the territory covered by the patent from making, using, offering for sale, importing, or selling the invention. Note that patent is not a right given to an inventor, but an absence of rights for others. Despite the global character of modern economy, intellectual property rights are national. This means that patent is registered in the national patent office and is valid in the territory of that country.

Once you get your patent, it is easy to learn about your invention since it is published. No one can keep patented invention unpublished. And besides, the patent needs to include a description of the procedure how to built your patented device or perform your patented method. It needs to be described to the level of details that results in relatively quick building of your patented gadget. If you keep some important elements secret, and a person skilled in the art of your invention cannot build the device or it does not work as you described in the patent, then your patent most likely is invalid.

The purpose of giving a patent to an inventor is to give him/her exclusivity for receiving benefits for the limited amount of time in exchange for full description of the invention, so it will become available to the public after the end of the patent life. So once you have obtained a US patent, it is fully disclosed in your patent description.

National character of IP means that your US patent does not preclude anyone to start production of your invention in, say, China, and selling it in, say, Germany.

Nobody can apply for a patent in any country for the invention that you patented in one country, but everyone can produce and sell in the territory of other countries without obtaining his own patent or licensing your patent.

9.6
International Patent Organizations

> And it comes from saying "no" to 1,000 things to make sure we don't get on the wrong track or try to do too much. We're always thinking about new markets we could enter, but it's only by saying "no" that you can concentrate on the things that are really important.
>
> *Steve Jobs*

It would be great to protect your IP in all countries of the world. However, it is probably not practical since it would be very expensive and time consuming. In order to simplify the patent examination procedure and coordinate efforts of

9.6 International Patent Organizations | 213

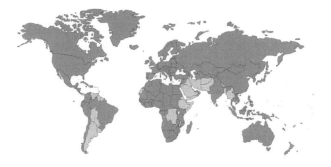

Figure 9.6 Countries, which are members of the PCT.

Figure 9.7 An example of World Intellectual property organization patent application.

national patent offices, especially in the case when the same patent is filed in a set of countries, a number of international organizations were created. The most important one is Patent Cooperation Treaty (PCT), which is an agreement for international cooperation in the field of patents between 144 countries. Figure 9.6 shows the world map, where PCT member are marked in dark color.

The PCT serves to coordinate efforts on filing, searching, and the examination of patent applications. It does not grant international patents. After the PCT examination, inventors can see the examination report and make a conscious decision on filing or not filing patents in a number of countries. The latter procedure is called *entering national stage*. The patent rights are granted by national offices.

World Intellectual Property Organization (WIPO) PCT patent publications have numbers that start with letters WO, Figure 9.7 shows "Metrology module for laser system" patent application by Carl Zeiss Laser Optics GmbH. One should keep in mind that those WO patent documents are patent applications, not yet patents.

9.7
Three Things Need to Be Done Before Creating Business

9.7.1
Prior Art Search

"Doubt is the father of invention."

Galileo Galilei

It is common that we often overestimate the greatness of our inventions and assume that nobody ever came to the same conclusions that we did. Unfortunately, in patent practice, we have situations over and over again, when the invention is pretty close to already known material. Otherwise, the whole field of patent examination would not exist. The examination is an officially certified comparison of the invention to the existing prior art. In order to avoid major disappointments, the inventor should perform thorough prior art search.

First of all, prior art search is performed to verify that the invention is novel. Patents are granted for novel ideas or nonobvious improvement of the existing technologies, materials, devices, and so on. Although, apparently, one can open another grocery store or landscape service, and it still will be called *creating business*, most likely it will be successful if this business will differ somehow from competitors. Novelty is a crucial element of success in any business.

Although you might think that after so many years in optical metrology, you know better than anyone who is doing what in the field, it is still better to get an official confirmation that your idea is novel. Patent office examiners are those experts who can officially certify that your innovation was never described in any other publication in the world, so it is novel.

Generally speaking, every new device and even material or method is a combination of known things. In order to protect your idea, you need to show the examiner that the combination is useful, it provides some additional features that were not attainable in the prior art. Let us analyze an example: if you created a novel optical system, where you replaced an optical laser emitting in the visible range with an ultraviolet one, you need to show that this replacement provides results that were not achievable using an optical laser. For instance, you can get higher resolution.

Another criterion for obtaining a patent is to prove that the advancement you made is nonobvious. If it is already known for every optical scientist that replacing optical (red) laser with UV laser provides essential improvement in the system resolution, then the examiner may reject your patent application as "obvious." In this case, you need to demonstrate that such replacement was not possible previously. For example, the photosensitive material that was used is not sensitive in UV range. So this new combination would not work unless you add, say, a novel processing mechanism.

This example explains criteria that the examiner will use to distinguish your invention from the prior art.

Obviously, you need to know the state of the art before starting your business and before applying for a patent. At least, each scientist has to perform it himself. Another option is to ask professional help of patent searchers to provide the results of prior art search.

A variety of databases can be used to perform the prior art search. I would suggest using Google Patent (*http://www.google.com/patents*) as a first step. In my view, it is the most convenient database to use and it has an excellent browser, which surfaces the most relevant patents, and besides you can download pdf copies of relevant patents right away, directly from this Web site.

My second choice would be US Patent and Trademark Office Web site (Table 9.1). Although it only includes US patents and applications, so the patent collection is limited, it is compensated by giving you options to search the database not just by keywords, but by using various patent characteristics (Figure 9.8). This provides you more flexibility in your search.

The work will not be completed without searching worldwide database of the European Patent Office database (*http://worldwide.espacenet.com/*). It houses close to 40 million patents from over 71 member countries.

Field code	Field name	Field code	Field name
PN	Patent Number	IN	Inventor Name
ISD	Issue Data	IC	Inventor City
TTL	Title	IS	Inventor State
ABST	Abstract	ICN	Inventor Country
ACLM	Claim(s)	LREP	Attorney or Agent
SPEC	Description/Specification	AN	Assignee Name
CCL	Current US Classification	AC	Assignee City
ICL	International Classification	AS	Assignee State
APN	Application Serial Number	ACN	Assignee Country
APD	Application Data	EXP	Primary Country
PARN	Parent Case Information	EXA	Assistant Examiner
RLAP	Related US App. Data	REF	Referenced By
REIS	Reissue Data	FREF	Foreign References
PRIR	Foreign Priority	OREF	Other References
PCT	PCT Information	GOVT	Government Interest
APT	Application Type		

Figure 9.8 Prior art search in US Patent Office database using various patent characteristics.

Table 9.1 Commonly used databases for the patent search and analysis.

	Databases	Web page
1	Google patent	http://www.google.com/patents
2	United States	http://patft.uspto.gov/
3	European	http://gb.espacenet.com/search97cgi/s97_cgi.exe?Action=FormGen&Template=gb/en/advanced.hts
4	Japanese	http://www19.ipdl.inpit.go.jp/PA1/cgi-bin/PA1INIT?1302804799012
5	Korean	http://eng.kipris.or.kr/eng/main/main_eng.jsp

If we are talking about patents in the optical field, I would highly recommend to perform additional search in Japanese and Korean Patent Libraries. Although typically we can only read their abstracts in English and the search is not simple, it is hard to overestimate the contribution of these countries in innovative approaches in modern optoelectronics and other high technologies. Table 9.1 shows most commonly used free of charge Web sites for patent search and analysis.

Two main techniques are typically used for patent search: a keyword search and a classification search. The keyword search speaks for itself. The only hidden trap there may be a multiple meaning of the same word, such as modulator. In this case, you should clarify what kind of modulator you are searching for. For instance, you may add "frequency adj modulator" or "quadrature adj modulator." Here, the connection word "adj" means "adjacent." Each search database has its own connective words that may be used. However, the standard set of "and, or, adj" exists almost everywhere.

Besides, it is very useful to search backward and forward citations of the most relevant patents that were found.

Performing patent search is definitely an art, although it is based on certain logical approaches. However, if you found a good patent searcher for your inventions, he/she should be treasured as jewelry. It is a kind of magic what they, patent searchers, do when they retrieve the most relevant patent out of a collection of millions of worldwide patents.

The results of prior art search are very useful for creating business. This knowledge allows formulating the distinction of your invention in comparison with others. It facilitates patent prosecution, and you will get your patent faster and will be less expensive, since it will save money otherwise spent on the stage of the patent examination. Moreover, the patent search gives you the names of your competitors, so you can track their business activity.

9.7.2
Patent Valuation

> Certainty? In this world nothing is certain but death and taxes
>
> *Benjamin Franklin*

Estimation of the profit that invention brings and the value of patenting is a complex subject with lot of uncertainties. Obviously, if the market size for your invention is small, it does not make sense to patent it. Before creating business, an inventor always tries to estimate the short-term and long-term profit range. Generally speaking, the value of patenting your invention is equal to profit with patenting minus profit without patenting. However, this simple estimation does not embrace risks that you are taking if you decide to proceed with production and selling without defending your intellectual property.

If your product has a significant market success, it can be counterfeited overnight. Even in such complicated high-tech industry as a product identification hologram, a fake hologram can be made within 24 h [3].

The absence of IP protection may lead to major reduction in your market niche because of abundance of counterfeit products imitating your invention.

And besides, if you did not patent your invention and did not disclose it in publication, then somebody else can patent it and block your business. Moreover, your company may get swamped in a lengthy law suit, and you cannot escape this procedure even if you have already stopped your business, since you will be a defendant, not a plaintiff.

An interested reader may learn more about patent valuation in [4].

9.8
Ownership Clarification

"I want it all ... "

Queen

According to the Bayh–Dole Act of 1980, US universities and other federal government-funded research facilities can have a control over their inventions and other intellectual property that resulted from such funding. Usually, it is stated in the employment agreement that your inventions belong to the employer (e.g., university). The university almost always retains rights in the invention even after it is licensed or sold; it is done to protect the continuity of the scientific research in this field. If your university owns the invention, then eventually you receive a certain portion of the profit. The size of this portion depends on the type of commercialization and on university rules [5]. We discuss those issues in the following sections.

If the invention is made under contract with the government agency (Department of Defense, National Institute of Health, Homeland security agency, etc.), they have rights to use your invention. The so-called "confirmatory license" provides the government with a royalty-free license to the invention. Note that the government always has rights on the part of the invention, which was developed during the contract and using the money of the contract. If, for example, you invented a new type of laser and later developed a cooling system of this laser under the contract,

Assignment: 1
Reel/Frame: 014636/0201   **Recorded:** 10/29/2003
Conveyance: CONFIRMATORY LICENSE (SEE DOCUMENT FOR DETAILS).
    Assignor: NORTHESTERN UNIVERSITY
    Assignee: AIR FORCE, UNITED STATES
                 26 ELECTRONICS PARKWAY
                 AFRL/IFOJ
                 ROME, NEW JERSEY 03441-4514
Correspondent: AFRL/IFOJ
                       JUDITH M. DECKER
                       26 ELECTRONIC PARKWAY
                       ROME NEW, 13441-4514

Figure 9.9 An assignment record in USPTO database showing the confirmatory license.

then the government will have certain rights on the cooling system, but not the laser itself. Typical record of the government confirmatory license is shown in Figure 9.9.

Government research is funded by taxpayers, and the government funding of the research often requires the product commercialization (production, sales, etc.) in the country of the invention.

If the invention is made under industry-sponsored contract, then the sponsor may have rights on your invention. Usually, it happens when the sponsor already has IP, and university research team continues testing or further develops what the sponsor owns.

There are special types of university funding, which exclude the possibility of inventing and patenting, for example, university bonds.

Students do not have employment agreements, so their IP rights in their inventions are determined based on university involvement in the invention. If the invention came from a school contest, extracurricular club, or individual initiative, the university keeps its hands-off. If the student invention came about under a professor's supervision, using school resources or grant money, then the university can assert an ownership right. Student invention ownership has become a popular topic for discussion in recent years. A number of undergraduates from various universities developed successful iPhone applications. For example, Tony Brown (University of Missouri) came up with an idea of the iPhone application NearBuy that later had more than 250 000 downloads within a year and a half. The app created by Brown and three other undergraduates won them a trip to Apple headquarters along with job offers from Google and other technology companies. The university allowed them to keep rights on their invention.

The patent assignment is registered at USPTO; you can check the assignment of all US patents and published patent applications at the Web site: *http://assignments.uspto.gov/assignments/q?db=pat*

In case of absence of record, the patent or patent application belongs to the inventors. Alternatively, it may belong to the inventors' employer, but those rights are regulated by the Contract Law.

We would like to discuss a special case of *joint ownership*. According to US legislation, each owner in this case has 100% of the invention, and he/she can commercialize it without asking permission of the other owner and without sharing the profit from the sales.

A detailed description of the inventor's right in university setting can be found in [6].

9.9
Patent Filing

> The work on satisfactory formulation of technical patents was a true blessing for me. It compelled me to be many-sided in thought, and also offered important stimulation for thought about physics.
>
> *Albert Einstein, Autobiography*

Once you have clarified that your invention is novel, it will bring profit, and you actually own the invention or your university owns it and the next step is to file a patent application.

The author's experience as a patent analyst for the market study and investments suggests that the number of patents should be at least six to be notable in various patent searches. A good size portfolio starts from 20 to 30 patents for a small business company.

If you are experienced in writing patents, then you can prepare and file the application yourself. It is not rocket science. However, we would suggest using professional help at least at the stage of the claims drafting. Claims are the heart of the patent; they disclose what you own, while the rest of the patent application including figures serves to support the claims. Claims are written according to certain rules and even its own nonstandard grammar. But this is not the most important thing about claims.

Claims are similar to possession of a land. You can negotiate with the patent office and "buy" as much "land" as you described in your patent application, but you cannot "buy" from the patent office the "land," which is already owned by somebody. You cannot step on somebody else's land. You have to formulate your claims around all existing prior art. So the first characteristic of claims is that they do not overlap with prior art.

It is important to get as much "land" as you can. The second characteristic of professionally written claims is that they should be as wide as possible. The main mistake that inventors make when they write the claims themselves: they meticulously describe each and every element of their device in claim 1. Such an approach limits the scope of the claims and, at the end, provides very limited protection of your invention.

Claim 1 should contain only elements that are absolutely necessary to perform the task. For example, if you invented a laser operating with a special type of integrated resonator, then you should claim this resonator features in your claim 1.

The fact that it is integrated is not important. The fact that it has certain type of couplers or a cooling mechanism should also be omitted from claim 1. You may add it later in the following claims.

If you mention a certain type of element that you use, for example, delta–beta switch, in your claim 1, then your competitor may create a similar device, but using another type of switch, and this device will not infringe your patent application. Therefore, the conclusion is the invention description in claim 1 must be as general as possible without stepping on prior art.

If you decide to use a patent professional for writing the claims, it would be very helpful if you make your best effort in describing your invention in written form and showing the difference between your invention and the prior art. This saves money and facilitates further prosecution of the patent application.

Patent is the proof of original work and a "keep off" sign for about 20 years. It allows to acquire larger market share compared to nonpatented invention sales. It allows longer development since it is protected from copying, and therefore provides a lower cost of the product development.

In academia, you should use Office of Technology Transfer (OTT) to proceed with your invention.

Ways to proceed with your patented invention:

- licensing to existing companies;
- start-up formation;
- or both.

The proportion is different for each university and depends on the specialization. MIT (Massachusetts Institute of Technology) shows that although only 35% of the licenses were granted to start-ups, they accounted for 77% of the investment and 70% of the jobs.

9.10
Commercialization

> You can't always get what you want
>
> *Rolling Stones*

In my view, the post-patent grant activity associated with selling the invention and getting a proper reward, which pleases the inventor and justifies all efforts, is the most complex part of the commercialization. It is a thorny path that often turns in a wrong direction and requires compromising and shrewd negotiation skills.

9.10.1
Licensing to Existing Companies

Typical license payment includes

- reimbursement and ongoing payment of patent prosecution costs;

- milestone payments;
- minimum annual royalties;
- a percentage royalty on sales.

Exclusive license gives the licensee sole and entire rights to operate excluding all others, including the right to sublicense. Nonexclusive rights give the licensee rights to make use, import, offer to sell, and/or sell the licensed technology; however, other companies may also obtain the similar license. Typically, nonexclusive license does not include a right to sublicense. Licensing does not constitute a change in ownership of the patent.

Licensing agreement for federally funded research may require US manufacturing and retained government rights.

The economic impact of US academia inventions licensing results in the investments of about $20 billion per year range, thus creating over 150 000 jobs per year.

Typically, OTT receives hundreds of invention disclosure forms (IDFs) per year, chooses to pursue about one-third of those and licenses half of these. Out of the latter 10% break even, 5% make some money and 0.01% generates a million dollars of revenue.

The revenue is usually split between the inventor(s), university, and inventor's laboratory, for example, 34% + 33% + 33%. The inventor's share varies from 25% to 70% depending on the university. In case of many inventors, revenue is typically shared equally unless inventors agree otherwise in writing. In some universities (e.g., Iowa State), first $100 000.00 of net revenue is given to the inventor(s).

9.10.2
Start-Ups

Proceeding with forming a start-up and initiating business activity based on your invention, often means route away from active scientific research and focus on the product development and on solving multiple business questions, such as getting funding, market penetration, advertisement, and others.

Although it is very exciting journey, it leads to uneasy life with thrilling successes and certain unavoidable mistakes. Roughly 5–10% of inventions meet the criteria necessary to become start-up companies.

Often, the TTO, and the faculty inventor wishing to create a start-up, are in a kind of "chicken and egg" situation. The start-up needs a secure commitment from the university that it has, or will soon obtain, full and exclusive rights to market a particular technology and the university seeks to grant licenses only to entities it believes have the capability of rapidly, and profitably, commercializing the technology. The university's contribution to a start-up is usually in the form of an exclusive license to a portfolio of intellectual property assets. The TTO also acts as a founder of the company, understanding that start-ups typically have little cash and no revenues. As a founder, the university typically accepts nonmarketable stock in lieu of cash as compensation.

In most cases, start-up companies seek exclusive licenses from the university to the intellectual property needed to form the core assets of the business. However, in nearly all cases, the work performed in the laboratory to generate the underlying IP continues well after the initial disclosure and subsequent patenting process. Additional discoveries are often more imperative to the commercialization of technology than the original invention. As a result, companies seek to secure rights to follow-on improvements via an option to future improvements. Options grant a company a period of time to evaluate an improvement for inclusion in the company's IP asset portfolio. If the company elects to exercise the option, this can be done for a small up-front fee and an agreement to incorporate the new technology into the company's License Agreement under the same terms and conditions as the original. If the company decides not to license the new technology, it is returned to the TTO for licensing elsewhere.

Successful story: In spring 2005, OmniVision acquired for US $30 million the company CDM Optics, a University of Colorado (CU)-based start-up, developing technology that greatly improves the clarity of images through a lens, using patented "depth-of-field" technologies from CU laboratories.

In the United States, the results of university inventions commercialization are summarized every year in reports by Association of University Technology Managers (AUTM) [7]. Just to give an idea about the scale of this activity, here are the numbers for the fiscal year 2009:

- 658 new commercial products introduced;
- 4374 total licenses executed;
- 596 new companies formed;
- 3423 start-up companies still operating as of the end of FY2009;
- $53.9 billion total sponsored research expenditures;
- 20 309 disclosures;
- and $2.3 billion total licensing income.

Patents filed: 18 214 total US patent applications; 12 109 new patent applications filed; and 1322 non-US patent applications.

9.11
Conclusions

> O summer snail,
> you climb but slowly, slowly
> to the top of Fuji
>
> *Haiku by Kobayashi Issa*

It is very hard to combine a research career with the commercialization of your scientific results. However, if business is your choice, then a number of precautions need to be done. The business is usually built slowly, step by step, and one of the first steps, in our view, includes clarification and ownership of your intellectual property.

It is needed, first of all, for your own product protection against somebody's lawsuit, where the competitor shows that you infringe somebody's patent and requests that you stop your business, compensate for what is already done and pay license fees. Being a defendant, not a plaintiff, you cannot abandon the suit and have just two options – to fight or to die. Going to the court and battling against your competitor is a full-time job, which takes lots of time and money. It redirects your attention from what you really want to do: to create new ideas and commercialize your inventions. IP provides you exclusivity in development of your product, which certainly saves time and money, and brings more profit at the end. If eventually you plan to sell your business, then the cost of the company with IP is many times higher than without IP. Up to 75% of the value of public companies is now based on their IP. You plan either to build a mass production or to sell the company, in any case, you need a strong IP portfolio.

The recently published book "Intellectual Property in academia: practical guide for scientists and engineers" edited by the author of this chapter offers detailed description of various aspect of IP protection and building business out of university inventions [8].

References

1. Reingand, N. and Osten, W. (2009) in *Fringe 2009, The 6th International Workshop on Advanced Optical Metrology* (eds W. Osten and M. Kujawinska), Springer, pp. 634–647.
2. (1990) Polaroid Corp. v. Eastman Kodak Co. U.S. District Court District of Massachusetts, decided October 12, 1990, case no. 76-1634-MA. Published in the U.S. Patent Quarterly as 16 USPQ2d 1481.
3. Dobert, D.J. (2010) Mixing It Up, March 21, 2010, http://www.atlco.com/blog/tag/holograms/ (accessed 01 May 2012).
4. Maiorov, M. and Spinler, S. (2011) in *Intellectual Property in Academia: Practical Guide for Scientists and Engineers* (ed. N. Reingand), Taylor and Francis, pp. 45–120.
5. Reingand, N. and Osten, W. (2010) in *Speckle 2010: Optical Metrology*, Proceedings of the SPIE, Vol. 7387 (eds A. Albertazzi Goncalves Jr. and G.H. Kaufmann), pp. 21-1–12-9.
6. Machi, P. (2011) in *Intellectual Property in Academia: Practical Guide for Scientists and Engineers* (ed. N. Reingand), Taylor and Francis, pp. 235–276.
7. Association of University Technology Managers (AUTM) (2009) Annual Report, 2009, http://www.autm.net/AM/Template.cfm?Section=Documents&Template=/CM/ContentDisplay.cfm&ContentID=5237 (accessed 01 May 2012).
8. Reingand, N. (2011) *Intellectual Property in Academia: Practical Guide for Scientists and Engineers*, Taylor and Francis, and CRC Press, ISBN: 978-1-4398-3700-9.

10
On the Difference between 3D Imaging and 3D Metrology for Computed Tomography

Daniel Weiß and Michael Totzeck

10.1
Introduction

Three-dimensional imaging denotes a set of techniques that allow the dimensional representation of the interior of an object without actually destroying it. These techniques comprise X-ray computed tomography (CT, [1]), optical coherence tomography (OCT), both conventional and with extended depth of focus [2] (EDOF [3, 4]), ultrasound tomography, confocal microscopy [5], and magnetic resonance imaging (MRI [6]). Figure 10.1 compares their typical penetration depths and resolution.

The importance of these techniques is rising with the complexity in geometry and composition of industrial fabricated parts and systems as well as biological samples.

However, it is one thing to get a 3D contrast and quite another to do actual dimensional 3D metrology. For the latter, we need to understand and correct geometrical distortions caused by the imaging process itself. The most prominent causes for these distortions are statistical and systematical errors of the imaging setup. So a comprehensive calibration is necessary. Furthermore, we have to cope with physical causes such as diffraction yielding fringing, refraction yielding displacements and aberrations, and dispersion effects. These, of course, depend strongly on the inspected material and require appropriate calibration methods. The relative contribution of these errors to the overall error depends on the applied imaging methods. For CT, as considered here, the calibration of the mechanical setup, as well as spectral dependence of the absorption of the material (beam hardening correction, cf. Section 10.3.3.2), yield the most important contributions.

This chapter is organized as follows. After a general consideration of 3D imaging and metrology, we present in Section 10.3 the principle of X-ray CT and review the means for obtaining a 3D metrology CT.

Optical Imaging and Metrology: Advanced Technologies, First Edition.
Edited by Wolfgang Osten and Nadya Reingand.
© 2012 Wiley-VCH Verlag GmbH & Co. KGaA. Published 2012 by Wiley-VCH Verlag GmbH & Co. KGaA.

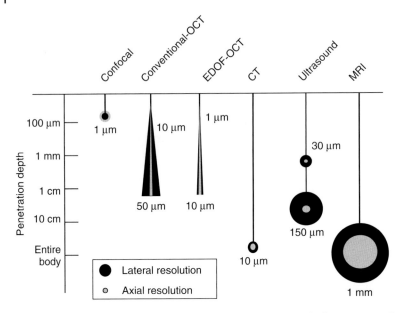

Figure 10.1 Resolution and penetration depth of 3D imaging methods. (Source: Ref. [2].)

10.2
General Considerations of 3D Imaging, Inspection, and Metrology

Any 3D imaging system yields in the first place a large amount of 3D data. This may be an irregularly spaced "point cloud" of vertices, for example, for optical surface triangulation methods, or it may be values on a regular grid. The usage we can make of 3D data is shown in Figure 10.2. We get a 3D image already by visualization of these data using standard software routines. Inspection searches for particular features, while for metrology we have to derive actual dimensional quantities from the volume data. These data can be used for comparison with a 3D computer-aided design (CAD) model or even to derive such a model in reverse engineering. In the remainder of this chapter, these issues are discussed in more detail.

10.2.1
3D Imaging

A 3D object is a volume assembly of properties. For a mathematical representation, we denote the properties by a vector dependent on all spatial coordinates (and generally of course also dependent on time) $\vec{P}(x, y, z)$. The task of 3D imaging is to produce a 3D representation $I(i, j, k)$ of one feature of a 3D object (here we consider actual volume data, not surface data).

$$\vec{P}(x, y, z) \mapsto I(i, j, k) \tag{10.1}$$

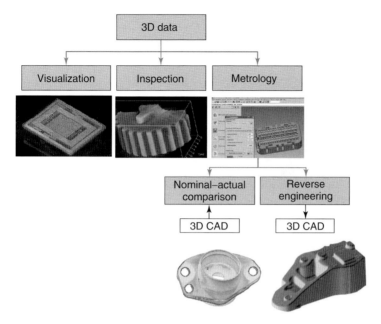

Figure 10.2 Usage of 3D data.

The result is a 3D "point cloud" of one feature, for instance, absorption, or refractive index. To obtain that, we need sufficient transparency of the object as well as sufficient 3D resolution of the imaging system. The geometry, however, might be distorted and it generally will be so. The data are usually obtained on a discrete grid of 3D volume elements, the voxels, numbered by 3D indices (i, j, k).

$I(i, j, k)$ can be visualized by standard software routines. Given the so-called "isothreshold," the "marching cubes" algorithm creates a triangulated surface approximating the isosurface of $I(i, j, k)$ at the voxel resolution [7], that is, a surface representing points of a constant value. This triangulated surface may then be used for visualization and modeling purposes. Alternatively, the visualization technique of "volume rendering" traces virtual rays through the 3D data. This enables transparency and other effects not possible when using the isosurface. As an example, Figure 10.3 shows the layer stack of a microelectromechanical system (MEMS) device measured with a 1300 nm OCT, visualized using volume rendering.

10.2.2
3D Inspection

The task of 3D inspection is to identify and visualize certain features of the object by automatic evaluation. It takes a 3D image and applies image processing routines to detect certain features. Usually this is done for quality control, for instance, the detection of delamination of layers or of faulty connections. The result is a vector

Figure 10.3 3D image of the layer stack of a MEMS device obtained with a 1300 nm OCT.

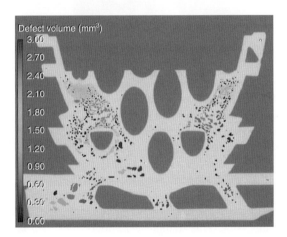

Figure 10.4 Automatic cavity detection in injection-molded polymer part.

of features $\vec{F}_{\text{img}}$. Prerequisites are discernible features and at least a simple object model. For inspection purposes, a high contrast must be achieved, while some geometrical distortion due to the imaging process is usually acceptable.

$$\begin{array}{ccc} \vec{P}(x,y,z) & \mapsto & I(i,j,k) \\ \downarrow & & \downarrow \\ \vec{F}_{\text{act}} & \leftrightarrow & \vec{F}_{\text{img}} \end{array} \qquad (10.2)$$

Applications comprise electronics [8], concrete [9], alloys [10], and polymer parts. As an example, Figure 10.4 shows the X-ray CT image of an injection-molded polymer part, where small cavities have been automatically detected and classified according to their size.

10.2.3
3D Metrology

The task of 3D metrology is to get an accurate 3D replica of the object geometry. The result is a set of dimensions $\vec{G}_{img}$, that is, for instance, the position and diameter of boreholes or the size of subparts.

$$\begin{array}{ccc} \vec{P}(x,y,z) & \mapsto & I(i,j,k) \\ \downarrow & & \downarrow \\ \vec{G}_{act}(x,y,z) & \leftrightarrow & \vec{G}_{img}(x,y,z) \end{array} \qquad (10.3)$$

The prerequisites for 3D metrology are a high reproducibility of the measurement system and a physical model for the relation between the measured quantity and the geometrical features (for instance, interfaces or edges). In general, in all quantitative measurement processes traceability is a must, but in practice often neglected. It requires a proper calibration standard as was done in [11] where a grating-like test object was used for calibration and accuracy assessment.

10.3
Industrial 3D Metrology Based on X-ray Computed Tomography

For a number of years, X-ray CT has been an established method for the non-destructive testing (NDT) of industrial parts. Such testing may be required for a production process that involves the possibility of creating inner defects, such as cavities in cast metals, for example, aluminum; or it can be used to check parts or assemblies with complex internal functionality. Depending on the circumstances, two-dimensional X-ray inspection may be sufficient, or complete three-dimensional object information may be required. For some applications, only a few samples of a casting process have to be tested subsequent to some change of the casting process parameters, and those samples can be investigated relatively leisurely in a dedicated measurement room. On the other hand, it might be necessary to test all produced parts "in-line," that is, as an integral part of the production line, and at the rate of production.

If complete 3D information is required, X-ray CT can be used to obtain a 3D image of the object under investigation within a time span ranging from a few seconds to several hours. Currently, the spatial resolution of this 3D image is limited for most objects by the detector resolution and lies between 1/1000 and 1/2000 of the object size; for small objects, the resolution may additionally be limited by the minimum effective source size of the X-ray source. Depending on the size and material of the object, the 3D image may show "artifacts," that is, apparent structures that have no counterpart in the real object. The 3D image is inspected manually or automatically for defects and errors arising from faulty processing or assembly.

10.3.1
X-Ray Cone-Beam Computed Tomography

10.3.1.1 Two-Dimensional Image Formation

X-ray cone-beam CT produces a 3D image of an object, based on a large number of 2D X-ray images of the object acquired under different viewing angles. This section describes the generation of these 2D images, consisting of the subsequent steps of X-ray generation, interaction with the object, and detection.

An X-ray tube is used to generate the X-rays. Inside the evacuated tube, a thin tungsten filament bent into a tip shape serves as a thermionic emitter of electrons. In comparison, a lanthanum hexaboride (LaB_6) crystal allows for a smaller electron source, albeit with a reduced emission current. The emitted electrons are accelerated, and the resulting electron beam is focused, by electrostatic and electromagnetic fields. The focused electron beam impinges on a metal target (the anode, usually also of tungsten), and the interaction of the electrons with the metal atoms creates both characteristic radiation, which is determined by the electron energy levels of the metal atoms and thus occurs only at specific energies, and a continuous spectrum of Bremsstrahlung. The maximum energy of the X-ray photons is given by the kinetic energy of the accelerated electrons and hence by the accelerating voltage of the X-ray tube; for material inspection, an acceleration voltage between 50 and 250 kV is used depending on the size and material of the object. For large objects composed of materials with high atomic number, linear accelerators may be used to obtain a kinetic energy of the electrons in excess of 1 MeV and corresponding photon energies of the X-rays.

Remark: While the CT community thinks in terms of energy E, the optics community works in terms of the wavelength λ. Both are simply connected by the photon energy

$$E = \frac{hc_0}{\lambda} \tag{10.4}$$

where h denotes the Planck's constant and c is the speed of light in vacuum. In this chapter, we use the energy notation.

The X-rays of energy spectrum $W(E)$ emitted from the tube anode leave by way of a vacuum window and impinge on the object. In general, the object is an inhomogeneous material with a spatial and wavelength-dependent absorption coefficient $\beta(E, x, y, z)$, which is just the imaginary part of the refractive index, that is

$$n = 1 - \delta + i\beta \tag{10.5}$$

with

$$\beta = \frac{n_a r_e \lambda^2}{2\pi} f_2^0(E) \tag{10.6}$$

with the electron radius r_e, the molecular density n_a (number of molecules per cubic centimeter), and the energy-dependent scattering coefficient f_2.

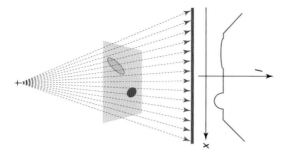

Figure 10.5 Schematic representation of the 2D image formation in X-ray cone-beam CT.

The X-ray image is a two-dimensional projection (Figure 10.5) resulting from the line integral of the local absorption integrated over all energies weighted with the spectral intensity

$$I = \int_{\text{spectrum}} W(E) \exp\left\{-4\pi \frac{E}{hc} \int_{d(x,y)} \beta(E,x,y,z) \mathrm{d}z\right\} \mathrm{d}E \qquad (10.7)$$

10.3.1.2 Imaging Geometries and 3D Reconstruction

In medical X-ray CT, the patient is translated along the body axis, while X-ray source and detector are moved on a circular trajectory around him. In industrial CT, detector and source are fixed while the object is rotated through 360° (circular source trajectory) and X-ray perspective projection images are acquired for 200–2000 angular positions. From these, the 3D reconstruction is done by cone-beam reconstruction. One of the most used methods for that is the so-called Feldkamp–Davis–Kress (FDK) method [1], described in 1984 by Feldkamp *et al.* [12]. It is a variant of the filtered backprojection algorithm. As the filtered backprojection it is based on the Fourier slice theorem, but additionally it takes the cone-beam geometry into account. It consists of three basic steps, which shall be only schematically stated here. A detailed description is found, for instance, in [1] and [13].

1) Convert X-ray intensity to integral absorption. Use negative logarithm (Lambert–Beer law) or more refined physical modeling.
2) Weigh and high-pass filter the absorption images to represent the cone-shaped geometry.
3) Backproject and accumulate by a filtered backprojection method modified to represent the cone-shape geometry.

Backprojection and accumulation dominate in practice the runtime (>95%). Because of that a suitable performance metric is "updates per second": for instance for 800 projections, a voxel needs 800 updates during reconstruction. Reconstructing a 1024^3 voxel volume from 1024 2D images requires 1.1×10^{12} updates.

10.3.2
X-Ray CT-Based Dimensional Metrology

10.3.2.1 Why CT Metrology?

For precision-machined parts, dimensional metrology is an important part of the process chain, ensuring compliance with ever-decreasing manufacturing tolerances. Competing with tactile and optical coordinate measurement machines (CMMs), X-ray CT offers several unique advantages making it the measurement method of choice for some applications: the 3D image permits measuring all outer and inner object features without exertion of a possibly deforming measuring force.

By comparing the CT-generated 3D image with the CAD model of the part, a nominal–actual value comparison can be created, that is, a color-coded representation of all surface deviations that permits an intuitive interpretation of the measurement results. The measurement precision of X-ray CT strongly depends on the size and material of the object, but is often much better than the spatial resolution of the 3D image, especially if many CT-measured surface points can be combined to determine a single measurement result, as in the case of extended geometrical elements.

10.3.2.2 Surface Extraction from 3D Absorption Data

X-ray CT yields the 3D distribution of absorption in the form of a point cloud, that is, unconnected voxel values. However, in dimensional metrology we deal with surfaces, of which the correct shape has to be determined (cf. Figure 10.6). So the main challenge is how to find the material–air interfaces with high accuracy, that is, how to extract a high-precision 2D surface from 3D voxel data.

A simple solution would be to apply a global threshold, as is done for edge determination from microscopical images. However, a global threshold is not good enough for high-precision measurements in the presence of artifacts. The solution is to use voxel gradients to extract interfaces with high subvoxel accuracy. Because the derivative enhances noise, suitable noise suppression is critical to obtaining a usable surface representation.

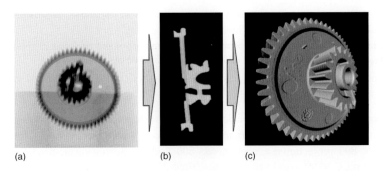

(a)   (b)   (c)

Figure 10.6 Transition from an X-ray projection (a), to 3D image (b), to surface data (c).

10.3.3
Device Imperfections and Artifacts

The following section deals with some common properties of a cone-beam X-ray CT device and how they influence the achievable measurement accuracy. In particular, we have to deal with the error sources listed in Table 10.1.

10.3.3.1 Geometrical Alignment of the Components
From a geometrical point of view, a cone-beam X-ray CT device may be said to consist of three simple geometrical objects: the X-ray source is effectively a point; the manipulator is reduced to a line, namely, the rotation axis of the rotary table, since at least for an FDK source trajectory, all linear manipulator axes will be stationary during acquisition of the projection set; and for the detector, the locus of photon detection is a plane. The original reconstruction algorithms stipulate an ideal geometry. For the FDK trajectory, the object rotation axis must be parallel to the detector plane and to the pixel columns of the detector. At the same time, the line projecting the source point onto the detector must intersect the rotation axis at a right angle (Figure 10.7).

10.3.3.2 Beam Hardening
The origin of beam hardening is that X-ray spectra are polychromatic and that the transmission of matter is energy dependent. As a result, the spectral composition of an X-ray depends on the traversed distance. This induces a problem because the mean absorption coefficient changes when the spectral composition of the X-ray beam changes (Figure 10.8). A detailed description and solution approach based on segmentation and reprojection can be found in [14].

Table 10.1 Error sources of CT metrology and their correction.

Property	Correction/calibration method
Nonideal manipulator (guide ways, bearings, drives, encoders, assembly)	"Computer-aided accuracy" (CAA): high-precision interferometric measurement of the reproducible manipulator errors, and correction in the control system
Geometric misalignment/unknown position of X-ray source, rotation axis, and detector	Precise alignment/measurement using special calibration tools (based on camera calibration)
(Thermal) movement of X-ray source (form errors and image scale error)	Source position monitoring and compensation
Beam hardening	Automatic determination of map "intensity $\rightarrow$ absorption" + gradient-based surface extraction
Reconstruction noise	Gradient-based surface extraction in combination with noise suppression

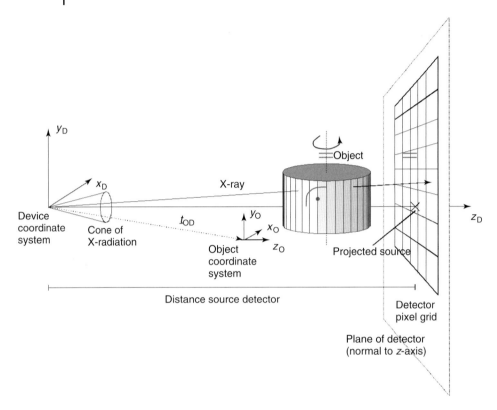

Figure 10.7 Component alignment important for high-accuracy CT metrology.

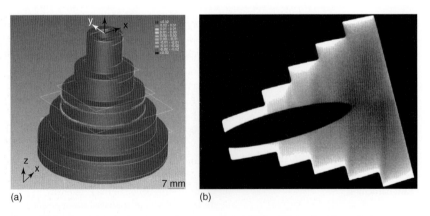

Figure 10.8 Beam hardening in an aluminum step cylinder (a): a slice through the reconstructed object (b) apparently shows inhomogeneous absorption although the material is homogeneous.

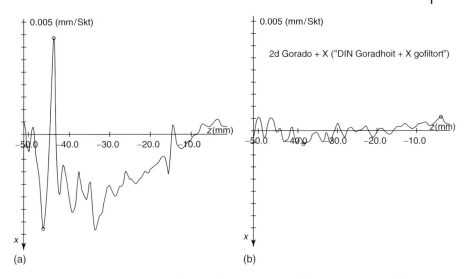

Figure 10.9 Beam hardening in an Al step cylinder: accuracy of dimensions. (a) Global threshold and (b) gradient-based surface determination. (Source: Ref. [15].)

It turns out that the gradient-based surface determination is significantly less sensitive to beam hardening than the global threshold. For a surface line extracted along the inner bore of the Al test cylinder shown in Figure 10.8, using a global threshold results in a surface error of 40 μm, while the error is reduced to 5 μm for the gradient-based surface (Figure 10.9).

Beam hardening correction also often improves the nominal–actual comparison.

10.3.4
Standards for X-Ray CT-Based Dimensional Metrology

The emergence of a new method for dimensional metrology, such as X-ray CT, confronts potential users with the problem of assessing the advantages and disadvantages of the new method when compared to existing approaches; they also want to compare the strengths and weaknesses of products offered by different manufacturers. Industry standards are therefore required both by device manufacturers and their potential customers. For coordinate measurement devices with CT sensors, the Association of German Engineers (VDI) has published guidelines VDI/VDE 2630 and VDI/VDE 2617 that explain how to apply the standard DIN EN ISO 10360 in order to perform acceptance testing and monitoring of length measurement and scanning errors. The guidelines are derived from industry-accepted procedures for tactile-probe CMMs and take into account the special conditions governing X-ray CT.

10.3.4.1 Length Measurement Error and Scanning Errors of Form and Size
According to the guideline, the main characteristics of an X-ray CT CMM are the length measurement error E and the scanning errors PF (form) and PS (size).

Pursuant to DIN EN ISO 10360, the length measurement error E describes the three-dimensional error behavior of the overall system, consisting of coordinate measuring device and sensor, in the entire measurement volume, while the scanning error describes the three-dimensional error behavior within a very small measurement volume. For an X-ray CT CMM, the scanning error is determined using precision-ground spheres with negligible surface roughness and form error, such as the ruby balls used for tactile-probe CMMs. The CT measurement of such a sphere produces a 3D image, which, in turn, is used to generate a number of individual surface points on the sphere (at least 25). To these surface points, an ideal sphere is fitted by minimizing the squared radial error between the fit sphere and the surface points. The form scanning error PF is then given by the range of the radial deviation of the surface points from the fit sphere (Figure 10.10a), the size scanning error PS by the difference between the diameter D_a of the fit sphere and the calibrated (true) sphere diameter D_r (Figure 10.10b).

Because the measurement precision of an X-ray CT CMM is strongly dependent on the object material and geometry, the length measurement error E may be determined in one of two ways. It is possible to measure gauge blocks such as are used for determining E for tactile-probe CMMs; however, because of the long X-ray penetration lengths inherent to the block geometry, the 3D images will likely exhibit beam hardening artifacts, so that measurement precision will be degraded.

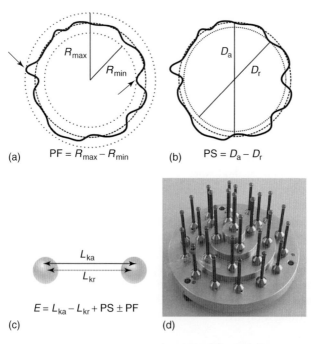

Figure 10.10 The scanning errors for (a) form and (b) size of a CT-measured sphere are combined with the center-to-center distance error of a sphere pair to form the length measurement error E (c); (d) many such sphere pairs may be measured simultaneously.

It is therefore permitted to choose a length gauge embodiment more suitable for X-ray imaging, namely, a combination of two spheres for which the center-to-center distance is determined and compared with the calibrated value. In this way, X-ray penetration lengths can be limited to twice the sphere diameter. However, since any symmetrical error in the surface determination (with respect to the sphere center) does not affect the center-to-center distance of the two spheres but should be represented in the length measurement error E, the scanning errors PS and PF are added to the center-to-center distance error in order to form the final value for E (Figure 10.10c).

According to DIN EN ISO 10360-2, five test lengths should be measured in seven spatial directions, with the smallest length not exceeding 30 mm and the largest covering at least 66% of the measurement volume diagonal. By suitable arrangement of a number of spheres, the required test lengths can be realized in a single test specimen and measured during a single CT measurement (Figure 10.10d). Usually, the manufacturer specifies a maximum permissible error (MPE_E) in the form of a length-dependent quantity such as $\pm(A + L/K)$, where A is a constant term and L is the measurement length scaled by a constant factor, thus permitting greater measurement errors for larger measurement lengths.

10.3.4.2 Dependence on Material and Geometry

In order to quantify the influence of the material and geometry of an object on the achievable measurement precision, the guideline also describes a test specimen in the form of a multitiered cylinder with a common central bore (Figure 10.8). For a given material, this specimen provides a broad range of X-ray penetration depths in a single object, and the influence of increasing penetration depth on the CT-determined cylinder radius and form can be observed. In this case, the specified characteristics are GR, the error of a CT-measured internal or external radius of one of the several tiers, GF, the corresponding form scanning error, and GG, the straightness of the axis of the central bore. These characteristics depend on the material of the test specimen and should therefore be specified for a number of relevant materials such as polymers, aluminum, and steel.

10.4 Conclusions

There is a significant difference between 3D imaging and 3D metrology. We have discussed this for CT: nonideal manipulators (guideways, bearings, drives, encoders, assembly) are accounted for by a precise one-time measurement of the reproducible errors. The positions and the alignment of X-ray source, rotation axis, and detector plane influence image artifacts and image scale error and must therefore be determined with special calibration tools. The thermal movement of the X-ray source requires source position monitoring and compensation. The effects of beam hardening and reconstruction noise are taken into account by a gradient-based surface extraction. Significant effort and a careful quantitative assessment of these error sources are necessary to transform an X-ray CT into a metrology tool.

VDI guideline 2630 describes the application of the standard DIN EN ISO 10360 to X-ray CT-based dimensional metrology. By looking at the errors involved in measuring form, size, and length, it provides an impartial way of assessing measurement performance. Adoption of this procedure by CMM manufacturers and customers alike will improve the acceptance of this new metrology technique.

References

1. Kak, A.C. and Slaney, M. (2001) *Principles of Computerized Tomographic Imaging*, IEEE Press 1988 and Society of Industrial and Applied Mathematics 2001.
2. Smolka, G. (2008) *Optical Coherence Tomograph: Technology, Markets, and Applications* 2008–2012, PennWell Corporation.
3. Ding, Z., Ren, H., Zhao, Y., Nelson, J.S., and Chen, Z. (2002) High-resolution optical coherence tomography over a large depth range with an axicon lens. *Opt. Lett.*, **27** (4), 243–245.
4. Leitgeb, R.A., Villiger, M., Bachmann, A.H., Steinmann, L., and Lasser, T. (2006) Extended focus depth for Fourier domain optical coherence microscopy. *Opt. Lett.*, **31** (16), 2450–2452.
5. Wilson, T. (1990) *Confocal Microscopy*, Academic Press.
6. Westbrook, C.T., Kaut Roth, and Talbot, J. (2011) *MRI in Practice*, 4th edn Wiley-Blackwell.
7. Lorensen, W.E. and Cline, H.E. (1987) Marching cubes: A high resolution 3D surface construction algorithm. *Comput. Graph.*, **21** (4), 163–169.
8. Teramoto, A., Murakoshi, T., Tsuzaka, M., and Fujita, H. (2007) Automated solder inspection technique for BGA-mounted substrates by means of oblique computed tomography. *IEEE Trans. Electron. Packaging Manuf.*, **30** (4), 285–292.
9. Suzuki, T., Ogata, H., Takada, R., Aoki, M., and Ohtsu, M. (2010) Use of acoustic emission and X-ray computed tomography for damage evaluation of freeze-thawed concrete. *Constr. Build. Mater.*, **24** (12), 2347–2352.
10. Suparta, G.B. and Handayani, N. (2009) Application of computed tomography to quality inspection of brass alloy. *Proc. SPIE – Int. Soc. Opt. Eng.*, **7522**, 75224C.
11. Kiekens, K., Welkenhuyzen, F., Tan, Y., Bleys, P., Voet, A., Kruth, J.-P., and Dewulf, W. (2011) A test object with parallel grooves for calibration and accuracy assessment of industrial computed tomography (CT) metrology. *Meas. Sci. Technol.*, **22** (11), 115502.
12. Feldkamp, L.A., Davis, L.C., and Kress, J.W. (1984) *J. Opt. Soc. Am. A*, **1** (6), Practical cone-beam algorithm. 612–619.
13. Turbell, H. (2001) Cone-beam reconstruction using filtered backprojection. PhD thesis. Linkopings Universitet, Linköping, February 2001.
14. Krumm, M., Kasperl, S., and Franz, M. (2008) Referenceless beam hardening correction in 3D computed tomography images of multi-material objects. 17th World Conference on Nondestructive Testing, Shanghai, China, October 25–28, 2008.
15. Weiss, D. and Lettenbauer, H. (2008) Optimized CT metrology through adaptive image processing techniques, XII. International Colloquium on Surfaces, Chemnitz, Germany, January 28–29, 2008.

11
Coherence Holography: Principle and Applications
Mitsuo Takeda, Wei Wang, and Dinesh N. Naik

11.1
Introduction

Coherence plays a fundamental role in holography and interferometry, and control of coherence is essential in an illumination system for microscopy and lithography. This chapter is intended to give an introduction to the concept, principle, and applications of a new technique of unconventional holography, called coherence holography [1], which was developed recently for synthesis and analysis of a spatial coherence function [2]. The basic principle of the new technique was inspired by the fact that the coherence function $\Gamma(\mathbf{r}_1, t_1; \mathbf{r}_2, t_2)$ and the optical field $u(\mathbf{r}, t)$ bear a formal analogy that both obey the same wave equation [3, 4]

$$\begin{pmatrix} \text{Coherence} \\ \text{function} \end{pmatrix} \quad \nabla^2_{i=1,2}\Gamma(\mathbf{r}_1, t_1; \mathbf{r}_2, t_2) = \frac{1}{c^2} \frac{\partial^2 \Gamma(\mathbf{r}_1, t_1; \mathbf{r}_2, t_2)}{\partial t^2_{i=1,2}}$$

$$\begin{pmatrix} \text{Optical} \\ \text{field} \end{pmatrix} \quad \nabla^2 u(\mathbf{r}, t) = \frac{1}{c^2} \frac{\partial^2 u(\mathbf{r}, t)}{\partial t^2} \quad (11.1)$$

where $\mathbf{r}$ is a position vector and t is the time. This formal mathematical analogy suggests that the basic principle of holography already established for the wave of the optical field can also be applied to the wave of a coherence function [5]. Just as a conventional computer-generated hologram (CGH) can create an arbitrary 3D optical field, a computer-generated coherence hologram (CGCH) can synthesize an optical field with a desired 3D coherence function. Furthermore, the time–space symmetry in the wave equation indicates that the role of a temporal coherence function can be replaced by a spatial coherence function [6]. This leads to a new concept of dispersion-free optical coherence tomography and profilometry [7], which makes use of a spatial (rather than temporal) coherence function of quasi-monochromatic light synthesized by a CGCH.

11.2
Principle of Coherence Holography

11.2.1
Reciprocity in Spatial Coherence and Hologram Recording

The principle of coherence holography can be explained on the basis of reciprocity in spatial coherence and holographic recording [2]. Referring to Figure 11.1, let us first consider recording of a conventional hologram. We note an arbitrary point P at r_P on a 3D object, which is recorded with a reference beam from a point source R at r_R. Suppose a point S at r_S on the hologram is located on one of the bright fringes resulting from constructive interference between the object and reference beams from points P and R. Then, the optical path difference (OPD) between the two beams becomes integer multiples of the wavelength λ

$$\mathrm{OPD}(r_P, r_R; r_S) = |r_P - r_S| - |r_R - r_S| = n\lambda \qquad (11.2)$$

Next, we illuminate the hologram with spatially incoherent quasi-monochromatic light. This generates an extended incoherent source with its irradiance distribution proportional to the intensity distribution of the hologram $I(r_S)$. Let us now consider a reciprocal process in which light emitted from point S on the source reaches points P and R to create fields $u(r_P)$ and $u(r_R)$, respectively. From reciprocity of wave propagation and because of Eq. (11.1), $u(r_P)$ and $u(r_R)$ are always in phase with the phase difference $2n\pi$, irrespective of the initial phase and the position of the point source S, as far as S is located on one of the bright fringes; this condition is automatically satisfied for all points on the hologram because no light is emitted from the location of the dark fringes. Thus, the incoherently illuminated hologram creates an optical field, which is in phase and has high coherence between these particular pair of points at P and R. If we fix one point as a reference at R and move the other point P over the 3D space with r_P as a variable, we can reconstruct the object as a 3D distribution of the mutual intensity $J(r_P, r_R) = \langle u(r_P) u^*(r_R) \rangle$, which can be detected with a suitable interferometer. In Figure 11.1, a Young's interferometer using an optical fiber is shown just for the conceptually simplest example.

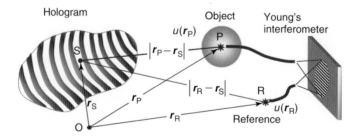

Figure 11.1 Schematic geometry for holographic recording and reconstruction of a coherence hologram.

11.2.2
Similarity between the Diffraction Integral and van Cittert–Zernike Theorem

Another explanation of the principle can be made on the basis of a formal analogy [6] between the diffraction integral and the van Cittert–Zernike theorem [3, 4]. Let us consider a conventional reconstruction process in which the hologram is illuminated from backward with a coherent spherical wave $\exp(-ik|\mathbf{r}_S - \mathbf{r}_R|)/|\mathbf{r}_S - \mathbf{r}_R|$ converging into the reference source point R. The reconstructed optical field, which forms a real image of the 3D object at point P, is given by the diffraction integral for the Fresnel–Huygens secondary waves

$$u(\mathbf{r}_P, \mathbf{r}_R) = \iint I(\mathbf{r}_S) \frac{\exp[ik(|\mathbf{r}_P - \mathbf{r}_S| - |\mathbf{r}_R - \mathbf{r}_S|)]}{|\mathbf{r}_P - \mathbf{r}_S| \cdot |\mathbf{r}_R - \mathbf{r}_S|} d\mathbf{r}_S \quad (11.3)$$

where $d\mathbf{r}_S$ denotes an element of the area at the position $\mathbf{r}_S$. It should be noted that Eq. (11.3) has exactly the same form as the formula for mutual intensity given by the van Cittert–Zernike theorem

$$J(\mathbf{r}_P, \mathbf{r}_R) = \iint I(\mathbf{r}_S) \frac{\exp[ik(|\mathbf{r}_P - \mathbf{r}_S| - |\mathbf{r}_R - \mathbf{r}_S|)]}{|\mathbf{r}_P - \mathbf{r}_S| \cdot |\mathbf{r}_R - \mathbf{r}_S|} d\mathbf{r}_S \quad (11.4)$$

except the difference that $I(\mathbf{r}_S)$ in Eq. (11.3) is the amplitude transmittance of the hologram, whereas $I(\mathbf{r}_S)$ in Eq. (11.4) is the intensity distribution of the spatially incoherent source. The implication of this formal analogy is that if a hologram, having intensity transmittance $I(\mathbf{r}_S)$ proportional to the recorded intensity, is illuminated with spatially incoherent light, an optical field will be generated, for which the mutual intensity between observation point P and reference point R is equal to the optical field that would be reconstructed if the hologram with the same amplitude transmittance were illuminated with a phase-conjugated version of the reference beam [1]. Unlike conventional holography, the reconstructed coherence image is not directly observable. It can be visualized only as the contrast and the phase of an interference fringe pattern by using an appropriate interferometer.

11.3
Gabor-Type Coherence Holography Using a Fizeau Interferometer

Detection of the coherence image by scanning the probe point of the Young's interferometer, as shown in Figure 11.1, is impractical. We used a simple optical geometry shown in Figure 11.2, which is in effect a Fizeau interferometer but we realized it with a Michelson interferometer in our initial experiment [1]. Light from an incoherently illuminated coherence hologram is guided into the interferometer to form interference fringes on the plane of the beam splitter as the result of interference between the fields at point R and point P. The fringe contrast reflects the mutual intensity between the two points given by the van Cittert–Zernike theorem in Eq. (11.4). Using the position vectors defined in Figure 11.2, the mutual

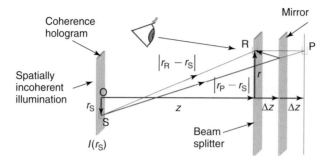

Figure 11.2 Schematic geometry for reconstruction of a coherence hologram.

intensity is now approximated by

$$J(\mathbf{r}_P, \mathbf{r}_R) \propto \iint I(\mathbf{r}_S) \exp[ik(|\mathbf{r}_P - \mathbf{r}_S| - |\mathbf{r}_R - \mathbf{r}_S|)] d\mathbf{r}_S$$

$$= J(\mathbf{r}, \Delta z) = \iint I(\mathbf{r}_S) \exp\left[ik(|z + 2\Delta z + \mathbf{r} - \mathbf{r}_S| - |z + \mathbf{r} - \mathbf{r}_S|)\right] d\mathbf{r}_S \quad (11.5)$$

$$\approx \exp\left[i\alpha(\Delta z)\right] \iint I(\mathbf{r}_S) \exp\left[-ik\left(\frac{\Delta z}{z}\right) \times \frac{|\mathbf{r} - \mathbf{r}_S|^2}{z}\right] d\mathbf{r}_S$$

where $\alpha(\Delta z) = k(2\Delta z)\left[1 - 2(\Delta z/z)^2\right]$ and the distance z between the hologram and the observation plane is assumed to be much larger than other parameters. It should be noted that the mutual intensity is given by the Fresnel transform of the incoherently illuminated hologram. If we record a Fresnel hologram with coherent light for an object at distance $\bar{z} = z^2/2\Delta z$ from the hologram and illuminate the hologram with spatially incoherent light from behind, we will observe on the beam splitter a set of interference fringe patterns whose fringe contrast and phase represent, respectively, the field amplitude and the phase of the original object recorded with coherent light. Just as a CGH can create a three-dimensional image of a nonexisting object, a CGCH can create an optical field with a desired three-dimensional distribution of spatial coherence function. This CGCH gives a new possibility of optical tomography and profilometry based on a synthesized spatial coherence function [7] and serves as a generator of coherence vortices [8]. An example of a Gabor-type (on-axis) CGCH for an object, the letter H, is shown in Figure 11.3a. Figure 11.3b shows a reconstructed coherence image, in which the letter H is displayed by the region of high-contrast fringes representing the designed high coherence area [1]. Figure 11.3c,d shows a coherence hologram for generating a coherence vortex and an interferogram of the coherence vortex, respectively [8]. While a conventional optical vortex has a singularity in the phase of the optical field at the location where the optical field vanishes, the coherence vortex has a singularity in the phase of the coherence function at the location where the coherence function becomes zero, which is indicated by the vanishing fringe contrast in Figure 11.3d.

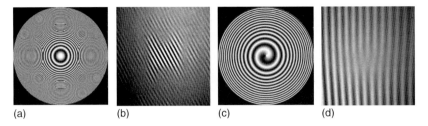

Figure 11.3 (a) Computer-generated Gabor-type coherence hologram for a letter, H. (b) Coherence image with the high-coherence region representing the letter H. (c) Computer-generated Gabor-type coherence hologram for generating a coherence vortex. (d) Interferogram of the coherence vortex with phase singularity in the coherence function.

11.4
Leith-Type Coherence Holography Using a Sagnac Interferometer

Although Gabor-type coherence holography has the advantage that it can be implemented with a relatively low-resolution spatial light modulator (SLM), it inherits the same drawbacks associated with a classic Gabor hologram in conventional holography. In this section, we introduce Leith-type coherence holography using a modified Sagnac radial shearing interferometer for the reconstruction of an off-axis coherence hologram [9]. Figure 11.4 shows the regimes of Gabor-type and Leith-type coherence holograms from the point of view of location of the object and the reference points. For a point object at P, the Gabor-type (on-axis) hologram is described by a Fresnel zone plate, which is suitable for controlling longitudinal coherence because $r_P - r_R$ is along the longitudinal (depth) direction, whereas

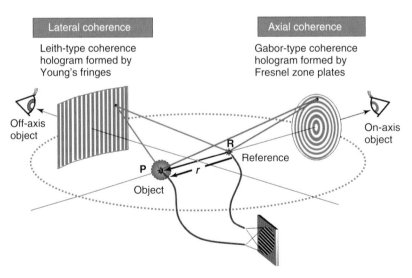

Figure 11.4 Recording geometry for a Gabor-type coherence hologram and a Leith-type coherence hologram.

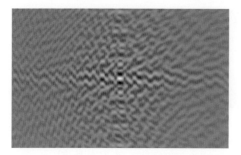

Figure 11.5 Computer-generated Leith-type coherence hologram.

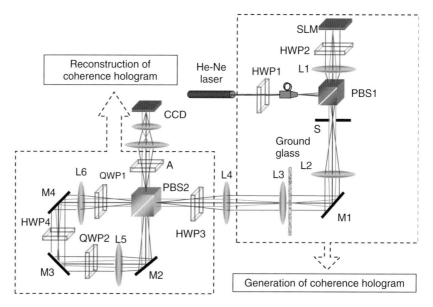

Figure 11.6 Experimental setup for generation and reconstruction of a Leith-type coherence hologram. (Please find a color version of this figure on the color plates.)

the Leith-type (off-axis) hologram is described by Young's parallel fringes, which is suitable for controlling lateral coherence because $r_P - r_R$ is along the lateral direction.

The coherence hologram used in the experiment is a computer-generated Fourier transform hologram of an off-axis binary object representing a Greek letter, Ψ. Figure 11.5 shows the CGCH for the Greek letter Ψ. A schematic illustration of the experimental setup for the reconstruction of the coherence hologram is shown in Figure 11.6 [9]. The experimental setup is functionally divided into two parts. The first part is a CGCH implemented by an SLM assembly to modulate the laser light intensity, followed by a rotating ground glass to destroy spatial coherence. The second part is a radial shearing interferometer to visualize a coherence image reconstructed from the coherence hologram. Linearly polarized light from a He-Ne laser passes through a half-wave plate1 (HWP1), which rotates the orientation of polarization to control the intensity of the beam illuminating an SLM through reflection from a polarized beam splitter1 (PBS1). A 5× microscope objective lens O and a lens L1 together serve as a beam expanding collimator. An HWP2 rotates the polarization of the collimated laser beam reflected from PBS1 to obtain maximum modulation efficiency at the SLM. The reflection-type liquid crystal on silicon-spatial light modulator (LCOS-SLM) (HoloEye Model LC-R1080) placed at the focal plane of lens L1 rotates the polarization of the incident collimated laser light according to the gray level of a CGH displayed on the SLM. The PBS1 functions as an analyzer for the light reflected from the SLM, transforming the localized polarization rotation introduced by the SLM into an intensity modulation. The effect of the SLM-induced discrete pixel structure in the modulated beam is eliminated by spatially filtering the higher order diffractions with a small circular aperture S placed in the rear focal plane of L1; subsequently, the hologram is imaged back onto a rotating ground glass by a relay lens L2. Thus, the hologram imaged onto the rotating ground glass acts, in effect, as a spatially incoherent but temporally coherent light source, which has an intensity distribution proportional to the gray level of the hologram. However, the actual ground glass does not behave as a perfect scatterer, and the light scattered by the rotating ground glass retains the directional characteristic of the incoming beam. A field lens L3 placed immediately behind the rotating ground glass directs the main lobe of the scattered light toward the optical axis of the subsequent optical setup, thereby enhancing the quality of the image at the interferometer output [10]. The field distribution of the incoherently illuminated hologram is Fourier transformed by lens L4 and introduced into the interferometer through an HWP3. PBS2 splits the incoming beam into two counterpropagating beams. The telescopic system with magnification $\alpha = 1.1$, formed by lenses L5 (focal length 220 mm) and L6 (focal length 200 mm), gives a radial shear between the counterpropagating beams as they travel through an interferometer before they are brought back together and detected by CCD. At any location r on the image plane, the interference is due to the superposition of fields at the locations $r\alpha$ and r/α of the original beam. Thus, at point r on the image plane, we have a cross correlation of the fields between two points separated by δr proportional to r scaled by the factor $(\alpha - 1/\alpha)$. The resulting interference gives a 2D correlation map that reconstructs the image as a coherence function represented by the fringe contrast. To detect the reconstructed coherence image represented by the fringe visibility, we need to introduce a phase shift into the common-path Sagnac interferometer. Because of its common path nature, the Sagnac interferometer is insensitive to the conventional PZT-based mechanical mirror movement. We therefore introduced the required phase shift by means of a geometric phase shifter [11]. Inside the interferometer we used two quarter wave plates (QWP1 and QWP2) to turn the linearly polarized light into a circularly polarized light and back to the linearly polarized light of orthogonal polarization with respect to the initial state of polarization. By rotating an HWP4, placed between QWP1 and QWP2, we introduced a geometric phase into the counterpropagating beams, which is used to find the fringe visibility. With an

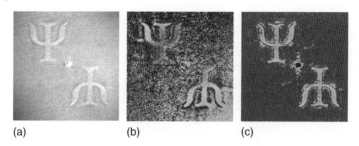

(a) (b) (c)

Figure 11.7 Reconstructed images of Leith-type coherence holography. (a) Raw intensity image resulted from shearing interference, (b) phase image, and (c) contrast image jointly representing the complex coherence function. (Please find a color version of this figure on the color plates.)

analyzer A with its axis kept at $45°$ to the orientation of the polarization of the two beams, interference between the two beams was achieved. Figure 11.7a shows the interference image captured by a CCD camera. Figure 11.7b,c shows the phase and the contrast of the interference fringes, respectively, which together represent the complex coherence function.

11.5
Phase-Shift Coherence Holography

In the previous section, a modified Sagnac radial shearing interferometer with a geometric phase shifter was employed together with phase-shift fringe analysis. The unique feature of the geometric phase shifter is that it can introduce phase shifts into the Sagnac common path interferometer, which is not possible by a conventional mirror shift with a piezoelectric transducer. However, the geometric phase shifter has a disadvantage. It introduces the phase shifts by mechanically rotating a wave plate, which makes the system complex and limits the speed of coherence measurement. As an alternative to the geometric phase shifter, a reconstruction scheme for coherence holography using computer-generated phase-shift coherence holograms has been proposed, which can reconstruct an object encoded into the spatial coherence function directly by using a set of incoherently illuminated CGHs with numerically introduced phase shifts [12]. Figure 11.8(a–d) shows four of the phase-shifted coherence holograms for a *smiley* as an object, each of which is phase shifted by an amount of $\pi/2$. The phase-shifted holograms are displayed on the SLM sequentially one after another, and the corresponding interference images are captured by the CCD. Figure 11.9(a–d) shows the phase-shifted interference images captured by a CCD camera, which correspond to the holograms shown in Figure 11.8. From these phase-shifted interference images, the fringe contrast and phase are obtained, as shown in Figure 11.10a,b, respectively, which jointly represent a complex coherence function. Note that an unwanted zeroth order correlation image is not observed. This is another advantage of this phase-shift holography.

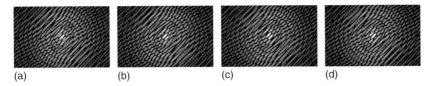

Figure 11.8 Phase-shifted coherence holograms. The increment of the phase shift is $\pi/2$.

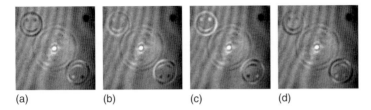

Figure 11.9 (a–d) Interference images corresponding to the phase-shifted holograms shown in Figure 11.8.

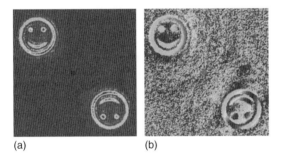

Figure 11.10 Reconstructed images show (a) the fringe contrast and (b) the fringe phase, which jointly represent the complex coherence function. (Please find a color version of this figure on the color plates.)

11.6
Real-Time Coherence Holography

In this section, we introduce yet another type of coherence holography, called real-time coherence holography, which is capable of real-time recording and reconstruction [13]. In real-time coherence holography, no use is made of devices, such as CCD and SLM, for recording and display of a hologram. Instead, a holographic fringe pattern due to the interference between object and reference beams is created directly on a rotating ground glass. The spatial coherence of the interfering field is destroyed completely by the rotating ground glass so that the hologram is represented in real time by the irradiance distribution of a spatially incoherent extended source. In effect, this extended source serves as an incoherently illuminated hologram that reconstructs the recorded object wave as a spatial coherence function. In this experiment, we adopted the geometry of a lensless Fourier transform hologram as shown in Figure 11.11, in which three letters U, E, and C serve as an object and

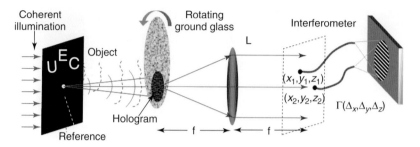

Figure 11.11 Lensless Fourier transform geometry for real-time coherence holography.

a point source as a reference. The field distribution of the incoherently illuminated hologram is Fourier transformed by lens L and introduced into an interferometer, which is a Sagnac interferometer although illustrated schematically as a fiber interferometer in Figure 11.11; this is just for simplicity. Instead of the phase-shift mechanism adopted in the previous sections, a spatial carrier frequency is introduced by giving a tilt between the interfering beams so that the fringe contrast and the phase are obtained from an instantaneous single-shot interferogram by means of Fourier fringe analysis [14]. Figure 11.12a shows the interferogram captured by a CCD camera at the interferometer output. A closer look at the interferogram reveals that high-contrast fringes are observed at locations corresponding to the letters U, E, and C, and the central area that corresponds to the zeroth order noise in conventional holography. Figure 11.12b shows the Fourier transform spectra of the interferogram. The spectrum around one of the first-order spectral peaks on the carrier frequency location was band-pass filtered, brought to center, and then inverse Fourier transformed to obtain the complex amplitude of the fringe pattern, which gives the coherence function $\Gamma(\Delta x, \Delta y, 0)$. The modulus and phase of the coherence function obtained by the Fourier transform method is shown in Figure 11.12c,d, which jointly demonstrates the validity of the proposed principle.

11.7
Application of Coherence Holography: Dispersion-Free Depth Sensing with a Spatial Coherence Comb

When the incoherent source (representing the coherence hologram in Figure 11.1) is at a distance very large compared to that between the two observation points P and R, the mutual intensity $J(\mathbf{r}_P, \mathbf{r}_R)$ in Eq. (11.4) can be approximated by a formula in the form of the Debye integral, which we call the McCutchen theorem [15]

$$J(\mathbf{r}_P - \mathbf{r}_R) \approx \iint I(\mathbf{r}_S) \exp\left[-ik\left(\frac{\mathbf{r}_S - \mathbf{r}_R}{|\mathbf{r}_S - \mathbf{r}_R|}\right) \cdot (\mathbf{r}_P - \mathbf{r}_R)\right] \frac{d\mathbf{r}_S}{|\mathbf{r}_S - \mathbf{r}_R|^2}$$

$$= \iint I(\mathbf{c}_S) \exp[-ik\mathbf{c}_S \cdot (\mathbf{r}_P - \mathbf{r}_R)] d\Omega_S \quad (11.6)$$

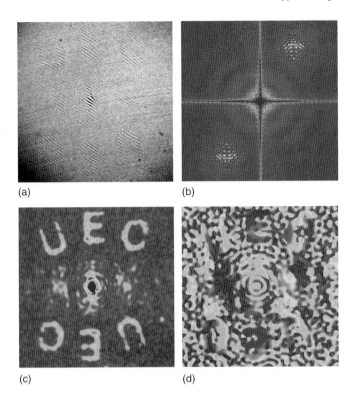

Figure 11.12 Reconstructed images. (a) Raw intensity image resulted from shearing interference in real time. Note that the region of high-contrast fringes represents the letters U, E, and C; (b) the Fourier spectrum of the interference image; and (c,d) are the contrast image and the phase image, respectively, which jointly represent the complex coherence function. (Please find a color version of this figure on the color plates.)

where $c_S = (r_S - r_R)/|r_S - r_R|$ is a unit vector pointing from point R toward the source element dr_S, $d\Omega_S = dr_S/|r_S - r_R|^2$ is an increment of the solid angle subtending the source element dr_S (Figure 11.13). The source distribution projected onto the unit sphere (c sphere) is called a *generalized source*. The MacCutchen theorem has a more simplified form of 1D Fourier transform

$$J(L) = \int_{-\infty}^{\infty} I(l) \exp(-ikLl) dl \qquad (11.7)$$

where $Ld = r_P - r_R$ and $l = c_S \cdot d$, with d being a unit vector in the direction of $r_P - r_R$, and $I(l)$ is the projection of the generalized source on a line oriented in the same d direction in the c sphere; $I(l)$ is called the *source distribution*. The McCutchen theorem gives an insight into the new concept of a spatial frequency comb and a spatial coherence comb, which can be applied for dispersion-free depth sensing [16]. Suppose a coherence hologram with a Fresnel zone plate (FZP)-like incoherent ring source structure, as shown in Figure 11.14. The thin ring sources

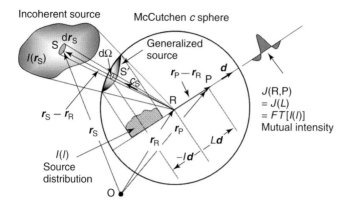

Figure 11.13 McCutchen's theorem for generalized source and mutual intensity.

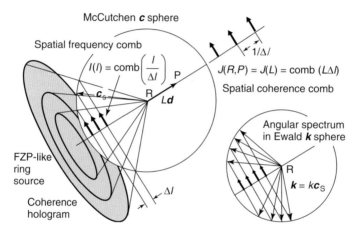

Figure 11.14 Spatial frequency comb generated by coherence holography.

are first projected onto the McCutchen c sphere and then onto a line connecting a reference point R and an observation point P to form a periodic impulse source distribution with period Δl, expressed by a comb function $I(l) = \text{comb}(l/\Delta l)$. This source distribution is called a spatial frequency comb because the directional unit vector c_s pointing to the ring source can be related to a k vector by $k = kc_s$, which represents an angular spectrum in an Ewald sphere, as shown in Figure 11.14. From the Fourier relation in Eq. (11.7), the mutual intensity of the optical field between points R and P (separated by L on the line parallel to d) becomes a comb function $J(L) = \text{comb}(L\Delta l)$, called a spatial coherence comb, and exhibits a selective high spatial coherence in the direction of d with period inversely proportional to Δl. By controlling the FZP-like ring source in the coherence hologram with an SLM, we can scan the coherence comb function in the depth direction d to obtain a 3D tomography image. Unlike a conventional optical frequency comb (which is composed of equally spaced multiple polychromatic line spectra), the

11.7 Application of Coherence Holography | 251

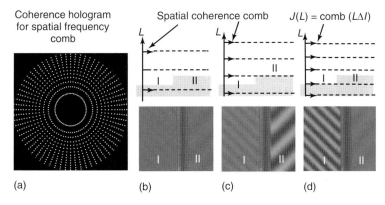

Figure 11.15 Sensing of the depth of block gauge surfaces by gating with a spatial coherence comb. (a) Source distribution for spatial frequency comb and (b–d) spatial coherence gating function is scanned by changing the interval of the spatial coherence comb.

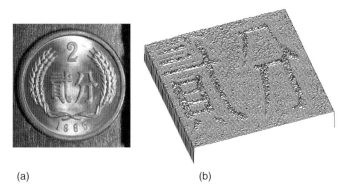

Figure 11.16 Spatial coherence profilometry using a spatial frequency comb generated by coherence holography. (a) A coin with a rough surface and (b) surface profile obtained by longitudinal coherence scanning.

spatial frequency comb is generated by monochromatic light and opens up a new possibility of spatial coherence tomography completely free from the problem of dispersion [7, 16].

An example of depth sensing with a variable longitudinal spatial coherence comb function created by an SLM-generated tunable spatial frequency comb is shown in Figure 11.15 [16]. Shown in Figure 11.15a is a coherence hologram with an FZP-like source distribution, in which each ring source is composed of the same number of point sources to equalize the comb heights. The interval of the spatial coherence comb is varied by SLM, and fringe contrast on the surface of two gauge blocks with a 400 μm height gap was observed with a Michelson interferometer. In Figure 11.15b, fringe contrast is low because the coherence comb does not match the surfaces of the gauge blocks, whereas high-contrast fringes are observed in Figure 11.15c,d because the coherence comb matches the surfaces I and II, respectively [16]. Figure 11.16 shows an example of surface profilometry based on axial spatial coherence scan by using a coherence comb generated by coherence holography [17]. The coin in Figure 11.16a has a rough surface and the position of the coherence comb matching is detected from the highest contrast of a random speckle pattern instead of a fringe pattern in the case of the polished surface. Figure 11.16b shows the surface profile of the coin obtained by the spatial coherence comb [17].

11.8 Conclusion

We reviewed the principle of coherence holography and the concept of the spatial frequency comb, which are inspired by the fact that a coherence function obeys the same wave equation as the optical field. The unique feature of coherence holography is that while an object is recorded with coherent light much in the same manner as in conventional holography, the recorded object is reconstructed as the 3D distribution of a spatial coherence function from the hologram illuminated with spatially incoherent light. The CGCH will open up a new possibility for dispersion-free depth sensing by controlling the 3D spatial coherence characteristics of the optical field.

Acknowledgments

We thank Joseph Rosen of Ben-Gurion University of the Negev and Zhihui Duan of Optoelectronics Co. Ltd. for their contributions in the initial stage of this research.

References

1. Takeda, M., Wang, W., Duan, Z., and Miyamoto, Y. (2005) Coherence holography. *Opt. Express*, **13**, 9629–9635.
2. Takeda, M., Wang, W., and Naik, D.N. (2009) Coherence holography: a thought on synthesis and analysis of optical coherence fields, in *Fringe* (eds W. Osten and M. Kujawinska), Springer-Verlag, Berlin and Heidelberg, pp. 14–21.
3. Mandel, M. and Wolf, E. (1995) *Optical Coherence and Quantum Optics*, Cambridge University Press, Cambridge.
4. Goodman, J.W. (1985) *Statistical Optics*, John Wiley & Sons, Inc., New York.
5. Lohmann, A.W., Mendlovic, D., and Shabtay, G. (1999) Coherence waves. *J. Opt. Soc. Am. A*, **16** (2), 359–363.
6. Takeda, M. (1997) in The philosophy of fringes: Analogies and dualities in fringe generation and analysis, in *Fringe '97 Automatic Processing of Fringe Patterns* (eds W. Jueptner and W. Osten), Akademie Verlag, Berlin, pp. 17–26.
7. Rosen, J. and Takeda, M. (2000) Longitudinal spatial coherence applied for surface profilometry. *Appl. Opt.*, **29**, 4107–4111.
8. Wang, W., Duan, Z., Hanson, S.G., and Miyamoto, Y. (2006) Experimental study of coherence vortices: Local properties of phase singularities in a spatial coherence function. *Phys. Rev. Lett.*, **96**, 073902.

9. Naik, D.N., Ezawa, T., Miyamoto, Y., and Takeda, M. (2009) 3-D coherence holography using a modified Sagnac radial shearing interferometer with geometric phase shift. *Opt. Express*, **17**, 10633–10641.
10. Liu, Z., Gemma, T., Rosen, J., and Takeda, M. (2010) Improved illumination system for spatial coherence control. *Appl. Opt.*, **49**, D12–D16.
11. Hariharan, P. and Maitreyee, R. (1992) A geometric-phase interferometer. *J. Mod. Opt.*, **39**, 1811–1825.
12. Naik, D.N., Ezawa, T., Miyamoto, Y., and Takeda, M. (2010) Phase-shift coherence holography. *Opt. Lett.*, **35**, 1728–1730.
13. Naik, D.N., Ezawa, T., Miyamoto, Y., and Takeda, M. (2010) Real-time coherence holography. *Opt. Express*, **18**, 13782–13787.
14. Takeda, M., Ina, H., and Kobayashi, S. (1982) Fourier-transform method of fringe-pattern analysis for computer-based topography and interferometry. *J. Opt. Soc. Am.*, **72**, 156–160.
15. McCutchen, C.W. (1966) Generalized source and the van Cittert-Zernike theorem: A study of spatial coherence required for interferometry. *J. Opt. Soc. Am.*, **56**, 727–733.
16. Duan, Z., Miyamoto, Y., and Takeda, M. (2006) Dispersion-free optical coherence depth sensing with a spatial frequency comb generated by an angular spectrum modulator. *Opt. Express*, **14**, 12109–12121.
17. Duan, Z., Kozaki, H., Miyamoto, Y., Rosen, J., and Takeda, M. (2004) Synthetic spatial coherence function for optical tomography and profilometry: influence of the observation condition, in *Interferometry XII: Techniques and Analysis*, Proceedings of SPIE, SPIE, Vol. 5531 (eds K. Creath and J. Schmit), pp. 236–243.

12
Quantitative Optical Microscopy at the Nanoscale: New Developments and Comparisons

Bernd Bodermann, Egbert Buhr, Zhi Li, and Harald Bosse

12.1
Introduction

In many scientific and technical applications small structures with dimensions in the micrometer or nanometer range are manufactured and used. In particular, the semiconductor industry produces ever smaller structures, for instance, on computer chips. In this respect, the functionality of these structures, for example, of a microelectronic element, is critically dependent on the exact size of the structure. For the *dimensional* characterization of micro- and nanostructures, microscopic methods are particularly applied: this metrological task is called quantitative microscopy.

For an accurate and reliable size measurement of micro- and nanostructures with an uncertainty level of a few nanometers or below, three different aspects are essential: the measurement conditions and parameter settings of the measurement system have to be carefully controlled and sufficiently stable to enable a high level of measurement reproducibility, the scale of the microscopic measurement has to be traceable to the SI unit, the *meter*, to enable the comparability of the results (both between different laboratories and between different measurement tools), and a detailed and comprehensive understanding and description of the imaging process is required to derive the demanded dimensional quantities of the structures from the microscopic image. This last aspect, in general, is particularly challenging.

Optical microscopy has been used for the imaging of small objects and the inspection of different kinds of surfaces ever since its early developments in the seventeenth century [1]. To measure the size of an object, a graduated scale or crosshair was placed in the image plane, the object under test was adjusted with respect to the scale and finally the number of tick marks was counted. For low-level dimensional microscopic measurements, this simple image-based method is still commonly used. Progress in manufacturing processes of optical lenses made it possible to improve the resolution and magnification performance of optical microscopes considerably. By the end of the nineteenth century, the performance of objective lenses had begun to approach the theoretical diffraction limit, which was observed and described by the work of Airy, Abbe, and Rayleigh [2, 3]. Owing to

Optical Imaging and Metrology: Advanced Technologies, First Edition.
Edited by Wolfgang Osten and Nadya Reingand.
© 2012 Wiley-VCH Verlag GmbH & Co. KGaA. Published 2012 by Wiley-VCH Verlag GmbH & Co. KGaA.

diffraction effects within the optical system, the optical image is not perfect – a point is imaged as a diffraction pattern (Airy disk) of finite size. The lateral resolution limit Δx of an optical microscope depends on the numerical aperture (NA) of the microscope objective and the wavelength λ used [4]

$$\Delta x = k \cdot \frac{\lambda}{\text{NA}} = k \frac{\lambda}{n \cdot \sin\left(\frac{\alpha}{2}\right)} \tag{12.1}$$

where $\text{NA} = n \cdot \sin(\alpha/2)$, n denotes the refractive index, and α is the angle of aperture. The factor k depends, for example, on the spatial coherence properties of the optical radiation used [5, 6]; typical values are between $k = 0.61$ and $k = 1.22$.

Equation (12.1) was already derived by Ernst Abbe, who, in 1873, published a model of microscopic optical imaging based on the two-stage Fraunhofer diffraction [7]. He established the scalar diffraction theory, which was used for a long time to describe the imaging performance of optical microscopes. This model was significantly improved by Berek [8] and Hopkins [9] by taking into account the degree of coherence of the illumination. However, these models still omit the influence of the light polarization on the microscopic image and approximate the structures to be imaged by planar, that is, two-dimensional structures. These approximations and this deficiency are no longer acceptable to meet the metrological requirements for current high-end nanostructures, such as those produced for semiconductor applications [10–12]. At present, the so-called rigorous diffraction models (as described, e.g., in [13]) take into account the three dimensionality of the object and consider the vector properties of optical fields. These models are used to calculate numerically the light–structure interaction, which is the basis for current sophisticated simulations of the microscopic images in quantitative microscopy.

For reliable and accurate microscopic measurements, one needs to compare in detail the measured images with the corresponding theoretical ones. Therefore, a proper understanding and simulation of the microscopic imaging process is indispensable for measurements with low uncertainties. These simulations have to be based on recognized and proven physical models, they have to consider all relevant experimental parameters (both of the instrument and the sample), and need to be calculated with the necessary numerical accuracy.

During the past decade, different fluorescence microscopic methods have been demonstrated and developed for imaging of biological samples, which partly reach very remarkable resolutions far below the Abbe limit. Among others, these include *stimulated emission depletion* (STED) *fluorescence microscopy* [14–16], *structured illumination microscopy* (SIM) [17, 18], and *photoactivated localization microscopy* (PALM) [19] or *stochastic optical reconstruction microscopy* (STORM) [20]. Remarkable resolution enhancement, the so-called super resolution (SR), is achieved in each case by a combination of nonlinearity (saturation for STED and SR-SIM, nonlinear switching of fluorescence molecules for PALM/STORM), and a dissection of the microscope image in time (raster scanning for STED, stochastic dissection for PALM/STORM, and deterministic dissection for SR-SIM). With these methods, lateral resolutions down to a few nanometers have been demonstrated [19, 21]. However, these super resolving fluorescence microscopic methods are suitable only for biological samples and are unfortunately virtually not applicable for quantitative microscopy, especially on nonbiological nanostructures.

Besides the well-known optical microscopy, ultrahigh-resolution electron microscopy and scanning probe microscopy (SPM) (such as atomic force microscopy (AFM) or tunneling microscopy) are also applied for nanometrology. All these microscopic methods have their assets and drawbacks and provide different but often complementing information about the structures to be measured. Therefore, comprehensive and reliable measurements of sophisticated structures with challenging requirements, for example, with regard to measurement uncertainty, often require the application of different microscopic methods. Here, we concentrate on quantitative *optical* microscopic methods, their applications and potential, and on current and future developments with an emphasis on activities at the Physikalisch-Technische Bundesanstalt (PTB).

In the following discussion, we especially describe recent developments in experimental realizations and appropriate modeling support used for the quantitative optical microscopy at the nanoscale. In particular, we discuss the very challenging field of the metrology of small features on photomasks. In addition, we present and discuss comparison measurements performed with other microscopic methods to validate the optical measurement results.

12.2
Quantitative Optical Microscopy

12.2.1
Metrological Traceability

The term *quantitative optical microscopy* means that the result of a measurement is traceable to the corresponding SI unit and that also a corresponding measurement uncertainty is given. For the realization of possible traceability routes as well as for the general procedures for calculating measurement uncertainties, internationally accepted guidelines exist such as the Bureau International des Poids et Mesures (BIPM) *mise en pratique* or the *Guide to the Expression of Uncertainty in Measurement* (GUM) [22]. If these guidelines are followed, optical microscopy can be used in a stringent quantitative way, which is necessary, for example, for applications in production quality control. The traceability of the measurement results ensures the international comparability of the results, which is an inevitable prerequisite to support globalized industrial production. Furthermore, it also ensures tool-to-tool comparability, which is of great importance, if different types of tools (e.g., electron optical and optical tools) are in use for one measurand.

If, for example, a dimension of a feature (e.g., its width) is extracted from a microscopic image, the length scale used to measure this dimension has to be traced back to the definition of the length unit, the *meter*. To achieve traceability in a microscopic measurement, a common approach is to use a reference standard, which has been calibrated before. To follow this approach, one needs to be sure

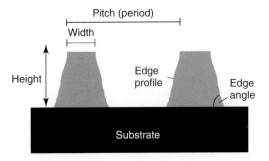

Figure 12.1 Schematic representation of important dimensional measurands of microstructure and nanostructure.

that the feature under test in the microscopic instrument behaves in the same manner as the corresponding feature of the standard. If this requirement cannot be met, appropriate correction factors or bias values to correct the measured values are needed. In practice, the determination of such correction values is often an issue of its own: one example is the measurement of the width of a chromium line on a quartz substrate (Figure 12.1) with an uncertainty of less than a tenth of the wavelength, where details of the chromium structure such as height, sidewall angle, or optical constants become important parameters to consider. In cases where no reference standard is available or where such a reference standard has to be calibrated for the first time (which is one of the main tasks of national metrology institutes such as PTB), the achievement of traceability is more difficult: one needs measuring instruments operated under well-controlled and stable conditions, the measurement procedure needs to be fully understood and is preferably based on fundamental physical laws, and all parameters that may influence the measurement need to be identified and their impact on the measurement needs to be quantified. These requirements often mean that rather slow but robust, instead of fast and possibly error-prone, measurement procedures are used.

Several standards for optical linewidth metrology have been developed, some of which are commercially available. One example is the photomask linewidth calibration standard SRM 2059 (Figure 12.2) from the US National Institute of Standards and Technology (NIST) [23], which contains patterns of clear and opaque isolated lines with nominal dimensions ranging from 0.25 to 32 μm, with an expanded uncertainty of below 26 nm.

In Europe, a high-quality photomask linewidth standard (Figure 12.9) was developed as a reference standard for different types of linewidth metrology instruments including optical transmission microscopy, low-energy scanning electron microscopy (SEM), and SPM [24]. This standard consists, among others, of isolated line structures in different tones and in x/y-orientation. Target linewidth values are currently going down to 100 nm, and the design and the minimum target linewidth values are continuously adapted to current technology nodes and the corresponding high-end mask production processes. Linewidth measurements performed using ultraviolet (UV) transmission microscopy and SEM agreed within

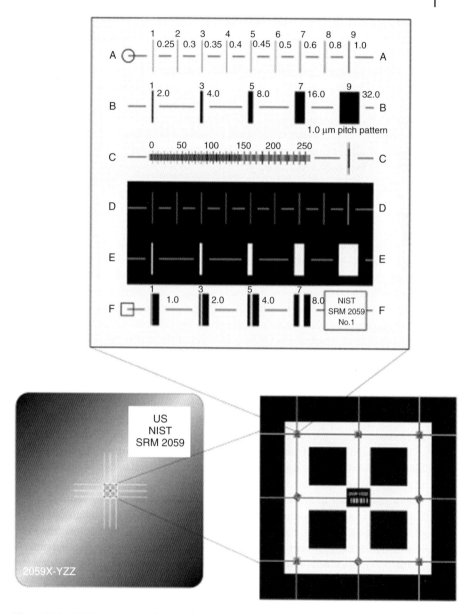

Figure 12.2 NIST SRM 2059 photomask linewidth standard.

a few nanometers [25]. In addition, a special standard has been developed for high-resolution deep ultraviolet (DUV) and confocal microscopy [26]. This standard provides structures in the submicrometer- to sub-100 nm scale and can be used for linewidth measurements and for resolution and astigmatism testing of these instruments.

Besides linewidth measurements, pitch measurements are also commonly performed to characterize resolution properties of optical microscopes. Therefore, many standards intended for optical microscopy contain also pitch structures such as one-dimensional or two-dimensional gratings. The calibration of these patterns can be carried out using metrological long-range SPMs [27] or optical diffractometry [28]. Uncertainties of the mean pitch down to about 10 pm can be achieved for gratings containing a large number of grating elements. International key comparison measurements of the pitch of two-dimensional gratings using SPMs and optical diffractometry revealed that the results are generally consistent and reliable to a very high accuracy [29].

12.2.2
Measurands and Measurement Methods

As mentioned above, quantitative microscopy measures dimensional quantities of micro- and nanostructures. The most important measurands are usually the widths and height of a structure and the distance or pitch between different structures (Figure 12.1). Other quantities of interest can be details such as edge angles or the edge shape (profile) or further lateral quantities such as the shape and diameter of a circular aperture. Since, in a classical microscope, two-dimensional images of the specimen are created, the characterization of the third dimension, the height (or depth) of a structure, requires special microscopic methods such as interference microscopy or confocal microscopy. State-of-the-art confocal measurement systems reach a depth resolution in the order of 1 nm and interference microscopes even significantly below 1 nm. In the following, we concentrate on microscopic systems and methods for measurements of lateral dimensions, in particular, of structure width and pitch.

Two different types of measurement methods for these parameters are in use: the image-based measurements and object scanning methods. In object scanning microscopy [30] (such as classical confocal microscopy), the specimen is scanned with a small probe and line by line an image of the object is recorded. Especially for the object scanning method, where the object is moved relative to the probe, in principle, a very low measurement uncertainty can be reached because the movement of the specimen can be measured directly (e.g., interferometrically) in the object space. In this way, a direct traceable calibration of the object dimensions is possible. In addition, the imaging geometry in the scanning microscope is fixed and near the optical axes, where the small aberrations of the imaging system, which are evident even in high-end microscopes, are usually minimal. Thus, the influence of these aberrations is significantly smaller than for imaging systems.

For an image-based measurement system, the complete specimen is imaged on a camera in the image plane, and the captured image is analyzed. For quantitative measurements, the image coordinates have to be calibrated carefully utilizing, for example, images of calibrated gratings. This calibration procedure is relatively complex: it depends on the sizes and distances of the camera pixels and on the lateral magnification of the microscopy. Therefore, an accurate calibration of the lateral magnification is necessary, which, in general, varies over the field of view because of residual aberrations of the imaging optics and may change slightly in time due to focus or thermal drifts. In addition, the position-dependent image sensitivity has to be taken into account. However, in industrial applications, image-based systems are preferably used because of their considerably higher measurement speed, their reduced sensitivity to mechanical noise (acoustic excitation and thermal drift), and due to the easier setup and the improved reproducibility as compared with scanning systems.

12.2.3
Image Signal Modeling

For the quantitative determination of measurements of, for example, the linewidths, a model-based evaluation is used [31]. This model-based analysis of the microscopic images is a crucial precondition for high-level quantitative measurements. The quality of the measurements and the achievable measurement uncertainty are vitally dependent on the accuracy of the numerical simulation of the light–structure interaction, on an adequate modeling of the imaging process, and also on the suitability of the applied models both for the structures and the optical system.

The modeling of the microscope image consists of two steps. In the first step, the diffracted electromagnetic fields are calculated. In the second step, the microscope image is computed from these electromagnetic fields by means of the given imaging parameters. Figure 12.3 shows a schematic representation of the components of the microscope simulation software used at PTB.

For the calculation of the diffracted electromagnetic fields, an appropriate geometric model of the structure has to be built up. If an electromagnetic plane wave irradiates this structured object, a part of the light is reflected or transmitted to

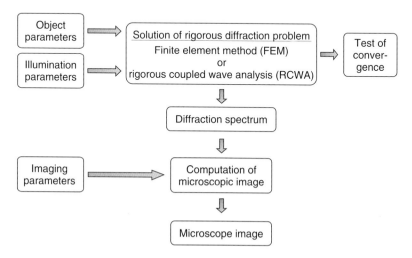

Figure 12.3 Schematic representation of the modeling of the microscopic image.

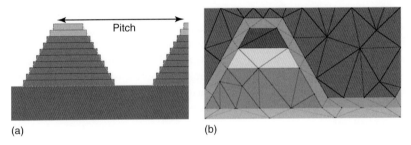

Figure 12.4 Representation of a trapezoid-shaped line structure using (a) the RCWA and (b) the FEM method, respectively.

the diffracted orders. The rigorous diffraction spectrum can be calculated using different theories [13]. All these theories solve the Maxwell equations numerically. For the numerical solution of the stationary Maxwell equation, both the rigorous coupled wave analysis (RCWA) and the finite element method (FEM) are used at PTB. The RCWA method [32] is based on a Fourier expansion of the permittivity function and electromagnetic fields. It was initially developed only for binary gratings. For such rectangular-shaped profiles, the Maxwell equations can be expressed as an eigenvalue problem. Arbitrary geometric profiles can be approximated by a staircase profile (Figure 12.4a), but for each staircase, an eigenvalue problem has to be solved.

The FEM method has the advantage that arbitrary geometries can be adapted (Figure 12.4b) and it shows fast convergence. Both methods calculate diffraction spectra of periodically continued structures. Single structures are approximated by increasing the period so that the images of adjacent structures do not overlap. Typically, the period is chosen in this case to be $\geq 10 \cdot \lambda$, where λ is the optical wavelength. The RCWA simulations are computed with the software package *MicroSim*, which has been developed at the *Institut für Technische Optik*, Stuttgart, Germany [33]. For the FEM method, either the software package *DIPOG* from the *Weierstrass Institute* (WIAS) in Berlin or *JCMWave* from the *Zuse Institute Berlin* (ZIB) is used [34, 35].

In the second step, the microscope image is computed from the complex diffraction spectrum by means of the given imaging parameter. The pupil approach according to Hopkins [9, 36] is applied: the entrance pupil is discretized and for each pupil point the electric and/or magnetic fields of all diffracted orders within the NA of the objective are summed up. For coherent illumination, the squared sum over the pupil points of the electric and magnetic fields gives the electric and/or magnetic field distribution in the image plane. For partly coherent illumination, the intensity distributions in the image plane corresponding to each pupil point have to be summed up over all pupil points [33].

The measurands such as the linewidth are determined by comparing the measured with the simulated intensity in the image plane of the microscope. Figure 12.5 shows the cross section of a rigorously modeled microscope image of a trapezoid-shaped chrome line using the RCWA method. The edge position is

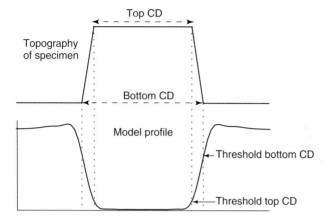

Figure 12.5 Rigorously modeled cross section of the microscopic image of a trapezoid-shaped line structure and evaluation of linewidth (CD) values using threshold criteria.

characterized via the intensity level at this position. This intensity value is called the *threshold value*. The corresponding edge position is derived from the measurement as the lateral position, for which the measured intensity agrees with the threshold value. The simulations can be adapted to infer different linewidth (critical dimension, CD) values from a measured signal profile, for example, the CD at a 50% height of the features or the CD at the top or the bottom level of the features.

12.2.4
Experimental Aspects

For the high-precision measurement of small features on photomasks we use different optical systems. In the following discussion, we briefly describe two different quantitative microscopic systems used at PTB for CD metrology: a UV transmission microscopic system and a novel type of dark-field microscope developed and set up recently at PTB for linewidth measurements on isolated line structures with enhanced resolution.

The UV transmission microscopic (Figure 12.6a) system has already been described in detail in [37]. It is designed, set up, and used for linewidth calibrations mainly on photomasks. The system is based on a modified commercial microscope (Zeiss Axiotron). The sample is imaged using Koehler illumination at a wavelength of 365 nm with a condenser NA of $NA_C = 0.2$ and a microscope objective with a lateral magnification of $m = 150$ and an NA of $NA_O = 0.9$. To realize a scanning microscope, a slit aperture is placed in the image plane in front of a photomultiplier. This aperture again is imaged demagnified into the object plane by the microscope objective. The linewidth measurement is based on the measurement of the transmitted light intensity, while the sample is moved highly precisely and interferometrically controlled in one dimension in the focal plane. The edge

(a) 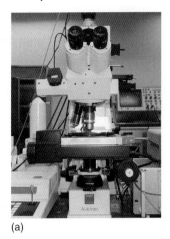 (b)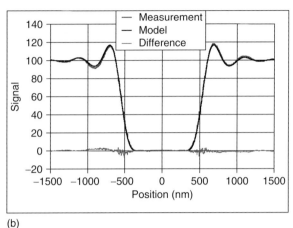

Figure 12.6 (a) Photo of the UV transmission microscope used at PTB for quantitative optical microscopy, primarily on photomasks, and (b) typical transmission intensity profile. (Please find a color version of this figure on the color plates.)

positions are deduced from the measured signal profile using a threshold criterion derived using the microscope imaging model described above (Figure 12.5). The width of the structure is obtained as the difference of the opposing edge positions. Figure 12.6b shows a typical transmission profile, measured on an opaque line feature on a photomask.

The dark-field microscope is based on a novel method, the *alternating grazing incidence dark-field* (AGID) *microscopy*, developed recently at PTB [38]. If the width of the structure to be measured is of the order of the wavelength, in conventional bright or dark-field microscopy the diffraction patterns of both edges overlap, leading to an apparent shift of the edge position (proximity effect). For even smaller linewidths this leads to the effect that the diffraction patterns of both edges cannot be resolved anymore. AGID microscopy has been developed to overcome this problem. This method is based on the alternating single-sided illumination of the specimen under grazing incidence. The resulting diffraction efficiencies of both edges of a line structure are asymmetric because of the asymmetric illumination. Depending on the polarization of the light, the diffraction pattern of one edge and therewith the proximity effect can be strongly suppressed. By illuminating the specimen successively from opposite sites perpendicular to the line structure, the edge positions can be detected independently and the difference of both positions yields the linewidth of the structure. For this method the classical resolution limit of a microscope is not applicable any more.

The setup realized at PTB for AGID microscopy is shown in Figure 12.7. Two temperature-stabilized UV laser diodes (374 nm, 8 mW) are used as radiation sources, combined with a UV microscope of the type INM200 from Leica Microsystems Semiconductor GmbH.[1] To adjust the polarization, $\lambda/2$ plates and

---

1) Now: KLA-Tencor.

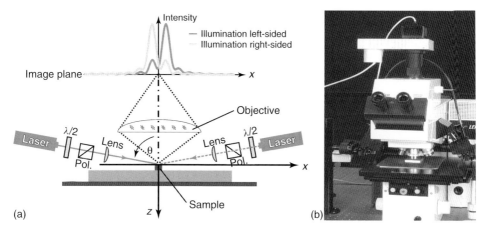

Figure 12.7 Experimental setup of AGID microscopy: (a) schematic diagram of the setup and (b) photograph of the system.

polarization beam splitters are used. A UV objective with an NA of 0.9 and a magnification of 300 is used. The setup is fully computer controlled. The specimen can be moved continuously or stepwise in z-direction. The lasers can be switched on and off independently and images are taken in each case by a 10 bit UV CCD camera. Figure 12.8 shows typical measured and modeled diffraction peaks for a 112 nm chrome line for left- and right-sided illumination in the plane of best focus (Figure 12.8).

To test the performance of AGID microscopy for quantitative linewidth measurements on chrome on glass (CoG) masks, we measured the CDs of several isolated Cr lines on a photomask (Figure 12.9) and compared the results with measurement results obtained by conventional CD-SEM metrology tools at PTB and from industrial partners (Figure 12.10). The linewidths were measured within the range of 100–1000 nm. The chrome lines had a trapezoid-shaped cross section with an edge angle of $\alpha = 71°$, thus resulting in a CD difference of ~46 nm between the top and the bottom CD. As an example, a top linewidth of 112 nm has been measured for the line with a nominal width of 200 nm. The measured value is in excellent agreement with top linewidth measurements carried out with three different CD-SEM tools, yielding values of 118, 114, and 112 nm, respectively.

These results demonstrate that AGID microscopy is capable of measuring the width of isolated lines down to about 100 nm, which is nearly a quarter of the illumination wavelength of 374 nm.

12.2.5
Measurement Uncertainty

As mentioned above, each quantitative traceable microscopic measurement involves the detailed analysis of the measurement uncertainty under consideration of the GUM [22]. This is, in particular, a challenging task if no suitable reference

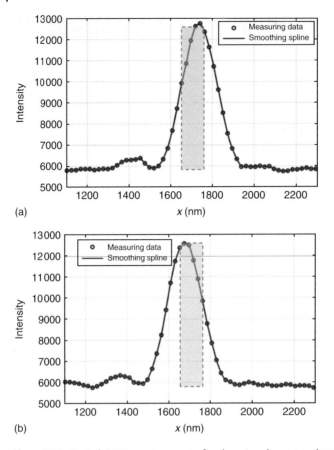

Figure 12.8 Typical AGID measurement of a chromium line on a photomask (top linewidth 112 nm, indicated by gray bar); dots: measured data, straight line: modeled profiles; (a) illumination left-sided and (b) illumination right-sided.

standard is available. As an example, Table 12.1 shows the uncertainty analysis for linewidth calibrations of a typical line feature on a standard antireflective CoG mask by means of the UV transmission microscope. For this example, an *expanded* (confidence interval of 95%) measurement uncertainty of $U = 19$ nm is obtained. This value contains statistical and systematic uncertainty contributions attributed to the measurement system such as, detector noise and nonlinearity, aberrations of the optical system, and residual uncertainties of instrumental parameters (e.g., of the objective NA). Significant uncertainty contributions are also always caused by the quality of the structures and the insufficient a priori information about the specimen. Examples are the knowledge of the optical parameters of the structure and substrate materials and also geometrical quantities such as structure height and edge angles. The example shows the significant contributions for a high-quality structure as produced in a current high-end lithography process.

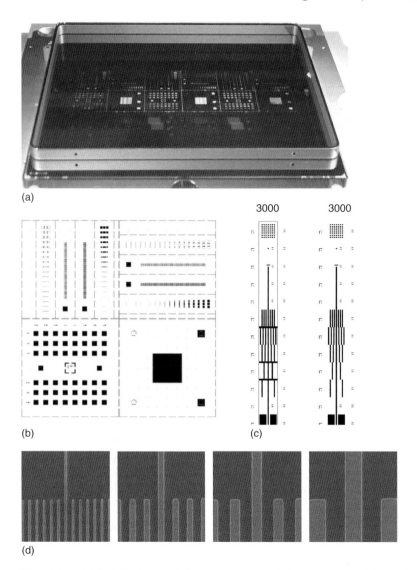

Figure 12.9 PTB's CoG photomask CD standard developed together with different European companies [24]: (a) photograph of the mask, (b) layout of one die on the mask, (c) layout of one CD test structure group, and (d) SEM images of the isodense transitions of the line feature measurement structures, nominal CD values from left to right: 100, 200, 300, and 500 nm (image size: 2.5 μm).

For lower quality structures, these uncertainty contributions can increase dramatically and further structure details as shown in Figure 12.11 such as line edge roughness, Cr edge run-out (the so-called footing), and corner rounding may also be significant sources of measurement errors.

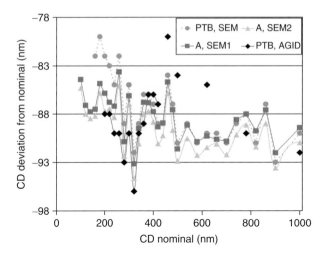

Figure 12.10 Comparison of CD measurements of isolated Cr lines on a state-of-the-art photomask using UV AGID measurements and conventional CD-SEM metrology tools.

A third group of contributions to the total measurement uncertainty results from the data analysis and the accuracy of the modeling. The latter may be affected, for example, by numerical errors, the influence of the approximations used in praxis in all rigorous diffraction calculations or simply by the use of a simplified (and in the worst case inadequate) structure model.

12.3
Comparison Measurements

Comparisons between different institutes, different instrumentation, and different methods, typically in the form of a round-robin test, are indispensable for the validation of measurement procedures and, in particular, for the conformation of uncertainty budgets. The latest international comparison on photomask line features was carried out about 15 years ago [39], the next international comparison between national metrology institutes, scheduled as Nano1, is currently under preparation. Within PTB, systematic comparison measurements have been performed between different linewidth metrology methods [40]. Figure 12.12 shows the results of measurements on CoG line features by the UV transmission microscope and by a metrological SEM, called the electron optical metrology system (EOMS) [41]. As for the measurement results of optical microscopy, the SEM measurement results also have to be supported by an adequate physical image simulation. For this task, Monte Carlo methods are often applied to describe the stochastic elastic and inelastic electron scattering events in the sample features to be measured [42].

Figure 12.12 shows that the application of suitable model-based image analysis allows one to achieve a good agreement of the measurement results of different

Table 12.1 Typical measurement uncertainty budget of a linewidth measurement using the UV transmission microscope; example of a Cr line with a linewidth of 1 μm.

Input parameter	Parameter uncertainty	Sensitivity coefficient	Standard uncertainty (nm)	σ^2	%
Tool-induced shift[a]	5 nm	1	5.0	25.0	27.8
Repeatability	4 nm	1	4.0	16.0	17.8
Sampling aperture	20 nm	0.14	2.8	7.8	8.7
Objective NA	0.05	44 nm	2.2	4.8	5.4
Photometer nonlinearity	2 nm	1	2.0	4.0	4.5
Defocus[b]	1.2 nm	1	1.2	1.4	1.6
Illumination wavelength	5 nm	0.14	0.7	0.5	0.5
Illumination NA	0.02	20 nm	0.4	0.2	0.2
Chrome thickness	2 nm	1.0	2.1	4.3	4.8
Layer structure	2 nm	1	2.0	4.0	4.5
Cr edge angle	4°	0.44 nm/°	1.8	3.1	3.5
Chrome k	0.2	4.8 nm	1.0	0.9	1.0
Chrome n	0.2	0.67 nm	0.1	0.0	0.0
Data filter[c]	3.7 nm	1	3.7	13.7	15.2
Modeling errors	2 nm	1	2.0	4.0	4.5
Sum			9.5 ($k=1$) 19.0 ($k=2$)	89.8	100

[a] Stray light, aberrations.
[b] Offset: +0.8 nm.
[c] Data interpolation at the threshold intensity level.

microscopic methods. On the other hand, the observed agreement is reduced for very small feature sizes below the 365 nm wavelength of the UV microscope. It is therefore promising to extend optical transmission microscopy to smaller wavelengths.

Another example of a comparison of characterizations with different microscopic methods including optical, SEM, and AFM on a special test mask is briefly discussed in the following paragraph. This test mask was manufactured to show the systematic variation of the sidewall characteristics of line features in different dies, as shown in Figure 12.13. The sidewall profiles can be described by a steep segment close to the top (edge angle $\alpha = 84°$), a flatter contour in the middle section (edge angle $\alpha = 45°$), and an exponential chrome contour at the bottom of the lines ("edge foot").

The top CD values were measured at PTB by means of AFM, SEM, and UV optical microscopy. For the model-based analysis of the UV optical measurements, three different geometrical models were used [43]. Firstly, a standard model of a double trapezoid (Cr and antireflection layer with the same edge angle), secondly, a triple trapezoid (two trapezoids for the Cr layer and one trapezoid for the antireflection layer), and thirdly, a quadruple trapezoid (three trapezoids for the Cr layer and one trapezoid for the antireflection layer) were used (Figure 12.14).

(a)

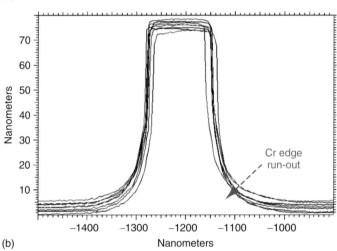

(b)

Figure 12.11 Examples for significant structure details on CoG masks: (a) SEM image of a Cr line with substantial line edge roughness and (b) AFM measurement result of a Cr line with considerable Cr edge run-out.

Figure 12.15 shows the results of the model-based determination of the top CDs in the different dies. The UV optical microscope measurements have been analyzed with the double, triple, and quadruple trapezoid models, and the results are compared to AFM and SEM top CD measurements. The quadruple trapezoid model approximates the profile geometries better than the simpler models such as triple or double trapezoid and yields results that are in excellent agreement with AFM and SEM measurements.

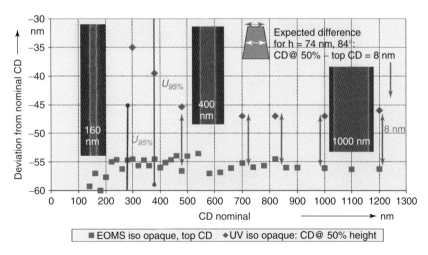

Figure 12.12 Linewidth (CD) comparison measurement results of opaque line features on an antireflective chrome on glass photomask. Shown are the results of UV transmission microscopy (CD at 50% of feature height) and SEM (EOMS, CD at the top of the features) and additional information gained from the atomic force microscopy (AFM) characterization of feature height and sidewall angle as well as three SEM images of line features of different nominal sizes.

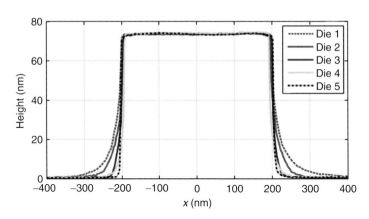

Figure 12.13 AFM measurements of the line features on the test mask.

12.4
Recent Development Trends: DUV Microscopy

The improvement of the lateral resolution of an optical microscope is highly desired in various research and industrial fields, especially in lithography and nanotechnology because of their rapid technology advances resulting in ever smaller structures being manufactured.

As is well known, there are several approaches to improve the resolution of an optical microscope. From the basic Abbe resolution law (Eq. (12.1)), one

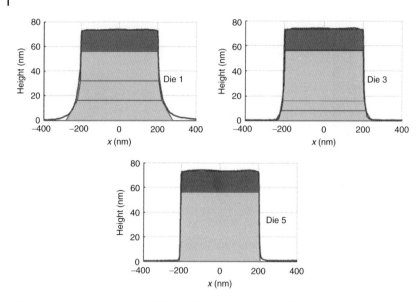

Figure 12.14 Approximation of the profiles used for the modeling of the UV microscopic images using a stack of four trapezoids. (Please find a color version of this figure on the color plates.)

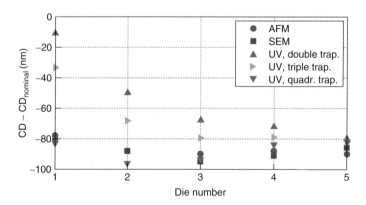

Figure 12.15 Deviations between measured and nominal top CD values determined optically by AFM and SEM. The optical measurements have been analyzed with three different models. The graph shows the results for a nominal 480 nm line structure as an example. The results for other linewidths are similar.

can immediately derive these options: reduction of the illumination wavelength, increase of the NA, and reduction of the factor k. Indeed, all these options are exploited in modern high-resolution optical microscopy. For instance, liquid immersion has been used to increase the NA above 1. Decreasing the factor k is exploited, for instance, in a confocal microscope where one can achieve an improved resolution of about 15% as compared with conventional bright field microscopy [44]. Further improvements of the lateral resolution of an optical microscope can

12.4 Recent Development Trends: DUV Microscopy

be obtained by employing several kinds of amplitude/phase pupil plates [45] or by structured illumination schemes. The above-described AGID microscopy is one example of the special modification of the entrance pupil.

However, the most effective way to further increase the lateral resolving power of an optical microscope is to employ light for illumination with shorter wavelength [46]. During the past few years, various UV microscopes, using wavelengths shorter than 400 nm, have therefore been developed and applied for different purposes [47–56], including CD metrology on photomasks. Commercial state-of-the-art DUV CD metrology systems are operating with a wavelength of around 248 nm [57] and 266 nm [58].

The system described in [57] shows an ultimate resolving power of about 140 nm using an objective with an NA of 0.9. By the application of water immersion at an illumination wavelength of 248 nm, a commercial optical CD metrology tool has been realized with an NA_O of 1.2 [59].

Significant progress in resolution can be achieved using DUV radiation with a wavelength of $\lambda = 193$ nm, as shown in Figure 12.16. This not only leads to a significant resolution enhancement but also enables at-wavelength metrology for current and future 193 nm lithography technologies. Carl Zeiss AG offers or is currently developing various 193 nm microscopy-based metrology systems such as the AIMS [60], the PROVE [61], the Phame [62], or the WL32CD [63] tool designed for different applications (aerial imaging, registration and overlay metrology, phase measurement and wafer-level CD metrology, respectively). However, a high NA CD metrology tool for photomasks using 193 nm radiation is currently not commercially available. Recently, the NIST has also started to develop and set up CD metrology tools operated with the optical radiation at a wavelength of 193 nm [64].

At PTB, a high-resolution DUV optical transmission microscope, the so-called CDM193, is currently under development [65–67], which is operated at a wavelength

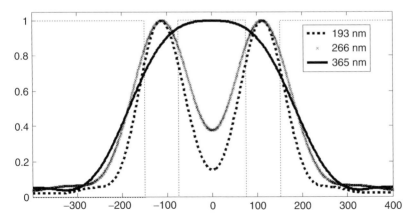

Figure 12.16 Modeled focal intensity profiles of an imaged double slit structure with a slit width of 75 nm and a slit separation of 150 nm on a Cr photomask under the condition of different illumination wavelengths (imaging parameters: $NA_O = 0.9$ and $NA_C = 0.6$).

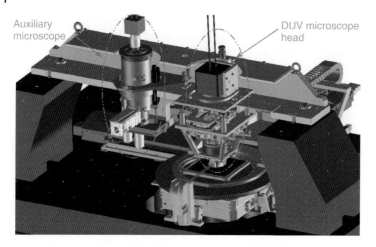

Figure 12.17 CAD (computer-aided design) drawing of the 193 nm DUV microscope in PTB.

of 193 nm. The whole system is designed to be highly stable, both mechanically and thermally in order to achieve the target expanded measurement uncertainty (with a confidence interval of 95%) of $U = 10$ nm (Figure 12.17). In the following section, the design and the current status of the CDM193 are described.

12.4.1
Light Source and Coherence Reduction

The image quality of an optical microscope is, to a large extent, determined by the quality of the microscope objectives in use. Although achromatic or even plan achromatic objectives are now quite common for a microscope with illumination light in visible spectrum, the design and construction of a near-perfect high NA objective applicable for DUV light with a wavelength down to 193 nm is quite challenging, since only a few optical materials, which are sufficiently transparent at 193 nm, are available.

In the CDM193, a monochromatic high NA objective (cf. Section 4.3) is used. Owing to the bandwidth of about 40 GHz of this microscope objective, an ArF excimer laser (Coherent OPTexPro-T) has been chosen as the light source, which is a pulsed laser system with a repetition rate of typically about 200 Hz. Equipped with a bandwidth narrowing module, the spectral bandwidth of the laser is as low as 40 GHz, and the laser offers a mean output power of up to 200 mW. The DUV laser light is delivered to the microscope by means of a multimode optical fiber, reducing the potential negative influence of laser acoustical noise and vibrations. Furthermore, by carefully choosing the fiber coupling method and the multimode fiber length, the lateral coherence length of the laser light could be reduced from 500 μm by about 1 order of magnitude [66].

12.4.2
Illumination System

The illumination and imaging system of the 193 nm DUV microscope realized at PTB features various imaging modes ranging from conventional bright field to special structured illumination schemes, which help to further enhance the flexibility and performance of the microscope. A schematic representation of the illumination system is shown in Figure 12.18: the light coming from the multimode fiber is collimated and sent to a beam homogenization unit. This unit consists of a diffractive optical element (DOE)-based diffuser [68], which has been specially designed for 193 nm wavelength, providing a homogeneous angular spectrum with a small diffusion angle of a few degrees. Together with the multimode fiber it is acting as an extended effective light source and offers a homogeneous illumination. This effective light source is imaged to the entrance aperture ($NA_C = 0.6$) of the condenser via a relay lens pair and the tube lens of the condenser. The arrangement of both diffusers yields a uniform illumination of both the entrance pupil and the field aperture of the condenser. In addition, the top surface of the object under test is homogeneously illuminated.

The functionality of the fundamental illumination system for the 193 nm DUV microscope has been investigated using ray-tracing calculations. The residual wave front aberrations were calculated to be less than $\lambda/10$ when the specimen

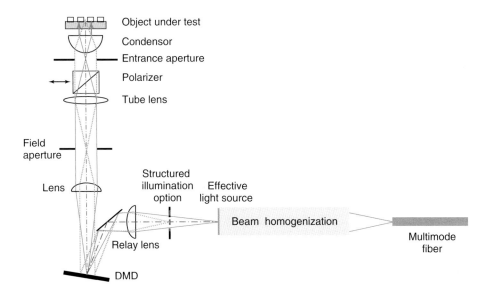

Figure 12.18 Illumination system of the 193 nm DUV microscope in which holographic diffusers are employed to evenly illuminate both the entrance aperture and the field aperture of the microscope; currently, the shown DMD is not yet installed; instead, a mirror replaces the DMD realizing conventional Koehler illumination.

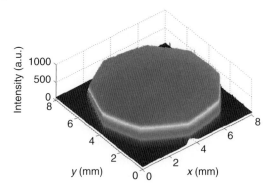

Figure 12.19 Intensity distribution measured in a conjugated plane of the entrance aperture of the condenser of the 193 nm DUV microscope.

was illuminated with the full condenser aperture. The actual performance of the fundamental illumination system for the DUV microscope has been experimentally investigated. With the DOE-diffuser-based homogenization unit, the root mean square homogeneity of the illumination light has been measured to be less than 1.4% in a conjugated plane to the entrance aperture of the condenser (Figure 12.19).

To enhance the system performance, the illumination system will also allow the use of different types of structured illumination. Two different general types of possibilities are introduced into the illumination scheme: structured aperture illumination and structured field illumination. A spatial light modulator (e.g., a digital micromirror device (DMD), specialized for a wavelength of 193 nm) will be placed in a plane conjugated to the condenser entrance aperture to realize the structure aperture illumination, as shown in Figure 12.18. With this option, one can precisely control the illumination direction on the specimen under test and, therefore, introduce different types of structured illumination such as dipole, quadrupole, or ring-shaped illumination, giving access to different resolution enhancement techniques.

12.4.3
Imaging Configuration

For imaging (Figure 12.20), the CDM193 employs a microscope objective developed by Carl Zeiss AG, Germany, featuring a high NA ($NA_O = 0.9$), a high Strehl ratio at the given wavelength (larger than 0.95 for a narrow bandwidth illumination of $\Delta\lambda < 0.005$ nm), a moderate working distance of 200 μm, and an adequate object field of 10 μm × 10 μm. In our microscope, a tube lens system with a nominal focal length of 1200 nm is used, yielding a lateral magnification of the microscope of 400. A DUV-enhanced CCD camera (Hamamatsu C8800) is located directly in the focal plane of the tube lens to acquire the image of the object under test. The CCD chip of the camera has 1000 × 1000 pixels with a pixel size of 8 μm, resulting in a pixel resolution of 20 nm. This resolution is about five times smaller than the actual

12.4 Recent Development Trends: DUV Microscopy

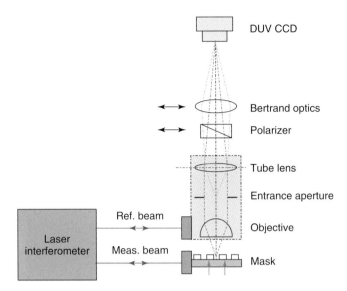

Figure 12.20 Imaging configuration of the DUV 193 nm microscope and the concept for ensuring the traceability of the microscope's measurement results.

resolution of the microscope, ensuring that the microscope's lateral resolution is fully exploited.

To analyze the Fourier spectrum of the structure under test, a Bertrand lens can optionally be inserted to image the back focal plane (the exit pupil) of the microscope objective directly onto the CCD camera. Traceability to the SI unit, the *meter*, will be accomplished by means of laser interferometry. A reference mirror is mechanically connected with the microscope objective, and the measurement mirror is fixed at the mask frame. A differential laser interferometer (Renishaw RLD) with subnanometer resolution is utilized to measure the relative displacement of the structures under test within the object plane of the microscope. In order to calibrate the actual magnification of the microscope, controlled by the interferometer, a structured specimen (e.g., a one-dimensional grid) will be moved by means of a piezo-stage within the object plane, and its image will be acquired by the CCD camera simultaneously. Thus, the microscope's actual magnification can be determined easily.

To investigate the basic performance of this system, a one-dimensional grating with line structures with a width of about 545 nm was imaged by the microscope. The measured profiles (Figure 12.21) show an edge steepness (68 nm between 20 and 80% intensity level), which is already near the theoretical limit of 63 nm. Owing to its very high resolution and designated small measurement uncertainty, the DUV microscope will be applicable and suitable for mask CD metrology for 32 nm technology and beyond. In addition, the microscope offers the important possibility of at-wavelength metrology for the currently applied 193 nm lithography technology: the CDM193 will thus ideally complement other techniques used for CD metrology of photomask features such as SEM, AFM, and scatterometry.

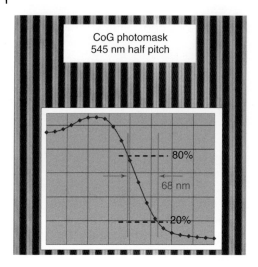

Figure 12.21 Microscopic image of a 1D grating with a nominal structure width of 545 nm demonstrating the resolution of the 193 nm DUV microscope; the measured edge profile is close to the theoretical limit.

12.5
Points to Address for the Further Development of Quantitative Optical Microscopy

In order to further develop quantitative optical microscopy as a valuable tool for metrology at the nanoscale, the following points should be taken into account and can be thus regarded as something like a "wish list."

As in other fields of metrology, it is necessary to use clear and unequivocal definitions of the measurand that has to be precisely determined. A lot of confusion is sometimes generated in the comparison of measurement results, measured by different methods, because the measurand has not been properly described.

An accurate simulation of optical images requires knowledge about the relevant optical material parameters of the structures to be analyzed as well as the respective substrates. If possible, these material parameters and their associated measurement uncertainties should be provided by the manufacturer.

Another important route to follow is the improvement and further acceleration of signal modeling algorithms, which are indispensable in optical microscopy. Here, a focus should be on an improved computation speed of the algorithms, a full 3D modeling, and – in addition to the so far often only "ideal" cases, which have been dealt with – also a complete treatment of configurations with "real" deviations, such as line edge roughness of the structures (sample characteristics) or aberrations of the microscope components (tool characteristics).

Up to now, very often only a part of the information from optical microscope images is used, namely, parts of the measured line profiles in the optimum focus position of the sample. It is highly recommended that full use is made of the information that is available from defocused images or from complete focus image series [69, 70].

In addition, the potential of scatterometry shall be further investigated and made use of because the theoretical models and implemented program packages for the description of the probe–sample interaction are identical for scatterometry and optical microscopy. Furthermore, in contrast to optical microscopy, scatterometry has proved the capability to characterize quantitatively nanostructures with sizes far below the Abbe limit of optical microscopy [71]. Scatterometry as an independent characterization method ("virtual imaging") provides valuable results for comparison measurements on repeated patterns of nanostructures [72, 73]. This also holds for nonoptical microscopic methods, such as SEM or SPM. We also see a great potential in the further development of sophisticated high-resolution optical microscopic methods, such as the special dark-field illumination AGID mode. Other potential applications of a structured field illumination for resolution enhancement have been discussed [74, 75]. Thorough exploitation of these different resolution enhancement techniques will enable the newly developed 193 nm CD metrology tool to be applicable also beyond the 32 nm node.

References

1. Hooke, R. (1665) *Micrographia*, Martyn & Allestry, London.
2. Lummer, O. and Reiche, F. (1910) *Die Lehre von der Bildentstehung im Mikroskop von Ernst Abbe*, Vieweg, Braunschweig.
3. Strutt, J.W. (1896) On the theory of optical images, with special reference to the microscope. *Phil. Mag.*, **42**, 167–195.
4. Hecht, E. (2001) *Optics*, Addison Wesley.
5. Menzel, E., Mirande, W., and Weingärtner, I. (1973) *Fourier-Optik und Holographie*, Wien, New York.
6. Marathay, S. (1982) *Elements of Optical Coherence Theory*, John Wiley & Sons, Inc., New York.
7. Abbe, E. (1873) Beiträge zur theorie des mikroskops und der mikroskopischen wahrnehmung. *Schultzes Archiv Mikr. Anat.*, **9**, 413–468.
8. Berek, M. (1926) Über Kohärenz und konsonanz des lichtes. *Z. Phys.* **36**, 675–688, 824–838; **37** (1926) 387–394; **40** (1927), 420–450.
9. Hopkins, H.H. (1953) On the diffraction theory of optical images. *Proc. R. Soc. Lond. A*, **217**, 408–432.
10. Nyyssonen, D. (1982) Theory of optical edge detection and imaging of thick layers. *J. Opt. Soc. Am.*, **72**, 1425–1436.
11. Nyyssonen, D. and Kirk, C.P. (1988) Optical microscope imaging of lines patterned in thick layers with variable edge geometry: theory. *J. Opt. Soc. Am. A*, **5**, 1270–1280.
12. Czaske, M. (1997) Strukturbreitenmessung auf photolithographischen masken und wafern im lichtmikroskop: theorie, einfluß der polarisation des lichtes und abbildung von strukturen im bereich der auflösungsgrenze. PTB-Opt-**55**, PhD thesis. Braunschweig.
13. Loewen, E.G. and Popov, E. (1997) *Diffraction Gratings and Applications*, Marcel Dekker, New York.
14. Hell, S.W. and Wichmann, J. (1994) Breaking the diffraction resolution limit by stimulated emission: stimulated-emission-depletion fluorescence microscopy. *Opt. Lett.*, **19**, 780–782.
15. Klar, T.A. and Hell, S.W. (1999) Subdiffraction resolution in far-field fluorescence microscopy. *Opt. Lett.*, **24**, 954–956.
16. Hell, S.W. (2007) Far-field optical nanoscopy. *Science*, **316**, 1153–1158.
17. Gustafsson, M.G. (2000) Surpassing the lateral resolution limit by a factor of two using structured illumination microscopy. *J. Microsc.*, **198**, 82–87.
18. Gustafsson, M.G. (2005) Nonlinear structured-illumination microscopy: wide-field fluorescence imaging with

theoretically unlimited resolution. *Proc. Natl. Acad. Sci. U.S.A.*, **102**, 13081–13086.

19. Betzig, E., Patterson, G.H., Sougrat, R., Lindwasser, O.W., Olenych, S., Bonifacino, J.S., Davidson, M.W., Lippincott-Schwartz, J., and Hess, H.F. (2006) Imaging intracellular fluorescent proteins at nanometer resolution. *Science*, **313**, 1642–1645.

20. Rust, M.J., Bates, M., and Zhuang, X. (2006) Sub-diffraction-limit imaging by stochastic optical reconstruction microscopy (STORM). *Nat. Methods (online)*, **3**, 793–796.

21. Rittweger, E., Han, K.Y., Irvine, S.E., Eggeling, C., and Hell, S.W. (2009) STED microscopy reveals crystal colour centres with nanometric resolution. *Nat. Photonics*, **3**, 144–147.

22. http://www.bipm.org (accessed 20 December 2011).

23. Potzick, J.E., Pedulla, J.M., and Stocker, M.T. (2002) New NIST photomask linewidth standard. *Proc. SPIE*, **4889**, 343.

24. Mirandé, W., Bodermann, B., Häßler-Grohne, W., Frase, C.G., Czerkas, S., and Bosse, H. (2004) Characterization of new CD photomask standards. *Proc. SPIE*, **5375**, 29–40.

25. Richter, J., Thorben, H., Liebe, R., Bodermann, B., Diener, A., Bergmann, D., Frase, C.G., and Bosse, H. (2007) Calibration of CD mask standards for the 65 nm mode: CoG and MoSi. *Proc. SPIE*, **6533**, 65330S.

26. Huebner, U., Morgenroth, W., Boucher, R., Meyer, M., Mirandé, W., Buhr, E., Ehret, G., Dai, G., Dziomba, T., Hild, R., and Fries, T. (2007) A nanoscale linewidth/pitch standard for high-resolution optical microscopy and other microscopic techniques. *Meas. Sci. Technol.*, **18**, 422–429.

27. Dai, G., Koenders, L., Pohlenz, F., Dziomba, T., and Danzebrink, H.-U. (2005) Accurate and traceable calibration of one-dimensional gratings. *Meas. Sci. Technol.*, **16**, 1241–1249.

28. Buhr, E., Michaelis, W., Diener, A., and Mirandé, W. (2007) Multi-wavelength VIS/UV optical diffractometer for high-accuracy calibration of nano-scale pitch standards. *Meas. Sci. Technol.*, **18**, 667–674.

29. Garnaes, J. and Dirscherl, K. (2008) NANO5 2D grating – final report. *Metrologia*, **45** (Tech. Suppl.) 04003, http://www.iop.org/EJ/abstract/0026-1394/45/1A/04003. (accessed 20 December 2011)

30. Geuther, H., Schröder, K.-P., Danzebrink, H.-U., and Mirandé, W. (1994) Rastermikroskopie im nah- und fernfeld an strukturen im submikrometerbereich. *Tech. Messen*, **61**, 390–400.

31. Colonna de Lega, X. (2012) Model-based optical metrology, *Optical Imaging and Metrology*, Wiley-VCH Verlag GmbH, Weinheim, pp. 281–302.

32. Moharam, M.G., Grann, E., and Pommet, D.G. (1995) Formulation for stable and efficient implementation of the rigorous coupled-wave analysis of binary gratings. *J. Opt. Soc. Am. A*, **12**, 1068–1076.

33. Totzeck, M. (2001) Numerical simulation of high-NA quantitative polarisation microscopy and corresponding near-fields. *Optik*, **112**, 399–406.

34. Program Package DiPoG: Direct and inverse problems for optical gratings http://www.wias-berlin.de. (accessed 20 December 2011).

35. Program Package JCMwave: http://www.jcmwave.com/. (accessed 29 March 2012)

36. Hopkins, H.H. (1977) Image formation with partially coherent light. *Photogr. Sci. Eng.*, **21**, 114–123.

37. Bodermann, B., Buhr, E., Ehret, G., Scholze, F., and Wurm, M. (2008) Optical metrology of micro- and nanostructures at PTB: Status and future developments. *Proc. SPIE*, **7155**, 71550V.

38. Ehret, G., Bodermann, B., and Mirandé, W. (2007) Quantitative linewidth measurement down to 100 nm by means of optical dark-field microscopy and rigorous model-based evaluation. *Meas. Sci Technol.*, **18** (2), 430–438.

39. Potzick, J. and Nunn, J. (1996) International comparison of photomask linewidth standards: United States

(NIST) and United Kingdom (NPL). *Proc. SPIE*, **2725**, 2725–2708.
40. Bodermann, B. and Bosse, H. (2007) An approach to validation of rigorous modeling in optical CD microscopy by comparison of measurement results with independent method. *Proc. SPIE*, **6617**, 66170Y-1–66170Y-10.
41. Häßler-Grohne, W. and Bosse, H. (1998) Electron optical metrology system for pattern placement measurements. *Meas. Sci. Technol.*, **9**, 1120–1128.
42. Frase, C.G., Gnieser, D., and Bosse, H. (2009) Model-based SEM for dimensional metrology tasks in semiconductor and mask industry. *J. Phys. D: Appl. Phys.*, **42**, 183001.
43. Ehret, G., Bodermann, B., Bergmann, D., Diener, A., and Häßler-Grohne, W. (2006) Theoretical modelling and experimental verification of the influence of Cr edge profiles on microscopic-optical edge signals for COG masks. *Proc. SPIE*, **6349**, 63494Y-1–63494Y-10.
44. Corle, T.R. and Kino, G.S. (1996) *Confocal Scanning Optical Microscopy and Related Imaging Systems*, Academic Press.
45. Yun, M., Wang, M. et al. (2006) Tunable transverse superresolution with phase-only pupil filters. *J. Opt. A: Pure Appl. Opt.*, **8**, 1027–1033.
46. Martin, L.C. and Johnson, B.K. (1928) Ultra-violet microscopy. *J. Sci. Instrum.*, **5**, 337–344.
47. Hignette, O., Woch, J., and Gotti, L. (1990) Large-bandwidth deep-UV microscopy for CD metrology. *Proc. SPIE*, **1261**, 79–90.
48. Eguchi, N., Oka, M. et al. (1999) New deep-UV microscope. *Proc. SPIE*, **3740**, 394–397.
49. Henderson, R. (1999) Beta test performance of the Leica LWM 250 UV CD measurement tool. *Proc. SPIE*, **3873**, 760–783.
50. Isomura, I., Tsuchiya, H. et al. (2002) A new inspection method for phase-shift mask (PSM) on deep-UV inspection light source. *Jpn. J. Appl. Phys.*, **41**, 4233–4237.
51. Doe, N.G., Eandi, R.D., and St. Cin, P. (2002) A new algorithm for optical photomask CD metrology for the 100 nm node. *Proc. SPIE*, **4689**, 35–44.
52. Har-Zvi, M., Liebe, R. et al. (2002) Inspection of EAPSMs for 193 nm technology generation using a UV-based 365 nm reticle inspection tool. *Proc. SPIE*, **4562**, 762–767.
53. Merriama, J., Jacoba, J.J., Broeke, D. et al. (2004) High-resolution actinic imaging and phase metrology of 193 nm CPL reticles. *Proc. SPIE*, **5567**, 894–904.
54. Lunde, S., Rouhani, S., Remis, J.P., Ruzin, S.E., Ernst, J.A., and Glaeser, R.M. (2005) UV microscopy at 280 nm is effective in screening for the growth of protein microcrystals. *J. Appl. Crystal.*, **38**, 1031–1034.
55. Shafer, D.R., Chuang, Y.-H., and Tsai, B.-M. (2008) Ultra-broadband UV microscope imaging system with wide range zoom capability. US Patent 7, 423, 805.
56. Kim, J., Song, H., Park, I., Carlisle, C.R., Bonin, K., and Guthold, M. (2011) Denaturing of single electrospun fibrinogen fibers studied by deep ultraviolet fluorescence microscopy. *Microsc. Res. Tech.*, **74**, 219–224.
57. Schlueter, G., Steinberg, W., and Whittey, J. (2002) Performance data of a new 248 nm CD metrology tool proved on COG reticles and PSM's. *Proc. SPIE*, **4562**, 379–385.
58. Doe, N., Eandi, R.D., and St. Cin, P. (2001) A new algorithm for optical photomask CD metrology for the 100 nm node. *Proc. SPIE*, **4562**, 225–236.
59. Hillmann, F., Scheuring, G., and Brueck, H.-J. (2008) New results from DUV water immersion microscopy using the CD metrology system LWM500 WI with a high NA condenser. *Proc. SPIE*, **6792**, 679215.
60. Guy, B.Z., Zait, E., Dmitriev, V., Labovitz, S., Graitzer, E., Böhm, K., Birkner, R., and Scheruebl, T. (2008) Mask CD control (CDC) using AIMS as the CD metrology data source. *Proc. SPIE*, **7028**, 70281C.
61. Klose, G., Beyer, D., Arnz, M., Kerwien, N., and Rosenkranz, N. (2008) PROVE: a photomask registration and

overlay metrology system for the 45 nm node and beyond. *Proc. SPIE*, **7028**, 702832.

62. Buttgereit, U., Birkner, R., Seidel, D., Perlitz, S., Philipsen, V., and De Bisschop, P. (2008) Phame® – phase measurements on 45 nm node phase shift features. *Proc. SPIE*, **7028**, 70282Z.

63. Beyer, D., Buttgereit, U., and Scheruebl, T. (2009) In-die metrology on photomasks for low k1 lithography. *Proc. SPIE*, **7488**, 74881N.

64. Sohn, Y.J., Quintanilha, R., Barnes, B.M., and Silver, R.M. (2009) 193 nm angle-resolved scatterfield microscope for semiconductor metrology. *Proc. SPIE*, **7405**, 74050R.

65. Ehret, G., Pilarski, F., Bergmann, D., Bodermann, B., and Buhr, E. (2009) A new high-aperture 193 nm microscope for the traceable dimensional characterisation of micro- and nanostructures. *Meas. Sci. Technol.*, **20**, 084010-1–084010-10.

66. Li, Z., Pilarski, F., Bergmann, D., and Bodermann, B. (2009) A 193 nm optical CD metrology tool for the 32 nm node. *Proc. SPIE*, **7488**, 74881J-0–74881J-10.

67. Bodermann, B., Li, Z., Pilarski, F., and Bergmann, D. (2010) A 193 nm microscope for CD metrology for the 32 nm node and beyond. *Proc. SPIE*, **7545**, 75450A.

68. Wurm, M., Bonifer, S., Bodermann, B., and Gerhard, M. (2011) Comparison of far field characterisation of DOEs with a goniometric DUV-scatterometer and a CCD-based system. *JEOS: Rapid Publ.*, **6**, 11015S.

69. Attota, R., Silver, R., and Dixson, R. (2008) Linewidth measurement technique using through-focus optical images. *Appl. Opt.*, **47**, 495–503.

70. Kerwien, N., Tavrov, A., Kauffmann, J., Osten, W., and Tiziani, H.J. (2003) Rapid quantitative phase imaging using phase retrieval for optical metrology of phase-shifting masks. *Proc. SPIE*, **5144**, 105–114.

71. Osten, W., Ferreras Paz, V., Frenner, K., Schuster, T., and Bloess, H. (2009) Simulations of scatterometry down to 22 nm structure sizes and beyond with special emphasis on LER. *AIP Conf. Proc.*, **1173**, 371–378.

72. Wurm, M., Bodermann, B., and Pilarski, F. (2007) Metrology capabilities and performance of the new DUV scatterometer of the PTB. *Proc. SPIE*, **6533**, 65330H.

73. Wurm, M., Pilarski, F., and Bodermann, B. (2010) A new flexible scatterometer for critical dimension metrology. *Rev. Sci. Instrum.*, **81** (2), 023701-1–023701-8.

74. Neil, M.A.A., Juskaitis, R., and Wilson, T. (1997) Method of obtaining optical sectioning by using structured light in a conventional microscope. *Opt. Lett.*, **22**, 1905–1907.

75. Poher, V., Zhang, H.X., Kennedy, G.T., Griffin, C., Oddos, S., Gu, E., Elson, D.S., Girkin, M., French, P.M.W., Dawson, M., and Neil, M.A. (2007) Optical sectioning microscopes with no moving parts using a micro-stripe array light emitting diode. *Opt. Express*, **15**, 11196–11206.

13
Model-Based Optical Metrology
Xavier Colonna de Lega

13.1
Introduction

The past two decades have witnessed a steady increase in the adoption of optical metrology for industrial applications, including quality control, process development, and production floor process control. Optical methods shine, in particular, where tight manufacturing tolerances dictate the use of high-sensitivity metrology tools. Adoption by the optics and semiconductor industries predates wider adoption by the data storage, automotive, flat panel displays, solar, and microelectromechanical systems (MEMS) industries. Similar trends are observed in biomedical and health-related domains.

This evolution is enabled in part by the rapid expansion of computer processing power. Indeed, we will show that while all optical metrology methods require modeling of the measurement process to solve the corresponding inverse problem [1, 2], there is a subset of advanced inverse problems that can only be solved through computation-intensive inverse models. This chapter surveys progressively more complex techniques and emphasizes the gains enabled by such inverse models compared to forward models. Our ambition is not to provide an exhaustive account of each development but rather to provide an overview of trends that have shaped the recent evolution of optical metrology, with a bias toward techniques that have found practical application in the industrial world.

A guiding thread throughout this chapter is coherence scanning interferometry (CSI), a versatile measurement technique that lends itself well to illustrating the benefits of model-based optical metrology.

13.2
Optical Metrology

An optical metrology technique derives quantitative estimates of a measurand [3] from light that interacts with the sample under test. This definition excludes optical inspection, defect detection, and visualization, which are qualitative in nature (we

Optical Imaging and Metrology: Advanced Technologies, First Edition.
Edited by Wolfgang Osten and Nadya Reingand.
© 2012 Wiley-VCH Verlag GmbH & Co. KGaA. Published 2012 by Wiley-VCH Verlag GmbH & Co. KGaA.

note in passing that these disciplines similarly benefit from inverse modeling or *identification* techniques [4]). The scope of this chapter is further restricted to the subset of noncontact methods that rely on free-space light propagation. Hence, we set aside the subset of optical metrology techniques based on guided light propagation in optical fibers and waveguides [5].

Noncontact optical methods provide information about a wide variety of physical characteristics of a sample volume or surface thanks to the many measurement principles and operating wavelengths available (from the deep ultraviolet to the far infrared). Practical applications are found in many phases of industrial product development and manufacturing. A nonexhaustive list follows, sorted by general application domains.

- Measurement of form, geometry (thickness, parallelism), slope, waviness, roughness, microroughness, texture, lay pattern: main applications in the automotive and technical glass industries
- Measurement of transmitted wave front, bulk optical material homogeneity, and birefringence: technical glass industries
- Characterization of material optical properties, thin-film stacks, dimensions, and registration of microstructures: main applications in the semiconductor, MEMS, LED, flat panel display, and data storage industries.

The first two categories in the list above rely mostly on the deterministic forward models discussed in Section 13.4, whereas the vast majority of applications in the semiconductor and related industries rely on the inverse models covered in Section 13.5.

Outside of industry, academia and institutional research laboratories use optical metrology in numerous domains, such as materials sciences (measurement of mechanical properties, e.g., Young's modulus or Poisson's ratio), the definition of fundamental physical constants or reference standards, and fundamental research (the largest interferometers are kilometer-sized gravitational wave detectors).

13.3
Modeling Light–Sample Interaction

13.3.1
From Light Detection to Quantitative Estimate of a Measurand

The chain of operations in an optical metrology measurement can be summarized as follows:

- An acquisition system detects and digitally samples the intensity of light that has interacted with the part under test. Multiple intensity values are measured as a function of one or more experimental parameters such as wavelength, polarization state, propagation direction, spatial coordinate (such as in the two-dimensional image of an object), temporal coordinate, optical delay (interferometers), and so on.

- A preliminary analysis based on a model of the experimental apparatus derives a refined estimate of the measured intensity (e.g., using calibration data) as well as other attributes of the detected light, such as amplitude, phase, polarization state, propagation direction, or optical path difference (OPD).
- A physical model of the light–sample interaction derives one or more characteristics of the sample under test from the refined light attribute(s). The complexity of this analysis step varies from essentially trivial (e.g., convert a measured phase into a traveled distance) to very complex as in scatterometry (derive from the properties of reflected light detailed information about a periodic optically unresolved structure). This step is the main topic of discussion in the upcoming sections.
- The measured sample characteristics are further analyzed according to the definition of a measurand in order to report a final metrology result. For example, numerical filters are applied to surface topography data to derive an estimate of surface roughness expressed as a power spectral density within a specific range of spatial frequencies. In another example, form data are fitted with a model surface to report a global characteristic such as cone angle or radius of curvature. Ideally, the measured value is qualified with an uncertainty estimate, taking into account bias and stochastic error sources [6].

Figure 13.1 summarizes the information flow outlined above and illustrates the two varieties of light–sample interaction models used for converting raw data into metrology data, as discussed in the following section.

13.3.2
Two Types of Light–Sample Interaction Models

Maxwell's equations [7] provide the mathematical framework to compute propagating and evanescent electromagnetic (EM) fields, given the material properties and geometry of the objects that reside within a volume of interest. Optical metrology relies on the dependence of such EM fields on a sample's properties to gain information about the sample. The practical inverse problem then consists in converting the measurable optical information back into sample properties. In some cases, the mathematics yields analytical solutions that can be readily inverted to describe physical parameters as functions of measured EM field. For instance, the motion of an optically smooth sample can be described as simply inducing a change in the phase of a wave front that reflects from the object surface. Assuming that the motion does not change the properties of the object surface, the potentially complex interaction of light at the surface can be summarized as a constant phase change on reflection. Measuring the variation of the phase of the reflected light in an interferometer then directly yields a measure of the object displacement (after scaling by the optical frequency).

We call forward model any analytical mathematical model that in a similar manner takes measured optical data as inputs and directly computes quantitative estimates of some characteristics of the sample under test. While such models can

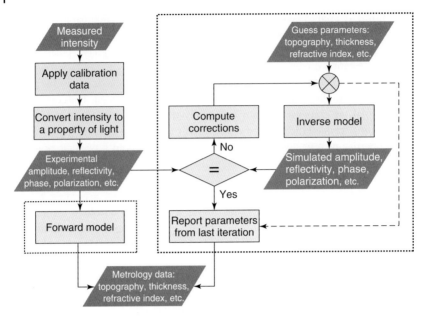

Figure 13.1 Data flow of an optical metrology measurement. Conversion of raw data into metrology data requires a model of the light–sample interaction. Such a model is called *forward* when the analysis procedure is direct (dotted box in lower left corner). Alternatively, an *inverse* model requires iteratively refining the parameters of a theoretical representation of the sample until the model predictions for the raw data match the experimental data (Section 13.3.2).

be complex in nature, they are deterministic and predictable: to a set of meaningful physical inputs corresponds a (usually) unique set of meaningful outputs.

More general inverse models provide algorithms that can only predict what optical data are expected to be measured for a given set of characteristics of the sample under test. For instance, the knowledge of the optical properties of a multilayer film stack allows computing its complex reflectivity as a function of wavelength or angle of incidence, for example, using the characteristic matrix formalism [8]. From the reflectivity we can further predict the intensity, or other derived property of light, that would be measured by an ideal instrument. However, this mathematical formalism does not generally provide a direct way to convert measured reflectivity into film stack characteristics (which would be a forward model by our definition). A practical analysis process thus requires iteratively modifying input metrology data, in this case refractive index and layer thickness, until the resulting simulated optical data match the experimental data within some tolerance. The convergence of this iterative process is not guaranteed in general nor is the uniqueness of the solution. Potential problems include cross talk between model unknowns, failure to converge when model parameters assumed known are in fact incorrect or when guess values for the unknown parameters are too far from the actual solutions, and measurement sensitivity that strongly depends on the model parameters. These pitfalls need to be properly accounted for in order to enable the broader metrology capabilities provided by inverse models.

Section 13.4 reviews a few examples of forward models and sets the stage for the discussion of inverse models in Section 13.5. Section 13.6 develops the topics of validation and confidence in inverse model techniques.

13.4
Forward Models in Optical Metrology

A Michelson interferometer generates an intensity pattern that exhibits sinusoidal variations with changes in the interferometer optical paths (Figure 13.2). This intensity variation is converted to a phase variation using one of a number of methods (heterodyne detection or phase shifting, e.g., [9]). One application is the measurement of displacement by linking one of the interferometer mirrors with the object to be monitored (e.g., a wafer stage in a photolithography tool [10]). Phase variations are converted into displacement data by applying a scaling factor proportional to the optical frequency (which amounts to a very simple forward model). The optical frequency acts in this case as a fundamental metrology metric and must be known to an uncertainty smaller than the uncertainty requirements of the application.

The same Michelson interferometer can conversely be used to characterize the optical frequency of a quasi-monochromatic light source [11], assuming in this case that the motion of one of the mirrors is known via some other means (e.g., another interferometer). In this case, the rate of change of phase with respect to displacement provides an estimate of the optical frequency – another simple forward model example. This is the operating principle behind Fourier transform spectroscopy, which can characterize simultaneously the multiple spectral components of a light source [12]. The forward model used in this case amounts to computing the Fourier transform of the interferometric intensity signal to yield the spectral distribution.

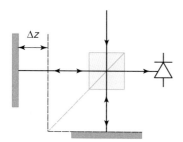
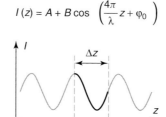

$$I(z) = A + B \cos\left(\frac{4\pi}{\lambda} z + \varphi_0\right)$$

Figure 13.2 A collimated beam of monochromatic light (e.g., from a laser) illuminates both arms of a simplified interferometer. The reflected light beams are recombined by the beam splitter and interfere at the detector. The measured intensity is a sinusoidal function of the difference in optical path in the two arms. Optical path variations are converted into displacement when the wavelength of light is known or into optical frequency when the displacement is known (see text).

The simple forward scaling model outlined above for the Michelson interferometer forms the basis of the data analysis applied to other interferometers used in the optics shop (Fizeau, Twyman-Green, Mach-Zehnder, see [13]), where phase is converted to OPD. When measuring the surface figure of an optic, the OPD is simply the distance between a reference surface and the optic under test. For the measurement of refractive index homogeneity in a transparent material, the OPD provides the required information about the location of the material interfaces and their indices [14]. A third class of applications is dedicated to the characterization of wave fronts transmitted or reflected by an optic. In this case, the OPD represents the departure of the test wave front from a reference wave front, which provides a measure of optical aberrations in the optical system.

Other types of instruments for wave front metrology include Shack–Hartman sensors or shearing interferometers that are sensitive to wave front slope [13]. These techniques trade the sensitivity and lateral resolution of traditional interferometers for simpler and more cost-effective instrument designs that have the ability to function over wide spectral ranges. Data analysis for Shack-Hartman sensors consists in the computation of the location of an array of dots created on a planar detector by an array of microlenses. Each dot corresponds to a small subregion of the wave front under test. The dot locations relate simply to the product of the microlenses' focal length and local wave front slope. Data analysis in a shearing interferometer requires computing the phase gradients of the interferogram (usually in two orthogonal directions) and integrating to recover the wave front shape.

A number of optical techniques used for form and strain measurement of diffusing objects (e.g., fringe projection [15], triangulation, speckle or holographic interferometry [16], and desensitized interferometry [17]) do not necessarily require collimating illumination and observation optics. This simplifies the optical design and allows covering a measurement volume larger than the aperture of the optical components. However, the use of diverging waves generally results in a field-dependent sensitivity, where sensitivity is the rate of change of a parameter such as optical phase as a function of a change of the physical parameter to be measured. This variation needs to be accounted for when converting the raw information into metrology data. Such a model of the measurement process is usually derived from a mix of known information about the system geometry and calibration procedures [15, 18].

Other metrology techniques require performing a coordinate transformation as part of the conversion of raw optical data into metrology data. For instance, low-coherence interferometers measure the form of inner and outer cylinders and cones [19, 20] by mapping these inherently three-dimensional surfaces onto two-dimensional imagers (CCD or CMOS cameras). The form data derived from the interferogram is mapped back to a three-dimensional coordinate system. This transformation is based on a priori information about the instrument and multiple calibration steps. The transformation of the raw optical data (optical path or phase) into a cloud of points describing a surface remains a forward model.

13.4 Forward Models in Optical Metrology | 289

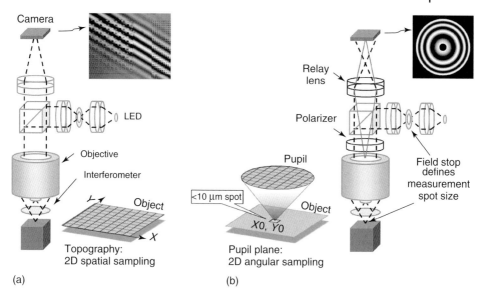

Figure 13.3 (a) Coherence scanning interferometer in surface profiling configuration. The light source is extended and/or spectrally broadband and generates localized interference fringes at sample locations that are in focus on the detector. An interference signal is captured at each detector element as the sample is mechanically scanned along the optical axis. (b) Same interferometer in pupil imaging mode for ellipsometry or scatterometry applications. Detector points correspond to different azimuthal and incidence angles in object space. The measurement spot is defined by a field stop.

CSI (also known as scanning white light interferometry, see [21]) is a versatile surface characterization technique. The limited coherence of the light source results in the observation of high-contrast fringes in a limited volume around object points that reside near the best focus plane of the instrument (Figure 13.3a). A typical CSI instrument records interferogram data as the object surface is scanned along the optical axis. The localization of the fringe pattern is used as a virtual contact probe when profiling optically rough object surfaces [22]. For optically smooth objects, both the location and the phase of the position of maximum interference contrast are used for subnanometer level topography measurements [23]. The phase is converted into surface height either using the assumed/known mean optical frequency of the light source or using the scan speed as the reference metric. The location of maximum contrast is used to determine the fringe order at each detector element (since phase is determined modulo 2π). However, phase and peak contrast location depend on both surface height and the optical properties of the materials present on the object surface [24, 25]. A proper model of the signal formation process accounts for the materials effects and provides in simple cases absolute topography and relational information [26]. These analysis steps fall within the forward models category as no iterative process comes into play to convert the raw optical data into topography data. Materials effects such as phase change on reflection are computed using known information about the optical properties [8, 27].

Digital holography [28] differs from analog holography in that the hologram is directly recorded on a digital detector (e.g., CCD camera). It otherwise provides the same capability for form and deformation metrology [29]. Hologram reconstruction is performed numerically and allows computing the light amplitude and phase distribution at any location within the object volume. This is an opportunity to reconstruct images that have better quality than what would be produced by the optical components alone. For example, one can correct for known optical aberrations [30, 31] and extend the range over which reconstructed information is in focus [32, 33]. This is achieved by modeling the propagation of the optical field through the measurement system and compensating for the physical imperfections of the system. Propagation is performed numerically as a multidimensional summation of elementary wave front components. In most cases, the computation is recast as a Fourier transformation – a linear operator.

The transfer function of an optical metrology instrument describes how faithfully it reproduces features of interest. For instance, for an instrument that measures surface topography, the transfer function describes how the spatial frequency components of the surface height are attenuated through the measurement process. For interferometers, the instrument transfer function (ITF) is identical to the optical transfer function [34] in the regime where surface features are smaller than the wavelength of light. The ITF can be readily measured using a calibration sample, such as a sharp surface discontinuity. This information is used to improve the accuracy of a measurement, such as when characterizing the power spectral density of an optical surface [35], which relates to the amount of light that the surface will scatter in an optical system.

To summarize, this section surveyed a few examples of optical metrology techniques that rely on forward models describing the interaction of light with the sample. The discussion is admittedly simplified and overlooks sometimes nontrivial implementation aspects. However, these methods all share this characteristic that the metrology parameters are derived from the measured light attributes via deterministic procedures that guarantee a meaningful result.

13.5
Inverse Models in Optical Metrology

13.5.1
Wave Front Metrology

The previous section gave an overview of wave front metrology techniques based on forward modeling. There is a class of methods that make use of inverse models for the same application. They are based on mathematical models that describe the propagation of light irradiance – "transport of intensity" – and allow recovering both amplitude and phase from that information. Data acquisition consists in recording intensity images at a few locations along the nominal propagation direction of the wave front. An iterative procedure that attempts to find the wave front characteristics (amplitude and phase) that match the measured intensity patterns is then applied [36]. The extreme simplicity of the experimental apparatus is the key attribute of these methods. The main trade-off compared to interferometers is the limitation to fairly low numerical aperture beams [37]. An interesting variation relies on an imaging system that exhibits longitudinal chromatism [38]. The acquisition of a single-color image provides the three intensity estimates required to reconstruct the wave front phase in the plane of best focus. This is used to image weak phase objects such as biological samples or reconstruct the topography of smooth reflective objects.

13.5.2
Thin-Film Structures Metrology

The development of thin-film metrology tools has tracked the development of semiconductor manufacturing processes, which rely in part on producing complex multilayered structures. These manufacturing and process control tools are used nowadays in other industrial fields, such as optical coatings, flat panel display devices, read/write heads used for data storage, MEMS, biomedical devices, and so on. The low cost of some low-end film metrology tools (e.g., compact reflectometers) further enables varnish or paint monitoring applications (automotive, glass panels for buildings) or hardening coatings on functional mechanical surfaces (e.g., diamond-like carbon coatings on bearing surfaces).

Ellipsometry [39] is arguably the most capable optical metrology tool for the characterization of thin material layers and likely one of the earliest examples of application of an inverse model for industrial metrology. The method relies on the measurement of the ellipsometric angles Δ and Ψ, which relate to the ratio of reflection coefficients of the sample for S and P polarized light via the equation: $\tan(\Psi) \exp(i\Delta) = r_P/r_S$. The two angles are derived from the variation of the measured reflected intensity for different polarization states of the illumination light (Figure 13.4). As mentioned in Section 13.3.2, models readily compute the reflection coefficients r_P and r_S (and thus predict the ellipsometric angles) using the optical properties and thickness of the layers in a given film stack. However, there are only a handful of cases for which analytical expressions allow to directly compute film stack parameters as a function of the measured ellipsometric angles. An example is the measurement of the thickness and refractive index of a single-layer film on a substrate of known optical properties [40]. In the general case, however, one is left with the application of an inverse model approach. This requires defining in a first step a complete numerical model of the film stack with the correct number of material layers and known or guess values for each layer's thickness and optical properties. Next, an iterative solver performs an optimization of the unknown model parameters until the computed ellipsometric angles match the measured values within some tolerance [41]. The most advanced ellipsometers perform such an analysis for multiple wavelengths and/or angles of incidence simultaneously in order to improve the likelihood of convergence of the iterative procedure and gain information about material dispersion.

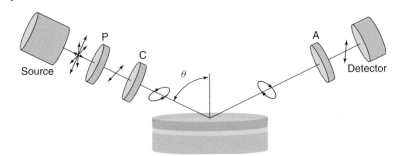

Figure 13.4 In a typical ellipsometer, the light from a collimated light source is first linearly polarized by a polarizer P and then converted to an elliptical polarization state by a birefringent compensator C before reflecting from the sample surface. A detector measures the intensity of the reflected light after transmission through an analyzer A. Measurements for different positions of the three elements P, C, and A yield the ellipsometry angles Δ and Ψ. The angle of incidence α is adjustable. Spectral ellipsometers repeat the measurement at multiple wavelengths to increase the amount of information collected about the sample.

The procedure just outlined is common to most inverse model techniques discussed in this section. A practical variation used when measurement throughput is critical consists in precomputing the values of the measured parameters (the angles Δ and Ψ in the case of ellipsometry) for a large number of possible values of the unknown parameters (e.g., layer thickness or refractive index). The experimental data are then rapidly matched to the precomputed values stored in a signal library. This approach is particularly relevant in the context of process control where information is available about the expected variability of the unknown parameters within the process window.

Contrary to ellipsometers, reflectometers only measure the reflectivity (there is no phase information) of a film stack as a function of wavelength or angle of incidence. Data analysis is performed in the same manner with an inverse model.

Some surface characterization tasks require measuring thickness variations of transparent material layers as well as form and roughness of the corresponding material interfaces, a combination of metrology capabilities that ellipsometers and reflectometers cannot provide. Low-coherence interferometry can be readily applied for such applications as each interface generates a localized interference signal when the sample is scanned along the optical axis of the interferometer. The experimental signal is modeled and decomposed as the combination of multiple copies of a single-interface signal template by a least-squares procedure [42]. The signal template is acquired experimentally. Figure 13.5 illustrates the principle of the technique for the measurement of film thickness profile and top surface topography. Note that the analysis procedure straddles the boundary between forward and inverse models as it relies on optimization procedures. Yet, the model is heavily constrained and convergence is guaranteed in practice.

The above method is generally limited to layer thicknesses on the order of a fraction of the coherence length of the broadband light source (e.g., about 1 μm

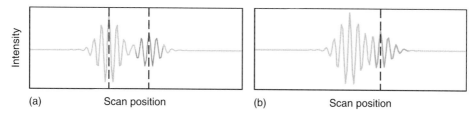

Figure 13.5 (a) Template signal matching for thick-film thickness measurement in CSI: the two best match signal positions provide an estimate of the optical thickness. (b) Template signal matching for topography measurement in the presence of a transparent film using a "leading-edge" template. (Please find a color version of this figure on the color plates.)

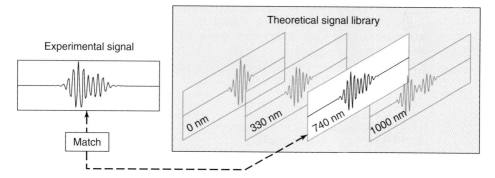

Figure 13.6 Inverse model film thickness measurement in CSI: experimental signals are matched to precomputed theoretical signals that depend on instrument calibration data and a model of the sample under test, in this example a layer of silicon dioxide on silicon.

for silicon dioxide layers on silicon). For thinner films, the interference signals originating at each interface overlap to create a distorted signal, the fine features of which contain signatures of the film properties [43, 44]. We can significantly extend CSI's film profiling capability by looking for a match between such experimental signals and collections of precomputed film signatures. This requires modeling the signal formation in the interferometer, taking into account spectral and numerical aperture effects as well as detailed optical properties of the instrument (dispersion, polarization-dependent effects, pupil-plane illumination, etc.) [45]. Figure 13.6 illustrates the process of comparing an experimental signal to model signals contained in a library. Figure 13.7 shows a map of photoresist thickness measured at a transistor gate on a thin-film transistor liquid crystal display (TFT LCD) panel. Figure 13.8 illustrates the method's capability when profiling a tapered silicon dioxide film on silicon with various interferometric objectives. Compared to the empirical approach of the previous paragraph, this inverse model method requires a priori information about the instrument and sample. However, the result is a capability improvement by more than an order of magnitude as the minimum film thickness that can be reliably measured drops below 50 nm. To put things in perspective, an ellipsometer will reliably measure films thinner

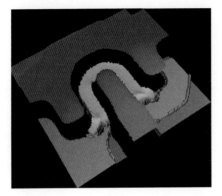

Figure 13.7 Inverse model film thickness measurement in CSI: map of photoresist thickness near a transistor gate on a TFT LCD panel. The horseshoe-shaped trench is nominally 400 nm thick and 5 μm wide at the bottom. (Please find a color version of this figure on the color plates.)

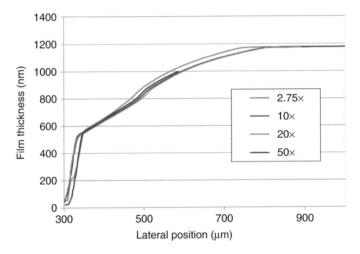

Figure 13.8 Inverse model film thickness measurement in CSI: thickness profiles of a silicon dioxide film on silicon measured at similar locations with Michelson and Mirau interference microscope objectives. (Please find a color version of this figure on the color plates.)

than 1 nm. However, an ellipsometer has no sensitivity to surface topography and measures film properties over an area typically many tens of micrometers wide, whereas the CSI instrument provides submicrometer lateral resolution with high-numerical-aperture objectives.

A further extension of low-coherence interferometry is the acquisition of interference data in the exit pupil of a high-numerical-aperture microscope objective (Figure 13.3b). Using polarized light and a Fourier decomposition of the experimental signals it is possible to recover ellipsometry angles over the spectral range of the light source and the illumination angle range of the objective [46]. These data are then processed as regular ellipsometry data to simultaneously measure thickness and optical properties of layered structures. Figure 13.9 shows example ellipsometry angles Δ and Ψ measured at multiple wavelengths and angles of incidence for a two-layer film stack. Combining this measurement mode with the model-based surface profiling technique outlined in the previous paragraph results in a metrology platform used for process control in the data storage industry [47]. The more sensitive pupil-plane technique provides estimates of material properties and nominal thickness of the film layers averaged over a ~10 μm wide measurement spot. The profiling technique uses this information to provide topography and film thickness profiles with submicrometer lateral resolution. The main benefit of this solution is the ability to perform subnanometer topography and film metrology with a single instrument.

As a last example, we refer the reader to chapter 10 in this book [48] for discussion of an inverse model technique used for profiling thick material stacks using optical coherent tomography – a technique related to CSI.

13.5.3
Unresolved Structures Metrology

Transistors in semiconductor devices are three-dimensional structures with overall dimensions smaller than the resolution of optical instruments operating at deep ultraviolet wavelengths. Consequently, such structures cannot be imaged optically. However, their characterization for process control is increasingly performed in production with optical techniques [49]. The main reason is the high measurement throughput of noncontact optical methods compared to atomic force microscopes or electron microscopes. The optical measurement is enabled in practice by the addition on the semiconductor wafers of periodic test structures used to monitor specific steps of the manufacturing process. For instance, shallow trench insulation (STI) is a process step where dielectric regions are created on the wafer to electrically isolate neighboring transistors. Test structures are developed in this case specifically for the purpose of monitoring the depth of the isolation regions, their width, their sidewall angle, and so on.

Scatterometry measurements are typically performed using advanced ellipsometers or reflectometers operating over large spectral bandwidths (150–1000 nm in some instances) [50]. The information they collect is the reflectivity of the test arrays, averaged over many periods. Consequently, these methods cannot detect local defects but only provide aggregate information about the geometry of the "average" structure. The light–material interaction is modeled by solving Maxwell's equations in the periodic structure, resulting in a computation burden orders of magnitude larger than that required by ellipsometry [51]. Similar to the inverse models used for film metrology, an iterative procedure optimizes the unknown parameters of a geometric model of the structure until the predicted reflectivities match the experimental data (typically for multiple wavelengths). Precomputed libraries of candidate signals are used when this optimization procedure cannot be

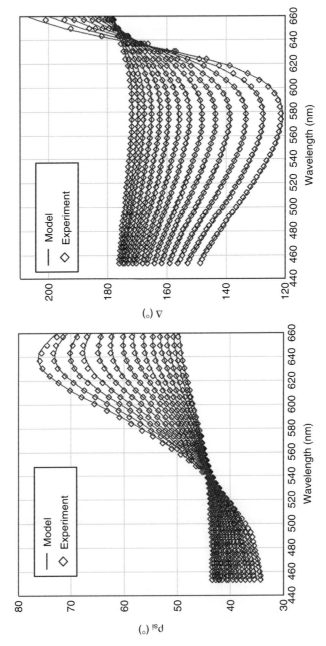

Figure 13.9 Ellipsometric angles Δ and Ψ measured using coherence scanning interferometry in the pupil for a 56 nm thick diamond-like carbon layer on 5 nm of alumina on a metal alloy. Each line in the plot represents a different angle of incidence, ranging from 20° to 46°.

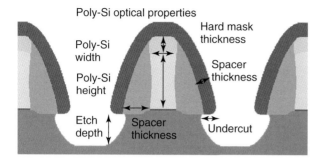

Figure 13.10 Example model geometry used in scatterometry to compute the reflectivity of a periodic optically unresolved array of transistor gate structures. The array pitch is less than 150 nm in this example. Seven geometry parameters and one refractive index parameter are optimized for this application.

performed fast enough for production line metrology (where the allowable measurement duration is on the order of 1 s per site). Figure 13.10 shows an example of a geometric model where seven parameters are optimized simultaneously to characterize transistor gate structures after etch.

Ellipsometers and reflectometers used in the semiconductor industry are single-point measurement instruments. Conversely, the scatterfield microscope creates an image of the wafer under test on a two-dimensional detector, as in conventional microscopy [52]. Multiple intensity images are acquired sequentially as a single source point is moved across the illumination pupil. This provides measurements of reflectivity as a function of angle of incidence, azimuth, and wavelength. More than one test structure can be present simultaneously in the field of view, which provides a benefit for overlay measurements that rely on multiple test pads.

It is also possible to use interferometry to collect information about unresolved structures. In this case, both amplitude and phase of the reflected light are measured by the instrument. In a conventional imaging mode, the step height between unpatterned and patterned areas correlates to a characteristic of the structures [53]. The mapping of measured step to geometrical parameter relies on a lookup table that is computed using the same modeling tools used for standard scatterometry. Multiple test pads are required in practice to measure more than one parameter.

The back pupil-plane CSI method described in [46] provides scatterometry data with enhanced sensitivity owing to the range of azimuthal angles provided by the measurement geometry [54]. The amplitude and phase of the reflected light are measured as a function of wavelength, angle of incidence, and azimuthal angle. The software used to solve Maxwell's equations for these applications is Microsim, a rigorous coupled wave analysis package [55]. A benefit compared to ellipsometers and conventional reflectometers is the reduced size of the measurement spot, which enables placing small test structures within the dies where semiconductor devices are printed, instead of between dies as is the common practice. Figure 13.11 shows

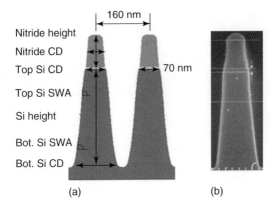

Figure 13.11 (a) Seven-parameter geometric model of a periodic shallow trench insulation structure measured using coherence scanning interferometry in the pupil. The test structure pitch is 160 nm, the depth of the silicon trench is ~300 nm, and the key critical dimension (CD) is 70 nm. Additional parameters include sidewall angle (SWA) and top or bottom CD. (b) Cross-sectional SEM of a similar test structure illustrating the agreement of the two metrology techniques.

the measured profile of an STI test structure with seven unknown parameters. An SEM cross section is shown for comparison. The 3-σ reproducibility of the length parameters in this model is better than 1 nm, as demonstrated during a three-day endurance test.

To conclude this section on unresolved structures metrology, it is worth noting that optical scatterometry is rolled out at an increasing pace for process control of semiconductor manufacturing. This requires the rapid development of more capable tools as semiconductor device features continue shrinking.

13.6
Confidence in Inverse Model Metrology

The previous sections draw a picture where inverse models enable new functionality for optical metrology tools. This, however, comes with some drawbacks. One of them is the computational burden that remains challenging for scatterometry and leads to the use of computer clusters to meet production line measurement throughput (most other inverse techniques can be tackled today with off-the-shelf workstations and/or Graphical Processing Units). Nevertheless, regular gains in computing power allow tackling more and more complex applications, such as three-dimensional unresolved structures.

Other limitations find their origin in the basic nature of inverse problem solvers, which usually rely on nonlinear iterative optimization procedures to find *a solution*, starting from some set of guess values. The end user, on the other hand, would like to know that the metrology tool provides the *correct solution*, within the measurement uncertainty expected for a given tool.

13.6.1
Modeling Pitfalls

A key step is setting up a starting model. This requires providing nominal material properties and dimensions of the object under test as well as choosing guess values for those parameters that will be measured. These choices strongly influence the robustness and likelihood of a successful measurement. This raises a number of questions both for the tool designer and for the end user:

- Is the model meaningful? That is, does the model capture all the required physical attributes that influence the light–sample interaction? A practical approach consists in starting with a model that is as simple as possible and gradually adding complexity as long as the model remains stable. For instance, a first-pass optimization estimates the thickness of unknown films in an ellipsometry measurement where each layer is modeled as a homogeneous domain. The model is then refined by the addition of interfacial layers to better account for material interface effects and roughness [41].
- Are model approximations adequate? In the case of scatterometry, the model of the structure of interest is typically broken into small rectangular slabs of materials. The larger the number of slabs, the more accurate the representation of the light interaction and the slower the computation. The algorithm needs to balance these two competing requirements (accuracy vs speed). In some cases, increasing the complexity of the up-front model analysis can provide significant convergence improvements [56]. In the context of wave front reconstruction from transport of intensity, there is a similar compromise between the lateral resolution of the reconstruction and the speed and convergence robustness.
- What is the required accuracy for the model parameters that are assumed known? The sensitivity analysis described in the next section can help define this requirement. It is not unusual to find multistep metrology procedures in the semiconductor industry where metrology results from previous process steps are fed forward for more accurate metrology in later manufacturing phases. Similarly, multisensor or sensor fusion approaches allow collecting complementary pieces of information that increase the confidence in the results from an inverse model [57].
- Is the modeling process robust? Is convergence guaranteed? Can ill-conditioned cases be detected? These questions relate to algorithm design and its ability to detect local minima, parameter correlations, and other obstacles to convergence toward the correct solution. Not surprisingly, specific inverse models call for specific solutions [58, 59]. Still, a common characteristic is that the likelihood of a successful measurement depends a lot on the quality of the model starting point. Significant robustness gains can be made when using the so-called grid search techniques where the optimization procedure is run multiple times with different seed values. This reduces the risk of the solver getting trapped in a local optimization minimum.

13.6.2
Sensitivity Analysis

A sensitivity analysis tool is a key component of an inverse model metrology solution. It relies on estimating the partial derivatives of the measured quantities (intensity, amplitude, phase, etc.) as a function of the model parameters. This information is then combined with a noise model of the measurement process to provide a report of the tool capability for the structure under test [60]:

- **Predicted measurement repeatability and bias for each model parameter**: these estimates typically assume no correlation in the various noise sources.
- **Parameter cross correlations**: this describes how a change in one model parameter can be exactly offset by a change in another. Strong correlations indicate that the instrument is not capable of accurately measuring these model components simultaneously. An example is the measurement of a thin layer of unknown optical properties with a fixed-angle laser ellipsometer. In this case, thickness and refractive index are heavily cross correlated. This limitation is usually overcome by repeating the measurement at different wavelengths or angles of incidence for which the correlation coefficient is different.
- **Uniformity of sensitivity over parameter range**: there are instances where an instrument suffers from a loss of sensitivity to some of the model parameters. For example, films with certain thicknesses will correspond to inflection points of the ellipsometric angles Δ and Ψ. Increasing the number of wavelengths is one of the means of restoring sensitivity to the instrument.
- **Optimum instrument configuration**: for an instrument where data can be acquired under a range of operating parameters (polarization, wavelength, propagation direction, etc.), the sensitivity analysis predicts those configurations that provide optimum sensitivity to the parameters of interest or minimize parameter cross correlations.
- **Uncertainty requirements for fixed input parameters**: the sensitivity analysis can also predict the biases induced in the unknown model parameters as a function of the model inputs assumed known. This provides an upper bound on the allowable uncertainty for these parameters and can inform the user of the need to characterize critical parameters beforehand.

13.6.3
Validation of the Overall Tool Capability

Ultimately, the capability of the optical metrology tool will be scrutinized by the metrologist. Repeatability and reproducibility tests provide some of the most important metrics of an instrument's performance, especially for process control applications. As mentioned earlier, an instrument's sensitivity can be heavily application dependent, which implies testing with varied samples. The outcome of repeatability tests is also used to improve the noise models used in a sensitivity analysis.

Measurement biases – the other key ingredient of an uncertainty budget – are best evaluated by comparing the results of multiple metrology tools. The total measurement uncertainty methodology analyzes such results by accounting for the uncertainty contributions of the metrology tools used as reference and is now common practice in the semiconductor industry [61]. These tests are performed with calibration standards and reference samples. A remaining difficulty is that measurement uncertainty is, by definition, sample dependent. In the case of inverse model instruments that have widely varying sensitivities, this implies deriving an uncertainty budget for each specific application.

13.7
Conclusion and Perspectives

Forward models provide a key data analysis step in many optical metrology tools in use today. There is, however, a growing breed of techniques based on inverse models. In some cases, they offer an alternative analysis technique that enables new instrumental approaches (such as transport of intensity for wave front measurement). In other cases, they enable a measurement capability not available with a forward model, such as in thin film or unresolved structure metrology. Their emergence is strongly linked to the steadily growing computation power available from commodity computer hardware.

The multiple application examples of CSI provided throughout this chapter illustrate how inverse models can significantly enhance and develop the metrology capability of a given optical technique. Indeed, there is no difference in the hardware required to measure the surface topography of a simple object, the thickness of a thick film, the optical properties and thickness of a multilayer stack, or even the detailed geometry of a periodic optically unresolved structure. The only difference between these use cases is the complexity of the data analysis and the amount of information provided by the user to seed the measurement. It is likely that similar developments will enhance the functionality of other existing optical metrology techniques.

Looking ahead, optical metrology will continue to gain acceptance in the industry as a result of ever tightening manufacturing tolerances, progress in tool functionality provided by advanced modeling, and progress in enabling technologies such as detectors (low-noise, high frame rate, and large format cameras), optics (aspheres, spatial light modulators, phase modulators, etc.), and light sources (high-brightness LED, supercontinuum lasers, etc.).

Inverse model optical metrology techniques do present some challenges, as seen in the last section. Their adoption will be more rapid if instrument designers can improve robustness and usability to the point where they become equivalent to that of more conventional forward-model-based instruments. The following list outlines areas of further development that will benefit any type of optical metrology tools:

- automated sensitivity analysis providing a comprehensive report of the expected capability of the instrument for a given application;
- automated system configuration optimization (choice of wavelength, illumination direction, polarization state, spatial sampling, etc.) to minimize parameter correlations and maximize sensitivity to parameters of interest;
- automated assessment of the quality of raw measurement data;
- estimated uncertainty computation for any given measurement;
- user input limited to providing information about the sample under test.

References

1. Bertero, M. and Boccacci, P. (1998) *Introduction to Inverse Problems in Imaging*, IOP Publishing, Bristol.
2. Osten, W., Ferreras Paz, V., Frenner, K., Lyda, W., and Schau, P. (2011) Different approaches to overcome existing limits in optical micro and nano metrology. *Proc. SPIE*, **8011**, 80116O.
3. ISO/IEC (2007) Guide 99:2007 *International Vocabulary of Metrology – Basic and General Concepts and Associated Terms (VIM)*, International Organization for Standardization, Geneva.
4. Osten, W., Elandaloussi, F., and Mieth, U. (2002) Trends for the solution of identification problems in holographic nondestructive testing (HNDT). *Proc. SPIE*, **4900**, 1187–1196.
5. Lopez-Higuera, J.M. (ed.) (2002) *Handbook of Optical Fiber Sensing Technology*, John Wiley & Sons, Ltd, Chichester.
6. ISO/IEC (2008) Guide 98-3:2008. *Uncertainty of Measurement – Part 3: Guide to the Expression of Uncertainty in Measurement (GUM: 1995)*, International Organization for Standardization, Geneva.
7. Born, M. and Wolf, E. (1999) *Principles of Optics*, 7th (expanded) edn, Cambridge University Press, Cambridge.
8. Born, M. and Wolf, E. (1999) *Principles of Optics*, 7th (expanded) edn, Cambridge University Press, Cambridge, pp. 58–63.
9. Malacara, D., Servin, M., and Malacara, Z. (1998) *Interferogram Analysis for Optical Testing*, Marcel Dekker, New York.
10. Gill, P. (1993) in *Optical Methods in Engineering Metrology* (ed. D.C. Williams), Chapman & Hall, London, pp. 153–177.
11. Michelson, A.A. (1995) *Light-Wave Analysis, in Studies in Optics*, Dover Publications, New York, pp. 34–45.
12. Davis, S.P., Abrams, M.C., and Brault, J.W. (2001) *Fourier Transform Spectrometry*, Academic Press, London.
13. Malacara, D. (ed.) (2007) *Optical Shop Testing*, John Wiley & Sons, Inc., Hoboken, NJ.
14. Deck, L.L. (2003) Fourier-transform phase-shifting interferometry. *Appl. Opt.*, **42** (13), 2354–2365.
15. Creath, K., Schmit, J., and Wyant, J.C. (2007) Moiré and fringe projection techniques, in *Optical Shop Testing* (ed. D. Malacara), John Wiley & Sons, Inc., Hoboken, NJ, pp. 667–742.
16. Jones, R. and Wykes, C. (1989) *Holographic and Speckle Interferometry*, 2nd edn, Cambridge University Press, Cambridge.
17. Boone, P.M. and Jacquot, P. (1991) Some applications of a HOE based desensitized interferometer in materials research. *Proc. SPIE*, **1554A**, 512–521.
18. Jones, R. and Wykes, C. (1989) in *Holographic and Speckle Interferometry*, 2nd edn, Cambridge University Press, Cambridge, pp. 81–82.
19. Albertazzi G., A. Jr., Viotti, M.R., Miggiorin, R.M., and Dal Pont, A. (2008) Applications of a white light interferometer for wear measurement of cylinders. *Proc. SPIE*, **7064**, 70640B.
20. de Groot, P.J. and Colonna de Lega, X. (2003) Valve cone measurement using

white light interference microscopy in a spherical measurement geometry. *Opt. Eng.*, **42**, 1232–1237.
21. de Groot, P.J. (2011) in *Optical Measurement of Surface Topography* (ed. R. Leach), Springer, Berlin, pp. 187–206.
22. Dresel, T., Häusler, G., and Venzke, H. (1992) Three-dimensional sensing of rough surfaces by coherence radar. *Appl. Opt.*, **31** (7), 919–925.
23. de Groot, P.J. and Deck, L. (1995) Surface profiling by analysis of white-light interferograms in the spatial frequency domain. *J. Mod. Opt.*, **42**, 389–401.
24. Biegen, J. (1994) Determination of the phase change on reflection from two-beam interference. *Opt. Lett.*, **19** (21), 1690–1692.
25. Harasaki, A., Schmit, J., and Wyant, J.C. (2001) Offset of coherent envelope position due to phase change on reflection. *Appl. Opt.*, **40** (13), 2102–2106.
26. de Groot, P.J., Colonna de Lega, X., Kramer, J., and Turzhitsky, M. (2002) Determination of fringe order in white-light interference microscopy. *Appl. Opt.*, **41** (22), 4571–4578.
27. Born, M. and Wolf, E. (1999) in *Principles of Optics*, 7th (expanded) edn, Cambridge University Press, Cambridge, pp. 63–70.
28. Goodman, J.W. and Lawrence, R.W. (1967) Digital image formation from electronically detected holograms. *Appl. Phys. Lett.*, **11** (3), 77–79.
29. Kreis, T. (2005) *Handbook of Holographic Interferometry*, Wiley-VCH Verlag GmbH, Weinheim.
30. Colomb, T., Montfort, F., Kühn, J., Aspert, N., Cuche, E., Marian, A., Charrière, F., Bourquin, S., Marquet, P., and Depeursinge, C. (2006) Numerical parametric lens for shifting, magnification, and complete aberration compensation in digital holographic microscopy. *JOSA A*, **23** (12), 3177–3190.
31. Ferraro, P., Grilli, S., Miccio, L., Alfieri, D., De Nicola, S., Finizio, A., and Javidi, B. (2008) Full color 3-D imaging by digital holography and removal of chromatic aberrations. *J. Display Technol.*, **4** (1), 97–100.
32. Ferraro, P., Grilli, S., Alfieri, D., De Nicola, S., Finizio, A., Pierattini, G., Javidi, B., Coppola, G., and Striano, V. (2005) Extended focused image in microscopy by digital holography. *Opt. Express*, **13** (18), 6738–6749.
33. Colomb, T., Pavillon, N., Kühn, J., Cuche, E., Depeursinge, C., and Emery, Y. (2010) Extended depth-of-focus by digital holographic microscopy. *Opt. Lett.*, **35** (11), 1840–1842.
34. Goodman, J.W. (1968) Frequency analysis of optical imaging systems, *Introduction to Fourier Optics*, McGraw-Hill, San Francisco, CA, pp. 138–145.
35. de Groot, P.J. and Colonna de Lega, X. (2005) Interpreting interferometric height measurements using the instrument transfer function, *Proceedings of Fringe 2005 – The 5th International Workshop on Automatic Processing of Fringe Patterns*, Springer, Berlin, pp. 30–37.
36. Fienup, J.R. (1982) Phase retrieval algorithms: a comparison. *Appl. Opt.*, **21** (15), 2758–2769.
37. Brady, G.R. and Fienup, J.R. (2009) Measurement range of phase retrieval in optical surface and wavefront metrology. *Appl. Opt.*, **48** (3), 442–449.
38. Waller, L., Kou, S.S., Sheppard, C.J.R., and Barbastathis, G. (2010) Phase from chromatic aberrations. *Opt. Express*, **18** (22), 22817–22825.
39. Azzam, R.M.A. and Bashara, N.M. (1987) *Ellipsometry and Polarized Light*, Elsevier Science B.V., Amsterdam.
40. Azzam, R.M.A. and Bashara, N.M. (1987) *Ellipsometry and Polarized Light*, Elsevier Science B.V., Amsterdam, pp. 315–332.
41. Woollam, J.A., Johs, B., Herzinger, C.M., Hilfiker, J., Synowicki, R., and Bungay, C.L. (1999) *Critical Reviews of Optical Science and Technology*, Proceedings of SPIE, SPIE – The International Society for Optical Engineering, Bellingham, Washington, Vol. CR72, pp. 3–28.
42. de Groot, P.J., Colonna de Lega, X., and Fay, M.F. (2008) Transparent film profiling and analysis by interference microscopy. *Proc. SPIE*, **7064**, 70640I.
43. Kim, S.W. and Kim, G.H. (2003) Method for measuring a thickness

profile and a refractive index using white-light scanning interferometry and recording medium therefor. US Patent 6, 545, 763, filed Mar. 23, 2000 and issued Apr. 8, 2003.

44. Tang, S., Freischlad, K., and Yam, P. (2007) Interferometry for wafer dimensional metrology. *Proc. SPIE*, **6672**, 667202.

45. de Groot, P.J. and Colonna de Lega, X. (2004) Signal modeling for low-coherence height-scanning interference microscopy. *Appl. Opt.*, **43** (25), 4821–4830.

46. Colonna de Lega, X. and de Groot, P.J. (2008) Characterization of materials and film stacks for accurate surface topography measurement using a white-light optical profiler. *Proc. SPIE*, **6995**, 69950P.

47. Colonna de Lega, X., Fay, M.F., de Groot, P.J., Kamenev, B., Kruse, J.R., Haller, M., Davidson, M., Miloslavsky, L., and Mills, D. (2009) Multi-purpose optical profiler for characterization of materials, film stacks, and for absolute topography measurement. *Proc. SPIE*, **7272**, 72723Z.

48. Weiss, D and Totzeck, M. (2011) On the difference between 3D imaging and 3D metrology for computed tomography, *Optical Imaging and Metrology: Selected Topics*, Wiley-VCH Verlag GmbH, Weinheim, pp. 223–236.

49. ISMI (2005) Unified Advanced Optical Critical Dimension (OCD) Scatterometry Specification for sub-90 nm Technology, International SEMATECH Manufacturing Initiative.

50. Raymond, C.J. (2001) in *Handbook of Silicon Semiconductor Metrology* (ed. A.J. Deibold), Marcel Dekker, New York, pp. 477–513.

51. Chu, H. and Wack, D. (2008) Forward solve algorithms for optical critical dimension metrology. *Proc. SPIE*, **6922**, 69221O.

52. Silver, R.M., Barnes, B.M., Attota, R., Jun, J., Stocker, M., Marx, E., and Patrick, H.J. (2007) Scatterfield microscopy for extending the limits of image-based optical metrology. *Appl. Opt.*, **46** (20), 4248–4257.

53. de Groot, P.J., Colonna de Lega, X., Liesener, J., and Darwin, M. (2008) Metrology of optically-unresolved features using interferometric surface profiling and RCWA modeling. *Opt. Express*, **16** (6), 3970–3975.

54. de Groot, P.J., Colonna de Lega, X., and Liesener, J. (2009) Model-based white light interference microscopy for metrology of transparent film stacks and optically-unresolved structures, *Proceedings of Fringe 2009 – The 6th International Workshop on Advanced Optical Metrology*, Springer, Berlin, pp. 236–243.

55. Totzeck, M. (2001) Numerical simulation of high-NA quantitative polarization microscopy and corresponding near-fields. *Optik*, **112** (9), 399–406.

56. Rafler, S., Götz, P., Petschow, M., Schuster, T., Frenner, K., and Osten, W. (2008) Investigation of methods to set up the normal vector field for the differential Method. *Proc. SPIE*, **6995**, 69950Y.

57. Kayser, D., Bothe, T., and Osten, W. (2004) Scaled topometry in a multi-sensor approach. *Opt. Eng.*, **43** (10), 2469–2477.

58. Fienup, J.R. and Wackerman, C.C. (1986) Phase-retrieval stagnation problems and solutions. *J. Opt. Soc. Am. A*, **3** (11), 1897–1907.

59. Quiney, H.M., Nugent, K.A., and Peele, A.G. (2005) Iterative image reconstruction algorithms using wave-front intensity and phase variation. *Opt. Lett.*, **30** (13), 1638–1640.

60. Silver, R., Germer, T., Attota, R., Barnes, B.M., Bunday, B., Allgair, J., Marx, E., and Jun, J. (2007) Fundamental limits of optical critical dimension metrology: a simulation study. *Proc. SPIE*, **6518**, 65180U.

61. Sendelbach, M. and Archie, C. (2003) Scatterometry measurement precision and accuracy below 70 nm. *Proc. SPIE*, **5038**, 224–238.

14
Advanced MEMS Inspection by Direct and Indirect Solution Strategies
Ryszard J. Pryputniewicz

14.1
Introduction

Recent and continuing advances in microelectromechanical systems (MEMS), also called microsystems or microdevices, are and will continue to dramatically influence the consumer, industrial, medical, and defense markets [1]. It is generally recognized that the microsystems technology will continue to be the focus of intense international competition and exciting new MEMS-based products will drive the pertinent markets. The new generation of MEMS chips will host mechanical devices, fluidic channels, chemical mixers and reactors, bioanalytical devices, photonic devices and circuits, and a myriad of others, all integrated with the electronics. It is believed that the success of the MEMS industry, to a large extent, will be determined by the availability of computer-aided design (CAD) and multiphysics simulation tools combined, usually, with state-of-the-art (SOTA) metrology [2, 3].

At present, these microsystems are used in a multitude of applications ranging from everyday use (e.g., automotive subsystems and household appliances) through national security applications to space exploration. For example, some of the microdevice-based subsystems are currently used in automobiles for fuel injection, tire pressure sensors, inertial measurement units (IMUs), and air bag deployment.

The automotive, aerospace, medical, as well as a number of other present-day applications of the microsystems, and the structures they enable, provide numerous and much diversified functions, which require sophisticated design, analysis, fabrication, testing, and characterization tools [4–8]. These tools can be categorized as analytical, computational, and experimental [9]. Solutions using the tools from any one category alone do not usually provide necessary information on the devices being developed; as a result, extensive merging, or hybridization, of the tools from different categories is used [10–13].

Methodologies used currently in development of MEMS and other microdevices can also be categorized as direct and indirect solution strategies [10] that relate to the above-identified categories of analytical, computational, and experimental solutions (ACESs). In general, analytical and computational solutions comprise the *direct*

14 Advanced MEMS Inspection by Direct and Indirect Solution Strategies

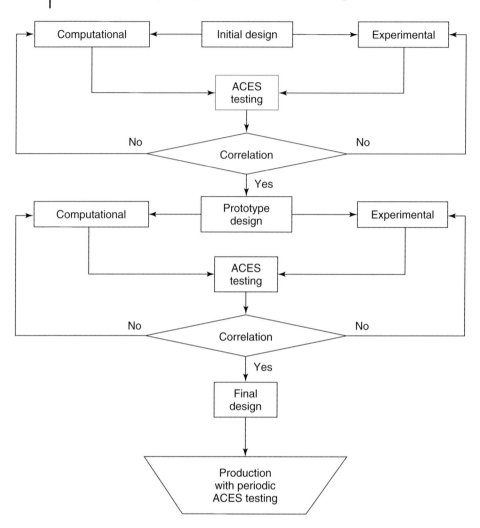

Figure 14.1 Multitier ACES methodology based on direct and indirect solution strategies.

strategy since they depend on well-known and well-understood long-established procedures, while experimental solutions are the basis of the indirect strategy since they, typically, require a number of assumptions and approximations and a priori knowledge to obtain the results [14–18]. One of the approaches employed in the development of microcomponents, as well as other complex structures of current interest, is based on a combined/hybrid use of direct and indirect strategies or ACESs methodology [9, 10, 19–21]. Such a hybrid methodology can be implemented, as shown in Figure 14.1.

The specific structure of the methodology shown in Figure 14.1 indicates three-tier process as follows: tier 1, evaluation of the initial design; tier 2, evaluation of the prototype design; and tier 3, production of the final design. Clearly, the methodology can also be implemented with either more or less than three-tier structure, depending on the development being made, as required in projects subject to the ever-present "golden rule." For the specific case of the three-tier structure displayed in Figure 14.1, once the initial design (tier 1) is arrived at, it is made to undergo computational and experimental assessments, with their results subjected to ACES testing according to the specific correlation (convergence) criterion. As long as the selected/imposed criterion is not satisfied, feedbacks loop back to computational and experimental assessments and testing with corresponding correlation is repeated until the expected convergence is reached. The result of the first tier is a prototype design incorporating any revisions arrived at during this stage of development. After arriving at the prototype design, tier 2 commences with evaluation of a short series of prototypes that are made to undergo computational and experimental assessments, with their results subjected to ACES testing according to appropriately defined correlation criterion, which can be different from that used for the tier 1 process. Again, as long as the tier 2 criterion is not satisfied, feedbacks loop back to computational and experimental assessments and testing with the imposed correlation is repeated until the convergence is satisfactorily achieved. During tier 2, the design can be further refined to make it more manufacturable resulting in the final design, which is used for production. During tier 3, the final design is produced/manufactured for delivery to customers. Typically, in this tier, testing is only performed when manufacturing deviations, which adversely affect the performance of the devices being produced, develop. Then, ACES testing is invoked and performed until the desired quality in the product is regained. The process may be repeated as needed. Depending on the specific product and customer requirements, correlation criteria during tier 3 can be appropriately selected to assure product reliability assessment as needed.

14.2 ACES Methodology

Effective development of structures, regardless of whether they are macro-, meso-, micro-, or nanosize, requires knowledge of design, analysis/simulation, materials, fabrication with special emphasis on packaging, and testing/characterization of the finished products [4, 5, 22]. These issues are best addressed via an integrated use of ACES methodology [9] or direct and indirect solution strategies [10].

The ACES methodology [20, 21] unifies/hybridizes the ACESs to obtain answers to problems where they would not be otherwise possible, to improve existing results, or to validate data obtained using other methodologies.

In the ACES methodology [9], *analytical tools* are based on exact, closed-form solutions. These solutions, however, are usually applicable to simple geometries for which boundary, initial, and loading (BIL) conditions can be readily specified. In addition, analytical solutions are indispensable to gain insight for overall representation of the ranges of anticipated results. They also facilitate determination of "goodness" of the results based on uncertainty analysis [23]. Computational tools, that is, finite element method (FEM), boundary element method (BEM), and finite difference method (FDM), provide approximate solutions as they discretize the domain of interest and the governing partial differential equations (PDEs). The characteristics of discretization, in conjunction with the BIL conditions, influence degree of approximation, and careful convergence studies should be performed to establish correct computational solutions and modeling [21]. It should be noted that both analytical and computational solutions depend on material properties. If material properties are well known, then solutions (typically) give correct results, provided convergence was achieved subject to properly specified BIL conditions; if material properties are not sufficiently known, in spite of having a good knowledge of other modeling parameters, erroneous results may be obtained [21, 24]. Experimental tools, however, in contrast to analytical and computational tools, evaluate actual objects, subject to actual/realistic operating conditions (including BIL conditions), and provide ultimate results while characterizing objects being investigated. The *experimental results* (typically) rely on the full-field-of-view (FFV) optoelectronic laser interferometric microscope (OELIM) system-based solutions, as discussed in Section 14.2.2.1.

With the development of more and more complex, but ever smaller, structures we find that there is a great number of design variables that affect their fabrication and performance. In the analysis of such structures, we find that experimental and theoretical models are equally important and equally indispensable for a successful development of viable, reliable, and low-cost (i.e., affordable) products. Although computational models can provide accurate simulations of specific designs of, for example, radio frequency (RF) MEMS switches [25–27], simple analytical models are sometimes preferred to develop an intuitive understanding of the behavior of these microswitches. In addition, the analytical models facilitate uncertainty analysis, which is invaluable for determination of the influence that variations in different process parameters (specified by tolerances and/or uncertainties) defining a microswitch have on the nominal results produced by the particular analytical model [28].

14.2.1
Computational Solution

Computational modeling of MEMS can be performed using commercial simulation tools [29, 30]. Parametric templates, utilizing, for example, Python scripting, for modeling MEMS can be developed and utilized for in-depth understanding of the designed/expected operation of microsystems. MEMS geometry, material properties, stress, contact forces, dynamic response, and other parameters can be investigated using the parametric templates to optimize performance. Atmospheric as well as packaging conditions (including vacuum), geometry of MEMS, and optimized pull-down voltage profiles can be modeled to understand and optimize the dynamic damping conditions of a packaged or unpackaged MEMS [31].

The coupled electrostatics-structures-flow simulations can also be performed using CFD-ACE+ software because it has the necessary multiphysics capabilities

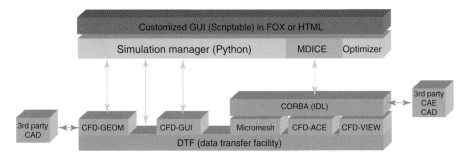

Figure 14.2 CFD-ACE+ coupling capabilities.

including flow, heat transfer, mechanics, and electrostatics [29]. All CFD-ACE+ capabilities are fully coupled to facilitate fast solution and determination of accurate results (Figure 14.2). According to this figure, CFD-ACE+ is a design environment for MEMS, which includes grid generation (CFD-GEOM); data visualization (CFD-VIEW); graphical problem setup (CFD-GUI); and implicit fully coupled fluidic, thermal, mechanical, electrostatic, and magnetic physical model solvers (CFD-ACE). The structural mechanics solutions are handled by FEM-STRESS, a finite element code, which is coupled implicitly with CFD-ACE+ for predictions of mechanical and thermal stress–strain fields in solid parts.

Using this software, physical models are solved on 2D, cylindrical, or 3D multidomain grids. Grids can be structured, unstructured, hybrid, or adaptive Cartesian. There is also a grid capability to track moving and deforming bodies and surfaces.

Strong links exist with the data transfer facilities (DTFs) for parallel execution on multiple machines. Physical models are implemented in a highly modularized code architecture that facilitates addition of future physical models. Therefore, CFD-ACE+ is well suited for analysis of MEMS and MEMS packaging with implicitly coupled modules that may be activated for analysis of various physical phenomena.

14.2.2
Experimental Solution Based on Optoelectronic Methodology

The optoelectronic methodology, as presented in this chapter, is based on the principles of optoelectronic holography (OEH) [13, 31, 32]. Basic configuration of the OEH system is shown in Figure 14.3. In this configuration, laser light is launched into a single-mode optical fiber by means of a microscope objective (MO). Then, the single-mode fiber is coupled into two fibers by means of a fiber-optic directional coupler (DC). One of the optical fibers coming out of the DC is used to illuminate an object, while the output from the other fiber provides reference against which the signals from the object are recorded. Both the object and the reference beams are combined by the interferometer (IT) and recorded by the system camera (CCD).

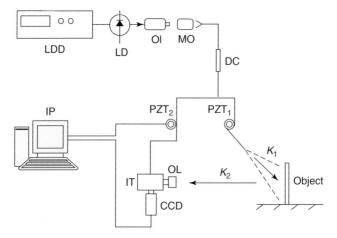

Figure 14.3 Single-illumination and single-observation geometry of a fiber-optic-based OEH system. LDD, laser diode driver; LD, laser diode; OI, optical isolator; MO, microscope objective; DC, fiber-optic-directional coupler; PZT$_1$ and PZT$_2$, piezoelectric fiber-optic modulators; IP, image processing computer; IT, interferometer; OL, objective lens. CCD is the camera, while K_1 and K_2 are vectors defining the directions of illumination and observation, respectively.

Images recorded by the CCD are processed by the system computer serving as the image processor (IP) to determine the fringe-locus function, $\Omega(x, y)$, constant values of which define fringe loci on the surface of an object under investigation. The values of Ω relate to the system geometry and the unknown vector L, defining displacements and deformations, via the relationship [33]

$$\Omega(x, y) = [K_2(x, y) - K_1(x, y)] \cdot L(x, y) = K \cdot L \tag{14.1}$$

where K is the sensitivity vector defined in terms of vectors K_1 and K_2 identifying directions of illumination and observation, respectively (i.e., geometry), of the OEH system as shown in Figure 14.3.

Quantitative determination of structural displacements/deformations due to the applied loads can be obtained, by solving a system of equations similar to Eq. (14.1), to yield [33]

$$L = \left[\tilde{K}^T \tilde{K}\right]^{-1} \left(\tilde{K}^T \Omega\right) \tag{14.2}$$

where $\tilde{K}^T$ represents a transpose of the matrix of the sensitivity vectors K. Equation (14.2) indicates that displacements/deformations determined from interferograms are functions of K and Ω, which have spatial, that is (x,y), distributions over the field of interest on an object being investigated. Equation (14.2) can be represented by a phenomenological relation [23, 34]

$$L = L(K, \Omega) \tag{14.3}$$

based on which the RSS-type (i.e., the *square root of the sum of the squares*) uncertainty in L, that is, δL, can be determined to yield

$$\delta L = \left[\left(\frac{\partial L}{\partial K} \delta K \right)^2 + \left(\frac{\partial L}{\partial \Omega} \delta \Omega \right)^2 \right]^{1/2} \quad (14.4)$$

where $\partial L/\partial K$ and $\partial L/\partial \Omega$ represent partial derivatives of L with respect to K and Ω, respectively, while δK and $\delta \Omega$ represent the corresponding uncertainties in K and Ω, respectively. It should be remembered that K, L, and Ω are all functions of spatial coordinates (x,y,z), that is, $K = K(x,y,z)$, $L = L(x,y,z)$, and $\Omega = \Omega(x, y, z)$, respectively, when performing partial differentiations required to solve (Eq. (14.4)). After solution of Eq. (14.4), the result indicates that δL is proportional to a product of the local value of L with the RSS value of the ratios of the uncertainties in K and Ω to their corresponding local values, that is

$$\delta L \propto L \left[\left(\frac{\delta K}{K} \right)^2 + \left(\frac{\delta \Omega}{\Omega} \right)^2 \right]^{1/2} \quad (14.5)$$

For typical geometries of the OEH systems used in recording of interferograms, the values of $\delta K/K$ are less than 0.01. However, for small deformations, the typical values of $\delta \Omega/\Omega$ are (usually) more than 1 order of magnitude greater than the values for $\delta K/K$. Therefore, the accuracy with which the fringe orders (based on which the Ω values are calculated [33]) are determined influences the accuracy in the overall determination of displacements/deformations [34] and, as such, is critical to implementation of this methodology for the development of MEMS and other structures, as discussed herein.

Therefore, to minimize this influence, a number of algorithms for determination of Ω were developed. Some of these algorithms require multiple recordings of each of the two states, in the case of the double-exposure method, of the object being investigated with introduction of a discrete phase step between the recordings [6, 35]. For example, the intensity patterns of the first and the second exposures, $I_n(x,y)$ and $I'_n(x, y)$, respectively, in the *double-exposure sequence*, can be represented by the following equations [13]

$$I_n(x, y) = I_o(x, y) + I_r(x, y)$$
$$+ 2\{[I_o(x, y)][I_r(x, y)]\}^{1/2} \cos\{[\Delta\varphi(x, y)] + \theta_n\} \quad (14.6)$$

and

$$I'_n(x, y) = I_o(x, y) + I_r(x, y) + 2\{[I_o(x, y)][I_r(x, y)]\}^{1/2} \cos\{[\Delta\varphi(x, y)]$$
$$+ \theta_n + \Omega(x, y)\} \quad (14.7)$$

where $I_o(x,y)$ and $I_r(x,y)$ denote the object and the reference beam intensities, respectively, with (x,y) indicating spatial coordinates, $\Delta\varphi(x, y) = \varphi_o(x, y) - \varphi_r(x, y)$ is the optical phase difference based on $\varphi_o(x, y)$, denoting random phase of the light reflected from an object, and $\varphi_r(x, y)$, denoting the phase of the reference beam, θ_n denotes the discrete applied nth phase step, and $\Omega(x, y)$ is the fringe-locus function relating to the displacements/deformations that an object incurred between the first and second exposures; Ω is what we need to determine. When Ω is known, it is used in Eq. (14.2) to find **L**.

In the case of the five-phase-steps algorithm with $\theta_n = 0, \pi/2, \pi, 3\pi/2,$ and 2π, the distribution of the values of $\Omega(x, y)$ can be determined using [13]

$$\Omega(x, y) = \tan^{-1}\left\{\frac{2[I_2(x, y) - I_4(x, y)]}{2I_3(x, y) - I_1(x, y) - I_5(x, y)}\right\} \tag{14.8}$$

Results produced by Eq. (14.8) depend on the capabilities of the illuminating, imaging, and processing subsystems of the OEH system used in the specific application. Developments in laser, fiber optic, CCD camera, and computer technologies have led to advances in the OEH methodology; in the past, these advances have almost paralleled the advances in the image recording media [36]. These developments resulted in educational procedures [5] and led to MEMS education alliance [37].

In response to the needs of the emerging MEMS technology, an OELIM system for studies of objects with micrometer-sized features was developed [38, 39], as discussed in Section 14.2.2.1.

14.2.2.1 The OELIM System

The OELIM configuration evolved into a modular optoelectronic station (Figure 14.4), developed especially for characterization of MEMS [40, 41]. In this station, the interchangeable optical subsystems (i.e., interferometric modules) were developed to have a long working distance to accommodate a chamber to conveniently place and hold MEMS being investigated/characterized while subjecting them

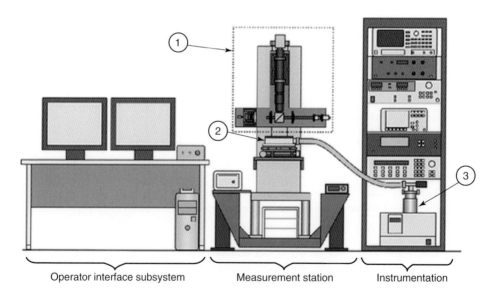

Figure 14.4 Schematic of a modular optoelectronic station for characterization of MEMS: (1) interchangeable interferometer module, (2) thermal/vacuum chamber, and (3) turbomolecular vacuum pump.

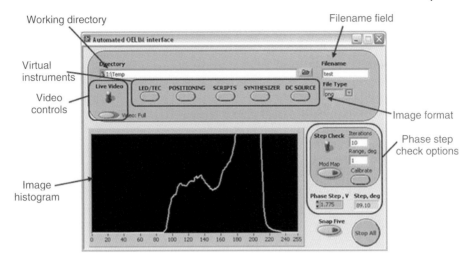

Figure 14.5 Front panel of an automated OELIM interface.

to various thermomechanical loads, including vibrations, under well-controlled temperature (for heating and cooling) as well as pressure/vacuum conditions.

A custom user interface has been developed for the modular optoelectronic station [41]. It is based on LabView software, with MatLab scripts embedded where needed. Communication with the hardware is achieved through IEEE 488 (GPIB), RS-232 (serial), and IEEE 1394 (Firewire) interfaces. The custom interface has been developed as a modular system. Nearly every piece of hardware is controlled by its own virtual instrument (VI). Also contained in the package are VIs that control various pieces of equipment to perform a scripted experiment, such as thermal cycling with data measurements. Finally, there are VIs whose sole function is to perform specific analysis, such as calculation of the measurement phase step. The main user interface, which is displayed on execution of the software and acts as the control panel for launching subprograms, is shown in Figure 14.5.

In the main interface (Figure 14.5), the user selects parameters such as the working directory, where data will be saved, format of images that are to be recorded, and the base file name for the images. During data acquisition, a series of phase-stepped images are recorded, with the phase step iteration number appended onto the base file name. The *image histogram* is displayed at all times on the front panel. This allows the user to ensure that there is no saturation. It may also be useful for optimizing modulation by adjusting the reference mirror position such that the width of the histogram is a maximum.

Pertinent features on the front panel open sub-VIs, each with its own interface, as detailed in Ref. [41].

OELIM measurements facilitate implementation of both direct and indirect solution strategies [10]. Use of the optoelectronic methodology is demonstrated on representative MEMS samples of contemporary interest, which are described in Section 14.3. Typical results obtained are presented in Section 14.4.

14.3
MEMS Samples Used

Although there are great many MEMS-based devices, in this chapter, only three representative MEMS of great contemporary interest are discussed: microgyroscope (also known as microgyro) as well as microaccelerometer, both facilitating development of miniaturized guidance systems, and RF cantilever-type microswitch, particularly suitable for development of various RF communication as well as remote sensing products and handheld portable information devices (also known as PIDs).

Figure 14.6 shows a representative MEMS gyroscope sample used in this study, based on a differential configuration consisting of dual proof masses, that is, shuttles, vibrating in a plane of the masses. Each of the masses is characterized by lateral dimensions of (typically) 300 μm × 300 μm and is supported by four folded springs, one in each corner of every mass [42]. One end of each of the folded spring(s) is attached to a post fixed on a die/substrate, this end of the spring is stationary and (ideally) does not experience any motions and/or deformations during functional operation of an MEMS gyroscope. The other end of the folded spring(s) is attached to the proof masse(s) and moves as the masses are actuated by the electrostatic comb drives. Each proof mass, in the configuration shown in Figure 14.6, is actuated by its own set of electrostatic comb drives.

The detail of a single, in this case left, proof mass/shuttle is displayed in Figure 14.7, while an enlarged view of the upper right section of this mass/shuttle, including its local suspension, is shown in Figure 14.8.

Functional operation of MEMS accelerometers, just like operation of microgyros, depends on motion of a proof mass in response to the applied acceleration, which is especially important in an application using a packaged sensor. The microaccelerometer proof masses (also known as shuttles) are suspended by folded springs attached to the shuttles in their corners, the same as practiced with microgyros. Because of advances in design, dual-axes microaccelerometers

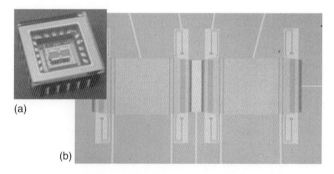

Figure 14.6 Microgyro: (a) a representative package and (b) a typical tuning fork gyroscope (TFG) configuration comprising dual proof masses/shuttles, each 300 μm × 300 μm in lateral dimensions.

14.3 MEMS Samples Used | 315

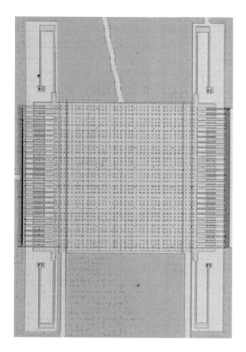

Figure 14.7 Left proof mass of the MEMS gyro shown in Figure 14.6b.

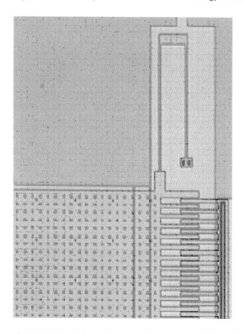

Figure 14.8 Enlarged view of the upper right section of the left proof mass shown in Figure 14.7.

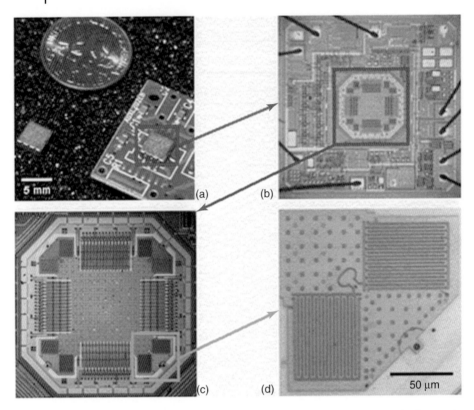

Figure 14.9 Dual-axes microaccelerometer: (a) overall view of a package, (b) the sensor – highlighted by the square in the center – and electronics integrated on a single chip, (c) the sensor consisting of a proof mass suspended by four pairs of folded springs – the square in the lower right corner highlights one of the spring pairs, and (d) view of the spring pair suspending the lower right corner of the proof mass. (Please find a color version of this figure on the color plates.)

were developed with suspension springs that have several folds (or turns) [43] (Figure 14.9). These multifold spring configurations allow compact design of a sensor, which facilitates a fast response.

On the basis of their topologies, the RF switches can be grouped into two categories [22]: (i) membrane type, that is, capacitive, and (ii) cantilever type, that is, resistive.

The microswitch considered in this chapter is a cantilever-type RF MEMS switch [44] (Figure 14.10). This figure shows a microcantilever-type contact of active length L fabricated parallel to a substrate in such a way that separation between the electrodes (one electrode is on a microcantilever and the other is directly below on top of a substrate) is d_e; for stable operation of a microswitch, d_e should be at least three times greater than the contact gap distance d_g. During functional operation, voltage applied to the electrodes induces an electrostatic

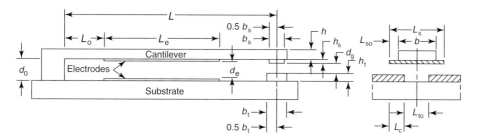

Figure 14.10 Geometry and pertinent dimensions of a cantilever-type microswitch.

force that activates/actuates a microswitch [45]. The electrostatic force bends a microcantilever causing the contacts to touch (i.e., by reducing the gap distance, d_g, to zero), which closes an electrical path (by making a cross bar, at the free end of a microcantilever, to bridge/close an "opening," defined by L_{to} in Figure 14.10, in a trace usually located below the free end of a microcantilever) and facilitates propagation of a signal. As the activation/actuation voltage is released, elasticity of the microcantilever is used to return it to its original, or open, position (i.e., making $d_g > 0$). Mechanically, the microcantilever of a switch behaves like a conventional cantilever [46]. In fact, traditional equations can be used to find microcantilever stiffness, natural frequency, pull-in voltage, and magnitude of the activation/actuation force. The switch fabrication methods are particularly important because they dictate the material type, surface finish, texture, and overall size of the microswitch components (especially electrical interfaces of a microswitch). Consequently, material properties have a direct influence on thermal management characteristics of a microswitch and its behavior under actual operating conditions [44]. In fact, fabrication tolerances and accuracy of material properties have profound influence on dynamics as well as the thermomechanical performance of microswitches [28].

14.4 Representative Results

The optoelectronic methodology, described in Section 14.2.2, was used to study displacements and deformations of three MEMS samples described in Section 14.3. The results of these studies are summarized in Sections 14.4.1–14.4.3, respectively.

14.4.1 Deformations of a Microgyroscope

Deformations of proof masses/shuttles were measured during the operation of a microgyroscope. In this application, interferograms were recorded stroboscopically, while microgyros were driven at their operating frequencies. To facilitate these recordings, the optoelectronic system was set up to be sensitive to the

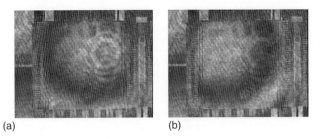

Figure 14.11 (a,b) Representative OELIM fringe patterns of the left shuttle, at different times in a vibration cycle, while a microgyro is operating at 10.1 kHz.

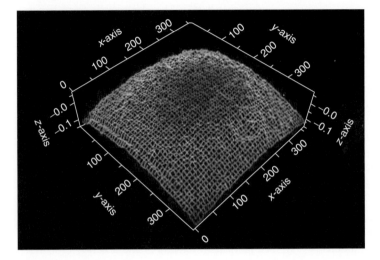

Figure 14.12 The out-of-plane 212 nm deformation component of the left shuttle based on the fringe patterns of Figure 14.11. (Please find a color version of this figure on the color plates.)

out-of-plane motions/deformations. Representative interferograms, corresponding to deformations of the left shuttle of a microgyro operating at 10.1 kHz, are shown in Figure 14.11. Observation of the fringe patterns of this figure clearly indicates asymmetry in deformations of the proof mass(es) of a microgyroscope. This can be related to structural design and suspension of the shuttles as well as to the way that electrostatic forces affect their motions and corresponding deformations. A representative display of deformations of the left shuttle of the microgyro, operating at 10.1 kHz, during a specific instant in a vibration cycle, is displayed in Figure 14.12, indicating deformations ranging up to 212 nm, which ideally should not exist. However, the deformations/motions were measured, in this case, to be orders of magnitude greater than typical (magnitudes of) motions of the proof masses due to the Coriolis forces [47]. Accuracy and precision of a microgyro depends on the quality of its suspension. This suspension is provided by folded springs attached, at one end, to proof masses/shuttles and, at the other end, to posts forming a part of a substrate, as illustrated in Figure 14.7.

Any deformations of the springs that are not in response to functional operation of a sensor will cause an incorrect (i.e., erroneous) output. Because of the nanoscale of these deformations and microsize objects over which they take place, it was not until the advancement of optoelectronic metrology that such deformations were quantified in the FFV [47].

14.4.2
Functional Operation of a Microaccelerometer

The functional operation of MEMS accelerometers was already mentioned in section 14.3. As the accelerometer of Figure 14.9 is actuated, its proof mass displaces during an operation cycle (Figure 14.13). This figure shows 1.48 μm displacement of a proof mass and corresponding deformations of the folded springs at the end of an actuation cycle [48]; because of a symmetric design of a sensor, only lower right corner, corresponding to the view shown in Figure 14.9d, is displayed. The displacements shown in Figure 14.13 are undesired and should be mitigated.

14.4.3
Thermomechanical Deformations of a Cantilever Microcontact

MEMS RF switches present a promising technology for high-performance reconfigurable microwave and millimeter wave circuits [49]. Low insertion loss, high isolation, and excellent linearity provided by MEMS switches offer significant improvements over an electrical performance provided by conventional positive intrinsic negative (PIN) diode and metal-oxide semiconductor field-effect transistor (MOSFET) switching technologies.

These superior electrical characteristics permit design of MEMS-switched high-frequency circuits not feasible with semiconductor switches, such as high-efficiency broadband amplifiers and quasi-optic beam steering arrays. In addition, operational benefits arise from low-power consumption, small size and weight, and integration capability of modern RF MEMS switches.

Prototype cantilever beams were fabricated and their dynamic characteristics were determined using OEH [50]. Figure 14.14 shows representative computational multiphysics results obtained for a 225 μm long Si cantilever contact. These results indicate that, as actuation conditions change, operational response of a switch also changes, as it should. For example, a cantilever vibrating at 160 kHz has the maximum amplitude of 500 nm, while the same cantilever vibrating at 1 MHz has the amplitude of merely 4 nm (Figure 14.15).

RF MEMS devices are currently used in telecommunications, wireless networking, global positioning systems, cellular, auto, and even toy industry. Microcomponents greatly reduce size and weight of many products, while reducing their cost, power consumption, and (simultaneously) improving performance and durability.

320 *14 Advanced MEMS Inspection by Direct and Indirect Solution Strategies*

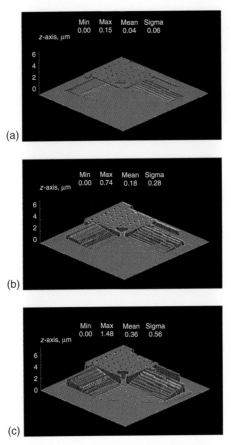

Figure 14.13 Optoelectronic holography measurements depicting displacements of a proof mass and deformations of the folded springs as a microaccelerometer is actuated: (a) 0.15 μm displacement of the proof mass, (b) 0.74 μm displacement of the proof mass, and (c) 1.48 μm displacement of the proof mass.

Although, at this time, MEMS brings a large number of advantages to a range of industries much can still be done to improve their reliability [51].

Ohmic-type (i.e., resistive/contact as opposed to capacitive) MEMS switches present a major reliability concern. Joule heat generated when an electrical signal is passed through a microswitch causes the electrical contacts to heat up, which leads to wear at an increased rate and, may even cause, self-welding of the electrical interfaces. Changing certain design parameters, for example, materials and dimensions, can enhance life expectancy of a microswitch. Modifying an integrated circuit (IC) comprising a microswitch is also a promising method for improving reliability of a microswitch because some modification/optimization can reduce the magnitude of the electrical signal passing through a microswitch, thus reducing joule heat.

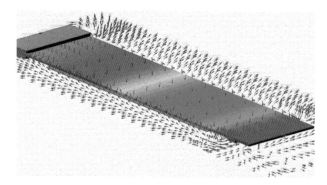

Figure 14.14 Computational multiphysics modeling of an RF MEMS switch closure at atmospheric conditions: 3D representation of air damping. (Please find a color version of this figure on the color plates.)

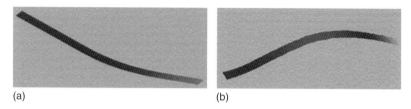

(a)　　　　　　　　　　　　　　(b)

Figure 14.15 OELIM measured dynamic characteristics of a 225 μm long Si cantilever contact: (a) the first bending mode at 160 kHz and 500 nm amplitude and (b) the second bending mode at 1 MHz and 4 nm amplitude.

On the basis of heat transfer analysis, the actual electrical interface/contact area of a microswitch, contact material, and an internal resistance of the contacts are parameters that greatly influence thermal management accompanying functional operation of RF MEMS contact switches. However, when two not sufficiently smooth bodies/interfaces are in contact (even without an electrical signal passing through them), a self-microwelding may occur. Microwelding is detrimental to reliable operation of a microswitch [52–54].

Completely smooth surfaces, however, experience strong van der Waals forces. van der Waals forces, along with capillary and electrostatic forces, are a cause of stiction in mechanical microdevices. Contact stiction that may occur in a microswitch causes permanent collapse of a microdevice. Surface roughness of contact interface is therefore an important design parameter that influences quality of electrical interfaces and thermal management issues relating to their operation. Quality of a surface finish determines the peak-to-peak distances on a given surface, which is why metal processing is so important; this surface roughness may be overcome, to a certain degree, by a sufficiently high contact force applied normal to the pertinent contacting interfaces [46]. Varying the stoichiometry of the used alloy/metal, the curing temperature, and application process, effects of surface roughness can be kept to a minimum, in order to avoid detrimental stiction.

Contact material is another important parameter that has to be carefully chosen in the design process of MEMS in order to assure long useful life. Most often, metals are chosen as the contact material of choice. Each metal, however, with its unique properties can greatly affect the reliability of a microswitch. The contact force in a microswitch has to be high enough to break through the film/oxide formed on the interfaces to assure a reliable electrical interconnection.

Heat generated by a current passing through a microswitch and across contacts (forming an electrical interface) is a source of thermal energy in a microswitch [22]. A representative microswitch was modeled using CAD/CAE software [55, 56]. The CAD model was then imported into the thermal analysis system (TAS) software [57], which is based on FDM of calculating heat distribution [58]. TAS is a software that uses resistors to represent finite elements that model a component in order to solve for the thermal distribution in a component being modeled.

The OELIM methodology can also be used to evaluate quality of motion of the proof masses of packaged MEMS operating at their resonance frequencies. This evaluation provides useful information for determination of motions of pertinent proof masses.

14.5
Conclusions and Recommendations

Advances in optoelectronic metrology were described with emphasis on static and dynamic measurements of absolute shape as well as displacements and deformations of objects according to direct and indirect solution strategies. Because of its scalability, the optoelectronic metrology is particularly suitable for applications over a great range of sizes of the objects, which need to be characterized with high accuracy and precision. As such, the optoelectronic metrology is very effective for applications from milliscale to nanoscale ranges. In fact, the optoelectronic metrology is more and more frequently the methodology of choice when it comes to testing and characterization of products fabricated by emerging technologies (ETs), such as, for example, MEMS.

MEMS is a revolutionary enabling technology that collocates, on a single chip, functions of sensing, actuation, and controls with computation and communication to affect the way people and machines interact with the physical world.

Need for remote and noninvasive measurements in the FFV providing data in three dimensions and in near real time that optoelectronic metrology is capable of will be ever increasing as the ETs evolve into mature technologies. This need will continue to be over multiscales ranging from milliscale to nanoscale and even down to *picoscale* as the "building blocks," out of which large structures will be made in the future, will be shrinking in size [59]. To be ready to satisfy testing and characterization demands that ETs will generate, development of metrology, especially optoelectronic metrology, should be continued.

The optoelectronic (e.g., OELIM) methodology can readily measure motions of the vibrating proof masses using stroboscopic laser illumination of MEMS.

These results can, in turn, be used to determine the quality (i.e., Q-factor) characterizing/defining measurement ability/sensitivity of modern vibrating/resonating/dynamic devices/sensors. Demands for high-quality measurements by such sensors will be increasing as technology ventures into ever smaller, yet more capable and affordable, devices. On the basis of recent evolution of microtechnology, exemplified herein by MEMS, and the role that direct and indirect solution strategies play in this evolution, it is believed that technologies that will emerge in the future will benefit from what is being done now to advance direct and indirect solution strategies.

All in all, using the direct and indirect (i.e., hybrid) strategies, computational results will facilitate design of experiments, results of which will have to be correlated with numerical data based on computations leading to ever more capable computer-based methodologies. It should be realized that correlation criteria will change and become more and more demanding as number crunching technology advances in the future, as we have witnessed this trend during the past few decades because of proliferation of computers. As a result, new, more capable hybrid methodologies will be introduced that will further accelerate development of ETs. It is inevitable that such development will lead to more accurate results that will be possibly based on the ETs of the future. One reason that this will be possible will be because of anticipated advances in computer hardware and accompanying software that will greatly facilitate evaluation of "what if" scenarios for a great number of parameters that will govern future devices and processes, their mutual interdependence, as well as interaction with surroundings in which they will operate. Such an evaluation will be performed in a relatively short time, and its results will readily indicate a set of parameters assuring optimum operation. The ACES process displayed in Figure 14.1 will facilitate evaluation of the "what if" scenarios for virtually any process. The major decision that will always have to be made before using the process will be to set the number of tiers to be employed and the correlation criteria to use in each tier. On the basis of past experience, this decision will depend on the "golden rule" as well as the technologies available at the specific time.

Acknowledgments

The author gratefully acknowledges support from all sponsors and thanks them for their permissions to present the results, of their projects, in this chapter. This work was also supported by the NEST Program at WPI-ME/CHSLT.

References

1. McWhorter, P.J. (1999) Intelligent microsystems: key to the next silicon revolution. MST News, Volume 4.

2. Athavale, M.M., Yang, H.Q., Li, H.Y., and Przekwas, A.J. (1998) A High-Fidelity Simulation Environment for Thermo-Fluid-Mechanical Design

of MEMS, Final Report for Baseline Project, DARPA Contract.
3. Pryputniewicz, R.J., Przekwas, A.J., Turowski, M., Furmanczyk, M., Hieke, A., and Pryputniewicz, D.R. (2003) Computational environment for predictive modeling and simulation of MEMS and MEMS packaging. Proceedings of the 5th Topical Workshop on MEMS, Related Microsystems, and Nanopackaging, Boston, MA, pp. 94–99.
4. Pryputniewicz, R.J. (2000) *Integrated Approach to Teaching of Design, Analysis, and Characterization in Micromechatronics*, Paper No. IMECE2000/DE-13, ASME - American Society of Mechanical Engineers, New York.
5. Pryputniewicz, R.J. (2001) *MEMS Design Education by Case Studies*, Paper No. IMECE2001/DE-23292, ASME - American Society of Mechanical Engineers, New York.
6. Pryputniewicz, R.J. (2007) Progress in MEMS. *Strain*, **43**, 1–13.
7. Osten, W. (2001) Optical microsystems metrology I. *Opt. Lasers Eng.*, **36** (2), 75–240.
8. Osten, W. (ed.) (2006) *Optical Inspection of Microsystems*, CRC Taylor & Francis, Boca Raton.
9. Pryputniewicz, D.R. (1997) ACES approach to the development of microcomponents. MS Thesis. Worcester Polytechnic Institute, Worcester, MA.
10. Osten, W. (2008) Digital image processing in optical metrology in *Handbook of Experimental Solid Mechanics*, Chapter 19 (ed. W.N. Sharpe Jr.), Springer, New York, pp. 481–563.
11. Pryputniewicz, R.J. (1994) A hybrid approach to deformation analysis. *Proc. SPIE*, **2342**, 282–296.
12. Furlong, C. and Pryputniewicz, R.J. (1998) Hybrid computational and experimental approach for the study and optimization of mechanical components. *Opt. Eng.*, **37**, 1448–1455.
13. Furlong, C. (1999) Hybrid, experimental and computational, approach for the efficient study and optimization of mechanical and electro-mechanical components. PhD Dissertation. Worcester Polytechnic Institute, Worcester, MA.
14. Hoffman, B. (1995) Ill-posedness and regularization of inverse problems - a review of mathematical methods in *The Inverse Problem* (ed. H. Lübbig), Akademie Verlag, Berlin, pp. 45–66.
15. Mieth, U., Osten, W., and Jüptner, W. (2001) Investigation on the appearance of material faults in holographic interferograms in *Proceedings of Fringe*, Elsevier, pp. 163–172.
16. Osten, W., Elandalousi, F., and Mieth, U. (2002) Trends for the solution of identification problems in holographic non-destructive testing (HNDT). *Proc. SPIE*, **4900**, 1187–1196.
17. Osten, W. and Jüptner, W. (1998) New light sources and sensors for active optical 3D-inspection. *Proc. SPIE*, **3897**, 314–327.
18. Elandalousi, F., Osten, W., and Jüptner, W. (1998) Automatic flaw detection using recognition by synthesis: practical results. *Proc. SPIE*, **3479**, 228–234.
19. Pryputniewicz, E.J. (2000) ACES approach to the study of electrostatically driven MEMS microengines. MS Thesis. Worcester Polytechnic Institute, Worcester, MA.
20. Pryputniewicz, R.J., Galambos, P., Brown, G.C., Furlong, C., and Pryputniewicz, E.J. (2001) ACES characterization of surface micromachined microfluidic devices. *Int. J. Microelectron. Electron. Pack. (IJMEP)*, **24**, 30–36.
21. Pryputniewicz, D.R., Furlong, C., and Pryputniewicz, R.J. (2001) ACES approach to the study of material properties of MEMS. Proceedings of the International Symposium on MEMS: Mechanics and Measurements, Portland, OR, pp. 80–83.
22. Pryputniewicz, R.J. and Furlong, C. (2002) *MEMS and Nanotechnology*, Worcester Polytechnic Institute, Worcester, MA.
23. Pryputniewicz, R.J. (1993) *Engineering Experimentation*, Worcester Polytechnic Institute, Worcester, MA.
24. Merz, T., Elandalousi, F., Osten, W., and Paulus, D. (1999) Active approach for holographic nondestructive testing of satellite fuel tanks. *Proc. SPIE*, **3824**, 8–19.

25. Stout, P. (1999) CFD-ACE+ a CAD system for simulation and modeling of MEMS. Proceedings of the SPIE, Paris, France.
26. Wilkerson, P.W., Kranz, M., and Przekwas, A.J. (2001) Flip-chip hermetic packaging of RF MEMS. MEMS4Conference, Berkeley, CA, August 24–26, 2001.
27. Przekwas, A.J., Turowski, M., Furmanczyk, M., Hieke, A., and Pryputniewicz, R.J. (2001) Multiphysics design and simulation environment for microelectromechanical systems. Proceedings of the International Symposium on MEMS: Mechanics and Measurements, Portland, OR, pp. 84–89.
28. Pryputniewicz, R.J., Pryputniewicz, D.R., and Pryputniewicz, E.J. (2007) *Effect of Process Parameters on TED-Based Q-Factor of MEMS*, Paper No. IPACK2007-33094, ASME - American Society of Mechanical Engineers, New York.
29. CFDRC (2004) CFD-ACE+ Multiphysics Software, *http://www.cfdrc.com* (accessed May 2010).
30. SRAC (1998) *COSMOS/M User's Guide*, Structural Research and Analysis Corporation, Santa Monica, CA.
31. Pryputniewicz, R.J. and Furlong, C. (2003) Novel optoelectronic methodology for testing of MOEMS. *Proc. Int. Symp. MOEMS Miniaturized Syst. III, SPIE*, **4983**, 11–25.
32. Brown, G.C. (1999) Laser interferometric methodologies for characterizing static and dynamic behavior of MEMS. PhD Dissertation. Worcester Polytechnic Institute, Worcester, MA.
33. Pryputniewicz, R.J. (1995) Quantitative determination of displacements and strains from holograms in *Holographic Interferometry*, Chapter 3, Springer Series in Sciences, Vol. 68, Springer-Verlag, Berlin, pp. 33–72.
34. Pryputniewicz, R.J. (1981) High precision hologrammetry. *Int. Arch. Photogramm.*, **24**, 377–386.
35. Pryputniewicz, E.J., Miller, S.L., de Boer, M.P., Brown, G.C., Biederman, R.R., and Pryputniewicz, R.J. (2000) Experimental and analytical characterization of dynamic effects in electrostatic microengines. Proceedings of the International Symposium on Microscale Systems, Orlando, FL, pp. 80–83.
36. Pryputniewicz, R.J. (1995) Hologram interferometry from silver halide to silicon and ... beyond. *Proc. SPIE*, **2545**, 405–427.
37. Pryputniewicz, R.J., Shepherd, E., Allen, J.J., and Furlong, C. (2003) University – National Laboratory alliance for MEMS education. Proceedings of the 4th International Symposium on MEMS and Nanotechnology (4th-ISMAN), Charlotte, NC, pp. 364–371.
38. Brown, G.C. and Pryputniewicz, R.J. (1998) Holographic microscope for measuring displacements of vibrating microbeams using time-average electro-optic holography. *Opt. Eng.*, **37**, 1398–1405.
39. Klempner, A.R., Hefti, P., Marinis, R.T., and Pryputniewicz, R.J. (2004) Development of a high stability optoelectronic laser interferometric microscope for characterization and optimization of MEMS. Proceedings of the 15th International Invitational UACEM Symposium, Springfield, MA, pp. 275–285.
40. Klempner, A.R. (2006) Development of a modular interferometric microscopy system for characterization of MEMS. MS Thesis. Worcester Polytechnic Institute, Worcester, MA.
41. Marinis, R.T. (2009) Development and implementation of automated interferometric microscope system for study of MEMS inertial sensors. PhD Dissertation. Worcester Polytechnic Institute, Worcester, MA.
42. Pryputniewicz, R.J. Tan, X.G., and Przekwas A.J. (2004) Modeling and measurements of MEMS gyroscopes. Proceedings of the IEEE-PLANS2004, Monterey, CA, pp. 111–119.
43. Kok, R., Furlong, C., and Pryputniewicz, R.J. (2003) Experimental modal analysis using MEMS accelerometers. Proceedings of the 30th Annual Symposium and Exhibition of IMAPS-NE, Boxboro, MA, pp. 116–123.
44. Pryputniewicz, R.J. (2007) *Thermal Management in RF MEMS Ohmic Switches,*

Paper No. IPACK2007-33502, ASME – American Society Mechanical Engineers, New York.

45. Pryputniewicz, R.J., Wilkerson, P.W., Przekwas, A.J., and Furlong, C. (2002) RF MEMS: modeling and simulation of switch dynamics. Proceedings of the 35th International Symposium on Microelectronics, Denver, CO, pp. 267–272.

46. Furlong, C. and Pryputniewicz, R.J. (2007) *Integrated Approach to Development of Microelectronic Contacts*, Paper No. IPACK2007-33345, ASME – American Society Mechanical Engineers, New York.

47. Pryputniewicz, R.J., Marinis, R.T., Klempner, A.R., and Hefti, P. (2006) Hybrid methodology for development of MEMS. Proceedings of the IEEE-PLANS2006 San Diego, CA.

48. Furlong, C. and Pryputniewicz, R.J. (2002) Characterization of shape and deformation of MEMS by quantitative optoelectronic metrology techniques. *Proc. SPIE*, **4778**, 1–10.

49. Mihailovich, R.E., Kim, M., Hacker, J.B.H., Sovero, A., Studer, J., Higgins, J.A., and DeNatale, J.F. (2001) MEM relay for reconfigurable RF circuits. *IEEE Microw. Wireless Compon. Lett.*, **11**, 53–55.

50. Pryputniewicz, R.J. (2008) Holography in *Handbook of Experimental Solid Mechanics*, Chapter 24 (ed. W.N. Sharpe Jr.), Springer, New York, pp. 675–699.

51. Zunino, J.L. III, Skelton, D.R., Han, W., and Pryputniewicz, R.J. (2007) Hybrid approach to MEMS reliability assessment. Proceedings of the International Symposium, on MOEMS-MEMS 2007: Reliability, Packaging, and Characterization of MEMS, San Jose, CA, Paper No. SPIE6563-03.

52. Tyco (2000) *Relay Contact Life*. Application Note 13C3236, Tyco Electronics Corporation – P&B, Winston-Salem, NC.

53. Tyco (2000) *Contact Arc Phenomenon*, Application Note 13C3203, Tyco Electronics Corporation – P&B, Winston-Salem, NC.

54. Kalpakjian, S. and Schmid, S.R. (2001) *Manufacturing Engineering and Technology*, Prentice-Hall, Upper Saddle River, NJ.

55. PTC (2003) *Pro/ENGINEER User Manual*, Parametric Technology Corporation, Needham, MA.

56. PTC (2003) *Pro/MECHANICA User Guide*, Parametric Technology Corporation, Needham, MA.

57. Rosato, D.A. (2002) Thermal Analysis System: User Manual, Version 6.1, Harvard Thermal, Inc., Harvard, MA.

58. Pryputniewicz, R.J., Rosato, D.A., and Furlong, C. (2003) Measurements and simulation of SMT components. *Microelectron. Int.*, **20**, 13–16.

59. Pryputniewicz, R.J. (2006) *Integrated Thermomechanical Design and Analysis with Applications to Micromechatronics*, Worcester Polytechnic Institute, Worcester, MA.

15
Different Ways to Overcome the Resolution Problem in Optical Micro and Nano Metrology

Wolfgang Osten

15.1
Introduction

A visible trend in the implementation of new technologies and creation of new products is the continuous reduction of feature sizes – a trend that is tangibly expressed by the term nanotechnology. Meanwhile, the critical dimensions (CDs) of structures written in silicon are considerably smaller than the wavelength of the applied light source, and the International Technology Roadmap for Semiconductors (ITRS) of SEMATECH [1] shows that this trend is to be sustained for the next 15 years. Beyond the reduction of feature sizes, both the degree of integration and the functionality are drastically increasing. A well-known example is the smartphone where in a minimum space, original telephone features are combined with many multimedia functions such as photo, video, and audio. But all improvement has its price. In the same way as the feature sizes are decreasing, the theoretical and practical constraints of making them and ensuring their quality are increasing. Consequently, modern production and inspection technologies are confronted with a bundle of challenges.

Important barriers for optical imaging and sensing coping with reduced feature sizes are the diffraction-limited lateral resolution, the limited depth of focus (DOF), and the limited space-bandwidth product of optical sensors. The observation of these physical and technical limitations is of increasing importance, not only for microscopic techniques but also for the application of 3D measurement techniques on wafer scale level. Consequently, the search for resolution-enhanced technologies becomes more and more important. A further challenge is the fast, reliable, and near-to-production detection of imperfections and material faults. This means that inline metrology/defectoscopy is a must for future production systems. Only the real-time feedback of the inspection results in the production process can contribute to a consistent quality assurance in processes with high cost risk. Moreover, the improvement of the robustness and flexibility of optical measurement systems, the assurance of the traceability, and the certified assessment of the uncertainty of the measurement results are ongoing challenges.

Optical Imaging and Metrology: Advanced Technologies, First Edition.
Edited by Wolfgang Osten and Nadya Reingand.
© 2012 Wiley-VCH Verlag GmbH & Co. KGaA. Published 2012 by Wiley-VCH Verlag GmbH & Co. KGaA.

Optical metrology still remains the measurement technique with the most potential, especially when dealing with systems having reduced feature sizes. Other techniques, such as atomic force microscopy, scanning near-field microscopy, or electron microscopy are time consuming, require partly special preparation of the specimen, and are often invasive. In contrast, optical metrology is a nondestructive, noninvasive, and fast working areal technique with many advantages over other technologies. However, some problems are still waiting for a solution. For instance, the 3D measurement of objects with high aspect ratio areas and noncooperative surfaces (corrupted, shiny, or translucent) requires special attention.

On the other hand, there are a lot of new approaches and tools that are tackling these challenges. New brilliant light sources with tunable properties, opto-electronic sensors with improved space-bandwidth product (SBP) and spectral sensitivity, spatial light modulators with both an ability for amplitude as well as phase control, and last but not least, the continuously growing computer power – especially taking use of the amazing graphics processing unit (GPU) performance – are a good basis for meeting the mentioned challenges. Furthermore, new approaches for the efficient inspection of extended surfaces having critical features in the micro- and/or nanoscale such as model-based feature reconstruction and new sensor fusion strategies deliver very promising results.

This chapter starts with a discussion of the physical and technical limitations in optical metrology. Afterwards, we try to give a systematic overview on the meanwhile huge diversity of approaches to enhance the resolution in optical imaging and metrology. On the basis of this, we present some promising strategies such as active wave front control, model-based metrology, superresolution (SR) using negative index material (NIM) lenses, and multi-scale sensor fusion.

15.2
Physical and Technical Limitations in Optical Metrology

Here we consider optical metrology as a far-field technique that is used for the measurement of object features acquired by imaging sensors. Measurement tasks that are based on optical imaging mainly suffer from the following four limitations that are closely related to each other:

- the lack of information for the unique solution of the inverse problem (the so-called ill-posedness);
- the diffraction-limited lateral resolution;
- the limited DOF;
- the limited SBP of optical imaging systems.

In the following sections, these limitations will be discussed briefly.

15.2.1
Optical Metrology as an Identification Problem

The problems to be solved in optical metrology are very similar to those in computer vision. Both disciplines process image like input and use methods of image analysis to derive a symbolic description of the image content or to reconstruct various physical quantities from the acquired intensity distribution. According to the well-known paradigm of Marr [2], computer vision is the development of procedures for the solution of the inverse task of the image formation process. This statement describes nothing else as the task to conclude from the effect (e.g., the observed intensity $I(i,j)$ in the pixel (i,j)) to its cause (e.g., the coordinates of the measured point $P(x,y,z)$ in space or its displacement due to any mechanical stress). In other words, in optical metrology also, an inverse problem has to be solved. From the point of view of a mathematician, the concept of an inverse problem has a certain degree of ambiguity which is well illustrated by a frequently quoted statement of J.B. Keller [3]: *"We call two problems inverses of one another if the formulation of each involves all or part of the solution of the other. Often for historical reasons, one of the two problems has been studied extensively for some time, while the other has never been studied and is not so well understood. In such cases, the former is called direct problem, while the latter is the inverse problem."* Both problems are related by a kind of duality in the sense that one problem can be derived from the other by exchanging the roles of the data and the unknown: the data of one problem are the unknowns of the other, and vice versa. As a consequence of this duality, it may seem arbitrary to decide what is the direct problem and what is the inverse problem. Following Bertero and Boccacci [4], for physicists and engineers, however, the situation is something different because the two problems are not on the same level: one of them, and precisely the one called the direct problem, is considered to be more fundamental than the other and for this reason, is also better investigated. Consequently, the historical reasons mentioned by Keller are basically physical reasons.

Processes with a well-defined causality such as the process of image formation are called direct problems. Direct problems need information about all quantities that influence the unknown effect. Moreover, the internal structure of causality, all initial and boundary conditions, and all geometrical details have to be formulated mathematically [5]. This includes the well-known initial and boundary value problems, which are usually expressed by ordinary and partial differential equations. Such direct problems have some excellent properties, which make them so attractive for physicists and engineers: If reality and mathematical description fit sufficiently well, the direct problem is expected to be uniquely solvable. Furthermore, it is in general stable, that is, small changes of the initial or boundary conditions also cause small effects only. Unfortunately, numerous problems in physics and engineering deal with unknown but nonobservable values. If the causal connections are investigated backward, we come to the concept of inverse problems. Based on indirect measurements, that is, the observation of effects caused by the quantity we are looking for, one can try to identify the missing parameters. Such problems, also called identification or reconstruction problems, are well known in imaging and optical metrology. Extensively studied examples are, for instance, the reconstruction of the phase distribution from observed intensities [6] and the reconstruction of object features from measured spectra in scatterometry [7]. Unfortunately, inverse problems have usually some undesirable properties: they are, in general, ill-posed, ambiguous, and unstable. The character of ill-posedness is addressed briefly.

The concept of ill- and well-posedness was introduced by Hadamard [8] into the mathematical literature. He defined a Cauchy problem of partial differential equations as well-posed if and only if for all Cauchy data there is a uniquely determined solution depending continuously on the data; otherwise the problem is ill-posed. In mathematical notation, an operator equation

$$F(x) = y, x \in D(F) \tag{15.1}$$

is defined with the linear or nonlinear operator $F : D(F) \subseteq X \to Y$ acting in Banach spaces X and Y with norms $\|.\|$ [5]. F is the mathematical expression for the cause-to-effect-map under consideration, x represents the solution to be found, and y the quantity that can be observed in general with a certain inaccuracy y_Δ only, where $\|y - y_\Delta\| \leq \Delta$. The domain $D(F)$ contains all permissible solutions of the inverse problem. Equation (15.1) is defined as well-posed if the following three Hadamard conditions are satisfied [8]:

1) $F(x) = y$ has a solution $x \in X$ for all $y \in Y$ (*existence*).
2) This solution x is determined uniquely (*uniqueness*).
3) The solution x depends continuously on the data y, that is, the convergence $\|y_n - y\| \to 0$ of a sequence $\{y_n\} = \{F(x_n)\}$ implies the convergence $\|x_n - x\| \to 0$ of corresponding solutions (*stability*).

If at least one of these conditions is violated, then the operator equation is called *ill-posed*. Simply spoken, ill-posedness means that we have not enough information available to solve the problem unambiguously.

In order to overcome the disadvantages of ill-posedness in the process of finding an approximate solution to an inverse problem, different techniques of regularization are used. Regularizing an inverse problem means that instead of the ill-posed original problem a well-posed related problem has to be formulated. The key decision of regularization is to find out an admissible compromise between stability and approximation [5]. The formulation of a sufficiently stable auxiliary problem means that the original problem has to be changed accordingly radically. As a consequence, one cannot expect that the properties of the solution of the auxiliary problem coincide with the properties of the original problem. But convergence between the regularized and original solutions should be guaranteed if the stochastic character of the experimental data is decreasing. In case of noisy data y_Δ, the identification of unknown quantities can be considered as an estimation problem. Depending on the linearity or nonlinearity of the operator F, we then have linear and nonlinear regression models, respectively. Consequently, least-square methods play an important role in the solution of inverse problems [9, 10]:

$$\|F(x) - y_\Delta\|^2 \overset{x \in D(F)}{\to} \min \tag{15.2}$$

with $y_\Delta = y + \Delta$, Δ – perturbations.

For the solution of inverse and ill-posed problems, it is important to apply a maximum amount of a priori knowledge and predictions about the physical quantities to be determined, respectively. In general, the question whether the measured data contain enough information to determine the unknown quantity clearly has to be answered. In case of direct problems where the data result from the integration of unknown components, data smoothing happens. Consequently, the direct problem is a problem directed toward a loss of information and its solution defines a transition from a physical quantity with a certain information content to another quantity with a smaller information content. For instance, in the process of image formation, this statement implies that the image is much smoother than the corresponding object. Consequently, after integration, the information about certain properties of the object is lost and therefore, different causes (causes used as synonym for the object) may result in almost the same effect. One example may serve again for explanation. In holographic nondestructive testing, characteristic fringe patterns are observed. These patterns are the basis for the identification of the usually nonobservable faults. The response of the fault to the applied load is smoothed since only the displacement on the surface gives rise to the observed fringe pattern. Depending on the structural properties of the object under test, the diversity of occurring faults, and the applied loading, a variety of fringe patterns can be observed. However, because of the mentioned integration process, these fringe patterns show a limited topology [11, 12]. Therefore, fixed recognition strategies based on known relations between the fault and the observed fringe pattern are only successful if the boundary conditions of the test procedure are limited in an inadmissible way. Since such conditions cannot be guaranteed, generally more flexible recognition strategies – we call them active strategies [13, 14] – have to be applied (Section 15.3.1.1). Most "classical" regularization procedures [9, 10] refer to a spatial neighborhood and are applied in passive methods of image analysis. In contrast to these passive methods, active approaches in optical imaging and metrology prefer another way to handle the difficult regularization problem by creating a temporal neighborhood where additional information continuously is provided (e.g., by changing the system configuration). Moreover, the direct and indirect problem systematically combined within a so-called model-based reconstruction strategy [14] (see section 15.3.1.2).

15.2.2
Diffraction-Limited Lateral Resolution in Optical Imaging

In general, optical resolution is a measure to evaluate an optical system with respect to the finest detail that can pass through without being distorted. The diffraction-limited lateral resolution in optics, that is, the minimum distance δx between two adjacent point images that can be distinguished optically is determined according to Abbe [15] by the quotient of the wavelength λ and the numerical aperture (NA)

$$\delta x = \kappa_1 \frac{\lambda}{\mathrm{NA}} \qquad (15.3)$$

The constant factor κ_1 is determined by experimental parameters such as illumination, signal-to-noise ratio (SNR) of the detector, a priori information, and the optical transfer function of the involved components. The value $\kappa_1 = 0.61$ corresponds to the well-known Rayleigh criterion for incoherent illumination with a 26% dip between the adjacent point images. To improve the resolution in optical lithography, immersion scanners with NA = 1.35 operating at a wavelength of 193 nm have been used for several years [16]. Using polarized illumination and multiple patterning technology [17, 18] they are capable of printing down to a 20 nm half pitch – the so-called CD – at full scan speed [19].

The resolution problem in imaging can also be discussed on the basis of the propagation of the angular spectrum [20]. With the interpretation of the propagation phenomenon as a linear space-invariant filter, the transfer function $H(u,v)$ of the free space can be written as

$$H(u,v) = \begin{cases} \exp\left[i2\pi \frac{z}{\lambda}\sqrt{1-(\lambda u)^2 - (\lambda v)^2}\right] & \text{for } \sqrt{u^2+v^2} \leq \frac{1}{\lambda} \\ 0 & \text{else} \end{cases} \qquad (15.4)$$

with z being the propagation distance and u and v being the spatial frequencies. According to Eq. (15.4), the free space acts like a low-pass filter with the cutoff frequency

$$u \leq \frac{1}{\lambda} \qquad (15.5)$$

The measurable information coming from structures having dimensions smaller than the wavelength is strongly attenuated. Consequently, the interpretation of the imaging process as a diffraction problem or as a direct problem leads to the same finding: the low-pass spatial bandwidth constraint of optical imaging systems resulting in a systematic loss of information about the original object. Therefore, the inverse process, that is, the reconstruction of the object from its image, shows all the difficulties that we have briefly discussed in Section 15.2.1.

The limited lateral resolution of imaging sensors has serious consequences for the limitation of the depth resolution of 3D-measuring techniques also [21]. In case of a triangulation-based sensor, the resolution δz of the measured height values z can be written as [21, 22]

$$\delta z = \kappa_1 \frac{C \cdot \lambda}{\mathrm{NA} \cdot \sin\theta} \approx \kappa_2 \frac{\delta x}{\sin\theta} \qquad (15.6)$$

where C is the speckle contrast characterizing the SNR in the point image [23] and θ is the triangulation angle used. In case of an areal triangulation sensor (e.g., a fringe projection system), it is useful to relate the resolution δz to the lateral extension x of the examined area F

$$\delta z_F = \frac{\delta z}{x} \qquad (15.7)$$

The derived quantity δz_F is a measure of the so-called area-related resolution that is often used for the characterization of the measurement system with respect to its performance on extended measuring areas. Because of the limited lateral resolution and the limited SBP (see Section 15.2.4) of imaging sensors the area-related resolution of current-triangulation-based 3D measurement systems only covers a range between 10^{-4} and 10^{-5}.

15.2.3
Diffraction-Limited Depth of Focus in Optical Imaging

The diffraction-limited DOF can be found using

$$\text{DOF} = \kappa_3 \frac{\lambda}{\text{NA}^2} \tag{15.8}$$

with κ_3 is the so-called process parameter [24]. Following Eq. (15.8), the DOF decreases drastically with the increase in the size of the NA. Consequently, there is a trade-off between the lateral resolution and the DOF, which has to be taken into account carefully if 3D measurement systems are designed.

15.2.4
Space-Bandwidth Product of Optical Imaging Systems

The SBP is another fundamental quantity for judging the performance of an optical imaging system. Following Max v. Laue, the SBP of a system is defined as a pure number that counts the degrees of freedom of the system. Lohmann et al. [25] define the SBP as the locations x and y and the regions of spatial frequencies $u = \sin \alpha / x$ and $v = \sin \beta / y$ where the signal is nonzero. The conventional definition of the SBP for a bandlimited function $g(x,y)$ can be written as [20]

$$\text{SBP} = 16 L_x L_y B_u B_v \tag{15.9}$$

where $2L_x$ and $2L_y$ are locations in space where the sampled function $g(x,y)$ delivers valid values and $1/2B_u$ and $1/2B_v$ are the distances where adjacent samples are taken according to the sampling theorem [26]. The sampling theorem states that if a signal is bandlimited, it can be represented by its sampled values if they are placed at equidistant intervals of $1/2B_u$ and $1/2B_v$. Bandlimited means that the spectrum of the signal is nonzero only over a limited region in the frequency space (u,v). Consequently, there are cutoff frequencies B_u and B_v such that $|u| \leq B_u$ and $|v| \leq B_v$, respectively. Thus for a discrete signal representation, the conventional definition of the SBP is given by the number of equidistant sampling points that accurately represent the signal. Simply spoken, for an optical signal that is sampled by a discrete sensor such as a CCD chip, $4L_x L_y$ is the area of the chip and $1/B_u$ is the pixel pitch. Therefore, the SBP can also be taken as the number of pixels that are used for the correct sampling of the function $g(x,y)$. Correct sampling means that the continuous function $g(x,y)$ can be reconstructed precisely from its sampled values.

15.3
Methods to Overcome the Resolution Problem in Optical Imaging and Metrology

Methods to overcome the mentioned limitations in optical imaging and metrology are known under the concepts of resolution enhancement [27], diffraction-limited resolution [28], super-resolution (SR) [29, 30], image restoration [31], image reconstruction [32, 33], and deconvolution [34]. For instance, SR is a frequently used term for many techniques that are dedicated to enhance the resolution of imaging systems considerably. However, only few of them really break the diffraction limit of optical systems (Section 15.3.2), while most of the SR techniques contribute only to a certain improvement of the resolution of digital imaging sensors. Figure 15.1 attempts to give a systematic overview of the applied methods. Here we mainly distinguish between methods based on direct imaging and those using reconstruction methods that are applied, for example, to spectral data captured with techniques such as scatterometry (Section 15.4). Most of the techniques mentioned in Figure 15.1 are described in the following text. For some special techniques, a reference is given in the caption of Figure 15.1.

It is far beyond the scope of this chapter to describe all existing solutions in detail. In the following sections, our focus is directed to some approaches that can help to overcome the above-mentioned difficulties. These approaches refer to

- new strategies for the solution of identification problems in optical imaging and metrology (active measurement strategies, model-based reconstruction strategies, and sensor fusion strategies) and

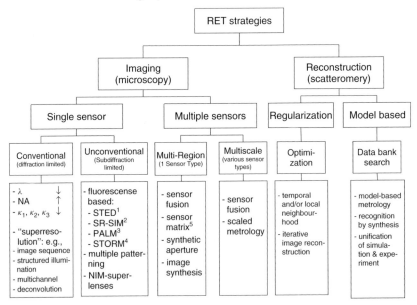

Figure 15.1 Overview of resolution-enhanced technologies applied in optical imaging and metrology. ([1]See [35], [2]see [36], [3]see [37], [4]see [38], [5]see [39].)

- methods that are directed to the improvement of the performance of a single imaging sensor, and methods that take advantage of the systematic combination of various sensors within a special inspection problem.

15.3.1
New Strategies for the Solution of Identification Problems

15.3.1.1 Active Measurement Strategies

Following the current trend in image analysis, more flexibility in the analysis strategy is obtained by combining the classical-data-driven bottom-up strategy with the so-called expectation-driven top-down strategy [40, 41]. The first strategy (Figure 15.2a) has been proved to be very efficient, but extra effort must be paid to obtain a high image quality and in most cases, a priori knowledge has to be added by operator interaction. Here the image formation process is considered as a rather fixed/passive data source and is not actively involved in the evaluation process. In contrast to this, the second strategy includes image formation as an active component in the evaluation process [12, 13]. Depending on the complexity of the problem and the state of evaluation, new data sets are actively produced by driving a feedback loop between the system components that are responsible for data generation (light sources, modulators, sensors, and actuators) and those that are responsible for data processing and analysis (algorithms, software, and processors) [42] (Figure 15.2b). Support from other sensors at different positions, recordings from different time instances, or the exploitation of different physical sensor principles, that is, sensor fusion approaches, are considered in that concept. The strategy that supervises the image analysis is connected to and controlled by the information gathered by sensors and by a knowledge base embedded in an assistance system [43, 44].

The implementation of an active feedback loop with the objective to control the image acquisition and formation process in a systematic way requires components

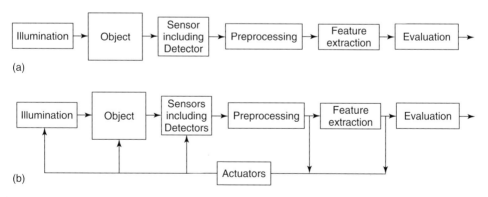

Figure 15.2 Different strategies for measurement and data processing in optical metrology. (a) Classical-data-driven bottom-up strategy and (b) combination of the data-driven bottom-up strategy with the expectation-driven top-down strategy by running an active feedback loop.

that are able to manipulate the decisive carrier of the information, that is, the light field and wavefront, respectively. Such actuator components – known as spatial light modulators – have been available since several years in various versions [42, 45]. They can be applied for the active control of all information channels of an electromagnetic wave such as the intensity, the phase, the angular spectrum, and the polarization. Detailed information about the application-adapted tuning of spatial light modulators (SLMs) is given by Kohler *et al.* [46–50].

15.3.1.2 Model-Based Reconstruction Strategies

As discussed in Section 15.2.1, the solution of the inverse problem of image formation needs the maximum amount of information that can be satisfied, particularly by the acquisition of high-precision measurement data, a priori knowledge, and model-based predictions about the physical quantities to be determined [51]. Model-based predictions are especially useful when nonresolved features have to be measured [52] (Chapter 13). The required data can be calculated by the simulation of the light–structure interaction for all relevant measurement constellations by using powerful simulation tools such as the rigorous coupled wave analysis RCWA [53]. The principle of such a strategy is shown in Figure 15.3. The starting point is a physical model of the measurement and image formation process. All known influences that contribute to the image formation process as the measurement principle, the instrument transfer function [54], the aberrations, and sensor features should be considered here. Such a model delivers the possibility of creating a database filled with templates that are representative of the addressed measurement constellations. As a kind of forward strategy, this simulation branch stands for the solution of the direct problem. In contrast to the measurement process in which the full complexity of interactions is always given simultaneously, this forward calculation opens the unique possibility to play with every single parameter independently. The other branch, namely, the real measurement of the object under test delivers those data that can be compared with the calculated templates. Such a kind of template matching opens the way to identify object

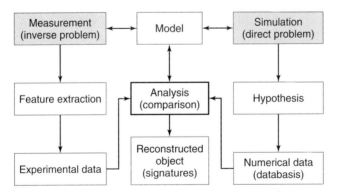

Figure 15.3 Model-based measurement strategy: combination of the direct and inverse problem within a feedback loop.

features, which cannot be measured directly. Before running such a feedback loop, a sensitivity analysis is recommended to learn about the impact of crucial system parameters [7].

As mentioned above, the reconstruction of the object data from the measured image data requires the solution of the inverse problem as a kind of backward strategy. Because of the limited information content of the measured data, the reconstructed object parameters match with the real parameters to a certain degree only. A comparison between the measured and simulated data gives reason for both improving the physical model for the simulation and changing the measurement procedure with respect to a more reliable data reconstruction. Such a complex reconstruction problem can be regularized by the systematic addition of new data taken from other information channels of the optical field (phase data, polarization data, and spectral data) [55, 56]. Consequently, the combination of both strategies, the measurement and the simulation branch, offers an elegant approach for the reconstruction of those parameters that cannot be observed directly. Exemplary implementations are presented in Section 15.4.1.

15.3.1.3 Sensor Fusion Strategies

Optical metrology offers a lot of efficient tools for the measurement of surface features across a wide scale range. However, as discussed in Section 15.2, there are some serious disadvantages of optical imaging sensors, such as diffraction-limited resolution. New strategies are needed especially in such cases where the dimensions of the object under test and the size of its features to be measured are in conflict. A representative example is the inspection of microelectromechanical systems (MEMS) at wafer scale with respect to manufacturing tolerances and local faults at micro/nanoscales, which requires, in principle, high-resolution measurement across a wide inspection field (Section 15.4.2). To judge the capacity of a particular sensor, we introduced the term area-related resolution δz_F (see Eq. (15.7)) as a figure of merit. δz_F is usually limited because of the diffraction problem, the SBP, and the system aberrations. As a consequence, inspection processes of extended surfaces by using only a single sensor can either be quick with low resolution or slow when sampled with high resolution. To find a compromise for that problem, several new concepts for inspecting extended surfaces were proposed and implemented in recent years. All these concepts are based on the cooperation of various sensors and can be classified under the term sensor fusion.

One such concept, called multiscale sensor fusion, refers to the systematic and sequential linking of different sensor types, which are gathering data about the object in successive scales [22, 43, 57–59]. Common inspection strategies usually follow a linear measurement chain with a single sensor. The multiscale strategy is a hierarchical and iterative approach starting with a coarse scale measurement of the specimen. A feature-based communication between different sensors is used to detect unresolved defects and to trigger higher resolution measurements. The new quality of this concept is characterized by the fact that the data acquisition in a certain scale is controlled by the measurement results obtained in the previous scale. Both the type of sensor to be used and the measurement area to be examined are specified by the respective preceding scale. Thus, step by step, the area of interest is reduced, while the boundary conditions for high-resolution sensors are improved. Such a multiscale sensor fusion concept (Figure 15.4a) relies on two assumptions. First, in usual industrial inspection processes only limited sections of an extended surface have to be measured with extreme resolution. Consequently, these sections have to be identified efficiently and only at these areas slow and expensive sensors should be applied. The other assumption is based on the hypothesis that deviations from the specification can already be recognized in low-resolution data sets, although the searched fault is not yet resolved sufficiently enough with respect to its unique classification. Such features that deliver neither a clear description nor a certainty about the presence of a measurement aberration but give indications for their presence are called as indicators [60, 61]. An improved description of the type and location of these features is a matter of the next measurement step following a sensor and position modification. Consequently, the involved sensors are not acting independent of each other. Rather their type, task, and position are determined by the results from the previous scale. Controlling the entire process is executed by a sophisticated assistance system that has the necessary knowledge about the available inspection task, sensors, actuators, and information processing routines [62] (Figure 15.4b).

Another concept to overcome the often unfavorable ratio of the lateral extension of the object to its feature sizes was proposed by Gastinger *et al.* [39]. In contrast to the above-discussed multiscale measurement, this concept refers to a multiregion measurement. The constraint resulting from the limited-area-related resolution of a single sensor is resolved here by the segmentation of the entire surface in many patches. To each patch only one sensor is assigned. In comparison to the multiscale approach, the processing time for such a sensor matrix is much shorter because the single-shot principle can be implemented. Furthermore, there is no additional overhead concerning sensor selection and positioning, indicator evaluation, and system control. However, such an implementation has a preassigned structure that is adapted to a defined inspection task. Thus, it has limited flexibility to be modified for function even when boundary conditions change (new object classes, changes in feature size, tolerances, etc.). An exemplary implementation was made for micro-opto-electronic-micro-systems (MOEMS) testing at wafer scale. The system consists of a 5 × 5 array of identical Twyman–Green or white-light interferometers. The diameter of a standard wafer is 8 in., while the structures to be investigated have typical lateral sizes between 3 mm and 10 μm. However, by using the interferometric measurement principle, the restrictions of triangulation sensors with respect to the area-related resolution δz_F come not into play.

15.3.2
Different Approaches for Resolution Enhancement of Imaging Systems

Refering to our overview given in Figure 15.1, the applied methods for resolution enhancement of optical imaging and measurement systems can be divided into two classes: conventional approaches with the objective to meet the resolution limit

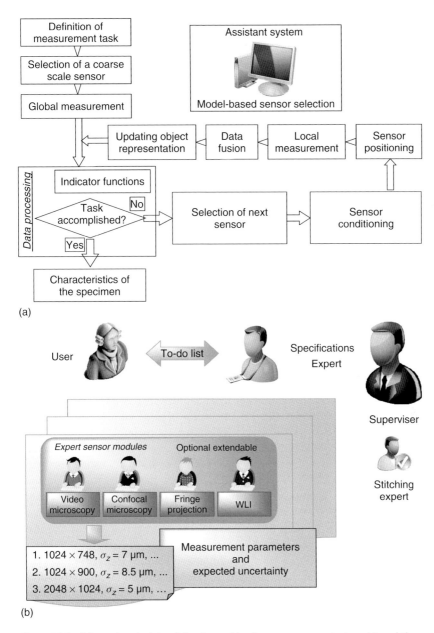

Figure 15.4 Schematic principle of the the multiscale measurement strategy (a) and the assistance system (b). (a) Multiscale measurement strategy and (b) core structure of the assistance system. WLI, white light interferometer.

(Eq. (15.3)) as best as possible and unconventional approaches with the objective to increase the resolution beyond the resolution limit.

15.3.2.1 Conventional Approaches to Achieve the Resolution Limit

Equation (15.3) offers the following three straightforward approaches for the enhancement of the lateral resolution [56]:

- decreasing the wavelength λ, that is, transition from the visual band across the UV band to the extreme UV region: visible (VIS) (550 nm) → UV (365 nm) → Deep UV (DUV) (248 nm) → Vacuum UV (VUV) (193 nm) → Extreme UV (EUV) (13,5 nm), (see, for instance, ITRS of SEMATECH [1] and recent VUV implementations in digital holography [63]);
- increasing the NA, by application of immersion and near-field methods [16], by taking advantage of the full angular spectrum by 4Pi-microscopy [64], by using methods of image synthesis [65], or by using synthetic aperture approaches that extend the collected spatial frequency content toward the $2/\lambda$ limit of the transmission medium [28];
- decreasing the process factor κ_1, which is possible by several means such as improved image acquisition techniques, optimized illumination as off-axis interference lithography, dark-field illumination [66], dark-beam scanning [67], and the simultaneous use of several information channels as in phase shifting polarization interference microscopy [55].

Advances in optical design and technology enable substantial improvements in the performance of optical systems. Meanwhile, it is state of the art in photolithography to write structure features in silicon that are much smaller than the wavelength of the used light [1]. However, the characterization of these subwavelength features is still a challenge [7]. At present, two main directions are pursued with respect to the further increase of the performance of metrology tools. One is taking advantage of the utilization of all information channels of the optical field. In addition to the intensity and phase, the polarization transfer due to the light–structure interaction and the complete angular spectrum of the scattered light field are evaluated [55, 56, 64, 68, 69]. The other direction refers to the implementation of reconstruction strategies in contrast to conventional imaging technologies. Nonresolved dimensional features such as the CD, the CDU (critical dimension uniformity), the side wall angle, the layer height, and material features (refractive index and dielectric constant, n- and k-value, respectively) are no longer measured directly but are calculated from the measured spectra by model-based reconstruction strategies [51]. With respect to the regularization of the ill-posed problem, additional data about the structure can be added. To make scatterometry data also sensitive for 3D-structures, conventional technologies can be upgraded by high-precision depth-scanning technologies such as white-light interferometry [52, 70].

15.3.2.2 Unconventional Approaches to Break the Resolution Limit

Meanwhile, many interesting approaches are published that are able to increase the resolution far beyond the classical resolution limit. Here, we classify these approaches as unconventional because they are based on fluorescence imaging and nonlinear transfer (Figure 15.1). Methods based on fluorescence imaging refer to special material properties and cannot be simply applied to all kinds of objects. Their main application fields are biology and medicine. Multiphoton microscopy [71, 72], stimulated emission depletion (STED) microscopy [35, 73], fluorescence microscopy by spatially structured illumination (SR-SIM – superresolution structured illumination microscopy) [36, 74], photoactivated localization microscopy (PALM) [37], and stochastic optical reconstruction microscopy (STORM) [38] are widely used. Other techniques such as multiple patterning [17, 18] use the nonlinear behavior of special photoresists in optical lithography and make a significant contribution by combination with immersion technologies for the unexpected decrease of the feature sizes down to 20 nm currently written in silicon with 193 nm photolithography. Here we discuss more extensively another unconventional approach: the application of meta-materials for making a lens that enables SR beyond the diffraction limit.

The Meta-Material Superlens Various activities are directed to the investigation of meta-materials [75] with respect to their application in the so-called superlenses [76, 77] that allow sub-λ imaging. Because surface plasmon polaritons (SPPs) [78] have a dominant influence on the unique properties of meta-materials, also called negative index materials (NIM), bulk NIMs can be replaced by resonantly coupled surfaces that allow the propagation of SPPs. Recently, it has been shown that a metallic meander structure is perfectly suited as such a resonant surface because of the tunability of the short-range surface plasmon polariton (SRSPP) and long-range surface plasmon polariton (LRSPP) frequencies by means of geometrical variation [79–82]. Furthermore, the passband between the SRSPP and LRSPP frequencies of a single meander sheet, induced by two Fano-type resonances, retains its dominant role when being stacked. A stack consisting of two meander structures (Figure 15.5) can mimic perfect imaging known from Pendry's lens [83] within this passband region. The meander structure under investigation is shown schematically in Figure 15.5a. It is basically a thin metal film corrugated on both sides with a thickness t, a meander depth D, and the periodicity P_x. To achieve inversion symmetry along the propagation direction of the incident light, the condition $W_r = P_x/2 - t$ has to be fulfilled. As soon as multiple meander structures are stacked (Figure 15.5b), the distance between two consecutive structures is labeled $D_{spa,i}$ from top to bottom. Silver is used as basic material because of the low absorption in the visible region. The dielectric function was derived from the Drude model and should be complemented by the Johnson and Christy [84] material data if higher frequencies are of interest. The structure is placed in vacuum and the incident light is s-polarized. The electromagnetic response is calculated with the RCWA simulation tool MICROSIM [53, 85], whereat 100 diffraction orders are considered. Schau et al. [80] have shown that such meander stacks can transfer energy resonantly over large distances with a high transmission and might enable subwavelength imaging. In Figure 15.6a, the transmittance of such a double meander structure as a function of the distance

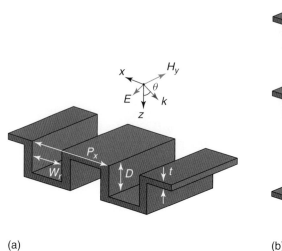
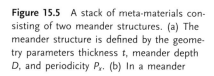
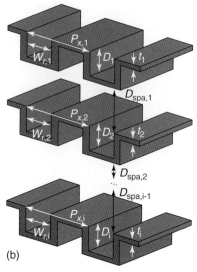

Figure 15.5 A stack of meta-materials consisting of two meander structures. (a) The meander structure is defined by the geometry parameters thickness t, meander depth D, and periodicity P_x. (b) In a meander stack, the geometry parameters of each sheet are labeled t_i, D_i, and $P_{x,i}$ from top to bottom according to the direction of the incident light. The distance between two meander structures is defined as $D_{spa,i}$.

between the meander sheets D_{spa} and the frequency f for different meander depths D is shown. One can observe that the plasmon passband of the single meander structure governs the transmission passband in the double and also (not shown) in the multimeander structure. The evolution of the passband also has an effect on the effective refractive index of the structure, resulting in interesting negative refraction and photon tunneling properties for imaging devices. Furthermore, the passband between the SRSPP and LRSPP frequencies is almost independent of the distance between the two meander structures, which allows low-loss energy transfer over large distances.

Within the passband of the structure in Figure 15.6a near-field imaging similar to Pendry's perfect lens can be achieved. In Figure 15.6b, a transverse magnetic (TM)-polarized wave ($f = 600$ THz and $\lambda = 500$ nm) is perpendicularly incident from the top and creates a point source behind a 100 nm wide subwavelength hole. As expected, the meander sheets ($P_x = 400$ nm, $t = 20$ nm, $D = 50$ nm, and $D_{spa} = 200$ nm) behave as coupled resonant surfaces and create a near-field deep subwavelength image of the slit behind the double meander structure. This first image has an FWHM of 148 nm, which is shown in the upper left inset of Figure 15.6b. However, the slit can also be imaged by a second focus with an FWHM of 202 nm almost two wavelengths behind the stack (lower right inset). This can still be considered as subwavelength imaging but occurs in a more usable distance behind the lens.

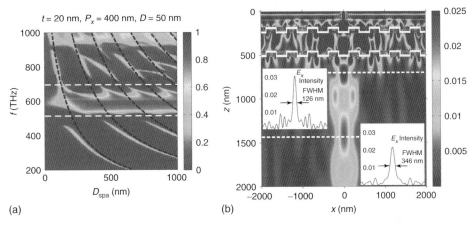

Figure 15.6 RCWA simulation of the transmission of a double meander structure as a function of the frequency f and the distance between the single structures D_{spa}. (a) The dashed black lines represent the predicted FP modes, while the dashed white lines indicate the SRSPP/LRSPP frequencies determined by dispersion diagram analysis (not shown). Ranging up to more than a micrometer, the passband is almost independent of D_{spa}. (b) Electric field intensity behind a slit with $w = 100$ nm and a double meander stack with $P_x = 400$ nm, $t = 20$ nm, $D = 50$ nm, and $D_{spa} = 200$ nm. The inset on the upper left side shows the electric field intensity along the x-axis at the focus plane. The inset on the lower right shows the electric field intensity at a distance of about two wavelengths ($\lambda = 500$ nm) behind the structure. (Please find a color version of this figure on the color plates.)

15.4
Exemplary Studies on the Performance of Various Inspection Strategies

15.4.1
Model-Based Reconstruction of Sub-λ Features

15.4.1.1 The Application of Scatterometry for CD-Metrology

In recent years, scatterometry could establish itself as a powerful tool for CD metrology besides CD-SEM (critical dimension scanning electron microscopy). The measured spectra are compared with simulated data that can be obtained using Maxwell equations solvers [85]. Performing a best match search for the measured spectra in a precomputed library of simulated spectra allows reconstructing the structure parameters.

The term scatterometry summarizes a rather great variety of techniques, all of which have in common the fact that they investigate the diffraction spectrum from a periodic array of nanostructures. Different methods are applied such as spectroscopic ellipsometry [86, 87], normal incidence reflectometry [88], 2θ-scatterometry [89], or angular resolved Fourier scatterometry [90]. A quite comprehensive overview of different techniques is given by Raymond [91, 92]. Imaging of diffraction spectra from the exit pupil of an imaging optics is well known as *conoscopy* [93].

Its application to scatterometry has been given various terms such as Fourier (transform) scatterometry [94] or scatterfield microscopy [95].

Besides the measurement technique itself the simulation and reconstruction is of great importance. A simple library search is largely applied; for 1D periodic structures, a real-time simulation with different techniques for quick scanning of the parameter space such as Recursive Random Search, Genetic Algorithms, or Levenberg-Marquardt Regression is applied [93]. Meanwhile, 2D periodic structures have been extensively investigated with scatterometry [96–99]. However, the technique still has serious problems such as enormous computation times and insufficient modeling background. Furthermore, scatterometry is currently confronted with two major problems: on the one hand, the feature size (CD) is continuously getting smaller and on the other hand, the structure inhomogeneity is getting larger. Therefore, we direct our attention in the following sections to the application of scatterometry for the inspection of rough sidewalls and to the investigation of the sensitivity of scatterometry tools with respect to the applied parameter settings in case of decreasing feature size [7, 100–102].

Line Edge Roughness and the Investigation of Its Influence on CD Metrology Line edge roughness (LER), that is, the roughness of sidewalls along line structures printed by lithography, has been of minor importance so far. Keeping present and future nodes (2010: 32 nm, 2011: 22 nm, and 2013: 16 nm [1, 103]), it is clear that a roughness amplitude in the range of a few nanometers becomes more and more crucial. Different authors have reported about investigations on LER in scatterometry. Boher *et al.* [90] investigated the influence of LER on the signals measured in a Fourier scatterometer. However, this method makes use of higher diffraction orders of the grating and hence is not applicable to present dense line structures. Quintanilha *et al.* [104] presented an experimental study that yielded a negligible influence of LER on scatterometry measurements for the considered samples and tool configurations. Germer [105] investigated the effect of linewidth and line position variation from trench to trench, that is, in the direction orthogonal to the lines. In a further study [106], he investigated line height variations and shape variations along the line cross section. He observed the spectra being seriously affected by the variations. Yakobovitz *et al.* [107] showed that measuring LER amplitudes above a certain threshold is possible using spectroscopic ellipsometry.

However, there is a strong need to understand more precisely the influence of LER on CD measurement and its measurability with scatterometry. Most of the work done so far avoids the costly modeling of LER using true 3D structures. A restriction to resist structures keeps the computation time bearable, since the RCWA algorithms converge much faster for dielectric gratings than for conductive gratings. In [7, 108], the first approach to consider true 3D roughness was proposed. In that work, reasonable computation times could be achieved because of a purely periodic modeling of the LER instead of a truly random model. Such a model, however, leads to diffraction because of the roughness rather than scattering and hence is a kind of oversimplification.

15.4 Exemplary Studies on the Performance of Various Inspection Strategies | 345

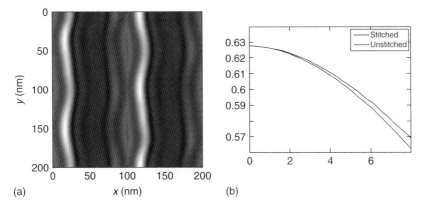

Figure 15.7 Exemplary near-field calculation for the case of a relatively large LER for Silizium bei 400 nm (almost grazing incidence and p-polarization). (a) Near-field and (b) validity examination: field amplitude of the zeroth diffraction order for the resulting far field of the stitched and unstitched cases.

There are methods to break large unit cells for rigorous computations into smaller ones, which are commonly known as *domain decomposition techniques* or field stitching [109]. However, in order to be fully rigorous, these methods have to comprise an overlap region and hence are not able to reduce computation time but only break a large problem into single feasible problems. Therefore, we skip the overlap region and introduce Kirchhoff boundaries [101]. Figure 15.7a shows an exemplary near-field calculation for the case of a relatively large LER amplitude of 4 nm peak to peak and a wavelength of 400 nm. The illumination is a single plane wave with almost grazing incidence. The leaps in the field amplitude are small but clearly recognizable. Figure 15.7b illustrates the field amplitude of the zeroth diffraction order of the resulting far field for the stitched and unstitched cases. The considered line structures for LER simulations are characterized by a mid-CD, a height, and a sidewall angle (SWA) (Figure 15.8a,b). The height of the lines is 100 nm. The nominal values for the SWA are $87°$. A modeling of the oblique sidewalls with only four layers yields a deviation of $<1\%$ from the best guess value using plenty of layers. The same is true for a restriction to five Fourier modes in both, x- and y- directions. This way the time for the RCWA computation of the 2D periodic gratings is reasonable. Provided that only roughness on the subwavelength scale occurs, it is obvious to introduce an effective medium approximation (EMA) layer sidewise to the line [110]. For the refractive index of this layer, the arithmetic mean of air and the resist material is chosen. A more advanced EMA model should include artificial birefringence [111]. The thickness of the EMA layer is named a_{EMA}. Figure 15.8c,d show two possible models for purely sinusoidal LER. The roughness pitch of such a sinusoidal roughness is denominated with Λ_{LER} and the peak to peak depth with a_{LER}. The first model (c) is called *"braid"* and describes a pure linewidth variation, whereas the second one (d) is called *"wave"* and describes pure LER. a_{LER} is gradually increased from zero in order to study the error of the stitched field amplitude as a

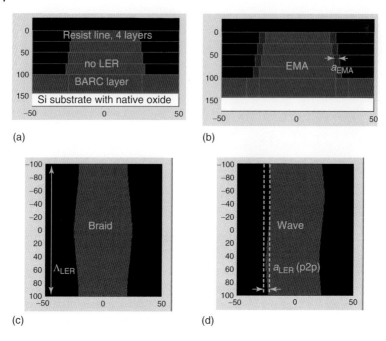

Figure 15.8 Structure models used in the LER simulation study. (a) Line structure for LER simulation without LER, (b) line structure for LER simulation with an EMA layer named a_{EMA}, (c) model for purely sinusoidal LER: braid (Λ_{LER}: roughness pitch), and (d) model for purely sinusoidal LER: waive (a_{LER}: peak to peak (p2p)).

function of a small perturbation of the periodic continuation. As can be seen, the errors remain acceptable for roughness depths of a few nanometers. The "wave" model assumes a variation of the line position only. For both Fourier scatterometry and spectroscopic ellipsometry the "braid" model is preferable for performing extensive simulations. The computation time can be reduced approximately by a factor of 2 because of the higher symmetry. Referring to the braid model we pursue the following approach: spectra with various LER parameters are calculated and considered as pseudomeasured spectra. Moreover, we simulate spectra without presence of LER and with a very simple type of LER in the form of an EMA layer. The latter spectra are considered as simulated spectra with and without LER. a_{EMA} is the free parameter. Then, reconstructions of different parameters such as CD and SWA are performed. The reconstruction errors of the parameters of interest as well as the reconstructed EMA layer thickness a_{EMA} can be considered as measures for the influence of LER in CD metrology. Figure 15.9 shows reconstructions of CD and SWA. The roughness pitch Λ_{LER} is varied over a wide range from 4 nm to 12.8 µm. The peak to peak depth of the LER (a_{EMA}) is chosen as 3 nm, which is believed to be a typical but rather large value. The reconstructed values are plotted as a function of Λ_{LER} whose axis features a logarithmic scale. The threshold value at which roughness diffraction begins is labeled by the gray vertical line. For graphs (a–c), the following findings can be stated: A floating EMA layer

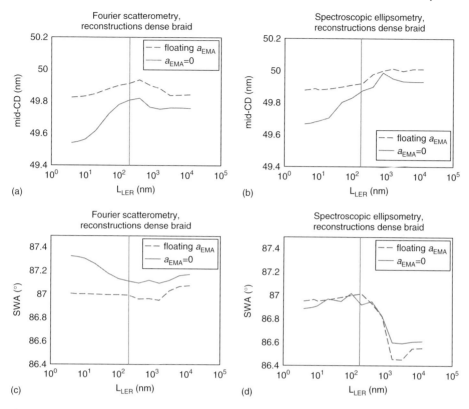

Figure 15.9 Reconstructions of CD and SWA using Fourier scatterometry and spectroscopic ellipsometry for vanishing and floating thickness of the EMA layer a_{EMA}. (a) Reconstructed CD (Fourier scatterometry), (b) reconstructed CD (spectroscopic ellipsometry), (c) reconstructed SWA (Fourier scatterometry), and (d) reconstructed SWA (spectroscopic ellipsometry). (Please find a color version of this figure on the color plates.)

highly reduces reconstruction errors even for the case of roughness diffraction. The influence of LER decreases with increasing Λ_{LER}. In graph (d) we observe a strong offset of the reconstructed SWA value for roughness pitches larger than 1 µm. There seems to be no obvious explanation for this behavior. Since the offset is observed for both modeling the lines with and without EMA layers, its origin cannot be assigned to the improper usage of the EMA model for large structures. The reconstruction errors in the case of Fourier scatterometry are slightly larger in the regime of optically smooth LER. However, keeping the offset in the SWA reconstruction in mind, there seems to be no clear advantage of spectroscopic ellipsometry. Summarizing the results, we can state that a floating EMA layer can clearly reduce reconstruction errors despite the presence of LER, even if the LER is on a strongly diffracting or scattering scale, that is, far outside the range of validity of EMA models. Further simulations of LER and investigations about its influence on the measured CD are given by Bergner *et al.* [112] and Bilski *et al.* [113]. Bilski *et al.* [113] could show that the influence of LER on scatter signatures is polarization dependent and proportional to the amount of LER.

Sensitivity Analysis of Scatterometry Tools with Respect to the Applied Parameter Settings As the size of semiconductor structures keeps decreasing the demands on scatterometry keep growing. Optimized measurement configurations can help to achieve maximum parameter sensitivity. These optimized configurations are also investigated with a simulation-based approach, performing a sensitivity analysis for different structures, parameters, and technology nodes. Sensitivity comparisons between different structure sizes and technology nodes allow to predict optimized measurement configurations for future industrial scatterometry tools. The sensitivity is analyzed for typical structure parameters such as height, CD, pitch, SWA, and so on, of simple resist line gratings having different pitch sizes.

To verify the correctness of the simulations, we use line gratings with different pitch periods that are produced with e-beam lithography. For the comparison between simulation and measurement, scatterometry spectra are measured with an industrial scatterometry tool [114]. In Figure 15.10, the results of the Fourier coefficients α and β over a wide wavelength range are shown. A very good agreement between simulation and measurement can be stated. Then we perform simulations for smaller structure sizes to be able to predict the evolution of the sensitivity for the next technology nodes. Here the sensitivity for further structure parameters such as the structure height, the CD, the pitch, the SWA, and different structure roundings is investigated by varying the measurement configuration for different wavelengths and incident angles. Figure 15.11a shows such a sensitivity plot for the CD parameter of dense resist lines. This sensitivity analysis was also performed for different structure sizes corresponding to actual and future technology nodes [7]. Figure 15.11b shows a simulation-based sensitivity prediction for the CD parameter at a typical fixed incidence angle used in industrial tools. The free parameter is the structure node. Such sensitivity predictions have been performed for different structure types and parameters [7]. With their help valuable predictions for optimized scatterometry configurations can be derived.

15.4.1.2 Model-Based and Depth-Sensitive Fourier Scatterometry for the Characterization of Periodic Sub-100 nm Structures

For spectroscopic ellipsometry, a fixed angle of incidence is chosen and the wavelength is varied while the ellipsometric angles (Δ, Ψ) or related quantities are measured at the position of the zeroth diffraction order [115]. In contrast, for 2θ scatterometry, the wavelength is fixed and the incidence angle θ is varied while measuring the diffraction efficiency of the zeroth order. These configurations can be realized with an experimental setup as schematically shown in

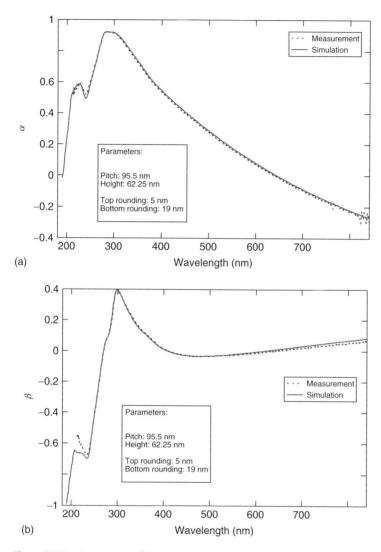

Figure 15.10 Comparison of the measurement is presented as dotted line and the corresponding simulation as continuous line made with MICROSIM of a resist line grating structure with a CD of 48 nm. (a) Fourier coefficient α and (b) Fourier coefficient β.

Figure 15.12. In Fourier scatterometry the angular response of a sample is transformed to spatial information, which can be imaged by a camera. Using a Köhler bright-field illumination with a high-NA objective the sample is illuminated by wide incident and azimuthal angle ranges simultaneously. In the imaged pupil plane, each incident and azimuthal angle corresponds to one specific position. This allows to analyze the angular response without the need of any mechanical scanning. Having all this information available in a single shot makes this

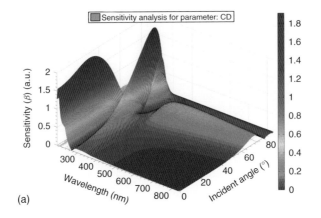

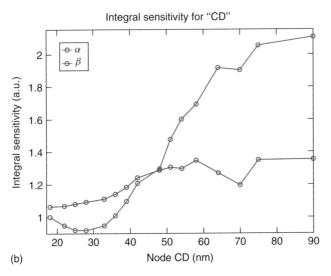

Figure 15.11 Sensitivity analysis of scatterometry tools. (a) Sensitivity toward the measurand β for a dense resist line grating (CD 48 nm) and the parameter CD dependent on the wavelength and the incident angle and (b) sensitivity trend (parameter CD, fixed incident angle) for shallow trench isolated (STI) line structures at different technology nodes. (Please find a color version of this figure on the color plates.)

method very promising for the solution of challenging tasks such as profile metrology on subwavelength periodic structures with noticeable LER (Section 15.4.1.1). Although Fourier scatterometry delivers 3D features of the inspected profile, it is necessary to improve the depth sensitivity of the method. Consequently, the combination with a more depth-sensitive method should contribute to the increase in reliability of the reconstructed 3D data. Scanning white-light interference microscopy is widely used for noncontact and high-speed 3D surface profilometry, providing phase-sensitive depth information with high resolution [116]. de Groot et al. [52] showed that imaging the pupil plane allows measuring the

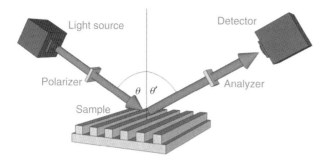

Figure 15.12 Classical scatterometry configuration. The used light source can have fixed or variable wavelengths and fixed or variable incidence angles.

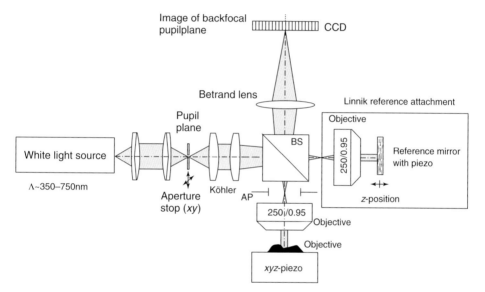

Figure 15.13 Schematic setup of depth-sensitive Fourier scatterometry by using a combination of white-light scanning interference microscopy and Fourier scatterometry. BS, beam splitter; AP, aperture.

incident-angle-dependent angular response and additionally using Fourier analysis of the white-light signal even allows wavelength-dependent measurements in one shot. We used white-light interference Fourier scatterometry for the investigation of sub-100 nm line gratings made with two-photon polymerization [70]. The measurement configuration is shown in Figure 15.13. The determination of the structure features follows the principle of model-based feature reconstruction explained in Section 15.3.1.2.

As the basis for the RCWA simulation of the scatterometric signals, we again use the software package MICROSIM [85]. The modeling of our light source is done by an illumination pupil sampled with an equidistant grid of points, which are defined by their NA coordinates NA_x and NA_y with $|NA| < 0.95$. Each point in the illumination pupil corresponds to one incident plane wave for which the diffraction spectrum is calculated with the RCWA. The calculated electric fields for each diffraction order have to be incoherently superposed in postprocessing in order to get the incoherent pupil image. In addition, to take into account the used white light, the calculation also has to be done for every wavelength at which the resulting fields are incoherently superimposed. For the case of having a reference branch, we need to model the beam splitter and to superimpose the electric field reflected by the reference mirror. This is done with help of the classical Jones matrix calculus. Each optical element can be separately described by its corresponding Jones matrix. The position of the reference mirror is taken into account by a phase term depending on the position z of the reference mirror. For an interferometric z-scan the diffraction caused by the object is rigorously calculated, while the reference branch and actual z-scan can be modeled fast, even though precisely, with the Jones calculus, getting a resulting pupil image for each z-position. With respect to the modeling of the complete experimental setup, we additionally included the spectral intensity distribution of the used white-light laser [117] and also the spectral response of the used CCD chip in our calculations. To estimate the sensitivity of the parameters of interest we use the formalism of uncertainty analysis as defined in the ICO Guide to the Expression of Uncertainty in Measurement [118]. With respect to its application for scatterometry Silver et al. [119] proposed more specified calculations. Three different types of structures are investigated here: (i) a simple periodic line grating of photosensitive material structured with two-photon polymerization on a glass substrate, (ii) the same structure but on a silicon substrate, and (iii) an e-beam-written photoresist line grating on a silicon substrate. The line width (CD) is 50 and 100 nm, respectively, while the period (pitch) is 100 and 200 nm, respectively. The height of the photoresist is in all cases 100 nm. The side wall angle of the structures is assumed to be 87°. A schematic description of the structure is given in Figure 15.14. To get an idea of how a pupil image in the back focal plane of the microscope objective looks like, Figure 15.15 shows the simulated Fourier scatterometry pupil images for an illumination with 410 nm and variation of the CD, height, and the side wall angle. The difference in the pupil images is

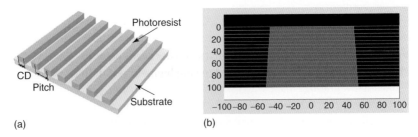

Figure 15.14 Schematic description of the investigated line structures. (a) Schematic line grating of photoresist on a substrate with a mid-line-width (CD) of 100 nm, period (pitch) of 200 nm, and a side wall angle (SWA) of 87° with 100 nm resist height (h) and (b) cross section with the staircase approximation used for the RCWA simulation.

15.4 Exemplary Studies on the Performance of Various Inspection Strategies | 353

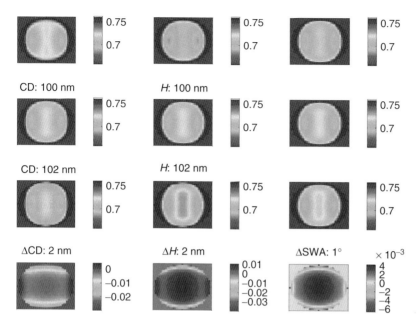

Figure 15.15 Simulated Fourier scatterometry pupil images for a photoresist line grating on silicon illuminated with 410 nm. The CD, height, and SWA are varied, and in the last row, the intensity difference is plotted to see the sensitivity toward that parameter. (Please find a color version of this figure on the color plates.)

an indicator of the sensitivity of the applied configuration. Further investigations with respect to the comparison of the 3σ uncertainty of data delivered by Fourier scatterometry at 410 nm with the data delivered by white-light interference Fourier scatterometry for the structure type "line grating" and the parameters CD, height, and side wall angle are made by Ferreras Paz et al. [70]. Table 15.1 shows exemplary results for a line grating of polymerized material on silicon. Especially for the parameter "height" the uncertainty is reduced significantly by more than a factor of 3 using white-light illumination and scanning the reference mirror. Also, the correlation between the different parameters especially CD with height and SWA is obviously reduced, which is important for library reconstruction (see Table 15.2).

A schematic overview of the experimental setup has been shown in Figure 15.13. A white-light laser as light source is used, giving higher intensity and a more flat intensity distribution for the used wavelength range as, for example, white-light light-emitting diode (LED) sources. Furthermore, a heat-protection filter is introduced to reduce the white-light spectrum to the visible range of about 400–800 nm because the used optical elements show smaller chromatic aberrations in this range. First the light coming out of the white-light laser fiber is collimated. The used white-light laser has an LP_{11} mode, which has to be homogenized with the help of two microlens arrays to get a homogeneous beam distribution [120].

Table 15.1 3σ-uncertainty for a line grating of polymerized material on silicon illuminated at 410 nm (first value) and for white-light illumination and scanning the reference mirror (second value) for dense lines with a CD of 50 or 100 nm.

Parameter	CD 50/Pitch 100 nm		CD 100/Pitch 200 nm	
CD	0.032 nm	0.038 nm	0.061 nm	0.074 nm
Height	0.164 nm	0.05 nm	0.071 nm	0.055 nm
SWA	0.162°	0.138°	0.122°	0.287°

Table 15.2 Covariance matrix containing the correlation coefficients for the different parameter combinations.

Parameter	CD		Height		SWA	
	50/100 nm	100/200 nm	50/100 nm	100/200 nm	50/100 nm	100/200 nm
CD	1 1	1 1	−0.90 −0.66	−0.87 −0.15	0.85 −0.21	0.85 −0.12
Height	−0.90 −0.66	−0.87 −0.15	1 1	1 1	−0.98 −0.84	−0.92 −0.86
SWA	0.85 −0.21	0.85 −0.12	−0.98 −0.84	−0.92 −0.86	1 1	1 1

The first value refers to Fourier scatterometry at 410 nm; the second value refers to the scanning white-light Fourier scatterometry.

A dispersion plate is allied to generate a diffuse and homogeneous light distribution for the Köhler illumination of the sample and also to get a uniform light distribution in the pupil plane that has to be imaged. The light enters a modified Leica DMR microscope where a beam splitter separates the light for the object and an optionally mounted reference branch. The reference branch consists of a Linnik-type interferometer setup, which is needed to accomplish the high NA of 0.95 of the used microscope objectives (PL APO 250×/0.95; PL APO is a special microscope objective lens from the company Leica (Wetzlar/Germany)). The high NA of 0.95 is needed to achieve incident angles of up to $\theta = \sin^{-1}(0.95) \approx 72°$. The z-position of the reference mirror can be scanned with help of a piezo actuator. The back focal planes of the object branch and the Linnik reference branch are then imaged together with the help of a Bertrand lens onto a frame transfer CCD (Hamamatsu C8000-10) to get the pupil plane images containing the scatterometric and interferometric information. Figure 15.16 shows the simulated and the measured pupil images for a complete scan of the reference mirror in case of an e-beam structured resist silicon grating. As one can see, simulated and measured pupil images look quite similar, but still some noise reduction, background subtraction, and calibration has to be performed until library-based comparison and feature reconstruction is possible with reasonable reliability. In recent work a silicon line grating with a CD = 200 nm and a period of 400 nm was measured in the sub-lambda regime [121]. The results for the model-based reconstruction of the profile parameters are in good agreement with the reference measurements made with AFM and SEM. The mean RMS-error bewteen the measured and simulated pupil image for the best matching scan is 6.5%.

15.4.2
High-Resolution Measurement of Extended Technical Surfaces with Multiscale Sensor Fusion

The objective of optical surface inspection is the measurement and description of the surface topography in different scales (global shape, waviness, and microstructure/roughness) and the detection of global and local deviations from the wanted shape (different kinds of surface defects). As critical areas, we denote those parts of the object where the resolution of the current sensor is not high enough to resolve the surface details sufficiently well with respect to the derivation of a reliable inspection result. To enable an efficient inspection process for such cases, the concept of multiscale sensor fusion [22] was introduced (Section 15.3.1.3). The new quality of this concept is characterized by the fact that the data acquisition in a certain scale is controlled by the measurement results obtained in the previous scale. Both the current type of sensor to be used and the current measurement area are specified by the respective preceding scale. Different features such as the fractal dimension, texture features, and the power spectral density are candidates for the indication of critical areas. Thus, step by step, the area of interest is reduced, while the boundary conditions for high-resolution sensors are improved.

The strategy described in Section 15.3.1.3 has been implemented into a system for the inspection of micro-optical and microelectromechanical components. To enable the system to select a suitable set of sensors for a given measurement task and to automatically explore the specimen, an assistance system was developed (Figure 15.4). This assistance system requires adequate knowledge about the measurement object and the possible defect types but allows a machine readable input for the inspection task. The implemented system provides detailed knowledge about the properties of different types of sensors to be activated sequentially in the multiscale inspection strategy, about the relevant object types, and about the surface defect classes. On the basis of this knowledge, the strategy for the current inspection problem is compiled automatically, taking into account both the sensor activation and the path control. Genetic algorithms are used to optimize the parameters and to compile a problem-adapted combination of various image processing functions with respect to a successful completion of the task [122]. For detailed information see Burla *et al.* [44, 62].

For the evaluation of the multiscale measurement strategy a demonstrator has been developed. The demonstrator is based on a modified Mahr MFU-100 measurement machine with a sensor mount for up to three different sensor systems (Figure 15.17). In the current configuration, three sensor types are mounted on the machine. In the coarse scale, a video microscope is used for the fast acquisition of the entire object. Various illumination types are realized for defect indication: dark field and scattered light illumination for the indication of scratches and dust particles, bright-field illumination for the recognition of shape defects,

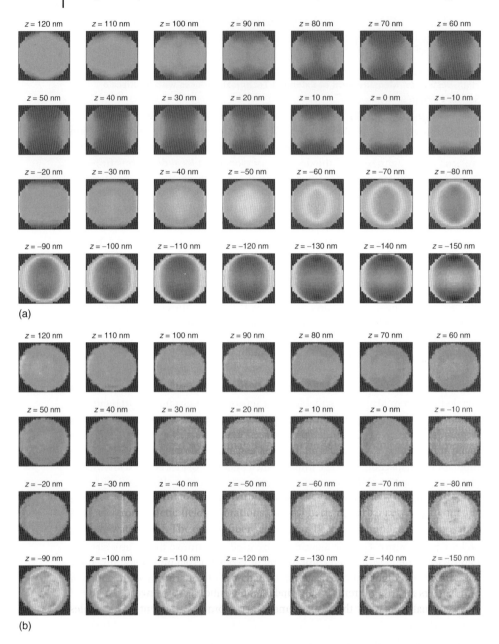

Figure 15.16 Complete scan of the reference mirror for an e-beam structured silicon grating of CD 200 and 400 nm pitch, with a height of 70 nm measured with white-light Fourier scatterometry: (a) measured results and (b) simulation results with the same parameters. (Please find a color version of this figure on the color plates.)

15.4 Exemplary Studies on the Performance of Various Inspection Strategies

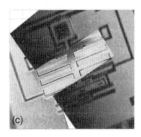

Figure 15.17 Mahr MFU-100 measurement machine adapted for multiscale metrology. (a) Mahr MFU-100 with three sensor types: video microscope, confocal sensor; (b) zoom of the sensor head; and (c) screen shot from the GUI (graphical user interface) of the control system.

and in case of inspection of micro-optical elements, a special illumination for the measurement of the point spread function. In the middle scale, a confocal microscope (CM) is used to measure 3D surface features. Depending on the task specification different magnifications and resolutions can be adopted. Finally, for high-resolution measurement, a chromatic confocal point sensor is used. Some relevant parameters of the used sensors are summarized in Table 15.3.

To assure a fast and precise position control of all sensor types, the tilts and positions for all axes are measured at 1 kHz with a resolution of 1 nm. This allows a fine positioning with submicrometer accuracy. The maximum velocity of the linear axes is above 200 mm s^{-1}, which also enables high-speed positioning [123]. The control program is implemented in Matlab. It is responsible for the completion of the measurement conditions in different scales, calculates new regions of interest, and triggers new measurements. To update the positions for a new measurement and to visualize results, the measurement data from the different sensor scales are registered into a global coordinate system. For that purpose, the sensor orientations and the position of the optical axis of the sensors with respect to the rotation axis of the MFU are carefully calibrated. A screenshot of the graphical user interface (GUI) along with the stitching of an example measurement is shown in Figure 15.17c.

For the proof of principle, the inspection system was tested on two different object classes: microlens arrays [124] and MEMS-based microcalibration devices. Here

Table 15.3 Parameters of different sensors implemented in the demonstrator for multiscale sensor fusion.

Parameter type	Video microscope	CM					Chromatic confocal sensor CCM		Siemens SiScan SC 400
Imaging data type	Color images (fovion chip)	Area-related topography and intensity map					3D point cloud		3D point cloud
Objective lens	55 mm, telecentric	5×	10×	20×	50×	100×	50×	63×	–
Numerical aperture	–	0.15	0.3	0.4	0.5	0.95	0.5	0.95	0.6
Field diagonal in μm	19000	3840	1920	960	384	192	Variable		Variable
Lat. resolution in μm	–	20	10	5	3	1	3	<2	<2
Ideal z-resolution (μm)	Not available	0.9	0.18	0.09	0.05	0.01	<0.05	<0.02	0.1
Working distance (mm)	100	20	10	12	10	0.3	10	0.16	6

CM, confocal microscope; CCM, chromatic-confocal-microscopy.

we show selected results for the inspection of microelectromechanical systems on the example of microcalibration devices [125, 126]. The inspection of MEMS is a relatively complicated task because of the wide diversity in layout and size and because of the fact that MEMS consist of various structures and elements with diverse design specifications (shape, surface properties, material properties, ...). Consequently, the description of all kind of MEMS by a single object class is an almost impossible task. Hence, for the verification of the multiscale principle, we concentrated on a special class of microcalibration devices as a kind of subset for the large diversity of MEMS (Figure 15.18).

The calibration devices are designed for the qualification of microdisplacement and microforce measurement systems. Figure 15.18a shows a sample of the in-plane displacement calibration devices and Figure 15.18b a sample of the out-of-plane displacement calibration devices. In the first case, a comb drive transforms a defined voltage into a defined in-plane movement of a mass held by springs and flexures. It can be used vice versa, transforming an elongation of the mass into a voltage. The mass of the moving part and the stiffness of the springs are specified so that the device can be used for dynamic and static calibration tasks. The overall size of the in-plane calibration device is 7 mm × 7 mm. It consists of the following critical components: optical detection areas, comb drives, springs and flexures,

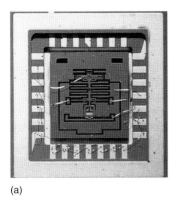 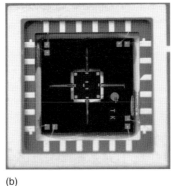

(a) (b)

Figure 15.18 Microcalibration devices developed by IMTEK in Freiburg in cooperation with ITO Stuttgart. (a) Photo of the in-plane displacement calibration device and (b) photo of the out-of-plane displacement calibration device.

Table 15.4 Specification of critical features and possible defects.

Component	Critical properties	Possible defects
Optical detection areas	Roughness	Scratches, digs, and bumps
	Reflectivity	Delamination
		Pollution
Moveable mass system	Volume	Deviation from calculated volume
	Mass	Blocking by pollution or unfulfilled edging
	Roughness	Deviation from specified roughness
Springs/Flexure	Stiffness	Deviation from specification
	Geometry	Cracks
	Resonance frequency	Broken springs
Comb drive	Dimensions	Broken combs and fingers
	Number of combs/fingers	Deflected combs
	(sidewall)Roughness: capacity (position), force (voltage)	Short cuts
	Roughness	Pollution
Contact area	Material	Short cuts
	Dimensions	Cracks
	Shape	Holes
Basic structure	Design	Shortcuts
		Moveable parts sticking
		Cracks or broken parts

contact areas and contact lines, the movable mass system, and the base structure. A list of selected critical components and possible defects is shown in Table 15.4. The springs for the plane device have a length of about 1.000 µm and a width of 10 µm. Each of the eight comb drives consists of up to 500 combs with a width of 10 µm and

Table 15.5 Parameters of the sensors suitable for measuring a micro-electro-mechanical systems (MEMS) waver.

Scale	Sensor	Subtask and defects	Detection algorithm
1.	Video microscope 84× bright field 84× dark field	Bright field: position of MEMS combs on waver Dark field: scratches and pollution	MEMS position: hough transformation Dark field: Local thresholding
2.	Confocal microscope (CM) 10× lens	Characterization of scratches and pollution Indicate missing or blocked combs Indicate cracks in flexures and springs	Fitting of standard geometrical elements (planes) Wavelet-transformation-based algorithm [28]
3.	50× CM or CCM 50×	Defect characterization	Fitting of standard geometrical elements (planes)

with a gap of 4 µm. Common defects for MEMS are cracks, bumps and scratches, delamination of layers, missing structures, short circuits, and broken wires. For the in-plane MEMS, additional defects are broken or blocked combs, deflections, and pollution on the optical windows. To find cracks or broken and deflected springs the sensor resolution has to be better than 10 µm. To find particle pollutions blocking the drive or any shortcuts between the combs the sensor resolution has to be better than 4 µm.

The configuration of the system is shown in Table 15.5. The video microscope uses a telecentric front lens and delivers the image data of the complete 6″ waver in a single shot. The bright-field images are used to identify the regions of the comb drives and springs on the MEMS waver. For the second scale, the CM is used to characterize larger cracks and scratches and to indicate defects in the comb regions. For high-resolution imaging 50× confocal lens or 50× chromatic-confocal-microscopy (CCM) lens are applied. The indicator detection algorithms are described in details by Lyda *et al.* [58]. Figure 15.19 shows typical measurement results of a MEMS waver in the three addressed scales.

15.5
Conclusion

In this chapter, we have tried to give a consistent description and classification of the existing limits in optical micro- and nanometrology. Based on this we have discussed several approaches to cope with the resolution problem, such as the active measurement strategy, the model-based reconstruction strategy, the multiscale sensor fusion strategy, and different ways for resolution enhancement of optical imaging and measurement systems. Finally, we have presented exemplary studies

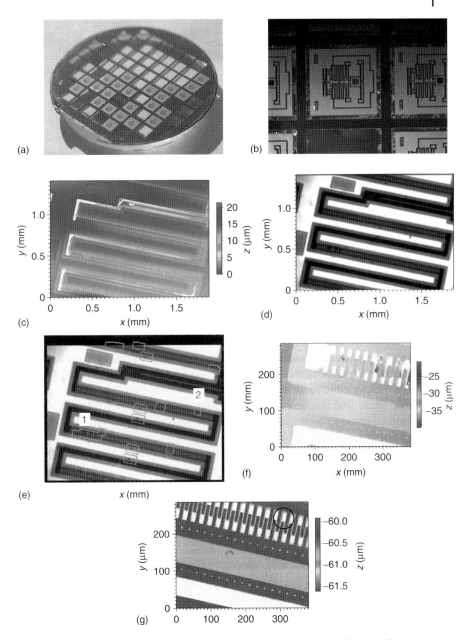

Figure 15.19 Inspection of a 6 in. wafer with in-plane microcalibration devices. (a) Photo of the wafer under test, (b) first scale measurement of MEMS structures made with a video microscope, (c) topography of the comb drive taken in the second scale with a 10× confocal microscope, (d) maximal intensity image of 10× confocal measurement, and (e) intensity image with highlighted region for possible defects, (f,g) third scale, high-resolution measurement with the 50× confocal microscope. (Please find a color version of this figure on the color plates.)

on the performance of these strategies using example of CD measurement for structures with LER; model-based feature reconstruction of sub-100 nm structures by a combination of scatterometry, white-light interferometry, and rigorous simulation; and the inspection of features of MEMS on wafer scale level by multiscale sensor fusion.

Acknowledgments

The author gratefully acknowledges the support of this work by the German Federal Ministry of Education and Research BMBF within the Grant No. 13N9432, the Baden-Württemberg Stiftung for financial support of the project OPTIM, the German National Science Foundation DFG for the support of the projects Os 111/18, OS 111/19, Os 111/22, and Os 111/28. Special thanks go to A. Burla, V. Ferreras-Paz, K. Frenner, T. Haist, W. Lyda, G. Pedrini, S. Peterhänsel, P. Schau, and Th. Schuster for their valuable contributions in the mentioned projects. Part of the work was already published in [127].

References

1. SEMATECH (1982) http://www.itrs.net/Links/2011ITRS/Home2011.htm.
2. Marr, D. (1982) *Vision: A Computational Investigation into the Human Representation and Processing of Visual Information*, Freeman and Company, San Francisco.
3. Keller, J.B. (1976) Inverse Problems. *Am. Math. Mon.*, **83**, 107–118.
4. Bertero, M. and Boccacci, P. (1998) *Introduction to Inverse Problems in Imaging*, IOP Publishing, Bristol.
5. Hofmann, B. (1995) in *The Inverse Problem* (ed. H. Lübbig), Akademie Verlag, Berlin, pp. 45–66.
6. Osten, W. (2008) in *Handbook of Experimental Solid Mechanics* (ed. W.N. Sharpe), Springer, pp. 481–563.
7. Osten, W., Ferreras Paz, V., Frenner, K., Schuster, T., and Bloess, H. (2009) Simulations of scatterometry down to 22 nm structure sizes and beyond with special emphasis to LER. *AIP Proc.*, **1173**, 371–378.
8. Hadamard, J. (1923) *Lectures on Cauchy's Problem in Partial Differential Equations*, Yale University Press, New Haven.
9. Tikhonov, A.N. (1963) Solution of incorrectly formulated problems and the regularization method. *Sov. Math. Dokl.*, **4**, 1035–1038.
10. Tikhonov, A.N. and Arsenin, V.Y. (1977) *Solution of Ill-Posed Problems*, Winston and Sons, Washington, DC (Transl. From Russian).
11. Mieth, U., Osten, W., and Jüptner, W. (2001) *Proceedings of Fringe*, Elsevier Science, pp. 163–172.
12. Osten, W. and Kujawinska, M. (2000) in *Trends in Optical Nondestructive Testing and Inspection* (eds P.K. Rastogi and D. Inaudi), Elsevier Science B.V., pp. 45–69.
13. Osten, W. (1998) Active optical metrology – a definition by examples. *Proc. SPIE*, **3478**, 11–25.
14. Osten, W., Elandaloussi, F., and Mieth, U. (2002) Trends for the solution of identification problems in holographic non-destructive testing (HNDT). *SPIE*, **4900**, 1187–1196.
15. Abbe, E. (1873) Beiträge zur theorie des mikroskops und der mikroskopischen wahrnehmung. *Arch. Mikrosk. Anat.*, **9**, 413–468.

16. Totzeck, M., Ullrich, W., Göhnermeier, A., and Kaiser, W. (2007) Pushing deep ultraviolet lithography to its limits. *Nat. Photonics*, **1**, 629–631.
17. Ebihara, A., Levenson, M.D., Liu, W., He, J., Yeh, W., Ahn, S., Oga, T., Shen, M., and M'saad, H. (2003) Beyond $k_1 = 0.25$ lithography, 70 nm L/S patterning using KrF scanners. *Proc. SPIE*, **5256**, 985–994.
18. Finders, J., Dusa, M., Vleeming, B., Hepp, B., Meenhoudt, M., Cheng, S., and Vandeweyer, T. (2009) Double patterning lithography for 32 nm: critical dimensions uniformity and overlay contol considerations. *J. Micro/Nanolith. MEMS MOEMS*, **8**, 011002.
19. http://www.electronics-eetimes.com/en/intel-and-micron-deliver-new-20-nanometer-process-for-8gb-mlc-nand-flash.html?cmp_id=7&news_id=222906887.
20. Goodman, J.W. (2005) *Introduction to Fourier Optics*, Roberts & Company Publishers, Greenwood Village.
21. Häusler, G. and Leuchs, G. (1997) Physikalische grenzen der optischen formerfassung mit licht. *Phys. Bl.*, **53** (5), 417–421.
22. Osten, W., Andrä, P., and Kayser, D. (1999) Hochauflösende vermessung ausgedehnter oberflächen mit skalierbarer topometrie. (High resolution measurement of extended technical surfaces with scalable topometry). *Tech. Messen*, **66** (11), 413–428.
23. Pedersen, H.M. (1975) On the contrast of polychromatic speckle patterns and its dependence on surface roughness. *Opt. Acta*, **22**, 14–24.
24. Born, M. and Wolf, E. (1999) *Principles of Optics*, Cambridge University Press, Cambridge.
25. Lohmann, A.W., Dorsch, R.G., Mendlovic, D., Zalevsky, Z., and Ferreira, C. (1996) Space-bandwidth product of optical signals and systems. *J. Opt. Soc. Am. A*, **13** (3), 470–473.
26. Shannon, C.E. (1949) Communication in the presence of noise. *Proc. Inst. Radio Eng.*, **37** (1), 10–21.
27. Schellenberg, F.M. (ed.) (2004) *Selected Papers on Resolution Enhancement Techniques in Optical Lithography*, SPIE Milestone Series MS178, SPIE, Bellingham, Washington.
28. Kuznetsova, Y., Neumann, A., and Brueck, S.R.J. (2007) Imaging interferometric microscopy – approaching the linear systems limits of optical resolution. *Opt. Express*, **15** (11), 6651–6663.
29. Zalevski, Z. and Mendlovic, D. (2004) *Optical Superresolution*, Springer, New York.
30. Zalevsky, Z., Borkowski, A., Marom, E., Javidi, B., Beiderman, Y., Mico, V., and Garcia, J. (2011) Recent advantages in the field of super resolved imaging and sensing. *Proc. SPIE*, **8082**, 80820G-1–80820G-10.
31. Castleman, K.R. (1996) *Digital Image Processing*, Prentice-Hall, Englewood Cliffs.
32. Long, D.G., Hardin, P.J., and Whiting, P.T. (1993) Resolution enhancement of spaceborne scatterometer data. *IEEE Tran. Geosci. Remote Sens.*, **32** (3), 700–715.
33. Early, D.S. and Long, D.G. (2001) Image reconstruction and enhanced resolution imaging from irregular samples. *IEEE Trans. Geosci. Remote Sens.*, **39** (2), 291–302.
34. Jähne, B. (2005) *Digital Image Processing*, Springer, Berlin, p. 489 ff.
35. Hell, S.W. and Wichmann, J. (1994) Breaking the diffraction resolution limit by stimulated emission: stimulated-emission-depletion fluorescence microscopy. *Opt. Lett.*, **19** (11), 780–782.
36. Gustafsson, M.G., Shao, L., Carlton, P.M. et al. (2008) Three-dimensional resolution doubling in wide-field fluorescence microscopy by structured illumination. *Biophys. J.*, **94** (12), 4957–4970.
37. Betzig, E., Patterson, G.H., Sougrat, R., Wolf Lindwasser, O., Olenych, S., Bonifacino, J.S., Davidson, M.W., Lippincott-Schwartz, J., and Hess, H.F. (2006) Imaging intracellular fluorescent proteins at nanometer resolution. *Science*, **313** (5793), 1642–1645.
38. Rust, M.J., Bates, M., and Zhuang, X. (2006) Sub-diffraction-limit imaging by stochastic optical reconstruction

microscopy (STORM). *Nat. Methods*, **3**, 793–796.

39. Gastinger, K., Johnsen, L., Kujawinska, M., Jozwik, M. et al. (2010) Next generation test equipment for micro-production. *Proc. SPIE*, **7718**, 77180F-1–77180F-16.

40. Aloimonos, J. (1994) Reply, what i have learned. *Image Underst.*, **60** (1), 74–85.

41. Liedtke, C.-E. (1998) Intelligent approaches in image analysis. *Proc. SPIE*, **3478**, 2–10.

42. Osten, W. and Jüptner, W. (1999) New light sources and sensors for active optical 3D-inspection. *Proc. SPIE*, **3897**, 314–327.

43. Kayser, D., Bothe, T., and Osten, W. (2004) Scaled topometry in a multi-sensor approach. *Opt. Eng.*, **43** (10), 2469–2477.

44. Burla, A., Haist, T., Lyda, W., and Osten, W. (2010) An assistance system for the selection of sensors in multi-scale measurement systems. *Proc. SPIE*, **7791**, 7791–7718.

45. Osten, W. (2006) Progress in total light control: components, methods, and applications. *Proc. SPIE*, **6341**, 63410G-1–63410G-12.

46. Kohler, C., Schwab, X., and Osten, W. (2006) Optimally tuned spatial light modulators for digital holography. *Appl. Opt.*, **45** (5), 960–967.

47. Kohler, C., Haist, T., Schwab, X., and Osten, W. (2008) Hologram optimization for SLMbased reconstruction with regard to polarization effects. *Opt. Express*, **16** (19), 14853–14861.

48. Haist, T., Zwick, S., Warber, M., and Osten, W. (2006) Spatial light modulators – versatile tools for holography. *J. Hologr. Speckle*, **3**, 125–136.

49. Zwick, S., Haist, T., Warber, M., and Osten, W. (2010) Dynamic holography using pixelated light modulators. *Appl. Opt.*, **49** (25), F47–F58.

50. http://www.holoeye.com/spatial_light_modulators-technology.html Vialux GmbH; http://www.vialux.de/pdf/dlp_overview_en.pdf.

51. Osten, W. and Kerwien, N. (2005) Resolution enhancement technologies in optical metrology. *Proc. SPIE*, **5776**, 10–21.

52. de Groot, P., Colonna de Lega, X., and Liesener, J. (2009) Model-based white light interference microscopy for metrology of transparent film stacks and optically-unresolved structures, *Proceedings of Fringe*, (eds W. Osten and M. Kujawinska) Springer, Heidelberg, pp. 236–243.

53. Totzeck, M. (2001) Numerical simulation of high-NA quantitative polarization microscopy and corresponding near-fields. *Opt. - Int. J. Light Electron Opt.*, **112**, 399–406.

54. de Groot, P. and Colonna de Lega, X. (2005) *Proceedings of Fringe*, Springer, Heidelberg, pp. 30–37.

55. Totzeck, M., Jacobsen, H., and Tiziani, H.J. (2000) Edge localization of sub-wavelength structures using polarization interferometry and extreme-value criteria. *Appl. Opt.*, **39**, 6295–6305.

56. Osten, W. and Totzeck, M. (2004) Optics beyond the limits: The future of high precision optical metrology and implications for optical lithography. Proceedings of the ICO 2004, Tokyo 2004, pp. 589–592.

57. Osten, W., Kayser, D., Bothe, Th., and Jüptner, W. (2000) High resolution measurement of extended technical surfaces with scalable topometry. *Proc. SPIE*, **4101A**, 166–172.

58. Lyda, W., Burla, A., Haist, T., Zimmermann, J., Osten, W., and Sawodny, O. (2010) Automated multi-scale measurement system for MEMS characterization. *Proc. SPIE*, **7718**, 77180G.

59. Lyda, W., Burla, A., Haist, T., Gronle, M., Osten, W., and (2012) Implementation and analysis of an automated multiscale measurement strategy for waver scale inspection of micro electromechanical systems. *Int. J. Precis. Eng. Manuf.*, **13** (4) 483–489. doi: 10.1007/s12541-012-0063-x

60. Burla, A., Lyda, W., Osten, W., Regin, J., Westkämper, E., Zimmermann, J., and Sawodny, O. (2010) Reliability analysis of indicator functions in an automated multiscale measuring system. *Tech. Messen*, **77** (9), 493–499.

61. Burla, A., Haist, T., Lyda, W., and Osten, W. (2011) Fourier descriptors for defect indication in a multiscale and multisensor measurement system. *Opt. Eng.*, **50**, 043603. doi: 10.1117/1.3562319
62. Burla, A., Haist, T., Lyda, W., Aissa, A., and Osten, W. (2011) Assistant system for efficient multiscale measurement and inspection. *Proc. SPIE*, **8082**, 808202–808801–10.
63. Faridian, A., Hopp, D., Pedrini, G., Eigenthaler, U., Hirscher, M., and Osten, W. (2010) Nanoscale imaging using deep ultraviolet digital holographic microscopy. *Opt. Express*, **18** (13), 14159–11164.
64. Hell, S.W. and Stelzer, E.H.K. (1992) Fundamental improvement of resolution with a 4Pi-confocal fluorescence microscope using two-photon excitation. *Opt. Commun.*, **93**, 277–282.
65. Yuan, C., Situ, G., Pedrini, G., Ma, J., and Osten, W. (2011) Resolution improvement in digital holography by angular and polarization multiplexing. *Appl. Opt.*, **50** (7), B6–B11.
66. Ehret, G., Bodermann, B., and Mirandé, W. (2007) Quantitative linewidth measurement down to 100 nm by means of optical dark-field microscopy and rigorous model-based evaluation. *Meas. Sci. Technol.*, **18** (2), 430–438.
67. Tavrov, A., Kerwien, N., Berger, R., Tiziani, H., Totzeck, M., Spektor, B., and Shamir, J. (2003) Vector simulation of dark beam interaction with nano-scale surface features. *Proc. SPIE*, **5144**, 26–36.
68. Totzeck, M., Jacobsen, H., and Tiziani, H.J. (1999) Phase – shifting polarization interferometry for microstructure linewidth measurement. *Opt. Lett.*, **24** (5), 294–296.
69. Kerwien, N. (2007) About the influence of polarization effects for microscopic image formation. Dissertation. University of Stuttgart.
70. Ferreras Paz, V., Peterhänsel, S., Frenner, K., Osten, W., Ovsianikov, A., Obata, K., and Chichkov, B. (2011) Depth sensitive Fourier-Scatterometry for the characterization of sub-100 nm periodic structures. *SPIE Proc.*, **8083**, 80830M-1–80830M-9.
71. Denk, W. Strickler, J., and Webb, W. (1990) Two-photon laser scanning fluorescence microscopy. *Science*, **248** (4951), 73–76.
72. Svoboda, K. and Yasuda, R. (2006) Principles of two-photon excitation microscopy and its applications to neuroscience. *Neuron*, **50** (6), 823–839.
73. Rittweger, E., Han, K.Y., Irvine, S.E., Eggeling, C., and Hell, S.W. (2009) STED microscopy reveals crystal colour centres with nanometric Resolution. *Nat. Photonics*, **3** (3), 144–147.
74. Carl Zeiss MicroImaging GmbH (2010) Superresolution Structured Illumination Microscopy (SR-SIM), White Paper, Carl Zeiss BioSciences, Jena 2010, http://www.zeiss.de/highres_visual_d (accessed 2011).
75. Veselago, V.G. (1968) The electrodynamics of substances with simultaneously negative values of ϵ and μ. *Sov. Phys. Usp.*, **10**, 509–514.
76. Liu, Z., Lee, H., Xiong, Y., Sun, C., and Zhang, X. (2007) Far-field optical hyperlens magnifying sub-diffraction-limited objects. *Science*, **315**, 1686.
77. Salandrino, A. and Engheta, N. (2006) Far-field subdiffraction optical microscopy using metamaterial crystals: theory and simulations. *Phys. Rev. B*, **74**, 75103.
78. Zayats, A.V., Smolyaninov, I.I., and Maradudin, A.A. (2005) Nano-optics of surface plasmon polaritons. *Phys. Rep.*, **408**, 131–314.
79. Schau, P., Frenner, K., Fu, L., Schweizer, H., Giessen, H., and Osten, W. (2011) Design of high-transmission metallic meander stacks with different grating periodicities for subwavelength-imaging applications. *Opt. Express*, **19** (4), 3627–3636.
80. Schau, P., Frenner, K., Fu, L., Schweizer, H., Giessen, H., and Osten, W. (2011) Rigorous modeling of meander-type metamaterials for sub-lambda imaging. *Proc. SPIE*, **8083**, 808303–808308.
81. Schweizer, H., Fu, L., Hentschel, M., Weiss, T., Bauer, C., Schau, P.,

Frenner, K., Osten, W., and Giessen, H. (2011) Resonant multimeander-metasurfaces: a model system for superlenses and communication devices. *Phys. Status Solidi B*, 1–7. doi: 10.1002/pssb.201084212

82. Fu, L., Schau, P., Frenner, K., Osten, W., Weiss, T., Schweizer, H., and Giessen, H. (2011) Mode coupling and interaction in a plasmonic microcavity with resonant mirrors. *Phys. Rev. B*, **84**, 235402-1–235402-6.

83. Pendry, J.B. (2000) Negative refraction makes a perfect lens. *Phys. Rev. Lett.*, **85**, 3966–3969.

84. Johnson, P. and Christy, R. (1972) Optical constants of the noble metals. *Phys. Rev. B*, **6**, 4370–4379.

85. (2012) MICROSIM ITO Software Package for the Rigorous Simulation of Light-Structure Interaction. http://www.uni-stuttgart.de/ito/Forschung/HMS/Modellierung%20und%20Simulation/modelling_and_simulation.html.

86. Niu, X., Jakatdar, N., Bao, J., Spanos, C., and Yedur, S. (1999) Specular spectroscopic scatterometry in DUV lithography. *Proc. SPIE*, **3677**, 159–168.

87. Niu, X., Jakatdar, N., Bao, J., and Spanos, C. (2001) Specular Spectroscopic DUV Scatterometry. *IEEE Trans. Semicond. Manuf.*, **14** (2), 97–111.

88. Ziger, D.H., Adams, T.E., and Garofalo, J.G. (1997) Linesize effects on ultraviolet reflectance spectra. *Opt. Eng.*, **36** (1), 243–250.

89. Zaidi, S.H., Prins, S.L., McNeil, J.R., and Naqvi, S.S.H. (1994) Metrology sensors for advanced resists. *Proc. SPIE*, **2196**, 341–351.

90. Boher, P., Petit, J., Leroux, T., Foucher, J., Desières, Y., Hazart, J., and Chaton, P. (2005) Optical Fourier transform scatterometry for LER and LWR metrology. *Proc. SPIE*, **5725**, 192–203.

91. Raymond, C.J. (2001) Scatterometry for semiconductor metrology, in *Handbook of Silicon Semiconductor Metrology* (ed. A. Diebold), Marcel Dekker, New York, pp. 477–513.

92. Raymond, C.J. (2005) Overview of scatterometry applications in high volume silicon manufacturing. *AIP Conf. Proc.*, **788**, 394–402.

93. Raymond, C.J. (2004) Overview of scatterometry applications in high volume silicon manufacturing. *Proc. SPIE*, **5375**, 564–575.

94. Kubota, H. and Inoué, S. (1959) Diffraction images in the polarizing microscope. *J. Opt. Soc. Am A*, **49** (2), 191–198.

95. Silver, R.M., Barnes, B.M., Attota, R., Jun, J., Stocker, M., Marx, E., and Patrick, H.J. (2007) Scatterfield microscopy for extending the limits of image-based optical metrology. *Appl. Opt.*, **46** (20), 4248–4257.

96. Bischoff, J., Hutschenreuther, L., Truckenbrodt, H., Bauer, A., Haak, U., and Skaloud, T. (1999) Scatterfield microscopy for extending the limits of image-based optical metrology. *Proc. SPIE*, **3743**, 49–60.

97. Quintanilha, R., Thony, P., Henry, D., and Hazart, J. (2004) 3D-features analysis using spectroscopic scatterometry. *Proc. SPIE*, **5375**, 456–467.

98. Quintanilha, R., Hazart, J., Thony, P., and Henry, D. (2005) Application of spectroscopic scatterometry method in hole matrices analysis. *Proc. SPIE*, **5752**, 204–216.

99. Reinig, P., Dost, R., Mört, M., Hingst, T., Mantz, U., Moffit, J., Shakya, S., Raymond, C., and Littau, M. (2005) Metrology of deep trench etched memory structures using 3D scatterometry. *Proc. SPIE*, **5725**, 559–569.

100. Schuster, T., Rafler, S., Osten, W., Reinig, P., and Hingst, T. (2007) Scatterometry from crossed grating structures in different configurations. *Proc. SPIE*, **6617**, 661715-1–661715-9.

101. Schuster, T., Rafler, S., Ferreras Paz, V., Frenner, K., and Osten, W. (2009) Fieldstitching with Kirchhoff-boundaries as a model based description for line edge roughness (LER) in scatterometry. *Microelectron. Eng.*, **86** (4–6), 1029–1032.

102. Schuster T. (2010) Simulation of light diffraction on crossed gratings and its application for scatterometry. Dissertation. University of Stuttgart.

103. Iwai, H. (2009) Roadmap for 22 nm and beyond. *Microelectron. Eng.*, doi: 10.1016/j.mee.2009.03.129
104. Quintanilha, R., Hazart, J., Thony, P., and Henry, D. (2005) Influence of the real-life structures in optical metrology using spectroscopic scatterometry analysis. *Proc. SPIE*, **5858**, 58580C-1–58580C-12.
105. Germer, T.A. (2007) Modeling the effect of line profile variation on optical critical dimension metrology. *Proc. SPIE*, **6518**, 65180Z-1–65180Z-9.
106. Germer, T.A. (2007) Effect of line and trench profile variation on specular and diffuse reflectance from a periodic structure. *J. Opt. Soc. Am. A*, **24** (3), 696–701.
107. Yakobovitz, B., Cohen, Y., and Tsur, Y. (2007) Line edge roughness detection using deep UV light scatterometry. *Microelectron. Eng.*, 84, 619–625.
108. Schuster, T., Rafler, S., Frenner, K., and Osten, W. (2008) Influence of line edge roughness (LER) on angular resolved and on spectroscopic scatterometry. *Proc. SPIE*, **7155**, 71550W. doi: 10.1117/12.814532
109. Layet, B. and Taghizadeh, M.R. (1996) Analysis of gratings with large periods and small feature sizes by stitching of the electromagnetic field. *Opt. Lett.*, **21**, 1508–1511.
110. Choy, T.C. (1999) *Effective Medium Theory, Principles and Applications*, Oxford University Press.
111. Turunen, J., Kuittinen, K., and Wyrowski, F. (2000) Diffractive optics: electromagnetic approach. *Prog. Opt.*, **40**, 341–387.
112. Bergner, B.C., Germer, T.A., and Suleski, T.J. (2010) Effective medium approximations for modeling optical reflectance from gratings with rough edges. *J. Op. Soc. Am. A*, **5**, 1083–1090.
113. Bilski, B., Frenner, K., and Osten, W. (2011) About the influence of line edge roughness on effective-CD. *Opt. Express*, **19** (21), 19967–19972.
114. (2012) Horiba Scientific UVISEL – Spectroscopic Ellipsometer. http://www.horiba.com/scientific/products/ellipsometers/.
115. Tompkins, H.G. (ed.) (2005) *Handbook of Ellipsometry*, William Andrew Inc., Norwich.
116. de Groot, P. and Deck, L. (1995) Surface profiling by analysis of white light interferograms in the spatial frequency domain. *J. Mod. Opt.*, **42**, 389–401.
117. LEUKOS, (2012) SP-UV-8-OEM, http://www.leukos-systems.com/IMG/pdf/LEUKOS-SPuv-2.pdf.
118. International Organisation for Standardization 98-3:2008 (2008) *ISO/IEC Guide 98-3:2008: Uncertainty of Measurement – Part 3: Guide to the Expression of Uncertainty in Measurement*, ISO, Genf 2008, ISBN: 92-67-10188-9.
119. Silver, R., Germer, T., Attota, R., Barnes, B.M., Bunday, B., Allgair, J., Marx, E., and Jun, J. (2007) Fundamental limits of optical critical dimension metrology: a simulation study. *Proc. SPIE*, **6518**, 65180U–65180U–17.
120. Harder, I., Lano, M., Lindlein, N., and Schwider, J. (2004) Properties and limitation of beam homogenizers made from microlens arrays. DGaO-Proceedings.
121. Ferreras Paz, V., Peterhänsel, S., Frenner, K., and Osten, W. Solving the inverse grating problem by white light interference Fourier scatterometry *in Light - Sci. Appl.*(submitted for publication).
122. Burla, A., Haist, T., Lyda, W., and Osten, W. (2012) Genetic Programming Applied to Automatic Algorithm Design in Multi-scale Inspection Systems. *Opt. Eng.*, **51** (6).
123. Lyda, W., Zimmermann, J., Burla, A., Regin, J., Osten, W., Sawodny, O., and Westkämper, E. (2009) Sensor and actuator conditioning for multi-scale measurement systems on example of confocal microscopy. *Proc. SPIE*, **7389**, 738903-1–738903-11.
124. Osten, W. (2008) Some answers to new challenges in optical metrology. *Proc. SPIE*, **7155**, 715503-1–715503-16.
125. Pedrini, G., Gaspar, J., Wu, T., Osten, W., and Paul, O. (2009) Calibration of optical systems for the measurement of microcomponents. *Opt. Lasers Eng.*, **47** (2), 203–210.

126. Pedrini, G., Gaspar, J., Schmidt, M., Alekseenko, I., Paul, O., and Osten, W. (2011) Measurement of nano/micro out-of-plane and in-plane displacements of micromechanical components by using digital holography and speckle interferometry. *Opt. Eng.*, **50** (10), 101504-1–101504-10.

127. Osten, W., Ferreras-Paz, V., Frenner, K., Lyda, W., and Schau, P. (2011) Different approaches to overcome existing limits in optical micro and nano metrology. *Proc. SPIE*, **8011**, 80116O-1–80116O-30.

16
Interferometry in Harsh Environments
Armando Albertazzi G. Jr

16.1
Introduction

There are many applications for interferometry where the specimen cannot be transported to the laboratory, where an appropriate interferometer should be operated in a quiet and stable environment. In those cases, the interferometer must be moved to the specimen site. Sometimes, the specimen is located in a harsh environment, where the interferometer must operate correctly. This chapter brings an analysis of the main factors that may disturb the interferometer performance or even make impracticable its use. It also brings a reflection about requirements and some of the current solutions already available to make possible an interferometer to successfully and confidently operate in harsh environments. Examples of successful environmental resistant interferometers are given and analyzed. Finally, conclusions are drawn on the present together with a quick walk into the future.

16.2
Harsh Environments

A quick definition for a harsh environment is *anywhere not "indoors."* However, this definition is not complete. It can be extended to *a place where any condition of extremes relative to the human condition applies: temperature; humidity; atmosphere (including pressure); radiation; shock and vibration; and erosive flows or corrosive media, whether indoors or not* [1].

Figure 16.1a shows an example of a harsh environment. It fulfills the above definition since it is exposed to temperature and atmospheric variations, humidity, sun radiation, vibration, wind, and dust. Figure 16.1b shows an interferometric measurement in progress in this same harsh environment. Residual stresses in a buried pipeline are measured by a portable digital speckle pattern interferometer (DSPI) using the incremental hole-drilling method [2]. The device is firmly attached to the pipeline using strong rare-earth magnets and an appropriate isostatic fixture.

Optical Imaging and Metrology: Advanced Technologies, First Edition.
Edited by Wolfgang Osten and Nadya Reingand.
© 2012 Wiley-VCH Verlag GmbH & Co. KGaA. Published 2012 by Wiley-VCH Verlag GmbH & Co. KGaA.

(a) (b)

Figure 16.1 (a) An example of a harsh environment. (b) An interferometric measurement in progress in this harsh environment. (Source: Courtesy *Photonita Optical Metrology*, Brazil.)

It is not only necessary to the interferometer to survive in the harsh environment but also fundamentally to give confident results.

Several other examples of harsh environments can be found in agricultural plants, farms, industrial plants (food, oil, pharmaceutical, chemical, and manufacturing), explosion areas, medical, transport and oceanography, and so on.

16.3
Harsh Agents

Temperature, humidity, atmosphere and pressure, shock and vibration, and radiation are among the main harsh agents affecting interferometric measurements, as shown in Figure 16.2. Their principal effects on the measurand or measurement performance are presented and discussed in the following sections. Although corrosion and erosion can be deadly harsh agents for interferometric measurements, they are not discussed here.

16.3.1
Temperature

Temperature usually has a strong influence on any kind of physical measurement. Since the mechanical, chemical, optical, or electrical properties of most materials are temperature dependent, the performance of a measurement system, made of components with different materials, is usually also temperature dependent. If the temperature varies while a measurement is in progress, it can be sometimes very difficult to say if the measurement indication changed because of temperature variations or because of changes in the measured quantity (measurand). On the other hand, the measurand itself can also be modified by temperature, as is the case of thermal deformation of mechanical components. Temperature can also

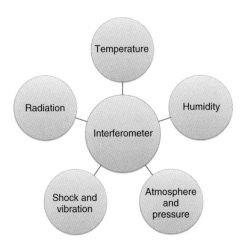

Figure 16.2 Main harsh agents for interferometric measurements.

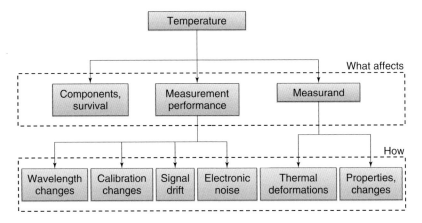

Figure 16.3 Effects of temperature on interferometric measurements.

influence the life and survival of components. All those effects are also present in optical measurement systems. Figure 16.3 synthesizes the principal effects of temperature on interferometric measurements.

The influence of temperature on the measurement performance of an interferometer can be affected by, at least, four main effects: wavelength changes, sensitivity changes, signal drift, and electronic noise. The wavelength of some light sources can be directly affected by temperature. That is particularly so in the case of nonstabilized diode lasers. Wavelength changes in interferometric systems can result in sensitivity changes, affecting the ratio between the fringe counting and the physical quantity variation. A signal drift can be associated with a wavelength change as well as with thermal dilation of the active parts of interferometer. High temperature can affect the amount of electronic noise of camera sensors and also has a strong influence on the life, or even survival, of electronic components.

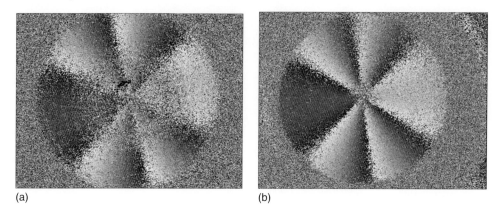

(a) (b)

Figure 16.4 Phase difference patterns obtained from DSPI measurements and a nonstabilized diode laser. (a) Obtained with a wavelength sensitive interferometer. (b) Obtained with an achromatic interferometer.

Figure 16.4 shows two phase difference patterns from the same displacement field measured by DSPI [3]. In both cases, a low-cost and nonstabilized diode laser was used as light source. Since ambient temperature was not stabilized, the laser wavelength was continuously changing with temperature. Figure 16.4a was obtained from a wavelength-sensitive configuration and Figure 16.4b with an achromatic configuration. The phase structure of Figure 16.4a is somehow blurred because of wavelength instability during the measurement. Figure 16.4b shows sharper phase structure since it is not sensitive to laser wavelength.

Temperature can also produce some changes in the measurand itself. Thermal deformation of the specimen will affect its dimensions. Nonuniform temperature fields, or a combination of materials with different thermal expansion coefficients, can produce distortions in the specimen, bringing about a departure from its original geometric shape. Optical, electrical, or chemical properties can also be affected.

16.3.2
Humidity

High humidity levels are also the source of problems for optical measurements, especially when there is moisture condensation. Figure 16.5 shows a schematic of the main effects of humidity on the components, measurement performance, and the measurand. High humidity levels may cause malfunction of a component. It can also change the effective wavelength of the light source, bringing about measurement errors. Moisture condensation can blur the imaging system and optical components as well as affect the surface reflectivity of the specimen.

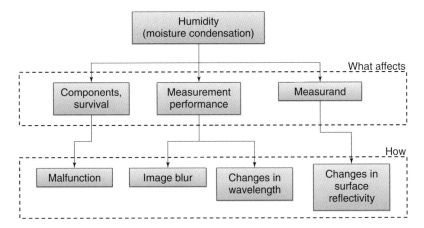

Figure 16.5 Effects of humidity on interferometric measurements.

16.3.3
Atmosphere and Pressure

Atmosphere and air pressure can bring about additional disturbances in an interferometric measurement. The main causes and effects are represented in Figure 16.6. Dust, droplets, and particles in suspension absorb light and make the air more turbid, reducing the image quality. However, the most drastic effect of dust and droplets is contamination of optical components and must be avoided. Pressure variation produces variations of the refractive index of the media, which also change the effective wavelength of the light source. Thermal currents, produced by natural convection, can bring about severe instability in interferometric measurement since they produce local and rapid variations of the refraction indexes of the surrounding

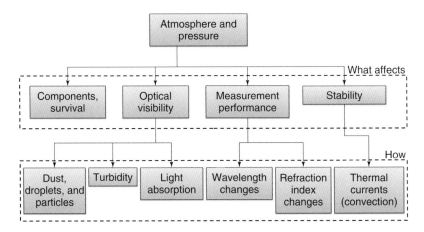

Figure 16.6 Effects of atmosphere and pressure on interferometric measurements.

atmosphere. The effects of such instability on interferometric measurement results are uncontrolled phase shifts and motions in the fringe patterns, which disturbs the fringe processing, drastically worsening the measurement uncertainty.

16.3.4
Shock and Vibration

Mechanical vibrations and shocks are usually associated with relative motion between the parts of an interferometer or between the interferometer and the measured specimen itself. Several effects of shock and vibrations are represented in Figure 16.7. Severe shock levels may compromise the survival of components and, of course, the interferometer itself. The relative motions brought about by moderate shocks or mechanical vibration can affect the measurement performance by blurring the interferogram, raising the level of random errors and reducing the lateral and principal resolution. Vibrations and shock can also affect the relative alignment of the interferometer parts and the alignment between the interferometer and the specimen.

16.3.5
Radiation and Background Illumination

Interferometers are mainly affected by two kinds of radiations: electromagnetic fields and background illumination, as shown in Figure 16.8. Strong electromagnetic fields can compromise the life, or even the survival, of electronic components such as camera sensors. Moderate electromagnetic fields can worsen the signal-to-noise ratio of some electronic components. Background illumination may affect some interferometer performance reducing the contrast of the measurement signal, deteriorating the signal-to-noise ratio, bringing about a higher level of random errors, and worsening the resolution.

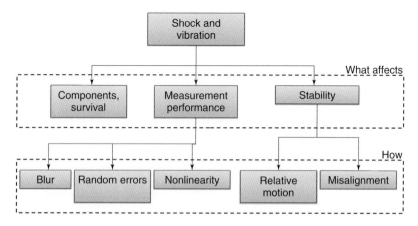

Figure 16.7 Effects of shock and vibrations on interferometric measurements.

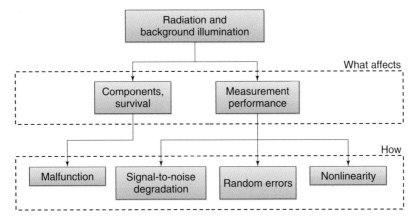

Figure 16.8 Effects of radiation and background illumination on interferometric measurements.

16.4 Requirements for Portable Interferometers

In order to successfully operate outside a laboratory, a portable interferometer must fulfill some requirements. Figure 16.9 shows a synthesis of six of them. They must be robust, flexible, compact, stable, friendly, and cooperative. In addition, as a product, a portable interferometer must be affordable to assure a commercial success. Each of the requirements is discussed in the following sections [2, 4].

16.4.1 Robustness

The interferometer must be robust against environmental demands. Its conception and construction must be designed to coexist with some degree of mechanical vibrations, temperature variations, voltage fluctuations, dust, moisture, and daylight. Since it is generally impossible to stop mechanical vibrations of the

Figure 16.9 Requirements for a portable interferometer for harsh environments.

specimen, the interferometer design must be optimized to minimize the relative motion between its parts and the interferometer and the specimen itself. The configuration and measurement principle should be ideally insensitive to temperature variations, or an efficient compensation strategy may be used. The electronics have to be resistant against voltage fluctuations or battery operated and protected against short circuits. If possible, the optical parts must be assembled in a sealed enclosure to avoid dust, moisture, and droplet contaminations and frequently protected from daylight. The mechanical structure has to be able to protect the interferometer up to some degree from direct damage caused by falling objects or the incidence of extraneous agents.

16.4.2
Flexibility

The interferometer conception must be flexible enough to allow it to be adjusted and installed in a variety of specimen geometries. Ideally, the interferometer has to be installed in flat or curved surfaces and in magnetic or nonmagnetic materials. It should have proper alignment devices to allow easy positioning and effective alignment between the interferometer and the specimen.

16.4.3
Compactness

It is highly desirable that interferometers have a compact design. A compact size is not only important to make it fit in anywhere but also a fundamental issue to achieve stiffness and mechanical robustness against shock and mechanical vibrations. A robust design is also an important feature to make it easily transportable to the measurement site.

16.4.4
Stability

Long-term stability is desirable for any portable measurement device. Ideally the system may remain trustable and traceable everywhere and every time. No further calibration would be required. This can only be true if the response of the interferometer has no time or temperature dependency, keeping stable its metrological parameters.

16.4.5
Friendliness

The interferometer has to be easy to use. It should be easy to transport, install, adjust, and operate. It is also highly desirable that it is able to present on-demand clear results. All these qualities are important for saving time during the measurements and to keep the system ergonomic for the user.

16.4.6
Cooperativeness

The interferometer should help the user to organize the experimental task. The software should guide the user step by step to complete the measurement. It should collect and organize the experimental data in a clear and permanent way. The software may also document the test and produce a technical report.

16.5
Current Solutions

There are some possible solutions to make interferometers to work outside a laboratory. They may be grouped into three main groups, as represented in Figure 16.10: (i) isolation, (ii) robustness, and (iii) simultaneous isolation and robustness. Isolation of the interferometers from the harsh agents is a natural solution. However, the interferometer must be able to receive an optical signal from the measurand. The second group is related to making the interferometer robust enough to successfully operate in harsh environments with no isolation. Finally, the third group is a combination of both: isolation and robustness.

16.5.1
Isolation

There are solutions for isolation that have been used for a long time. Some of them are discussed in the following sections.

16.5.1.1 Atmosphere Isolation
Encapsulation is a very common infield solution to isolate the instrument from atmospheric contamination produced by dust, rain, droplets, water sprays, and even for underwater operation. The parts to be isolated are kept within a hermetic

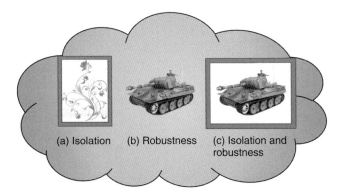

Figure 16.10 Three groups of solutions for interferometry in harsh environments.

container, usually with a transparent and efficiently cleanable optical window. There is also the possibility of using a nonhermetic container with an open window. In this case, clean and dry air is continuously blown inside the container resulting in a positive pressure, producing a permanent airflow through the window from inside to outside, and preventing dust and particles from entering the container.

16.5.1.2 Temperature Isolation

To isolate an optical system from excessive room temperature, it can be kept inside a container built with thermal insulating materials. Active cooling by air conditioning or chilled water circulation can be added in severe cases. Thermoelectric cooling, using the Peltier effect, is another attractive active cooling possibility. A closed-loop temperature control is needed in cases of strong temperature dependence of the optical system.

16.5.1.3 Radiation Isolation

Faraday cages are effective ways to isolate optical systems from strong electromagnetic fields. A Faraday cage or a Faraday shield is an enclosure formed by conducting material or by a mesh of such material. Such an enclosure blocks out external static electric fields.

The interferometer can be isolated from ambient light by keeping it inside a closed container, forming a darkroom. Although this solution is effective, a large darkroom cannot be very practical for applications outside the laboratory. However, a compact interferometer can be packed inside a small and closed metallic container, producing simultaneous isolation against ambient light, electromagnetic fields, and atmospheric effects.

Another possibility of drastically reducing the influence of ambient light is by the use of narrow band-pass interference filters in front of the imaging system. This is particularly useful for laser-based interferometers. The interference filter is tuned in the laser wavelength. The amount of ambient light intensity in such a wavelength is usually negligible when compared with the intensity coming from the area illuminated with the laser light. In this case, a portable darkroom may not be necessary.

16.5.1.4 Vibration Isolation

Passive and active dumping elements are widely used for vibration isolation in laboratory environments. There is a large variety of optical tables equipped with pneumatic or elastomeric isolators. Passive damping is very effective for seismic vibration isolations (coming from the ground underneath the table) but is affected by varying tabletop loads. Actively damped tables incorporating sensors and electronics to both sense and actively damp tabletop vibrations, become effective for isolation of both kinds of vibrations. They are usually not portable elements.

Active stabilization is another approach that can be used to reduce vibration effects in optical systems. It can make the system more complex and expensive but can provide a useful solution for portable devices. The main concept is presented in Figure 16.11. The optical system is equipped with a sensor capable of detecting the

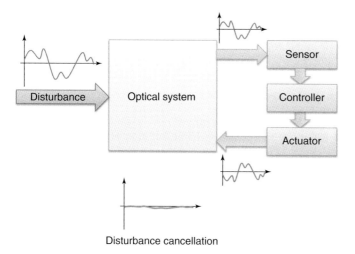

Figure 16.11 Principle of active stabilization.

presence of a disturbance that produces mechanical vibrations. The disturbance signal is sent to an appropriate controller that produces a reactive signal with a sign opposite to the input signal. This reactive signal is sent to an actuator and it is applied to the optical system. The superimposing of both signals produces a disturbance cancellation effect in the optical system.

There are several possible ways to capture the disturbance signal. Additional photodiodes, complementary metal-oxide-semiconductor (CMOS) cameras with the ability of fast readings in small areas, or additional interferometers are some possibilities. The actuator can be formed by a piezoelectric-driven mirror or optical fibers stretched by a piezoelectric element, changing diode laser injection current, acoustic optical modulators, among others [3, 5, 6, 28]. Commercial fringe stabilization systems are able to actively stabilize vibrations from 0 to 500 Hz up to $\lambda/20$.

Another very interesting application class is active disturbance cancellation by adaptive optics. It is frequently used in astronomy to reduce the effects of atmospheric instabilities. An appropriate wave front sensor is used to dynamically estimate the distortions in the optical field due to external disturbances. An active optical element is then modified inside a feedback loop to compensate the effects of the disturbance, producing an undisturbed image. Deformable mirrors, driven by a set of pistons, or spatial light modulators, are examples of convenient adaptive optical elements.

16.5.2
Robustness

In the Oxford dictionary, the adjective robust has some meanings: (i) *strong and healthy; vigorous*; (ii) *sturdy in construction (of an object)*; (iii) *able to withstand or overcome adverse conditions (of a process, system, organization, etc.)*; (iv) *uncompromising and forceful*; and (v) *strong and rich in flavor or smell (of wine or food)*. Therefore, a robust interferometer should be able to reliably operate in adverse conditions.

The robustness of an interferometer can be distributed in two components: physical properties and measurement strategies. They are discussed in the following list.

Robustness in physical properties is understood here to be related to a sturdy construction. It involves a few aspects:

- **Rigid/stiff construction**: A rigid/stiff mechanical design is very important to reduce relative motions between the interferometer internal parts, increasing the interferometer's ability to work under moderate vibration levels.
- **Stiff clamping**: Stiff clamping ability is also very important to minimize the relative motions between the interferometer and the specimen surface during the measurement.
- **Compact**: Compact units are usually stiffer and easier to be firmly attached to the specimen surface.
- **Monolithic**: Another very attractive approach is using monolithic constructions. In this case, all active optical components are carved and integrated in the same solid transparent media, usually fused silica [7, 8].

The measurement strategies can strongly influence the success of an interferometer performance in nonideal environments. They can be connected to measurement principle and/or to the data analysis approach.

- **Common or quasi-common path interferometer**: In these configurations, the two interferometer arms are almost equal [9]. Atmospheric or vibration disturbances produce undesirable phase shifts in both interferometer arms by nearly the same amount. Therefore, the phase difference between the arms, which is usually related to the quantity of interest, is minimally affected by the disturbances. Lateral shearing interferometers are examples of quasi-equal path interferometers. Self-referencing interferometers are also robust since they generate the reference wave locally with respect to the signal wave so that the reference and signal waves experience common aberrations and path length changes and thus naturally maintain constant relative phase.
- **Robust algorithm**: There are vibration-tolerating algorithms already developed for phase-shifting calculation [10–12]. They may use sinusoidal phase-shifting algorithms and harmonic analysis [10] or a combination of fast and slow image acquisition hardware [11] or pure mathematic compensations [12].
- **Fast/one shot**: The most promising approach for interferometry in the harsh environment is based on a single-shot measurement [13–21]. The main idea is to use an exposure time short enough to freeze the effects of vibrations and atmospheric disturbances. Several approaches, involving application of carrier fringes [13, 14], polarization and pixelated arrays [15–19], or simultaneous acquisitions with two wavelengths [20, 21], have been reported. The vibration effects are simply frozen. The effects of convection currents can be averaged out by combining several repeated measurements.

Features	Harsh agents disturbances				
	Temperature	Atmosphere	Vibration	Humidity	Radiation
Physical properties					
• Rigid/stiff construction			☺		
• Stiff clamping			☺		
• Compact	😐	😐	☺	😐	
• Monolithic	😐	☺	☺	😐	
Measurement strategies					
• Common/quasi-common path	😐	☺	☺	😐	
• Robust algorithms	😐	😐	☺		
• Fast/one shot	😐	☺	☺		😐
• Self-compensating	☺	😐	😐		😐

Figure 16.12 Relationship between robustness features and harsh agents' disturbances.

- **Self-compensating**: It is desirable that the measurement principle, or working algorithm, be self-compensating for long-term temperature or wavelength fluctuations. There are some possibilities using achromatic interferometers [22].

The relationships between the physical properties and measurement strategies with their effectiveness against the main harsh agents are presented in Figure 16.12. A smiling face (☺) means strong effect. An indifferent face (😐) means medium effect. The other unpainted areas correspond to weak or nonexisting effects.

Self-compensating strategies are the most effective solution for dealing with long-term temperature variations. A fast/one-shot measurement is excellent to stop vibrations and atmospheric effects. When possible to use, common or quasi-common path configurations are a very good and simple solution to automatically reduce vibration and atmospheric disturbances. From the physical principle point of view, monolithic constructions are also very attractive to reduce vibrations and atmospheric disturbances. The chances of an interferometer to successfully operate under vibration are higher if it is stiff, compact and can be clamped very stiffly on the specimen surface.

16.6
Case Studies

Seven examples of interferometers for application outside the laboratory are given and discussed in the following sections. The information presented here was obtained from technical publications as well as from commercial web pages. For additional examples, see Ref. [23].

16.6.1
Dantec ESPI Strain Sensor (Q-100)

The first example is a compact electronic speckle pattern interferometer (ESPI) produced by Dantec Dynamics, as shown in Figure 16.13 [24]. The system is really very compact in size: $55 \times 55 \times 59\,mm^3$ and weighs only 370 g. It has diode lasers, dedicated optics, and a compact camera integrated in a single piece. It is installed on the specimen surface by means of an external fixture with magnetic feet and springs that firmly press the unit against the specimen surface, as shown in Figure 16.13b. It is able to measure all the three displacement fields on an area of $25 \times 35\,mm^2$, as well as in-plane strain fields in Cartesian coordinates, in a nonlaboratory environment.

The robustness of this system is mainly due to its compactness, rigid/stiff construction, and the ability to be stiffly clamped on the specimen surface, reducing the relative motion between the specimen surface and the device.

16.6.2
Monolitic GI/DSPI/DHI Sensor

The second example is a very compact integrated interferometric sensor developed by the Warsaw Technology University [8]. It has a monolithic construction, carved in a glass or poly methyl methacrylate (PMMA) block, as shown in Figure 16.14. Figure 16.14a shows a sensor configuration suitable for either grid interferometry (interferometric moiré) or DSPI. Figure 16.14b is a configuration for digital interferometric holography. The LSDM (light source and detector matrix) is formed by a light source, a beam-forming optics, the imaging optics, and the detector

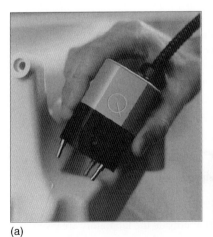
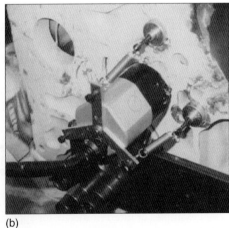

(a) (b)

Figure 16.13 (a) Compact ESPI strain sensor. (b) The sensor installed on a specimen. (Source: Courtesy *Dantec Dynamics*, Denmark.)

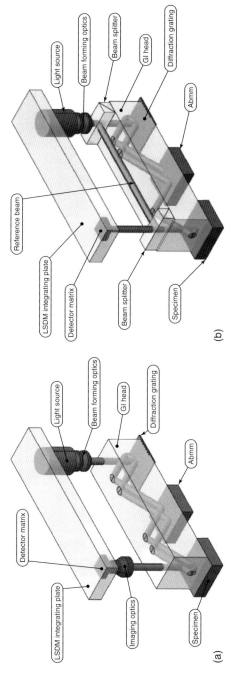

Figure 16.14 Multipurpose monolithic interferometric sensor: (a) Grating interferometer (GI)/DSPI sensor and (b) digital holographic interferometry (DHI) sensor. (Source: Courtesy M. *Kujawinska et al.*, Poland.)

matrix. The main interferometric head (IH) is made of a glass or PMMA block with diffraction grating integrated. Designed to guide object illuminating beams, formed by reference diffraction grating, in a proper way. The Active Beam Manipulation Module is an micro-opto-electro-mechanical-based (M(O)EMS) phase shifter enabling usage of temporal phase-shifting algorithms ameliorating interference phase decoding and beams for digital holography switching.

The robustness of this sensor comes from its very small size and monolithic construction. Since the parts can be very stiffly integrated, the relative motions between the internal parts of the interferometer are very small. When stiffly attached to the specimen surface, the system becomes resistant to vibrations. There is no sensitivity to atmospheric disturbances at all. The use of diffraction gratings to produce symmetric double illumination results in an achromatic DSPI sensor. Therefore, the system is insensitive to laser wavelength variations, allowing the use of compact and inexpensive diode lasers as light source.

16.6.3
ESPI System for Residual Stresses Measurement

The third example is a DSPI developed by the Federal University of Santa Catarina (Brazil) [2, 4, 22]. The DSPI system has a special diffractive optical element (DOE) that produces a circular double illuminated region with radial in-plane sensitivity. The double illumination angles are produced by the especial DOE in such a way that the sensitivity of the system is not dependent on the laser wavelength [22]. Figure 16.15a shows the DSPI system used for residual stress measurement in a buried pipeline located in the site in Figure 16.1. The interferometer is quite compact and stiffly attached to the specimen surface by a combination of four rare-earth magnets and three sharp conical tips.

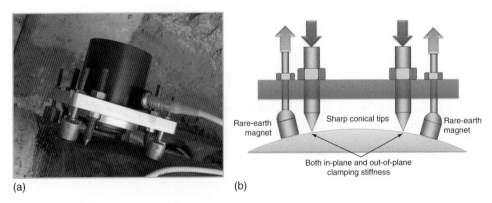

Figure 16.15 A DSPI system for infield residual stresses measurement. (a) The actual device. (b) The clamping system. (Source: Courtesy *Photonita Optical Metrology*, Brazil.)

Figure 16.15b shows a detail of the clamping system. Three feet with sharp conical tips are kept in contact with the specimen surface. Their heights are adjusted in such a way that the system becomes approximately parallel to the area of interest on the specimen. The three-feet set results in a very rigid and isostatic configuration. The sharp tips stick in the specimen surface drastically restricting the in-plane movement. The four rare-earth magnets produce a vertical force that pushes the system against the specimen surface, restricting the out-of-plane movement. This combination results in a very stiff clamping.

The system has been commercially used for residual stresses measurement in pipelines since 2006. Its robustness is mostly due to its compactness and stiffness, stiff clamping scheme, and the achromatic configuration that allows the use of cheap and nonstabilized diode laser as light source. Therefore, mode hopping of the diode laser and long-term instability are not problems.

16.6.4
Pixelated Phase-Mask Dynamic Interferometer

The fourth example is a pixelated phase-mask interferometer developed by the American company 4D Technology. The system is practically insensitive to mechanical vibrations and air disturbances. Figure 16.16a shows the actual device named PhaseCam 6000. Figure 16.16b presents its working principle [15–18].

Each arm of the pixelated interferometer uses orthogonally polarized light. A pixelated phase mask, placed just in front of the sensor of a high-resolution camera, produces phase-shifted images that depend on the polarization direction. The pixelated mask is constructed in such a way that each four neighbor pixels have polarizations oriented in different angles, producing four 90° phase-shifted signals. By an appropriate algorithm, the information is combined and a wrapped phase value is obtained for each camera pixel. Only one shot is needed to a complete and instantaneous phase measurement.

The robustness of the system comes from the one-shot measurement ability. Since a relatively high-power laser is used, the exposure time is short, in the range of 30 µs, which is enough to freeze any relative motion. The system is insensitive to vibrations between the interferometer and the specimen. Disturbances produced by air motion are removed by averaging a set of successive phase maps acquired in a few seconds. The drawback of this parallelized approach is the reduction of the spatial resolution.

Parallelized approaches are very attractive options to make interferometers less sensitive to mechanical vibrations or air motion. The pixelated mask approach is a very interesting one, but it is not the only one. For example, in Ref. [25], the authors develop a parallelized interferometer for testing aspheric surfaces using microlens arrays to make use of multiple test beams propagating under different angles through the interferometer.

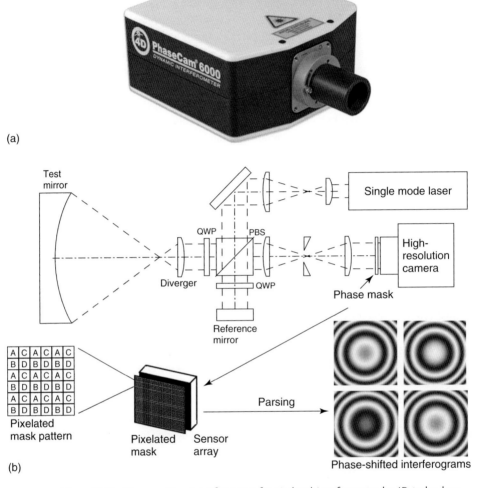

Figure 16.16 Twyman–Green configuration for pixelated interferometer by 4D technology. QWP, quarter wave plate; PBS, polarizing beam splitter. (Source: Courtesy *4D Technology*, USA.)

16.6.5
Digital Holographic Microscope

The fifth example is a digital holographic microscope produced by the Swiss company Lyncée Tec, shown in Figure 16.17. It measures both geometry and motion of small size surfaces from single acquisition (no scan) by off-axis digital holography [20, 21]. The usual half-wavelength measurement range is increased

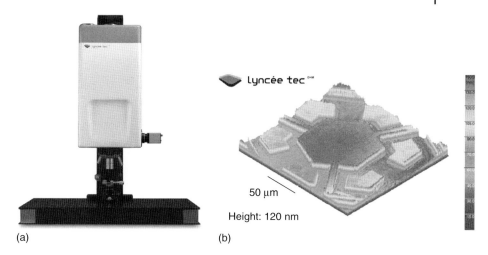

Figure 16.17 (a) Digital holographic microscope by Lyncée Tec. (b) A measurement example of a vertical displacement mirror. (Source: Courtesy *Lyncée Tec*, Switzerland.)

up to 15 μm by multiplexing on the same hologram the carrier frequencies for two wavelengths by introducing different references waves angles.

Apart of usual DHM (digital holographic microscopy) operation described above, DHM is compatible with stroboscopic techniques. Stroboscopic synchronization can be used in conjunction with DHM allows measurement of high repetitive vertical displacements up to 25 MHz with a unique feature: a vertical deformation range larger than several tens of micronmeters. It is obtained thanks to a large coherence length of the laser source and to laser pulses length down to 7.5ns.

Since only one shot is required, and the acquisition time is very short, the system freezes all the relative motion between the specimen and the microscope. This makes the system robust against vibrations. Figure 16.17b shows a measurement example of a vertical displacement mirror.

16.6.6
Shearography

The sixth example is a shearography measurement close to the sea and 180 km away from the coast. Shear interferometers are quasi-common path interferometers since the specimen image interferes with a laterally displaced one [9]. It is used for measurement of deformations in rough surfaces. It is resistant to moderate vibrations and it is not very much sensitive to atmospheric effects. The application shown in Figure 16.18 is to detect bonding defects in a protective jacket made of a composite material. The jacket is used to reduce corrosion effects in a steel pipeline (riser) used to bring gas from the bottom of the sea to a production plant. Shearography measurements were made to verify the proper adhesion between the protective jacket and the steel pipe. Thermal loading was used.

(a) (b)

Figure 16.18 (a,b) Shearographic measurement in a composite jacket of a gas riser 180 km away from the coast. (Source: Courtesy *Photonita Optical Metrology*, Brazil.)

The measurement site was affected by vibrations, temperature variations, wind, sunlight, and moisture. The shearography device was protected against wind, water sprays, and sunlight by a kind of tent. It was firmly attached to the pipe surface by traction belts, reducing the relative vibrations between the specimen and the interferometer. Since thermal loading was used, temperature effects were not a disturbing factor in this case.

The robustness behind this application comes from the measurement technique itself, which is a quasi-equal path interferometry.

16.6.7
Fiber-Optic Sensors

There is a large variety of interferometric fiber-optics-based sensors. In common, they are very resistant to harsh environments. They are especially immune to corrosion, electromagnetic fields, vibrations, atmospheric disturbances, and frequently they are waterproof. The optical fiber sensors can be located hundreds or thousands of meters away from the reading unit, allowing remote operation [26, 27].

Fiber Bragg grating sensors are strain transducers that transform a static or dynamic distance variation into a change in reflected wavelength of a prestressed Fiber Bragg grating. They can be bonded or spot welded to structures and components (metallic, concrete, etc.). These sensors are self-referenced and have very good long-term stability.

Fabry–Perot deformation sensors are static or dynamic strain transducers that transform a distance variation into a change in the path unbalance between two optical fibers. They can be surface mounted or directly embedded in concrete and mortars or fiber-reinforced plastics and other composite materials. They can be configured to be also insensitive to temperature variations. They can have a very high sensitivity and long lifetime. The robustness of this kind of optical fiber interferometric sensors is mainly due to its quasi-equal path configuration, monolithic construction, and very high degree of isolation.

16.7
Closing Remarks

16.7.1
Summary

This chapter discusses the main difficulties and some alternatives for bringing interferometers from laboratories to harsh environments. The difficulties are related to the ability of interferometers to survive and operate confidently in environments with unusual levels of temperature, humidity, pressure and atmospheric disturbances, mechanical shock and vibration, and all kind of radiations. The solutions can be grouped into three groups: isolation, robustness, and a combination of robustness and isolation. There are effective solutions already available for temperature, vibration, atmosphere, pressure, moisture, and radiation isolation. Robustness can be present in constructive aspects as well as in the measurement principle and strategy.

At present, only fiber-optics-based interferometers are truly operating in aggressive, harsh environments. However, important steps have been taken toward making the current interferometers increasingly resistant to moderate harsh environments.

Perhaps the most promising and effective way to making interferometers work in harsh environments is the one-shot measurement strategy. A unique and fast acquisition is able to stop mechanical vibrations and shocks as well as reduce the effects of atmospheric instabilities. That is the case of successful systems already commercially available as the pixelated phase-mask dynamic interferometers and digital holographic microscopes.

Quasi-equal path configurations are very simple and effective ways to reduce the influence of environmental instabilities. This makes it possible for shearography to be applied outside the laboratory without any very sophisticated apparatus. This also makes it possible for a fiber-optic sensor to be located hundreds or thousands of meters away from the light source and from the reading unit.

Miniaturized monolithic sensors, carved from solid optical materials, or formed by optical fibers have very high degree of isolation and insensitivity to atmospheric disturbances and contaminations, electromagnetic radiations, corrosion, and frequently they are waterproof. Monolithic sensors naturally solve a lot of problems and are very promising, efficient, and inexpensive solutions for harsh environment applications.

Robust measurement principles, insensitive to wavelength or temperature fluctuation, are important steps to increase the reliability of the measurement results outside controlled laboratory environments. Robust algorithms are also very promising approaches to reduce and compensate instability effects.

16.7.2
A Quick Walk into the Future

One-shot strategy looks to be the most promising approach for bringing interferometers to harsh environments. It is reality today for digital holography and for Fizeau, Twyman-Green, and Mach-Zehnder interferometers. One-shot interferometric moiré, one-shot DSPI/ESPI, one-shot shearography are in a wish list for the future.

Miniaturized monolithic sensors can become very robust against shock and vibration, and nonexpensive interferometers have also been used outside laboratories. The mass fabrication processes of monolithic sensors using customized optical elements are becoming less and less expensive nowadays. Applications involving long-term monitoring of buildings and bridges are today a reality. Underwater applications are today feasible using optical fiber-based interferometers. Specimen/structure-imbibed sensors should be used more and more. A large growth is expected in this area.

Noninvasive fast active stabilization systems are also in the trend. CMOS cameras are becoming faster, much powerful, and less expensive everyday, as well as parallel processing using GPCGU (general-purpose computation on graphics hardware). A combination of both resources, with specialized algorithms, brings several possibilities to build effective and nonexpensive active-phase stabilization systems to increase vibration and atmospheric tolerance of interferometers. The intensive use of GPCGU opens up new possibilities for very fast image processing by applying disturbance tolerating algorithms. A wish list includes also an open source scientific image processing library using CGCGU platforms.

Special optical configurations can be designed to be achromatic, that is, wavelength independent. As part of a wish list for the future, temperature-insensitive and self-calibrating interferometers have a good chance to succeed in harsh environments. Specimen-imbibed interferometers are also excellent candidates for operation in harsh environments. They can measure very interesting and valuable engineering quantities.

Also, as part of a wish list, the development of DOEs can bring new conceptual design and robust optical measurement systems for a variety of applications, including a new demanding field of underwater optical metrology.

References

1. DataPlex (2006) Maintaining high reliability for new electronic designs intended for harsh environments. Vol. 3, issue (2), http://www.dataplex.com/blog/ (accessed 01 April 2012).
2. Viotti, M., Albertazzi, A. Jr., and Kapp, W. (2008) Experimental comparison between a portable dspi device with diffractive optical element and a hole drilling strain gage combined system. Opt. Lasers Eng., **46**, 835–841.
3. Viotti, M.R. and Albertazzi, A. Jr. (2009) Industrial inspections by speckle interferometry: general requirements and a case study. SPIE 2009-Optical Metrology, Munich, Alemanha, 2009, Vol. 7389, pp. 73890G-1–73890G-15.
4. Albertazzi, A. Jr. and Viotti, M.R. (2009) In-field loading analysis of pipelines

using a radial digital speckle pattern interferometer combined with the hole drilling method. OPTIMESS 2009-4th International Conference on Optical Measurement Techniques for Structure & Systems, Antwerp, 2009, Vol. 1, pp. 1–6.
5. Santos, J.L., Newson, T.P., and Jackson, D.A. (1990) Electronic speckle-pattern interferometry using single-mode fibers and active fringe stabilization. *Opt. Lett.*, **15**, 573–575.
6. Liu, J., Yamaguchi, I., Kato, J., and Nakajima, T. (1997) Real-time surface shape measurement by an active interferometer. *Opt. Rev.*, **4** (1), 216–220.
7. Ledger, A.M. (1975) Monolithic interferometric angle sensor. *Appl. Opt.*, **14** (12), 3095–3101.
8. Kujawinska, M. *et al.* (2006) New generation of full-field interferometric sensors. Proceedings of the Symposium on Photonics Technologies for 7th Framework Program, pp. 463–466.
9. Malacara, D. (2007) *Optical Shop Testing*, 3rd edn, John Willey & Sons, Inc.
10. de Groot, P. (2009) Design of error-compensating algorithms for sinusoidal phase shifting interferometry. *Appl. Opt.*, **48** (35), 6788–6796.
11. Deck, L. (1996) Vibration-resistant phase-shifting interferometry. *Appl. Opt.*, **35** (34), 6655–6662.
12. Deck, L.L. (2009) Suppressing phase errors from vibration in phase-shifting interferometry. *Appl. Opt.*, **48** (20), 3948–3960.
13. Sykora, D.M. and de Groot, P. (2010) Instantaneous interferometry: another view. Proceedings of the International Optical Design Conference (IODC), Jackson Hole, WY, June, 2010.
14. McLaughlin, J.L. and Horwitz, B.A. (1986) Real-time snapshot interferometer. *Proc. SPIE*, **680**, 35–43.
15. Kimbrough, B., Millerd, J., Wyant, J., and Hayes, J. (2006) in *Interferometry XIII: Techniques and Analysis*, Proceeding of SPIE, Vol. 6292 (eds K. Creath and J. Schmit), p. 62920F.
16. Millerd, J.E. (2005) in *Fringe 2005* (ed W. Osten), Springer, New York, pp. 640–647.
17. Hayes, J. (2002) *Dynamic Interferometry Handles Vibration*, Laser Focus World, pp. 109–113.
18. Freischlad, K., Eng, R., and Hadaway, J.B. (2002) Interferometer for testing in vibration environments. *Proc. SPIE*, **4777**, 311–322.
19. Deck, L.L. (2004) Environmentally friendly interferometry. *Proc. SPIE*, **5532**, 159–169.
20. Kühn, J.E. (2008) Axial sub-nanometer accuracy in digital holographic microscopy. *Meas. Sci. Technol.*, **19**, 074007.
21. Kühn, J. *et al.* (2007) Real-time dual-wavelength digital holographic microscopy with a single hologram acquisition. *Opt. Express*, **15** (12), 7231–7242.
22. Viotti, M.R., Kapp, W., and Albertazzi, A. Jr. (2009) Achromatic digital speckle pattern interferometer with constant radial in-plane sensitivity by using a diffractive optical element. *Appl. Opt.*, **48**, 2275.
23. Osten, W., Garbusi, E., Fleischle, D., Lyda, W., Pruss, C., Reichle, R., and Falldorf1, C. (2010) in *Speckle 2010: Optical Metrology*, Proceeding of SPIE, Vol. 7387 (eds A. Albertazzi Gonçalves Jr. and G.H. Kaufmann), p. 73871G.
24. Dantec Dynamics, *ESPI Strain Sensor (Q-100)*, http://www.dantecdynamics.com/Default.aspx?ID=853 (accessed 13 April 2011).
25. Garbusi, E., Pruss, C., and Osten, W. (2008) Interferometer for precise and flexible asphere testing. *Opt. Lett.*, **33** (24), 2973–2975.
26. Rogers, A.J. (2008) Optical-fiber sensors, in *Sensors Set: A Comprehensive Survey* (eds W. Göpel, J. Hesse, and J.N. Zemel), Wiley-VCH Verlag GmbH, Weinheim, pp. 355–398.
27. López-Higuera, J.M. (ed.) (2002) *Handbook of Optical Fibre Sensing Technology*, John Wiley & Sons, Inc., 828 pp. ISBN: 978-0-471-82053-6.
28. Young, P.P., Priambodo, P.S., Maldonado, T.A., and Magnusson, R. (2006) Simple interferometric fringe stabilization by charge-coupled-device-based feedback control. *Appl. Opt.*, **45**, 4563–4566.

17
Advanced Methods for Optical Nondestructive Testing
Ralf B. Bergmann and Philipp Huke

17.1
Introduction

Nondestructive testing (NDT) is indispensable for quality assurance in production and for maintenance purposes [1, 2]. Commonly, NDT is related to the measurement of the state or a property of an object and the comparison to a nominal or desired condition. The nominal condition can be defined either by the layout or by a measurement of a reference object. Generally, the measurement is influenced by the material parameters and certain tolerances have to be introduced allowing for a definition of margins of tolerable deviations. If a detected deviation exceeds this margin, it is commonly called a *defect* and the object is discarded. There exists a wide range of different defects related to, for example, shape, internal structure, or composition of the material. Material-related defects are a frequent source for defective products or component failure in service, and it is therefore desirable to detect these defects at an early stage in the production chain or in service.

Standard techniques such as ultrasound (US) or eddy current are well established for a large variety of applications. However, optical methods for NDT can be very useful, since they provide contactless, fast, and sensitive measurements. This chapter presents a discussion on a range of selected techniques such as thermography, reflectometry, speckle shearography, and laser ultrasound (LUS) and their application to optical nondestructive testing (ONDT). A compilation of their properties and a comparison to conventional techniques such as US and X-ray computed tomography (X-ray CT) give an overview on suitable areas of application. Following the presentation of results of advanced ONDT techniques, this chapter describes possible directions of further development.

17.2
Principles of Optical Nondestructive Testing Techniques (ONDTs)

ONDT techniques can be divided into two: (i) time- or depth-resolved and (ii) integrating techniques. Examples for the first case are phase-resolved thermography

Optical Imaging and Metrology: Advanced Technologies, First Edition.
Edited by Wolfgang Osten and Nadya Reingand.
© 2012 Wiley-VCH Verlag GmbH & Co. KGaA. Published 2012 by Wiley-VCH Verlag GmbH & Co. KGaA.

and LUS, which allow for a volumetric measurement. The distance in the in-depth direction is calibrated using time gating, which can be achieved by a time-resolved excitation, for example, pulsed thermal excitation or US waves. Integrating techniques, as the second case, observe the behavior of the surface of an object under the influence of a suitable load. Transient thermography, shearography, and reflectometry are examples for this case.

Concerning the second case, sophisticated detection schemes were investigated that measure the object in its original state and in an altered state by applying a load. Accordingly, these techniques are based on the following three factors:

1) Material or object properties (including defects)
2) Loading (excitation of a defect-related signal)
3) Measurement technique (with its sensitivity to factors 1 and 2).

This section is subdivided accordingly and concludes with a comparison of selected ONDT techniques as well as X-ray CT and US with respect to defect resolution.

17.2.1
Material or Object Properties

Material parameters such as Young's modulus, optical properties such as transparency, and those related to the production processes such as local stress, strain, or roughness have to be regarded individually for every object under test. For a particular ONDT technique, their influences on the measurement have to be determined, often by experiments. However, the effect of the defect on a solid can be classified in a suitable way.

Figure 17.1 shows the location of defects with respect to the surface that may serve as a useful classification for defect detection scenarios. In this contribution, we concentrate on nondestructive materials testing and distinguish (i) surface defects such as scratches, grooves, pits, open pores, dents, or bumps, which can, in

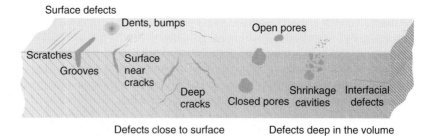

Figure 17.1 Defects may be classified as (i) surface defects such as scratches, grooves, dents, or bumps; (ii) defects close to the surface, for example, cracks, (open) pores, or cavities; and (iii) defects deep within the volume of a sample, for example, again cracks, pores, or cavities and various types of interfacial defects, for example, kissing bonds, which are particularly difficult to detect.

most cases, easily be detected by optical (microscopy) inspection; (ii) defects close to the surface; and (iii) defects deep within the volume of the material such as cracks, pores, shrinkage cavities, and interfacial defects such as debonding. Defects under the categories (ii) and (iii) belong to the typical domain of tasks for detection by material-related NDT techniques. Additional difficulties for detection may arise, for example, from interfacial defect types such as kissing bonds, since in this case, no voids occur and the disturbance only consists of a lack of chemical binding. In addition to the cases described above, there are many other applications of NDT such as the detection of inclusions and leakage in compound structures. The suitability of NDT techniques for the detection of defects depends, however, not only on their depth below the surface but also on a number of other parameters such as the defect size, their orientations toward the surface, the shape of the surface, or the volume of the sample as well as its response to the applied load.

17.2.2
Application of Thermal or Mechanical Loads for NDT

Surface-related optical metrology techniques show a high resolution perpendicular to the surface under investigation (vertical resolution). Techniques based on interferometry employed for surface inspection offer a vertical resolution in the nanometer range, while the resolution parallel to the surface (horizontal resolution) is limited to the micrometer range due to Abbe's resolution limit. In order to turn optical metrology techniques [3] into useful ONDT techniques capable to detect hidden defects, a suitable load has to be applied so that a defect-related signal is generated.

Figure 17.2 shows the reaction of an originally unloaded sample owing to exposure to heating or vibration. In addition to these commonly applied loads, other types of loads, for example, mechanical forces, may be applied. Even if the defect itself has macroscopic dimensions, the resulting deformation is typically very shallow and on the order of micrometers or even less. As a consequence, the load and the optical metrology technique have to be chosen carefully considering the object and material properties, since the high resolution usually obtained in optical

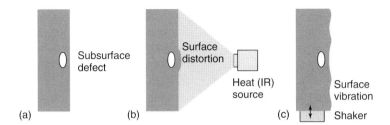

Figure 17.2 Detection of hidden defects using a suitable thermal or vibrational load, schematic representation, not drawn to scale. (a) Sample with defect within the material before loading. (b) Sample after thermal excitation (e.g., by illumination using an infrared (IR) source). (c) Sample during vibrational excitation (e.g., using a mechanical shaker).

metrology does not necessarily turn directly into useful resolution for defects below the surface.

The following four subsequent steps prove useful for such a combined ONDT technique:

- Measurement of the object in its initial state (e.g., surface shape)
- Loading (thermal, vibrational, pressure, vacuum, etc.)
- Measurement of the object under load or thereafter
- Retrieval of surface deviation (e.g., deformation, vibration, and thermal flux).

In general, the defect-related signal enables an in-plane localization. In order to obtain information about the depth and the geometric features of the defect detected by the ONDT technique, additional boundary conditions have to be included. The appearance and evaluation of defect-related deformations have been investigated by a number of authors, see, for example, Refs [4–7], and are therefore not dealt here.

17.2.3
Selected Measurement Techniques Suitable for Optical NDT

Figure 17.3 presents simplified schemes of whole-field measurements with interferometry, shearography, and reflectometry suitable for the measurement of

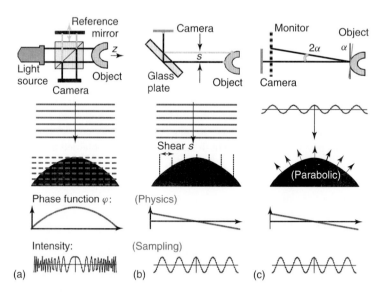

Figure 17.3 (a) Conventional interferometry: phase function φ proportional to wave front. High surface gradients result in high fringe density that may produce aliasing. (b) Shearography: phase function φ proportional to difference quotient. Path difference results from positions separated by shear s. (c) Fringe reflection/deflection: phase function φ proportional to surface slope. Local surface slopes determine the intensity variation.

surface profiles. All these principles end up measuring fringe patterns that can then be evaluated to obtain quantitative surface profiles. An extended discussion on digital image processing for optical metrology was recently presented by Osten [8].

Interferometry (Figure 17.3a) can be used to measure phase differences between the wave fields coming from the reference arm and the object arm of the setup. Therefore, high surface gradients lead to narrow fringe patterns. If the patterns become too narrow, they cannot be resolved by the CCD camera any more and impede the measurement, as indicated in the lower part of Figure 17.3a. In practical applications, interferometry using a light source with a wide band width, the so-called white light interferometry (WLI) [9, 10], is used for metrology because of its ability to measure unambiguously surface topographies several orders of magnitude higher than the wavelength of light. The influence of surface roughness on the measurement has been described in detail in the literature, see, for example, Refs [11, 12]. The effect of high surface slopes can be compensated by tailoring the reference wave due to a suitably shaped reference mirror [13].

Shearography [14], with its principle shown in Figure 17.3b, measures a phase function proportional to the difference quotient of the deformation. The difference quotient itself depends on the shear s. Therefore, the spatial frequency of the measured intensity variation can be adjusted by choosing an appropriate shear. In the case of fringe reflection [15], as shown in Figure 17.3c, the phase function is proportional to the surface slope and can be adjusted by choosing an appropriate spatial frequency of the fringe pattern of the illumination source (monitor).

As can be seen from Figure 17.3, reflectometry and shearography have a high tolerance towards steep surface curvature. The parameter space with respect to surface properties such as surface roughness or surface angle is dealt with in Ref. [16]. The next chapter describes the methods and their properties in detail.

17.2.4
Comparison of Properties of Selected NDT Techniques

Figure 17.4 shows a selection of NDT techniques according to their resolution, location of defects (as classified in Figure 17.1), and the geometrical complexity of an object. The last feature does not consider the smoothness or roughness of an object's surface, but its increasing complexity from simple flat surfaces up to complex-shaped objects containing voids and inner surfaces. A recent classification approach for measurement systems with respect to their freeform capability is described in Ref. [17]. For an overview on three-dimensional shape measurement techniques, see, for example, [18]. Although this chapter is dedicated to ONDT techniques, it is useful to compare them to conventional US and X-ray CT [19].

The technique of X-ray CT is an extremely versatile tool for NDT, since it enables the measurement of geometrically very complex objects containing inner volumes and allows for a resolution on the order of 1 μm [19]. Employing phase contrast information, structures on the 100 nm length scale can be resolved [20].

As for all techniques, the resolution limit depends on the objects under test by means of geometry and material properties and cannot be predicted for arbitrary

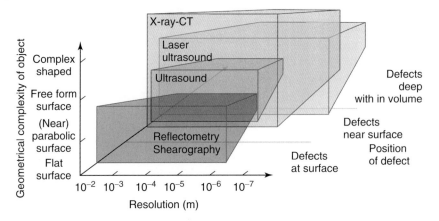

Figure 17.4 Schematic representation of the parameter space of selected NDT techniques according to geometrical complexity of the object, resolution, and location of defects: X-ray CT, laser- and conventional ultrasound, shearography, and reflectometry. The schematic representation compiles the main characteristics of the given approaches for cooperative materials or surfaces, respectively. Note that the order of magnitude of measurement resolution of the techniques is given, which, however, is not necessarily obtained when used as NDT technique. For further details, see text.

situations. Accordingly, the resolution decreases with the depth of the defect and the complexity of the surface as indicated in Figure 17.4. Drawbacks are limitations for large objects and the strong material-dependent penetration depth, which impedes its application in some materials or material systems. In addition, the application of X-ray CT in arbitrary environments suffers from the necessity of the shielding of ionizing radiation.

US testing is the established NDT method in many applications [1]. US has a good penetration for almost all materials and enables a resolution on the order of 50 μm [2]. A high resolution of around 1 μm may be obtained using acoustic microscopy [21]. In general, limitations arise from the need for a coupling medium (such as water or gel) and the difficulty to apply US testing to samples with a complex geometry. The advantages of LUS over conventional US appear especially attractive if one considers that it allows for contact-free measurement, tolerates more complicated sample geometries, allows for a pointlike excitation, and easily enables a separation of the location of excitation and detection and therefore extends the applicability of NDT far beyond conventional US. In addition, using LUS with very short pulse duration, a resolution down to the submicrometer range may be achieved under certain conditions that are dealt with later. The two main drawbacks of LUS are the need of hazardous laser radiation and equipment cost. Safety issues can be resolved by a proper housing or protective personal equipment. Unfortunately, capital investment into laser equipment suitable for LUS is presently still far beyond those for a phased-array system.

17.3
Optical Methods for NDT

17.3.1
Thermography

Thermography [22–25] is based on the thermal flow or heat flux in materials related to temperature gradients. As every object emits blackbody radiation corresponding to its temperature, thermal flow can be detected by monitoring this infrared radiation. It tells much about the performance of a system, process, or object [26]. Infrared imaging is based on the whole-field detection of radiation in the infrared range of the electromagnetic spectrum (3–14 µm). Most infrared cameras are based on a cooled detector consisting of a narrow bandgap semiconductor. For ONDT, this method can be used passively or actively using an excitation source. Passive thermography is of little use for material-related NDT because most materials exhibit a homogeneous temperature distribution that reveals no inner features [27].

In contrast, active thermography is an imaging method measuring the response behavior of an object to energy provided by an excitation source [26]. It is based on alteration of the thermal flow as a consequence of inhomogeneities or defects. Depending on the excitation, the defects may become thermal sources as well. Excitation sources may be optical, mechanical, or electrical. Optical excitation can be achieved with halogen- or flashlamps. Mechanical excitation is based, for example, on an US transducer, which produces elastic waves that emit heat energy during interaction with a defect. Another often used method is the excitation with eddy currents that experience larger losses at defects where the electrical energy is turned into heat. Another distinguishing feature is the detection principle. Transient thermography measures the cooling behavior of a solid, while the phase-resolved thermography is based on the analysis of heat waves. All sources mentioned above enable a time-gated or modulated excitation, and the respective methods are named optical lock-in thermography (OLT), ultrasound lock-in thermography (ULT), and induction lock-in thermography (ILT) [28, 29]. Generally, the phase-resolved methods show a higher resolution in defect depth and defect geometry and are therefore better suited for ONDT.

The lower detection limit of the respective phase-resolved method is given by the excitation frequency and the material constants determining propagation velocity (elastic modulus, stress, strain, etc.) and damping (thermal conductivity) of the heat pulse. As an example, a measurement of a gearwheel is shown in Figure 17.5.

17.3.2
Fringe Reflection Technique (FRT)

The fringe reflection technique (FRT) [15] is a whole-field measurement technique that uses the image of fringes being reflected and hence distorted by the objects's

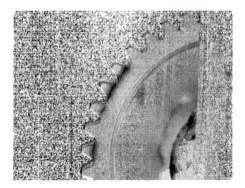

Figure 17.5 Measurement of a defective gearwheel with ultrasound lock-in thermography (ULT). The cracks in the region of the sprockets are clearly visible. (Source: Courtesy of EdeVis GmbH (C. Spiessberger).)

surface. It enables shape measurements with a resolution in the nanometer range. This capability is increasingly important [17] for macroscopic objects such as coatings on cars or mirrors also. The FRT demonstrates some remarkable properties: the axial resolution (that points into the direction perpendicular to the surface) does not depend on the size of the object and can be in the nanometer range, while the object height may range over multiple millimeters. Thus, the measurement dynamics is larger than $1 : 10^6$. The following sections presented here originate from the work of Bothe *et al.* [15].

17.3.2.1 Principle of FRT

Figure 17.6 shows a basic capture set-up of a fringe reflection system comprising basically three hardware components: A monitor that displays parallel sinusoidal fringes (Figure 17.6a), the object under investigation showing the mirrored and by its shape distorted fringe patterns and a CCD camera to recording the fringes (Figure 17.6b).

Figure 17.6 Basic setup of a system based on the fringe reflection technique (FRT): (a) a monitor generates a sinusoidal fringe pattern that is (b) reflected by the object, in this case a car window. The fringe pattern is captured by a CCD camera (here fixed on top of the monitor). (Source: Taken from Ref. [16].)

The object, in this case a car window (Figure 17.6b), is reflective, but the fringe pattern is visible even when the surface is only partly specular. A flat mirror would cause an undistorted image of the fringe pattern. Any tilt of the surface or a part of it would displace the fringes. A complex surface will deform the former straight parallel fringes according to the local angle of the surface against the normal or in other terms according to the surface gradient (Figure 17.3c). The reflected fringe pattern is measured using a phase-shifting method [30] and evaluated by means of an image processing system [31]. By field integration of the gradients [32], the shape can be evaluated with an axial resolution down to the nanometer range [15].

17.3.2.2 Experimental Results

As FRT is very sensitive to surface angle changes, it can be used for NDT, when combined with a load onto the object that generates surface angle changes in defective areas. In contrast to shearography, which works fine on scattering surfaces, FRT requires at least a partially specular surface as discussed in Ref. [16].

Figure 17.7 shows measurements on a sample of lacquered wooden marquetry. The sample shown in Figure 17.7a has been prepared so that the surface delaminates from the object base in the regions marked with dashed lines within the frame of measured area. For investigation, the sample is heated by an infrared lamp, which generates a temperature increase of $\Delta T \approx 10$ K. Consecutively, the passive cooling phase is monitored by FRT and locally reveals the change of surface

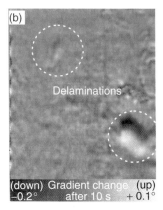

Figure 17.7 Use of fringe reflection for ONDT: (a) photo with reflected fringes on the lacquered part of a wooden marquetry sample. Sample size depicted in vertical direction is 200 mm. The measurement area and the delaminations are marked. (b) Measurement of surface gradient change after heating and 10 s of passive cooling. The surface angle changes during the first 10 s within a range from −0.2 to +0.1°. The changes clearly show the areas of the delaminations. (Source: Taken from Ref. [16].)

angles inside the marked measurement area. The result shown in Figure 17.7b demonstrates that FRT is able to detect the delaminated areas. Owing to the transient heat flow, the defects are best measured at particular time intervals after heating. Thus, the detection capabilities of NDT based on heat loading benefits from a series recording of the gradients.

17.3.3
Digital Speckle Shearography

Digital speckle shearography measures shape or deformation of an objects surface [14, 33]. It also belongs to the whole-field measurement techniques. Numerous examples demonstrate the potential of shearography for nondestructive inspection of technical components with respect to material defects and structural imperfections [34–37]. Recent significant developments reported here go back to the work of Falldorf [38].

17.3.3.1 Principle of Shearography
Besides the glass plate shown in Figure 17.3b, a variety of approaches using, for example, gratings or birefringent components have been proposed to create a shear [39–41]. An advanced system, developed at BIAS, is now commercially available and makes use of the birefringent properties of a liquid crystal spatial light modulator (SLM) as the shearing element [42]. In this configuration, no mechanically moving parts are required to set the shear, the system thus combines flexibility, high accuracy, and high reproducibility at the same time. As one of its major benefits compared to other interferometric techniques, the system has, due to the small differences in the relative path lengths of the light, rather low demands regarding the stability of the setup and the coherency of the light [43, 44].

This configuration, which is described in detail in Ref. [45], produces two laterally shifted images of the same surface on the camera. The reflected light coming from the surface of the sample is captured by the lens objective and is separated by the $\lambda/2$-plate in two orthogonal polarized beams. The SLM is situated in the Fourier plane of lens L_1. The SLM affects only one polarization and therefore shifts one of the beams laterally. The SLM is also positioned in the Fourier plane of lens L_2, which projects the light coming from the SLM via the polarizer A on the CCD camera (Figure 17.8). A double-image incident on the CCD originated by a certain point on the surface is obtained creating the desired shear. If the surface is rough, both intensity distributions feature a noisy appearance called *speckle effect* [46].

17.3.3.2 Experimental Results
Various kinds of samples ranging from artwork [47] to industrial objects [48] have been investigated by shearography as an NDT tool. A typical thermal loading regime gently increases the surface temperature of the object by $\Delta T \approx 5$ K.

Figure 17.9 shows an example of an investigation on a wood inlay [49]. The size of the sample shown in Figure 17.9a is $15 \times 15 \,\text{cm}^2$. A delamination has been

17.3 Optical Methods for NDT | 403

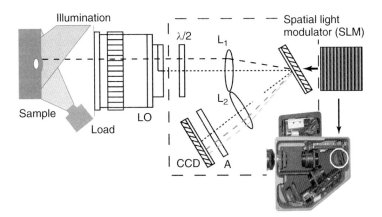

Figure 17.8 Schematic setup of advanced digital speckle shearography: the sample containing the defect is illuminated by a laser source. The load is applied, for example, by an infrared light source. The shearography setup contains the lens objective (LO), a λ/2 plate (λ/2), the lenses L_1 and L_2, the spatial light modulator in the Fourier plane of both lenses creating a shear by dividing the incoming beam into an ordinary and extraordinary beam, a polarizer (A), and the CCD camera. A picture of the complete sensor head shows the position of the SLM marked by a circle. For further explanation, see text.

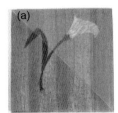 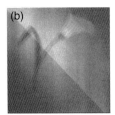 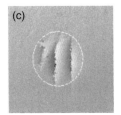

Figure 17.9 (a) Photograph of a wood inlay with a size of 15 cm × 15 cm. (b) Image of the wood inlay with shear of 2.5 mm in x- and y-direction. (c) Unwrapped phase image after filtering. Three cracks are identified, the encircled area indicates location and size of the delamination. (Source: Measurement at BIAS, sample supplied by M. Stefanaggi, Laboratorie de Recherche des Monuments Historique (LRMH), Champy sur Marne, France; see Ref. [49].) (Please find a color version of this figure on the color plates.)

purposely introduced in the central part of the sample. A laser with a wavelength of 532 nm and a power of 400 mW serves for sample illumination. Figure 17.9b shows the image of the wood inlay with a shear set to 2.5 mm both in x- and y-directions. An infrared lamp serves as a load and increases the surface temperature by about 5 °C within 10 s. Figure 17.9c depicts the phase image of the sample after filtering and phase unwrapping. The investigation reveals the presence of the delamination (encircled) with a size of about 60 mm and three cracks.

17.3.4
Laser Ultrasound

US has always been a primary method for NDT, because of its high resolution, its large depth of measurement, and foremost its ease of use. Phased-array techniques have been developed as a consequent follow-up to single transducers, enabling the three-dimensional in-depth search for defects [50, 51]. However, objects showing poor accessibility, for example, due to high temperature or complex shape, remain a challenge to conventional US. These limitations can be overcome with laser-assisted generation and detection of US [52]. The technique provides absolute measurement values during contactless, remotes detection of in-depth defects, and gains rising attention [53]. Part of the work presented here originates from the successful investigation and implementation of the LUS technique by Kalms [54–56] over the past decade.

17.3.4.1 Principle of Operation

Figure 17.10 shows an LUS setup consisting of two parts: laser-assisted generation of US waves in a sample and a suitable detection scheme. The generation beam is a short-pulsed laser that excites a transient thermoelastic stress zone that creates elastic waves. These waves are reflected at the rear side and induce surface vibrations once they reach the front side. The detection scheme shown in Figure 17.10 includes a detection beam, which is frequency modulated by the surface vibration in the

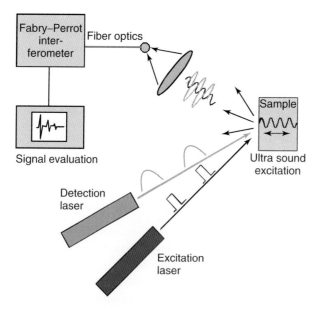

Figure 17.10 Schematic laser ultrasound setup in the impulse-echo method. For explanation, see text.

picometers- to nanometers range. This vibration can be detected by optical or electrical demodulation showing a signal comparable to US NDT with single transducers but with much higher frequency. Typical detection schemes use a pump–probe setup [57], a vibrometer, or for two-dimensional detection of surface acoustic waves a holography setup [58] or a shearography system as described above [59].

In cooperation with the aeronautic industry, an LUS setup was realized at BIAS [60]. It uses a CO_2 laser with a repetition rate up to 500 Hz for US generation. The pulse width can be varied from 50 to 100 ns. The detection system consists of a scanning laser and an interferometer coordinated with it. The laser wavelength of the detection system is 1064 nm, and the energy per pulse is about 90 mJ with a pulse length of 70 µs. A confocal Fabry-Perot interferometer converts the optical input signal into an electrical output signal with an amplitude proportional to the detected ultrasonic signal.

Typically, the generation of US waves in a medium takes place with a short-pulsed laser or a laser with a steep slope at the beginning of its pulse. A variety of lasers ranging from KrF-Laser [52], (frequency doubled and frequency tripled) Nd:YAG-Laser [61], (ring-) dye-Laser [62], Ruby-Laser, or CO_2-Laser [52, 60, 63, 64] have been used. The available pulse energy ranges from nanojoules to a few millijoules [52, 65], while the pulse widths range from femtosecond to a few microseconds. The choice of the appropriate laser depends on the available energy that can be used for generation and hence, the light–matter interaction plays a major role.

For NDT, the thermoelastic effect is most often used for generation of US waves. If the beam of the generation laser interacts with the object's surface, the energy is absorbed, transmitted, or reflected. A good portion of the absorbed energy is converted into heat. The fast rising temperature builds up local stress in the media that dissipates through elastic waves carrying a part of the induced energy away. Generally, a bunch of waves with different characteristics are excited that can be divided into two categories: surface and bulk waves. Surface and coupled waves such as lamb waves are useful for surface or near-surface defect detection. However, owing to the complexity in detection and signal analysis, they are rarely used. One especially interesting example is the lattice defect detection described in Ref. [57].

Bulk waves consist of two different types of waves, the shear waves, which have a transversal character, and the longitudinal waves (including the central bulk wave) [66]. These waves can be converted into each other through reflection at a boundary layer with a refractive index jump. In general, the longitudinal waves travel much faster than the transversal waves, and accordingly, the separation of the waves at a certain detection point can be carried out via the specific run time. The excitation and propagation of different waves depends on laser beam characteristics, for example, energy, pulse width, and beam profile, as well as on material constants, for example, speed of sound, damping, E-modulus, and homogeneity [66].

If we presume an "ideal" medium with a defect-free crystal lattice, no boundaries, no scattering, and no dispersion, the excited spectrum of US waves

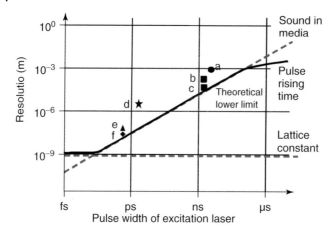

Figure 17.11 Lower limit of axial resolution of laser ultrasound compared with experimental data. (Source: Literature sources: a: [51], b: [65], c: [64], d: [60], e: [62], f: [61]. Taken from Ref. [16].)

would be a continuum of frequencies with the highest frequency given by $f_{max} = 1/(\text{pulse width})$. Furthermore, we assume that the detection limit is low enough for detection and frequency-dependent damping can be neglected. Under these conditions, the minimum wavelength excited by the laser pulse would be $\lambda_{sound, min} = v_{sound}/f_{max}$.

Basically, the detection limit or the axial resolution for single defects corresponds to the wavelength λ_{sound}, even though a high density of small defects would produce a significant and measurable scattering [67]. Naturally, a lower limit is given by the lattice constant for acoustical lattice vibrations. A comparison of the "ideal" resolution limit and literature results is shown in Figure 17.11. However, typically, the evolution of the thermal stress zone takes more time than the pulse width, so that generally the available detection frequencies are below the maximum.

17.3.4.2 Experimental Results

Figure 17.12a shows an example of a complex-shaped carbon-fiber-reinforced plastic (CFRP) component as required by aircraft industry investigated by LUS. The object is investigated in a position with the flat sides having an angle of 45° relative to the optical axes of the generation and detection lasers using the impulse-echo mode. Scanning is carried out using an x–y translation stage. Figure 17.12b shows a photo of the back portion of the component with inserted flat bottom holes. Five artificial flaws are situated in the region of the radius and two are located on the flat side. Each flaw is marked with the size of its diameter. Figure 17.12c shows the result of the inspection in a time of flight scan. All flaws are identified – the five flaws in the radius as well as the flaws on the flat side. Besides the unequal depth of the inserted drilling holes, the thickness along the area of curvature also varies.

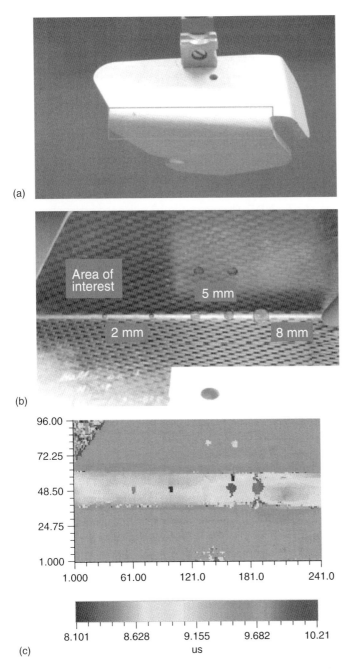

Figure 17.12 (a) Photograph of a CFRP component, (b) rear side with inserted flaws, and (c) C-scan, for explanation, see text. (Source: Taken from Ref. [16], first published in Ref. [64].) (Please find a color version of this figure on the color plates.)

17.4
Conclusions and Perspectives

A wide range of ONDT techniques have been successfully developed and improved up to the present date. The ONDT of materials and objects was classified here into time-/depth-resolved and integrating methods that observe the surface behavior of an object under a suitable excitation or load. A short introduction and overview of thermography, FRT, digital speckle shearography, and LUS was given with some examples.

The FRT is an appropriate method to fulfill the requirements of industrial demands as a robust and fast measurement method. It has a resolution in the nanometer regime with respect to surface distortions and has an extraordinary large dynamic range, also for NDT. The method implies that the surface has to be partly reflective, at least. It has to be pointed out that the technique applies white light. Thus, the users do not need protection against laser radiation. Shearography is also capable of high-precision optical metrology and NDT. As surface gradients are measured, the same discussion as for reflectometry applies. However, in contrast to fringe reflection, shearography may be used both on specular and rough surfaces. Owing to the use of a common path and the absence of mechanical adjustment of the shear in the setup developed at BIAS, it is very rugged. To make shearography safer to use, light sources with a lower degree of coherence are under investigation. For industrial inspection, an automated and rugged fault detection is desired.

With regard to NDT, both methods are, in contrast to LUS, techniques that rely on the response of the defective object to a suitable load applied to it. Therefore, it is always necessary to adapt the load to the properties of the object under investigation.

Finally, for certain applications, LUS surpasses the applicability of conventional US for NDT. US waves are excited using a thermoelastic generation with the help of a short-pulsed laser. The bulk and the surface waves are detected at certain distances of the excitation point enabling the discrimination of the waves by their runtime. Distinct advantages of laser-based US over conventional US are as follows: (i) the measurement is contactless and does not require any coupling medium, (ii) the generation of US is independent of the direction of incident beam, (iii) it allows for investigation of objects with complex geometry, (iv) it is suitable for a large range of materials, (v) it allows for a (large) distance to the test object and therefore inspection of moving and even glowing or hot objects, and (vi) it can detect much smaller structures than conventional US.

Accordingly and in response to the industrial requests, currently smaller, cheaper, and robust systems are under development. Constant effort is also put to the signal analysis in order to improve extraction of information as much as possible. For example, the sound velocity depends on the stress and strain and the differences are measurable. In addition, a retrieval of complete tomographic information would be desirable.

Acknowledgments

The authors would like to thank Th. Bothe, C. Falldorf, C. von Kopylow, O. Focke, M. Kalms, and R. Klattenhoffat BIAS for their valuable contributions. We also acknowledge the contributions of Wansong Li, VEW Corp. in Bremen, for his significant input into applications. Long-term collaborations with industrial partners, mainly VEW Corp. (VereinigteElektronikwerkstätten), Satisloh Corp., and Airbus Corp. have substantially stimulated this work.

Financial support of the following organizations is gratefully acknowledged: Part of this work was performed within the EU projects EVK4-CT-2002-00096 and GRD1-2000-25309, the German BMWi (Federal Ministry of Economics and Technology) projects KF0086401WM4 and 20W9904, and the DFG (German Research Foundation) projects Ju142/54-1 and Ju142/60-1.

References

1. Moore, P.O. (series ed.) (1998–2010) *Nondestructive Testing Handbook*, 3rd edn, vols. 1–10, The American Society for Nondestructive Testing (ASNT), Columbus, OH.
2. Bergmann, R.B. and Zabler, E. (2006) in *Handbuch der Mess- und Automatisierungstechnik in der Produktion* (eds H.-J. Gevatter and U. Grünhaupt), 2nd edn, Springer, Berlin, pp. 363–410.
3. Yoshizawa T. (ed.) (2009) *Handbook of Optical Metrology*, CRC Press, London.
4. Osten, W., Jüptner, W., and Mieth, U. (1993) Knowledge based evaluation of fringe patterns for automatic fault detection. SPIE Interferometry VI: Applications, pp. 256–268.
5. Kreis, T., Jüptner, W., and Biedermann, R. (1995) Neural network approach to holographic nondestructive testing. *Appl. Opt.*, **34**, 1407.
6. Osten, W., Elandalousi, F., and Jüptner, W. (1996) Recognition by synthesis – a new approach for the recognition of material faults in HNDE. *SPIE*, **2861**, 220–224.
7. Furlong, C. and Pryputniewicz, J. (1996) Hybrid, experimental and computational, investigation of mechanical components. *SPIE*, **2861**, 13–24.
8. Osten, W. (2008) in *Springer Handbook of Experimental Solid Mechanics* (ed. W.N. Sharpe), Springer, Berlin, pp. 481–563.
9. Thian, S.C.H., Feng, W., Wong, Y.S., Fuh, J.Y.H., Loh, H.T., Tee, K.H., Tang, Y., and Lu, L. (2006) Dimensional measurement of 3D microstructure based on white light interferometer. *J. Phys.: Conf. Ser.*, **48**, 1435–1446.
10. Bergmann, R.B., Drabarek, P., Kallmann, U., Schmidtke, B., and Bauer, J. (2006) in *Photonik – Grundlagen, Technologie und Anwendung* (eds E. Hering and R. Martin), Springer, Berlin, pp. 263–281.
11. Häusler, G. and Leuchs, G. (1997) Physikalische grenzen der optischen formerfassung mit licht. *Phys. Bl.*, **53**, 417–422.
12. Dresel, T., Häusler, G., and Venske, H. (1992) Three-dimensional sensing of rough surfaces by coherence radar. *Appl. Opt.*, **31**, 919.
13. De Groot, P. and Colonna de Lega, X. (2003) Valve cone measurement using white light interference microcopy in a spherical measurement geometry. *Opt. Eng.*, **42**, 1232–1237.
14. Hung, Y.Y. (1979) Image-shearing camera for direct measurement of strains. *Appl. Opt.*, **18**, 1046–1051.
15. Bothe, T., Li, W., Kopylow, C., and Jüptner, W. (2004) High resolution 3D shape measurement on specular surfaces by fringe reflection. *Proc. SPIE*, **5457**, 411–422.

16. Bergmann, R.B., Bothe, T., Falldorf, C., Huke, P., Kalms, M., and von Kopylow, C. (2010) in *Proceedings of SPIE Interferometry XV: Applications*, vol. 7791 (eds C. Furlong, C. Gorecki, and E.L. Novak), SPIE, Bellingham, WA, pp. 1–15.
17. Savio, E., De Chiffre, L., and Schmitt, R. (2007) Metrology of freeform shaped parts. *CIRP Ann.– Manuf. Technol.*, **56** (2), 810–835.
18. Chen, F., Brown, G.M., and Song, M. (2000) Overview of three-dimensional shape measurement using optical methods. *Opt. Eng.*, **39** (1), 10–22.
19. Bergmann, R.B., Bessler, F.T., and Bauer, W. (2006) Non-destructive testing in the automotive supply industry – requirements, trends and examples using x-ray CT. 9th European Conference for Non-Destructive Testing, Berlin, DGZFP Proceedings BB 103-CD, Th.1.6.1.
20. Dierolf, M., Menzel, A., Thibault, P., Schneider, P., Kewish, C.M., Wepf, R., Bunk, O., and Pfeiffer, F. (2010) Ptychographic X-ray computed tomography at the nanoscale. *Nature*, **467**, 436–440.
21. Zakutailov, K.V., Levin, V.M., and Petronyuk, Y.S. (2010) High resolution ultrasonic ultrasound methods: microstructure visualization and diagnostics of elastic properties of modern materials (review). *Inorg. Mater.*, **46**, 1655–1661.
22. Busse, G., Wu, D., and Karpen, W. (1992) Thermal wave imaging with phase sensitive modulated thermography. *J. Appl. Phys.*, **71**, 3962–3965.
23. Rantala, J., Wu, D., and Busse, G. (1996) Amplitude modulated lock-in vibrothermography for NDE of polymers and composites. *Res. Nondestr. Eval.*, **7**, 215.
24. Dillenz, A., Zweschper, T., and Busse, G. (2000) Elastic wave burst thermography for NDE of subsurface features. *Insight*, **42** (12), 815–817.
25. Maldague, X. and Marinetti, S. (1996) Pulse phase infrared thermography. *J. Appl. Phys.*, **79** (5), 2694–2698.
26. Maldaque, P.V.X. (1995) *Nondestructive Evaluation of Materials by Infrared Thermography*, Springer-Verlag, ISBN: 3-540-19769-9.
27. Dillenz, A. (2001) MP Materialprüfung Jahrgang, **43** (1–2), 30–34.
28. Zöcke, C.M. (2009) Quantitative analysis of defects in composite material by means of optical lock in thermography. Dissertation. Naturwissenschaftliche Technische Fakultät III der Universität des Saarlandes und der EcoleDoctorale MIM der Université Paul-Verlaine/Metz.
29. Riegert, G. (2007) Induktions-Lockin-Thermografie ein neues Verfahren zur zerstörungsfreien Prüfung. Dissertation. Institut für Kunststofftechnik der Universität Stuttgart.
30. Burke, J., Bothe, T., Osten, W., and Hess, C. (2002) Reverse engineering by fringe projection. *Proc. SPIE*, **4778**, 312–324.
31. (2010) Description of the fringe processor. http://www.fringeprocessor.com, (accessed 02 April 2012).
32. Li, W., Bothe, T., Kopylow, C., and Jüptner, W. (2004) Evaluation methods for gradient measurement techniques. *Proc. SPIE Int. Soc. Opt. Eng.*, **5457**, 300–311.
33. Osten, W., Kalms, M., Jüptner, W., Tober, G., Bisle, D., and Scherling, D. (2000) A shearography system for the testing of large scale aircraft components taking into account non-cooperative surfaces. *Proc. SPIE*, **4101B**, 432–438.
34. Yang, L., Chen, F., Steinchen, W., Hung, M.Y. (2004) Digital shearography for nondestructive testing: potentials, limitations, and applications. *J. Hologr. Speckle*, **1**, 69–79.
35. Hung, Y.Y. and Ho, H.P. (2005) Shearography: An optical measurement technique and applications. *Mater. Sci. Eng.*, **49**, 61–87.
36. Zhang, J. and Geng, R. (2008) Studies on digital shearography for testing of aircraft – composite structures and honeycomb-based specimen. Proceedings of the 17th Conference on Nondestructive Testing, Shanghai.
37. Moser, E. (2008) Detection of state-of-the-art shearography systems. Proceedings of the 17th Conference on Nondestructive Testing, Shanghai.
38. Falldorf, C. (2011) Measuring the complex amplitude of wave fields by means

of shear interferometry. *J. Opt. Soc. Am. A*, **28** (8), 1636–1647.

39. Mallick, S. and Robin, K.L. (1969) Shearing interferometry by wavefront reconstruction using a single exposure. *Appl. Phys. Lett.*, **14** (2), 61–63.

40. Nakadate, S. (1997) Phase shifting speckle shearing polarization interferometer using a birefringent wedge. *Opt. Lasers Eng.*, **26**, 331–350.

41. Falldorf, C., von Kopylow, C., and Jüptner, W. (2007) Proceedings of 3DTV (eds IEEE CNF), Kos, Greece, Conference, doi: 10.1109/3DTV.2007.4379400.

42. Falldorf, C., Osten, S., von Kopylow, C., and Jüptner, W. (2009) *Opt. Lett.*, **34** (18), 2727–2729.

43. Falldorf, C., Kolenovic, E., and Osten, W. (2003) Speckle shearography using a multiband light source. *Opt. Lasers Eng.*, **40**, 43–52.

44. Harder, I., Schwider, J., and Lindlein, N. (2005) DUV-Shearing interferometer with reduced spatial coherence. *DGaO Proc.*, **106**, A11.

45. Falldorf, C., Klattenhoff, R., Gesierich, A., von Kopylow, C., and Bergmann, R. (2009) in *Lateral Shearing Interferometer based on a Spatial Light Modulator in the Fourier Plane, Fringe 2009: 6th International Workshop on Advanced Optical Metrology* (eds W. Osten and M. Kujawinska), Springer, Heidelberg, pp. 93–98.

46. Goodman, J.W. (1975) in *Laser Speckle and related Phenomena* (ed. J.C. Dainty), Springer, Berlin, pp. 9–75.

47. Falldorf, C., Klattenhoff, R., von Kopylow, C., and Jüptner W. (2008) Digital speckle shearography applied to artefacts, in *Handbook on the Use of Lasers in Conservation and Conservation Science* (eds M. Schreiner, M. Strlic, and R. Salimbeni), COST Office, Brussels, Belgium, CD-ROM, pp. 10.

48. Kalms, M. and Osten, W. (2003) Mobile shearography system for the inspection of aircraft and automotive components. *Opt. Eng.*, **42**, 1188.

49. BIAS (2006) Laser Multitask Non-Destructive Technology in Conservation Diagnostic Procedures (LaserAct), Conclusive Report, European Project (EVK4-CT-2002-00096, January 2, 2003–January 31, 2006), Partner 2: BIAS.

50. Kröning, M., Ribeiro, J.G.H., and Vidal, A. (2008) Progress in NDT system engineering through sensor physics and integrated efficient computing. Proceedings of the 17th World Conference on Non Destructive Testing, pp. 1–15, doi: 10.1.1.119.787

51. Schmerr, W.L. and Song, S.-J. (2007) *Ultrasonic Nondestructive Evaluation Systems: Models and Measurements*, Springer, Berlin.

52. Monchalin, J.-P., Neron, C., Bussiere, J.F., Bouchard, P., Padioleau, C., Heon, R., Choquet, M., Aussel, J.-D., Durou, G., and Nilson, J.A. (1998) *Adv. Perform. Mater.*, **5** (1–2), 7–23.

53. Monchalin, J.-P. (2004) Optical and laser NDT: a rising star. Proceedings of the 16th World Conference on Non destructive Testing.

54. Kalms, M., Focke, O., and van Kopylow, C. (2008) in *Proceedings of SPIE 9th International Symposium on Laser Metrology*, vol. 7155 (ed. by C. Quan and A. Asundi), SPIE, Bellingham, pp. 71550E–715501.

55. Kalms, M. and Bergmann, R.B. (2009) Applications of laser ultrasound NDT methods in aircraft industry. The 13th Asia-Pacific Conference on Non-Destructive Testing.

56. Kalms, M., Peters, C., and Wierbos, R. (2011) Assessment of carbon fiber-reinforced polyphenylene sulfide by means of laser ultrasound. SPIE Conference on Nondestructive Characterization for Composite Materials, Aerospace Engineering, Civil Infrastructure, and Homeland Security (ed. H. Felix Wu), Proceedings of SPIE, Vol. 7983, San Diego, pp. 79830B-1–79830B-8.

57. Hess, P. and Lomonosov, A.M. (2010) *Ultrasonics*, **50**, 167–171.

58. Telschow, K.L., Deason, V.A., Schley, R.S., and Watson, S.M. (1999) Imaging of lamb waves in plates for quantitative determination of anisotropy using photorefractive dynamic holography, in *Reviews of Progress in Quantitative Nondestructive Evaluation*, Vol. 18 (eds by D.O. Thompson and D.E. Chimenti), Springer.

59. Huke, P., Focke, O., Falldorf, C., von Kopylow, C., and Bergmann, R.B. (2010) Contactless defect detection using optical methods for non destructive testing. Proceedings of the NDT in Aerospace.
60. von Kopylow, C., Focke, O., and Kalms, M. (2007) Laser ultrasound – a flexible tool for the inspection of complex cfk components and welded seams. *Proc. SPIE*, **6616**, 66163J, doi: 10.1117/12.732043
61. Hao, H.-Y. and Maris, H.J. (2001) Experiments with acoustic solitons in crystalline solids. *Phys. Rev. B*, **64**, 064302.
62. Thomsen, C., Grahm, H.T., Maris H.J., and Tanc, J. (1986) Surface generation and detection of phonons by picosecond light pulses. *Phys. Rev. B*, **34** (6), 4129–4138.
63. Blouin, A., Kruger, S.E., Levesque D., and Monchalin, J.-P. (2008) Applications of laser-ultrasonics to the automotive industry. Proceedings of the 17th World Conference on Non Destructive Testing, Shanghai.
64. Focke, O., Kalms, M., and Jueptner, W. (2006) NDT with laser ultrasonic of complex composites. *Non Destr. Test. Aust.*, **43** (4), 123–129.
65. Blouin, A., Neron, C., Campagne B., and Monchalin, J.-P. (2008) Applications of laser-ultrasonics and laser-tapping to aerospace composite structures. Proceedings of the World Conference on Non Destructive Testing, Chinese Society for Non-Destructive Testing.
66. Scruby, C. and Drain, L. (1990) *Laser Ultrasonics: Techniques and Application*, Chapter 5, Adam Hilger, Bristol.
67. Seale, M.D. and Smith, B.T. (1996) *Review of Progress in QNDE*, vol. 15A, Plenum Press, New York, pp. 261–266.

18
Upgrading Holographic Interferometry for Industrial Application by Digital Holography

Zoltán Füzessy, Ferenc Gyímesi, and Venczel Borbély

18.1
Introduction

After inventing the laser, there was a significant expectation toward realizing the full potentials of holography – particularly of holographic interferometry (HI) – in solving measuring tasks both in and outside the laboratory walls. In the 1960s, the efforts to make HI a versatile measuring tool met with full success worldwide. Methods and techniques were developed for measuring deformation (displacement) [1], vibration [2], shape [3, 4], and refractive index distribution changes [5] of opaque and transparent objects. For many other applications of HI and for further details of related methods and techniques, the reader is referred to the well-known monographs [6–8]. It seemed that nothing could prevent HI from replacing traditional measuring tools such as dial indicators, inductive and capacitive transducers, 3D measuring machines, and so on used in the above-mentioned areas.

Being a noncontact, nondestructive full-field method with high sensitivity and accuracy, HI rightly claimed to be a widely used measuring tool in industry. However, the high sensitivity may result in nonresolvable fringe system, which can be considered as a disadvantage at quantitative evaluation of interferogram. For unambiguous fringe count, the fringe system should be observed as a whole – the full-field character appears as an unfavorable peculiarity in this respect. Nevertheless, the main practical problem was connected with the wet and lengthy chemical development of silver halide recording material, which was not alleviated by its high resolution and large size available.

On the basis of the research at the Holography Laboratory of the Department of Physics, mainly, this chapter shows that despite its inherent difficulties the HI does register successes both in development of new methods/techniques and in industrial applications as well. Digital holography is considered as a reliable version of holography that proved adaptable to the actual conditions and requirements (rapidity, flexibility, etc.) and so it can upgrade HI for industrial application. The quantitative evaluation of high-density fringe system is forced by new techniques developed at the laboratory.

Optical Imaging and Metrology: Advanced Technologies, First Edition.
Edited by Wolfgang Osten and Nadya Reingand.
© 2012 Wiley-VCH Verlag GmbH & Co. KGaA. Published 2012 by Wiley-VCH Verlag GmbH & Co. KGaA.

18.2
Representative Applications

The most dynamic period of analog HI in view of its industrial use was certainly the 1980s. This chapter starts with a brief survey of practical applications of HI in some representative countries – without attempting to be comprehensive – which is followed by a more or less detail discussion on industrial use of HI in Hungary.

Nagasaki Technical Institute has applied HI in factory environment at Mitsubishi Heavy Industries Ltd [9]. Vibration mode of steam turbine blades and compressor vanes was tested, and propagating transverse waves in diesel engines was studied. Vibration of a car body exited by electromechanical shaker was also investigated.

Since 1979, Ford Motor Company has been developing HI to supplement more conventional test methods to measure vehicle component vibration [10]. At early stage, an Apollo PHK-1 double-pulsed holographic laser system was used to visualize a variety of complex vibration modes, primarily on current production and prototype power train components. The holocamera (model PHK-1) developed by Rottenkolber GmbH and marketed by Apollo Lasers, Inc. in the United States was among the first commercial double-pulsed holographic laser systems to be offered with internal triggering controls to synchronize pulse timing to test object vibration.

Volkswagenwerk AG applied HI for studying noise and vibration relations of automotive components. Influence of engine suspension on vibrations of the engine itself and that of car body, the stiffness of the gear box, dependence of noise level on the velocity and speed of engine rotation, and vibration and noise transmission and vibration modes of the car body were investigated [11].

Parallel to the HI, several other optical methods, also, have been developed for industrial applications. Portable devices have been constructed using electronic speckle pattern interferometry (ESPI) to measure vibrations and deformations of car elements [12–14]; residual stresses have been investigated by digital speckle pattern interferometry (DSPI) [15]; and aircraft components, abradable seals in jet engines [16], local excitation of delaminations and disbonds [17], residual stresses [18], and so on have been studied by shearography.

18.3
Contributions to Industrial Applications by Analog Holography

18.3.1
Portable Interferometer in the Days of Analog Holographic Interferometry

The main common feature of the representative applications of HI is the fact that only a single interferogram was recorded during each measuring phase. So the interferometer in each case had one sensitivity vector that allowed determining one component of the displacement at deformation or vibration of the object under test. There was no attempt to evaluate the single interferogram quantitatively in those applications.

18.3 Contributions to Industrial Applications by Analog Holography

The aim of the authors was to create conditions for measuring all three components of the displacement vector produced by deformation/vibration of any technical object subjected to static/dynamic load in hostile environment. The way to do this is to have a portable holographic interferometer that records at least three interferograms with three sensitivity vectors and a reliable evaluation algorithm [19, 20].

At technical object investigation in factory environment, there is no unmoved point on the object surface, as usual. Therefore, special means have to be taken to avoid it by enhancing the redundancy of data. Four interferograms with different sensitivity vectors provide sufficient data for their unambiguous interpretation even in the case when the zero-order fringe cannot be identified.

Figure 18.1 shows two views of the portable measuring system HIM developed at the Department of Physics.

The interferometer is mounted on a tripod (1). It can be lifted up hydraulically (2) and rotated along horizontal and vertical axes. The beam forming optics is affixed on frame consisting of pipes and ribs (6). The main components are four measuring holocameras (4), ruby laser (7), optical system producing five reference beams and one object beam, path-length-matching optomechanical system, and control holocamera (5). Interferograms are recorded on thermoplastic film simultaneously with four measuring ones in order to control the quality of holograms in silver halide emulsions in some 10 s after exposures.

- Technical data: energy 30–1500 mJ imp^{-1}, impulse length 20 ns.
- Dimensions: the height of the two upper measuring cameras is 185 cm, the height of the ruby laser is 150 cm, length of the interferometer is 160 cm, width of the interferometer is 110 cm.
- Mass: interferometer 200 kg, capacitor 200 kg, controlling box 100 kg.

Figure 18.1 (a) Side and (b) front views of the interferometer (HIM).

416 | *18 Upgrading Holographic Interferometry for Industrial Application by Digital Holography*

(a) (b)

Figure 18.2 Investigation of MU 51 universal milling machine. (a) The interferometer (right) with the milling machine (left) at the factory and (b) the interferogram of the milling machine at static load.

In the period 1980–1995, typically two to three measurements took place annually by HIM, mainly in factories of Hungarian machine tool industry. In the following discussion, a few representative applications are presented.

Figure 18.2 shows the interferometer in front of the MU 51 universal milling machine (prototype) and an interferogram at static load. At static and dynamic loads (30 kg mass on the right edge of the table and single-way and conventional milling), the console, beam, and frame deformations were investigated.

An interferogram and its evaluation are shown in Figure 18.3. The object was the carriage support of RF4-75/2000 radial drilling machine.

The carriage support was dynamically loaded by no-load speed of the main spindle. At a given surface point, the amplitude of vibration was measured by piezoelectric vibrometer (1.4 µm) and compared with that of the interferometrically measured one (1.3 µm). The graph displays components of the displacement vector determined by quantitative evaluation of the interferogram along the line at markers on the central line 3, 6, 10, 15, 20, 25 (the x-axis is along the support, z is vertical; the carriage is near the marker 15). It can be seen that the displacement of the support in the y-direction around the marker 15 is slightly influenced by the carriage.

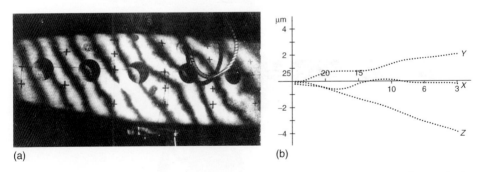

(a) (b)

Figure 18.3 The interferogram of the RF4-75/2000 radial drilling machine at dynamic load (a) and its evaluation (b).

(a) (b)

Figure 18.4 Investigation of the C-frame of LMC 250 CNC puncher: (a) the interferometer in front of the object and (b) the recorded interferogram.

Figure 18.4 shows the C-frame of LMC 250 CNC puncher and an illustrative interferogram at dynamic load. The dimension of the frame is 3×2 m; half of the surface was investigated in one laser shot. The fringes on the upper part of the interferogram are too dense to be observed. The number of recorded/evaluated interferograms in this application was $48 \times 4 = 192$.

As an example, the vibration behavior of automotive parts is illustrated in Figure 18.5.

The car under investigation was the type Watrburg-Tourist P 353 produced and prepared for the holographic investigation by the IFA-Kombinat Personalkraftwagen, Karl-Marx-Stadt, and the VEB Automobil Werke Eisenach (former DDR). An examination roller bed was provided by the Hungarian Transport Science Research Institute to simulate real dynamic loading conditions of the car. The vibration of the car was studied at different speeds of rotation of the engine (740–4200 rpm) and different simulated road conditions.

(a) (b) (c)

Figure 18.5 Investigation of the automotive parts: (a) the car under investigation, (b) the interferogram of the car body, and (c) the interferogram of the engine block of the car (c).

It may be concluded that HI has successfully been applied in machine tool and automotive industries in 70–90 years of the past century. Double-pulsed portable hologram interferometric measuring system such as HIM with computer-based evaluation of interferograms seemed to be a usable tool for 3D deformation investigation in factory environment. Nevertheless, there were meaningful difficulties in realizing the same. One of the inherent properties of analog holography is the lengthy wet chemical recording of the hologram, and industrial people were rightly averse to this. The high sensitivity of HI did not promote its widespread industrial applications. The measuring limit is relatively low, and so the measuring range of HI allows the inspection of moderate object deformations only. This is the case at shape measurement, too, where even a small surface gradient can result in local contour fringe density higher than the practical upper measuring limit. The goal is to cut out the dense fringe to make it resolvable, which is equivalent to extending the practical upper measuring limit of HI. The related procedure is named *fringe compensation*.

18.3.2
Difference Holographic Interferometry (DHI) – Technique for Comparison and Fringe Compensation

When a hologram storing more than one wave field is illuminated with coherent light, the reconstructed waves can interfere with one another. In 1965, the significance of this inherent property of holography was first and fully appreciated [1–6]. The common important peculiarity in all these applications, which should be emphasized, is the existence of correlation between the two interfering wave fronts, which is conditioned by the same microstructures of the comparable object surfaces. This unavoidable condition may be convincingly demonstrated by real-time interferometry: the stored wave front illuminates the very same object in its second state during the second exposure and interference fringes are observed. On the other hand, no interference fringes will be observed if the an object change would take place between the two exposures.

The conclusion states the well-known fact: by conventional holographic techniques, there is no possibility to compare two diffusely reflecting objects by purely optical way regarding their deformation, shape, or refractive index distribution changes, in the case of transparent objects, because of the different microstructures of the interfering wave fronts.

Two-exposure HI, when object under investigation is illuminated directly by laser light, produces interference pattern that displays the phase difference between the interfering waves. This implies that the phase difference between illuminating beams is zero.

On the other hand, if we would be able to create and store for later illumination two coherent wave fronts with a given phase difference, this stored phase difference may serve as a reference to which the actual phase difference can be compared. By this way, the direct optical comparison in HI will be solved.

18.3 Contributions to Industrial Applications by Analog Holography | 419

This is done by first recording a double-exposure interferogram of the first, named master object, and then the waves reconstructed from this hologram are used for illumination of the second, named test object. This holographic illumination of the test object is the crucial moment leading to realizing the direct optical comparison of the two objects. The interferogram obtained in such a way displays the difference between characteristics of the two objects (displacement, shape, or refractive index distribution change).

The applicability of holographic illumination was first demonstrated by Denby and his coworkers [21] in 1975 in ESPI. The authors emphasized the comparative feature of the technique and its ability to display the difference in shape of two nominally identical objects (turbine blades). The idea of holographic illumination was outlined for its possible application in HI by Neumann [22] in 1980 as a purely principled possibility. First experimental result of the author was published in 1985 [23]. In this chapter, particular stress was laid on fringe compensation property provided by comparative holography. The central role of holographic illumination and its deterministic role in comparative measurement by HI was demonstrated by a series of measurements performed by Füzessy and Gyímesi; the first paper was published in 1983 dealing with comparative deformation measurement of two nominally identical objects [24]. On the basis of the conception of Denby's group, the authors named the technique difference holographic interferometry (DHI). During the first 10 years, the authors developed different techniques/arrangements for realization of the holographic illumination, proved their effectiveness in the case of comparative deformation measurements, that of transparent object investigations, and applied DHI in two-refractive index contouring [25]. The wave optical theory of DHI has also been elaborated [26]. Problems and realization of comparative two-wavelength contouring are discussed in Ref. [27].

The DHI has enhanced the potentials of HI both in academic and industrial applications by its ability to compare two nominally identical objects by purely optical way and by providing a means for managing irresolvable fringe systems. The DHI also finds its place in digital version of HI, and so this combination is considered as a possible upgrading HI for application in macromechanical engineering industry.

The following short introduction to the principle of DHI highlights its power. Because of the easier terminology and unanimity, the technique will be introduced in terms of deformation inspection keeping in mind that all statements are in force both for contouring and transparent object investigations. The concept of holographic illumination states that the holograms belonging to the initial and final states of the master object are reconstructed by conjugate reference beams used at their recordings, and the object to be compared in its initial and final states is illuminated by reconstructed real images of the master object belonging to its initial and final states, respectively.

Owing to the deformation, there is a definite phase difference between the two reconstructed master wave fronts used for illumination, and, on the other hand, the deformation of the test object produces a given phase difference between the test object wave fronts. In the measurement, these two phase differences are compared, strictly speaking, they are simply subtracted. The possible deviation of these two phase differences results in an interference pattern that displays the deformation difference of the two nominally identical objects subjected to nominally identical loads. In the optimal case, the "interferogram" is fringe free. The DHI works even in the case of high master and test object deformations: in cases when the corresponding interference fringes are too dense to be resolved.

Master holograms can be recorded using single or two reference beams [25]. Applying the single reference beam technique, the test object in its two states is simultaneously illuminated by both holographic images of the master object. The technique is less noise sensitive, but the simultaneous holographic illumination causes disturbing background fringes, as usual. Recording master holograms by two reference beams of different directions, the holographic images are separately accessible and holographic illumination has great flexibility. The price that should be paid is that special steps should be taken to reach high fidelity of holographic illumination to project back master object waves with true phase difference. For that reason, an adjusting beam should be present in the arrangement.

According to what has been said above the difference interferogram, which displays the difference in the displacement fields of two nominally identical objects (master and test) subjected to nominally identical loads, is obtained in two steps: recording master holograms (Figure 18.6a) and making difference interferogram itself (Figure 18.6b).

Figure 18.6a contains the fundamental optical elements for recording master holograms. The master object O_M is illuminated from direction K_{IM} by I_M and observed from K_{OM} (holographic plate H_M). There are two reference beams R_1 and R_2 for recording master holograms. The holograms belonging to the initial and deformed states are recorded on the plate H_M. A spherical wave B is recorded at each exposure and serves as adjusting beam. The overlapping of the two reconstructed object images leads to an interference pattern with phase distribution $\Phi_M(\mathbf{r})$ given by

$$\Phi_M(\mathbf{r}) = (\mathbf{K}_{OM} - \mathbf{K}_{IM})\, \mathbf{L}_M = \mathbf{S}_M\, \mathbf{L}_M \tag{18.1}$$

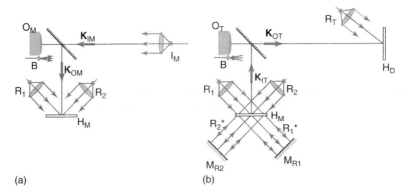

Figure 18.6 Optical arrangement for recording master holograms (a) and difference interferogram (b).

where L_M is the displacement vector, S_M is the sensitivity vector, and $|K_{OM}| = |K_{IM}| = 2\pi/\lambda$.

Figure 18.6b shows the main components of the experimental arrangement for recording the difference interference pattern displaying result of comparison directly (holographic plate H_D). The master object is replaced by the test object O_T and R_T is the reference beam for recording the difference interference pattern.

The mirrors M_R reverse the direction of the two reference plane waves R_1 and R_2 used at recording the master holograms, so the test object is illuminated from the direction $K_{IT}(-K_{OM})$ by reconstructed real images of the master object. A new holographic plate H_D is placed in the direction from where the master object was illuminated, so the observation direction for the test object is $K_{OT}(-K_{IM})$. In its ground state, the test object is illuminated by real image belonging to the initial state of the master object; the test object is subjected to a load, and in this state, it is illuminated by wave front belonging to the loaded state of the master object. The recorded double-exposure interferogram is the difference pattern itself that displays the result of comparison made by purely optical way.

The phase distribution due to the test object deformation is given by

$$\Phi_T(r) = (K_{OT} - K_{IT}) L_T = (K_{OM} - K_{IM}) L_T = S_T L_T \qquad (18.2)$$

with $K_{OT} = -K_{IM}$ and $K_{IT} = -K_{OM}$.

As it was stated earlier, the difference of phases (1) and (2) is recorded directly on the holographic plate H_D because of the holographic illumination. The comparison (subtraction) of the two phases results in

$$\Delta\Phi_T(r) = \Phi_M(r) - \Phi_T(r) = (K_{OM} - K_{IM})(L_M - L_T) \qquad (18.3)$$

This survey states that the output of the procedure is an interference pattern that displays the difference in the two displacement fields of the two objects. If there is no difference in the response of the two objects, the overlapping region is fringe free. Furthermore, Eq. (18.3) shows that the actual phase difference $\Delta\Phi(r)$ may be affected by unwanted changes of the sensitivity vector. It may change because of the difference in geometry of the arrangements used in comparative process and/or if the comparison is made not for corresponding points of object surfaces. The latter may take place as a consequence of inaccurate wave front reversal and/or inaccurate master test object replacement.

The accuracy of the wave front reversal (holographic illumination) is enhanced by adjusting beam B (Figure 18.7a,b). The beam B is a spherical wave at deformation inspection that comes from a point source. (At comparative two-wavelength contouring by DHI, the fidelity of holographic illumination is controlled in some more complicated way [27].) In this case, the unchanged point source is recorded twice, together with both states of the master objects using reference beams R_1 and R_2 (Figure 18.6a). The back-projected reference beams R_1^* and R_2^* (Figure 18.6b) reconstruct this point source in two copies that exactly coincide with each other and with their original position (with the position of the pinhole) at perfect wave front reversal. Thus, as first geometrical approach to the back projection alignment, this

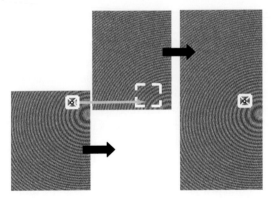

Figure 18.7 Steps of fitting image elements together (marker area in the first element, center-target area in the second element, and successful overlapping of the two elements).

coincidence is to be controlled and achieved with the accuracy provided just by the naked eye already.

After this, the observation of the interference of these two back-projected spherical waves takes place at the vicinity of the point images on a screen. The interferometrical approach means clearing out these fringe systems as much as possible by changing the back-projecting reference beams R_1^* and R_2^* properly. Usually, it is not possible to clear it completely but only to one fringe or to a fraction of a fringe. However, at this state of alignment, the wave front reversal may be considered as correct and the back-projected master fringes visible on the test object surface display the actual master object deformation with acceptable accuracy.

The inaccurate master/test object replacement influences the accuracy of the comparative measurement. The incorrect test object position does not destroy the operation of DHI – the only effect is the change of the shape of the difference fringe systems [28]. This shape change displays the fact that the actual wave phases are subtracted from each other – however, this subtraction takes place not between phases belonging to the corresponding surface points of the two objects but between phases belonging to displaced points. The corresponding phases are mismatched just with the positioning error of the test object.

If the fringe system of the master object is rare enough, then much larger positioning errors can be tolerated than in the case of denser master fringes.

It can be chosen as a rule of thumb that one-tenth of the spacing of the master fringe system is an accuracy (macroscopic and not interferometric) that is usually easy to comply with at the position change and which is sure not to influence the difference fringes visibly. But in any given case, an actual decision is to be drawn according to accuracy requirements and to the positioning possibilities available.

It may be concluded that DHI does contribute to the HI and promote its practical/industrial application providing a useful tool for direct optical comparison of two objects, as circumstances may require works as fringe compensation technique to make dense fringe system resolvable.

If the concept of DHI is applied in digital holography when digitally stored holograms can be transmitted via Internet, there is no need to have both samples to be located at the same place. The master laboratory can be at a given place of the earth, and the test object can be investigated at another site. The main feature of the technique that meets this challenge is generation, transmission, and reconstruction of the complete optical information belonging to an object (master), the displacement/shape of which is to be compared to that of a test object [29]. In this case, the faithful wave front reversal and the exact master-test object replacement need special attention.

18.3.3
Straightforward Way of Managing Dense Holographic Fringe Systems

It is a requirement at the development of the interferograms produced by HI measurements that the full fringe system has to be recorded and observed in one piece. Nowadays, interferograms are mostly read out electronically by CCD or CMOS cameras into a computer and they are evaluated by computers, as well. Therefore, the full image has to be imaged on the detector of a camera, which has limited size and resolution, whereas proper optics is used (or not). The image processing program of the computer has to be able to identify the fringes for evaluation. From the point of view of the CCD detector, or rather the data processing software, it is sensible to define the maximum fringe density or fringe number, which can be still resolved by the evaluation system. From hereon, this fringe density will be regarded the practical upper limit of HI and according to the above, and this is limited by the size and resolution of the CCD detector and by the capabilities of the evaluation program.

Therefore, according to the large sensitivity of HI, only those measuring regions can be investigated, which are relatively small and which are especially not too great with respect to the practical application requirements. In addition to this, an otherwise manageable fringe system can become dense beyond measure locally, in some places, because of a material fault causing rapidly changing deformation or because of some part of the object, which has significant surface gradient. This can be further aggravated the unwanted displacements occurring at loading simultaneously, the fringe system of which can make even denser the actual fringe system. Thus, the management of the dense fringe systems and the rarifying of these fringe systems in any forms mean the same as the extension of the practical upper limit of the HI measuring technique. Therefore, this has been a general goal since the beginnings of the applications of HI, as well.

One of the really effective methods of managing dense fringe systems is the so-called interferogram-puzzle technique in which the dense fringe system is recorded by scanning in parts, and the subinterferograms are put together in the memory of the computer to build up a large, complete hologram. This method does not require any accurate control of scanning because the correlation-based fittings of the subholograms recorded with overlapping parts make this completely unnecessary.

18.3.3.1 Upper Limit of the Evaluating Camera–Computer System

The full-field peculiarity of HI also requires full-field evaluation. The whole fringe system is to be observed and evaluated as a one – if absolute and not only relative values are to be evaluated. From this, it follows that the whole fringe system of the object investigated has to be captured by some detector connected to the computer – most usually by a CCD camera.

The whole fringe system as one is to be placed on the pixel matrix at the same time, and its densest parts, too, are to be resolved by the pixels. This is what limits the maximum number of fringes that can be dealt in one step.

In principle, 2 pixel fringe spacing could be enough for detecting a dark and bright fringe if only the maximum and minimum are to be found – but only if these extreme values just exactly meet one-one pixel. Of course, this cannot be really expected everywhere in a fringe system; therefore, larger fringe spacing is required. The minimum fringe spacing required on CCD depends on the structure of the fringe system – regarding its shape and the size of the disturbing speckles. In addition, it strongly depends on the software, which performs the task of identifying the fringes.

It is not enough to have the fringes displayed on the computer monitor in a visible form for the human eye – the computer "eye" has to see it as well. Special software is required for this – that is to recognize the fringes from the speckle noise and follow their shape. Although there are commercially available programs [30], the authors have preferred their own "fringe integrator program" (FRINT). This carries out fringe intensity integration along the fringes, and the same time it fits a cosine intensity distribution perpendicularly to them.

According to our experience, a minimum fringe spacing of $d_{min}^{CCD} = 3.6$ pixel, that is a maximum fringe density $\rho_{max}^{CCD} = 0.28$ fringe/pixel, can be chosen as a practical upper limit for the densest fringes resolvable by CCD – and simultaneously for the upper limit of HI as well, as it will be seen later. This has been found in the case of the quite general fringe system and with our FRINT fringe processing and identifying program. These limiting values may change somewhat for other fringe processing programs although not much. Nevertheless, it is important to note that they do not really depend on the type of the CCD, and most certainly, they do not depend on the pixel numbers of the CCD.

However, the maximum number of detectable fringes is exclusively determined by the pixel numbers of CCD and only by that – with $\rho_{max}^{CCD} = 0.28$ fringe/pixel regarded as a universal constant for all CCDs. If the smaller pixel number (usually the vertical N_v) is considered as the limiting one, then in an ideal case the maximum fringe number acceptable by a CCD in one step is simply

$$I_{max} \text{Fr}_{CCD}^{N_v} = \rho_{max}^{CCD} N_v = 0.28 \, N_v \tag{18.4}$$

In this context, the "ideal case" means that all the pixels of the CCD are covered by the fringe system and the fringe system density is maximum everywhere. This may happen perhaps never in practice for the following several reasons. The object image does not fill out exactly the CCD, and some extra space is left around it. There is a rim of the object, which does not move during the deformation – that is, only part of the object is active and covered by fringes. Finally, homogeneous equidistant fringe systems are very rare because they belong to pure rigid body rotations – without any deformation at the same time.

According to this, the actual maximum fringe numbers on a chosen object in a given deformation do not characterize correctly the capabilities of an interferometric technique – instead of that the ideal maximum fringe number is to be used for comparisons. The actual maximum fringe numbers are not correct for comparison even if always the same object and same type of deformation and the same imaging are used because of the changing inhomogeneity of the fringe systems with increasing deformations.

For a cheap bottom line CCD camera – what we have used – the pixel numbers are about 768 × 576 and thus $I_{max} Fr_{CCD}^{576} = 0.28 \times 576 \approx 160$ fringes. On the other end, for a very expensive top line CCD camera, pixel numbers are about 4000 × 4000 nowadays and thus $I_{max} Fr_{CCD}^{4000} = 0.28 \times 4000 \approx 1100$ fringes.

There is a factor of 7 between ideal maximum fringe numbers – but there is a factor of about 100 between their prices for the different cameras! Keeping this in mind, it is clear that any technique capable of extending the camera limitations to any extent is fully justified even if it remains below the extension factor of 7 – that is, it does not make the cheapest camera to overcome the power of the best. A much smaller achievement is worth any effort, too.

18.3.3.2 Measuring Range of Holographic Interferometry

HI has a basic theoretical upper limit that the displacement should be within the kernel of the focal distribution of the imaging optics or in other words: within the speckle size. Fortunately, this allows a very wide measuring range. If, for instance, $F = 36$ is used, this means that a maximum out-of-plane displacement of $L_{max}^{out} = 8\lambda F^2 \approx 5mm$ can still be measured. ($F = 36$ is not an extreme value yet, this belongs to an aperture of 5 mm in the focal plane of an objective of 180 mm focal length.) If the sensitivity is maximum at 514 nm argon ion laser light that is deformation is measured by 0.25 µm steps, this means that 20 000 fringes are possible in a monotonous increase of deformation, only. (And much more if the increase is not monotonous.) Compared to the 160 fringes of a bottom line CCD camera, it follows without doubt that the real limiting factor in HI comes from the camera.

This remains the final verdict when the two other limiting factors of HI are considered – which are related to each other. First, secondary requirement is that the interferometric fringes are to be localized on the object surface – to make possible to connect displacement values evaluated from the fringes to the object points. This requires a relatively small aperture where the actual size depends strongly on the magnitude of the in-plane component of the displacement. However, a small aperture increases the speckle size, and here comes the second secondary requirement: the fringes have to be seen between the speckles. Because the speckle size comes from the focal distribution of the imaging optics and it is usually its perpendicular section – this secondary requirement can be stated differently. Namely, the interferometric fringes have to be resolvable by the imaging optics – which is a natural requirement from the other side. Taking practical numbers, if $F = 36$ is used as before, which usually much more than usually required for the localization, the resolution limit is $r_{min} = 1.22\lambda F \approx 22\,\mu m$. This means that 45 fringe mm^{-1} fringe density is allowed on the object – resulting in fringe number of 1800 on a 40 mm object length. It easy to reach 10 000 fringes, if the object size is tripled to 120 mm and F number is decreased to its half, to $F = 18$, which is usually still an acceptable value for localization.

As the final conclusion, it can be stated that in HI the upper limit of the measuring range is imposed by the observing and evaluating camera–computer system and not by the optical technique itself.

18.3.3.3 PUZZLE Read-Out Extension Technique – for Speckled Interferograms

The maximum fringe number acceptable by a CCD in the ideal case is directly proportional to the vertical pixel number, and this is the only variable in the expression $I_{max}\mathrm{Fr}_{CCD}^{N_v} = \rho_{min}^{CCD} N_v = 0.28 N_v$ – thus, the increase of the pixel numbers can be the only extension possibility. However, this is not possible in reality on a CCD camera, therefore the CCD camera itself is to be virtually multiplied as a whole by using it multiple times in a single interferogram – that is, by scanning the interferogram with the CCD camera.

Scanning is a straightforward extension idea but it requires a very precise motion of the camera – and at least in two directions, all around within the image plane. If the CCD camera is used at its best, the accuracy of the motion has to be about the pixel size or even better and in addition on a scale up to 50–250 mm or even more.

Fortunately, computer-controlled image fitting can reduce the required accuracy of motion to about 1/10 of the CCD matrix size (instead of pixel size), that is, to about 0.5 mm. The elementary images have to be recorded with sufficient overlapping areas, and they are fitted together on these overlapping areas with the required pixel size accuracy. A proper computer program can do this if some markers can be found on the overlapping areas of the elementary images to be fitted together. This is like assembling a puzzle with very indistinctive details: fringes and most of all speckles and speckles.

In spite of their indistinctive characters, the speckle structure is unique enough to serve as a natural marker and even everywhere, that is, as a marker system on the whole holographic interferogram. Figure 18.7 illustrates the main steps of the fitting procedure. First, a marker area has to be chosen on the overlapping upper stripe of the first image element. Then, a center target area is to be chosen on the overlapping lower stripe of the second image element. After these choices, the center of the marker area is moved around in the center target area to find the best fit for the marker area pixels with the pixels of the second image element below them at their actual positions. The position of best fit can be selected out simply by continuously calculating the average of the absolute values of the intensity differences between the two image element pixels within the marker area – and finding the minimum.

To make use of this puzzle read-out (PRo) technique, a magnified image of the interferometric fringe system is to be produced – or at least, this image should be larger than the size of the CCD matrix. Usually, this is not a real magnification of the object yet but just a smaller demagnification of the object than the one which is used when a whole image is produced on the CCD matrix.

The optics used for the magnification may move together with the camera as a part of it, or the optics may be separated from the camera and it can produce a magnified aerial image in itself, which is scanned by the moving "naked" camera stripped from its optics. The second version has the advantage that smaller weight is to be moved around at scanning and the dimension of the movable part is smaller as well. However, this may also require large diameter optics if the object is large.

The advantage of the separated optics version is kept without its inherent disadvantage if the real image of the hologram is used to produce an 1 : 1 aerial image for scanning. The 1 : 1 imaging is already an about 10 times magnification for a 50 mm diameter object with respect to an image matching a CCD matrix. The PRo results in this chapter have been performed with this holographic real image version.

The maximum extension factor of PRo is approximately equal to the number of steps of capturing with CCD camera – if the magnification applied is really just minimum required to have still resolvable fringes on the CCD matrix and besides overlapping is negligibly small. Thus, the ideal fringe number becomes for mPRo-HI having "m" steps $I_{\max}\text{Fr}_{m\text{PRo-HI}}^{\sum N_v} = \rho_{\max}^{\text{CCD}} \sum N_v = 0.28 \sum N_v$, where $\sum N_v$ stands for mN_v to take into account the overlapping losses.

If the magnification applied is really just the minimum required to have still resolvable fringes on the CCD matrix, the maximum extension factor of PRo-HI is $I_{\max}\text{Fr}_{m\text{PRo-HI}}^{\sum N_v} = \rho_{\max}^{\text{CCD}} \sum N_v = 0.28 \sum N_v$. In the following experimental results, the magnification was larger (the holographic real image version was applied) and the deformation could not be pushed for the outmost extreme acceptable by the CCD matrix because of fatigue of the material of the model object.

To emphasize that these results belong to smaller maximum fringe density on the CCD than the maximum possible one, the "act" (actual) antecedent will be used in the fringe number expressions where needed Act $I_{\max}\text{Fr}_{m\text{PRo-HI}}^{\sum N_v} = \text{Act}\rho_{\max}^{\text{CCD}} \sum N_v$.

In the experiment, the object was a circular Al membrane of 52 mm diameter clamped rigidly on its perimeter. The diameter of the deformable part was 41 mm. The membrane was loaded at the center from behind by micrometer screw, causing a symmetric bulging and thus a circular fringe system. At the upper limit of HI at the camera used (768 × 576 pixels), the actual number of fringes occurred to be 2 × 49 half circles and the fringes were observable everywhere. At three times above the upper limit of HI, the fringes are already too dense to be captured by the camera as it is illustrated in Figure 18.8a, where six times computer-magnified part of the fringe system is shown – with clearly invisible parts in it.

However, as seen in Figure 18.8b, PRo even if not pushed to its outmost limit makes these fringes observable – with eight-step readout in one direction on the real holographic image. Figure 18.8c shows the result of Figure 18.8b completed with FRINT processing computer identification of the fringes.

The actual maximum fringe density on the CCD matrix during the steps has become only $\text{Act}\rho_{\max}^{\text{CCD}} = 0.09$ fringe/pixel – well below the possible maximum.

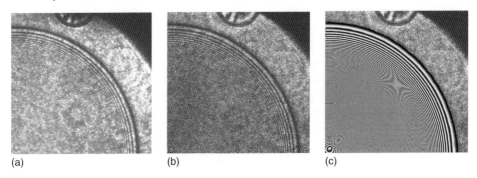

(a)   (b)   (c)

Figure 18.8 Three times above the upper limit of HI (actual fringe number 2×126 half circles): (a) one-step readout, (b) eight-step puzzle readout, and (c) eight-step puzzle readout and FRINT processing.

It gives an actual ideal fringe number on $\sum N_v = 4160$ pixel of the eight steps: $\mathrm{Act}I_{\max}\mathrm{Fr}_{\mathrm{8PRo-HI}}^{4160} = \mathrm{Act}\rho_{\max}^{\mathrm{CCD}}\sum N_v \approx 480$ fringes – which results an actual extension factor of 3 but it is still far from the possible maximum. By increasing the number of readout steps or by increasing the actual maximum density, further significant extension can be expected.

18.4
Contributions to Industrial Applications by Digital Holography

Analog holography uses high resolution (5000–10 000 line pairs mm^{-1}) and of almost arbitrary size (even up to the meter range) photo materials for recording the interference fringes of an aerial hologram and subsequently for recording a real hologram grating for the reference light wave to diffract on it at reconstruction.

Digital holography uses matrix detectors instead of photo plates, and the hologram recorded on the sensor is transferred to a computer. The diffraction of the light on the hologram grating is calculated from the diffraction integral, and the image is digitally "displayed" almost in real time. Digital holography, on the one hand, is exempt from the wet and time-consuming chemical processing, and, on the other hand, the result can be used in more ways and in more flexible manner than before.

In our days, CCD and CMOS cameras are used to record the holograms, although the size of the detector (5–40 mm) and their resolution (100–150 line pairs mm^{-1}) is still far away from these parameters of the photo materials.

The small size and the low resolution of the present-day CCD and CMOS chips restrict, on the one part, the resolution of the digital holographic image, and, on the other part, they restrict the angle of view of the holographic image, and with this the size of the object to be investigated. These disadvantages may be alleviated by virtually increasing the size and resolution of the detector target.

As for the limited size of the detector target, scanning the larger aerial hologram somehow with the small matrix detector is the most straightforward approach.

Although there are important other methods, as well, to build up some type of a synthetic aperture, this mostly means the relative motion of the camera and the object with respect to each other [31–36]. The synthetic aperture can be in the real hologram plane directly [33–36], somewhere before the real hologram plane in the Fourier plane of an intermediate optical system [31], or in a virtual calculation-transformed hologram plane [32]. In most cases [32, 35, 36], the camera is the only moving element. In [31], some part of the optical setup moves together with the camera, and in [32], separate fixed cameras are used. Besides, there is an opposite possibility of moving the object instead of the camera, as well [33]. Both on-axis [35, 36] and off-axis [31–34] optical arrangements are practiced. Phase-shifting procedure is applied in on-axis arrangement [35] or at moving reference case [31]. The stitching of the holograms is usually performed by correlation calculation on their overlapping side areas.

The authors present results in probably the simplest version of the synthetic aperture approach, which does not seem to have been directly exploited up to now. An optics-free small bottom line camera is moved directly in the aerial hologram plane in a conventional off-axis arrangement – up to 10 (or even 20) times to its original size. The increase in resolution is at a factor of 6 (or even 14).

18.4.1
Scanning and Magnifying at Hologram Readout

The virtual increase of the size of the CCD chip is to be achieved that way that a hologram being larger than the CCD chip size is scanned part by part by a moving camera – and the subimages are put together into the large hologram with the help of computer-aided image processing. This is nothing else, practically, than applying the PRo technique discussed previously. Nevertheless, there is a conceptual difference between fundamentals of the two techniques. In the case of measuring interferograms (macrointerference pattern), the fringe system is speckled with moderate visibility. The speckles are used as markers at correlation matching of subinterferograms. At recording holograms, however, the overlapping of object and reference waves produces a holographic interferogram (microinterference pattern) where the fringes are not speckled: the fringes are inside the speckles. Macro- and microinterference patterns are shown in Figure 18.9a,b, respectively.

The virtual increase of the resolution of the CCD chip is to be achieved by connecting the scanning with magnification [37].

Making use of these ideas, a robust digital holographic interferometer has been developed for deformation inspection and shape measurement (Figure 18.10). Both the resolution of the reconstructed image and field of view were increased roughly by 6. This increase of field of view makes possible to investigate object of 330 mm in diameter when the object–CCD distance was 1 m.

A spectacular demonstration of image resolution increase is shown in Figures 18.12 and 18.13. The object was an electronic panel with dimensions 120×120 mm. The object–CCD distance was 1 m in both cases. The CCD characteristics are pixel number 512×512, pixel size 7.4 μm, and sensor area 3.8×3.8 mm.

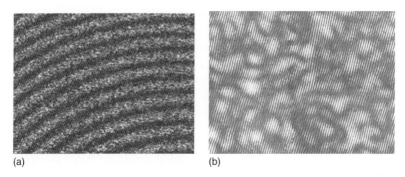

Figure 18.9 Macro- and microinterference patterns.

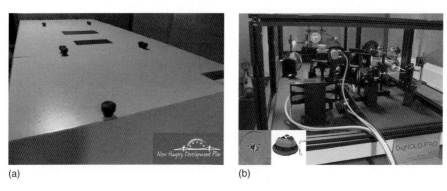

Figure 18.10 Digital holographic interferometer: (a) the robust interferometer with covering and (b) the robust interferometer during a measurement process.

Figure 18.11 shows reconstructed images of the hologram recorded with the naked CCD; the object was in this case the 32 × 32 mm part of the panel. In the case of Figure 18.11a, a single hologram was reconstructed, while Figure 18.11b shows the image reconstructed from a large hologram synthesized from 16 × 16 subholograms producing a virtual sensor area 16 × (3.8 × 3.8) mm.

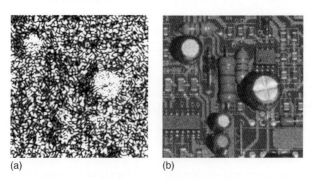

Figure 18.11 Increasing image resolution by scanning: holographic image reconstructed by 1 subhologram (a) and from 16×16 subholograms (b).

 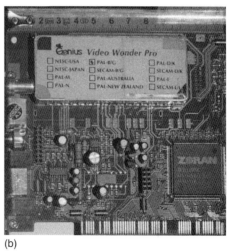

(a) (b)

Figure 18.12 Increasing field of view and image resolution by magnification and scanning: (a) pixel size is virtually decreased four times and (b) the sensor area is virtually increased 16×16 times with decreased pixel size.

The limited resolution of the camera does not allow recording hologram of the whole panel (120 × 120 mm) in the present geometry.

To increase the sensor resolution, the pixel size was optically decreased by 4× microscope objective resulting in virtual pixel size $7.4/4 = 1.85$ μm. The reconstructed image is shown in Figure 18.12a: the resolution is poor. However, increasing the sensor area by scanning at decreased pixel size and constructing large hologram from 16 × 16 subholograms, the resolution of the reconstructed image is acceptable (Figure 18.12b).

We believe that owing to the realized increase of the resolution and field of view, the potentials of digital holography in practical applications become comparable with those of its analog version.

18.4.2
Digital Holography for Residual Stress Measurement

Just in our days, after digital holography having grown up fully to classical holography, digital HI is already in the position to take full advantage of its four (3 + 1) unique capabilities – and it will most certainly do so in the near future. To summarize these properties:

1) fast and interferometrically same place (without changes of the hologram plates) sequential recordings of plenty of holograms if needed (with subsequent separate access to any two of them);
2) fast and convenient (numerical) reconstruction of the holograms;

3) extremely flexible manipulation of the (numerical) reconstruction with special regards to the evaluation process of the interferograms;
4) developer-friendly support at the development of new measurement methods: automatic (numerical) "archiving" of the results at any level with the constant possibility of restarting from anywhere if required.

These properties are totally missing in classical holography, and exactly these properties are needed in real industrial measurement situations.

Residual stress measurement with hole drilling can be one of the first spectacular examples for the long-awaited industrial breakthrough of HI – just because of these properties of digital HI. Nevertheless, already analog holography did the first step on this area [38, 39] and even with a portable device [40]. Besides this, other nonholographic optical methods, as well, have entered this field, mostly together with their digital versions. These are ESPI [41–43] including its digital version DSPI [15, 44, 45], shearography [18, 46], Moire interferometry [47–50], and digital image correlation [51, 52]. Digital HI solution was presented first by the authors just on this HoloMet 2010 conference (as will be mentioned and illustrated here below when treating the repositioning problems) and another already fully developed realization, too, has been recently reported [53] where the simultaneous measurements of out-of-plane and radial in-plane displacement fields are performed.

In the first, most direct approach, holography has to replace only the conventional strain gage method – first just by measuring the extension or contraction of the three gage stamps by HI. This means that the 2D deformation field has to be measured in the plane of the inspected object surface, and the displacement data of the end points of the gage stamps have to be evaluated and in the direction of the stamp orientations. Quite naturally, this can only be the first, the very direct step. HI can measure a complete 3D deformation field and proper finite element models can make better use of this than discrete and averaged out strain gage data can provide.

Nevertheless, the hole-drilling procedure is far from being natural in HI measurements between the two exposures. This type of loading can always be accompanied by unwanted rigid body motions even if the drilling takes place in the optical arrangement or it is performed out of it, after having removed the object from the optical arrangement. In the first case, the drilling push can deform or displace the holder of the object permanently and the vibration caused by the drilling can cause alterations in the whole optical arrangement everywhere. In the second case, the replacement of the object has to be performed with interferometric precision, which, as well, is not really possible to do, especially not with good repetition rate. Here comes the very important third property of digital HI, the extremely flexible manipulation of the (numerical) reconstruction with special regards to the evaluation process of the interferograms. With this, the unwanted fringes of the unwanted deformation accompanying the hole drilling can be removed perfectly in most cases. Because, in most cases, the unwanted fringes are oblique straight equidistant fringe systems, which are easy to identify on some reference surface supposed not to have any fringes on it – and can be subtracted from the actual valuable fringe system.

Even classic holography could do this with sandwich holography [54] where the two plates of the two holograms are fastened two each other with a gap between them; therefore, they act by shifting the images when they are rotated with respect to the common reference wave. Digital holography provides separate access to the two holograms and thus can apply even different reference waves at their reconstructions without any difficulty. Besides, plenty of other compensation possibilities can be performed directly, much more than the simple shifting of images. Such a compensation was already used in digital HI, although in a very special parallel case only, where positioning holographic illuminations was performed for comparative distant shape control (DISCO) [55].

Now the authors would like to show another application of the latter digital approach – for unwanted rigid body motion compensation, especially in the case of residual stress measurement with hole drilling (Figure 18.13). Here, the compensation takes place with respect to a reference surface attached to the object rigidly and which is supposed not to be influenced by stress relief at hole drilling. Therefore, unwanted repositioning fringes appear clearly and undisturbedly on the ring form reference area and can be compensated on the joint fringe system of the active central area.

Having solved the repositioning problem, the digital holographic method becomes really viable. The data are obtained everywhere on the surface at arbitrary distances from the rim of the hole, and there is no need to position the hole precisely. It can be applied in the case of curved or distributed surfaces (close to curves or corners) and requirements for surface quality are notably reduced. It provides prompt and unlimited repeatability (with redrilling of holes), possibility of monitoring changes of the stress field on the surface almost in real time, and so on. All of this seems to lay the foundation of a real industrial breakthrough in this area.

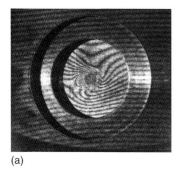

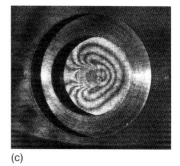

(a)   (b)   (c)

Figure 18.13 Compensation of unwanted rigid body motion in the case of residual stress measurement: (a) combined (distorted) fringe system, (b) fringe system of the repositioning error, and (c) undistorted deformation fringe system.

18.5
Conclusion and a Kind of Wish List

There is a long time need and expectancy to measure relatively large deformation and to compensate unwanted rigid body motions at the applications of HI – in macroscopic mechanical engineering industry. Digital holography can provide all of this already at its present state of development and most certainly will do it even better in the very near future. It can be foreseen that digital holography will end its "endless march" toward industrial applications in the following years. Nevertheless, some further developments could make this breakthrough into industry even faster and even more efficient. The wish-to-do-list should contain at least the following three main steps:

1) At compensation of unwanted rigid body motions, the different areas of compensation possibilities have to be investigated severally and more in details (compensation in the calculated image plane or in the 3D space, compensation in the back-calculated object plane, and compensation in the hologram plane – virtually, but physically).
2) At reconstruction of the digital holograms and holographic interferograms, problems connected to the larger size of the higher resolution holograms have to be investigated (resolution and distortion of images over the whole image area and localization of the interferometric fringes).
3) At the hardware side, camera motion mechanics and their compensation-oriented control have to be developed – even for 3D motion possibilities, as well.

Acknowledgments

The authors would like to thank to their former colleagues Béla Ráczkevi, György Molnár and to all the partners in the cooperating companies in the previous GVOP. 3.1.1.-2004-05-0403/3.0 project: Aladár Czitrovszky, Attila Tibor Nagy, Győző Molnárka, Abdelhakim Lotfi, Attila Nagy, István Harmati, and Dezső Szigethy. Special thank goes to Dezső Szigethy (Technoorg Linda Ltd. Co.) for supporting a current project in the same direction and to József Schmelzer for its programming contributions. The results of these projects have helped the authors of this chapter to come to the predictions above.

References

1. Aleksandrov, E.B. and Bonch-Bruevich, A.M. (1967) Investigation of surface strains by the hologram technique. *Sov. Phys. Tech. Phys.*, **12**, 258–265.
2. Powell, K.L. and Stetson, K.A. (1965) Interferometric vibration analysis by wave front reconstruction. *J. Opt. Soc. Am.*, **55** (12), 1593–1596.
3. Haines, K. and Hildebrand, B.P. (1967) Multiply-wavelength and multiply-source holography applied to contour

generation. *J. Opt. Soc. Am.*, **57**, 155–162.
4. Tsuruta, T., Shiotake, N., Tsujiuchi, J., and Matsuda, K. (1967) Holographic generation of contour map of diffusely reflecting surfaces by using immersion method. *Jap. J. Appl. Phys.*, **6**, 661–662.
5. Zaidel, A.N., Ostrovskaya, G.V., and Ostrovski, Y.I. (1969) Plasma diagnostics by holography (a review). *Sov. Phys. Tech. Phys.*, **13**, 1153–1164.
6. Vest, C.M. (1979) *Holographic Interferometry*, John Wiley & Sons, Ltd, Chichester.
7. Kreis, T. (1996) *Holographic Interferometry*, Akademie Verlag GmbH, Berlin.
8. Ostrovsksky, Y.I., Butusov, M.M., and Ostrovskaya, G.V. (1980) *Interferometry by Holography*, Springer-Verlag, Berlin, Heidelberg.
9. Murata, M. and Kuroda, M. (1983) Application of holographic intereferometry for practical vibration study. *SPIE Proc.*, **398**, 74–81.
10. Brown, G.M. and Wales, R.R. (1983) Vibration analysis of automotive structures using holographic interferometry. *SPIE Proc.*, **398**, 82–84.
11. Felske, A. (1986) Laser metrology for development and quality control of automobiles. *VDI Ber.*, **617**, 128–132.
12. Steinbichler, H. and Gehring, G. (1996) TV-holography and holographic interferometry: industrial applications. *Opt. Lasers Eng.*, **24**, 111–127.
13. Van der Auweraer, H., Steinbichler, H., Haberstok, C., Freymann, R., Storer, D., and Linet, V. (2001) Industrial Applications of Pulsed-Laser ESPI Vibration Analysis.
14. Krupka, R. and Ettemeyer, A. (2001) Brake vibration analysis with three-dimensional pulsed ESPI. *Exp. Tech.*, **25** (2), 38–41.
15. Viotti, M.R., Kapp, W.A., and Albertazzi A. (2010) A portable optical DSPI strain sensor with radial sensitivity using an axis-symmetrical DOE. *Proc. SPIE*, **7387**, Article Number: 73870B.
16. Krupka, R., Walz, T., and Ettemeyer, A. (2005) Industrial applications of shearography for inspection of aircraft components. *Proc. SPIE*, **5852**, 476–479.
17. Steinchen, W., Gan, Y., Kupfer, G., and Mackel, P. (2003) Non-destructive inspection and vibration analysis of disbonds in carbon fibre structures using laserdiode shearography. *Proc. SPIE*, **5144**, 885–893.
18. Hung, Y.Y. and Ho, H.P. (2005) Shearography: an optical measurement technique and applications. *Mater. Sci. Eng.*, **49**, 61–87.
19. Füzessy, Z., Ádám, A., Bogár, I., Gyímesi, F., and Szarvas, G. (1983) Hologram interferometric measuring system for industry. *SPIE Proc.*, **398**, 111–115.
20. Füzessy, Z. (1989) Application of holographic interferometry in machine tool industry. *VDI Ber.*, **761**, 29–35.
21. Denby, D., Quintanilla, G.E., and Butters, J.N. (1975) Contouring by electronic speckle pattern interferometry *Proceeding Strathclyde Conference*, Cambridge University Press, pp. 323–349.
22. Neumann, D.B. (1980) Comparative holography, tech. Digest, topical meeting on hologram interferometry and speckle metrology. *Opt. Soc. Am.*, Paper MB2-1.
23. Neumann, D.B. (1985) Comparative holography: a technique for eliminating background fringes in holographic interferometry. *Opt. Eng.*, **24** (4), 625–627.
24. Füzessy, Z. and Gyímesi, F. (1983) *SPIE Proc.*, **398**, 240–243.
25. Füzessy, Z. and Gyímesi, F. (1993) Difference holographic interferometry: technique for optical comparison. *Opt. Eng.*, **32** (10), 2548–2556.
26. Gyímesi, F. and Füzessy, Z. (1988) Difference holographic interferometry: theory. *J. Modern Opt.*, **35** (10), 1699–1716.
27. Gyímesi, F., Füzessy, Z., Borbély, V., and Ráczkevi, B. (2009) Analogue difference holographic interferometry for two-wavelength contouring. *Opt. Commun.*, **282** (2), 276–283.
28. Füzessy, Z., Ráczkevi, B., Borbély, V., and Gyímesi, F. (2003) Error factors of comparison in difference holographic interferometry. Proceedings of SPIE: Optical Measurement Systems for Industrial Inspection III, Munich, Germany, June 23-26, 2003, Vol. 5144, 115–123.

29. Osten, W., Baumbach, T., and Jüptner, W. (2002) Comparative digital holography. *Opt. Lett.*, **27** (20), 1764–1766.
30. BIAS (2002) BIAS Bremen Institute for Applied Beam Technology, Fringe Processor 3.0.
31. Le Clerc, F., Gross, M., and Collot, L. (2001) Synthetic-aperture experiment in the visible with on-axis digital heterodyne holography. *Opt. Lett.*, **26**, 1550–155211.
32. Massig, J. (2002) Digital off-axis holography with synthetic aperture. *Opt. Lett.*, **27**, 2179–2181.
33. Binet, R., Colineau, J., and Lehureau, J.C. (2002) Short-range synthetic aperture imaging at 633 nm by digtital holography. *Appl. Opt.*, **41**, 4775–4782.
34. Kreis, T., Adams, M., and Jüptner, W. (2002) Aperture synthesis in digital holography. *Proc. SPIE*, **4774**, 69–76.
35. Nakatsuji, T. and Matsushima, K. (2008) Free-viewpoint images captured using phase shifting synthetic aperture digital holography. *Appl. Opt.*, **47**, D136–D143.
36. Martínez-León, L. and Javidi, B. (2008) Synthetic aperture single-exposure on-axis digital holography. *Opt. Express*, **16**, 161–169.
37. Gyímesi, F., Füzessy, Z., Borbély, V., Ráczkevi, B., Molnár, G., Czitrovszky, A., Nagy, A.T., Molnárka, G., Lofti, A., Nagy, I., Harmati, I., and Szigethy, D. (2009) Half-magnitude extensions of resolution and field of view in digital holography by scanning and magnification. *Appl. Opt.*, **48** (31), 6026–6034.
38. Bass, J.D., Schmitt, D., and Ahrens, J. (1986) Holographic in situ stress measurements. *Geophys. J. R. Astr. Soc.*, **85** (1), 13–41.
39. Nelson, D.V. and McCrickerd, J. (1986) Residual-stress determination through combined use of holographic interferometry and blind-hole drilling. *Exp. Mech.*, **26** (4), 371–378.
40. Onishchenko, Y., Kniazkov, A., Shulz, J., and Salamo, G. (1999) The portable holographic interferometer for residual stress measurement and nondestructive testing (NDT) of the pipelines. *Proc. SPIE*, **3588**, 16–24.
41. Díaz, F.V., Kaufmann, G.H., and Möller, O. (2001) Residual stress determination using blind-hole drilling and digital speckle pattern interferometry with automated data processing. *Exp. Mech.*, **41** (4), 319–323.
42. Steinzig, M. and Ponslet, E. (2003) Residual stress measurement using the hole drilling method and laser speckle interferometry: Part I. *Exp. Tech.*, **27** (3), 43–46.
43. Schajer, G.S. and Steinzig, M. (2005) Full-field calculation of hole drilling residual stresses from electronic speckle pattern interferometry data. *Exp. Mech.*, **45** (6), 526–532.
44. Viotti, M.R., Albertazzi, A.G., and Kapp, M. (2008) Experimental comparison between a portable DSPI device with diffractive optical element and a hole drilling strain gage combined system. *Opt. Lasers Eng.*, **46** (11), 835–841.
45. Albertazzi, A., Viotti, M., and Kapp, M. (2008) A radial in-plane DSPI interferometer using diffractive optics for residual stress measurement. *Proc. SPIE*, **7155**, 715525-1–71552510.
46. Hung, M.Y.Y., Shang, H.M., and Yang, L. (2003) Unified approach for holography and shearography in surface deformation measurement and nondestructive testing. *Opt. Eng.*, **42**, 1197–1207.
47. McDonach, A., McKelvie, J., MacKenzie, P., and Walker, C.A. (1983) Improved moire interferometry and applications in fracture mechanics, residual stress and damaged composites. *Exp. Tech.*, **7** (6), 20–24.
48. Min, Y., Hong, M., Xi, Z., and Lu, J. (2006) Determination of residual stress by use of phase shifting moire interferometry and hole-drilling methods. *Opt. Lasers Eng.*, **44** (1), 68–79.
49. Cardenas-Garcia, J.F. and Preidikman, S. (2006) Solution of the moire hole drilling method using a finite-element-based approach. *Int. J. Solids Struct.*, **43** (22–23), 6751–6766.
50. Chen, J., Peng, Y., and Zhao, S. (2009) Comparison between grating rosette and strain gage rosette in hole-drilling combined systems. *Opt. Lasers Eng.*, **47**, 935–940.

51. McGinnis, M.J., Pessiki, S., and Turker, H. (2005) Application of three-dimensional digital image correlation to the core-drilling method. *Exp. Mech.*, **45** (4), 359–367.
52. Nelson, D.V., Makino, A., and Schmidt, T. (2006) Residual stress determination using hole drilling and 3D image correlation. *Exp. Mech.*, **46** (1), 31–38.
53. Viotti, M.R., Kohler, C., and Albertazzi, A. (2011) Simultaneous out-of-plane and in-plane displacements measurement by using digital holography around a hole or indentation. *Proc. SPIE*, **8082**, 80820E.
54. Abramson, N. (1974) Sandwich holo-gram interferometry: a new dimension in holographic comparison. *Appl. Opt.*, **13**, 2019–2022.
55. Baumbach, T., Osten, W., von Kopylow, C., and Jüptner, W. (2006) Remote metrology by comparative digital holog-raphy. *Appl. Opt.*, **45** (5), 925–934.

Color Plates

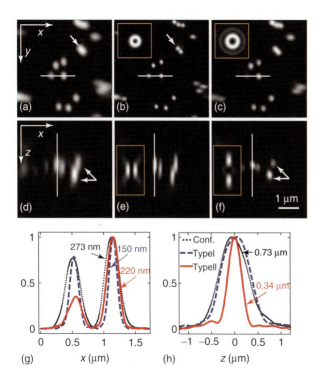

Figure 1.11 (a,b,c) Lateral and (d,e,f) axial fluorescence images of 200 nm beads with (a,d) confocal acquisition, (b,e) acquisition with a doughnut STED beam (type I), and (c,f) acquisition with a "bottle" STED beam (type II). (g,h) Normalized intensity line profiles of lateral and axial images, respectively, with specified FWHM. Insets show the corresponding PSFs. Note the differences in lateral and axial resolution between two STED imaging modes. Source: With permission from [71]. (This figure also appears on page 17.)

440 | Color Plates

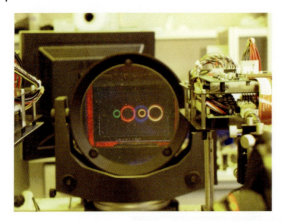

Figure 1.18 Photo of the display. Source: With permission from [88]. (This figure also appears on page 21.)

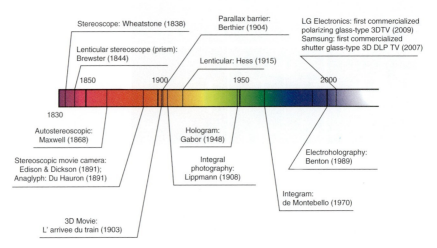

Figure 2.1 Progress of technical developments for 3D display and imaging. (This figure also appears on page 32.)

Color Plates | 441

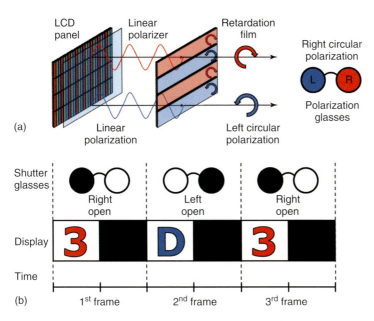

Figure 2.2 The principle of representative stereoscopic displays, using (a) polarizing glasses and (b) LC shutter glasses. (This figure also appears on page 34.)

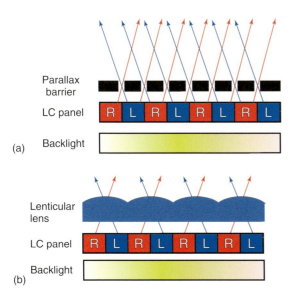

Figure 2.4 The principle and structure of (a) the parallax barrier method and (b) the lenticular lens method. (This figure also appears on page 36.)

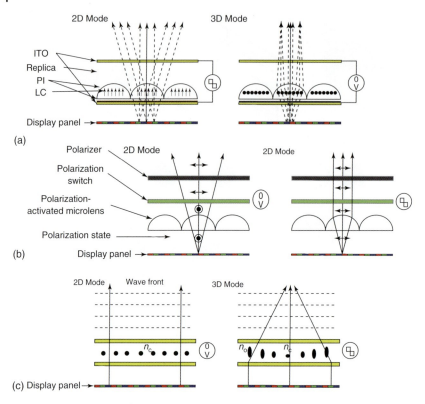

Figure 2.5 Three representative methods for achieving LC active lenticular lenses: (a) a surface relief method, (b) a polarization-activated lens method, and (c) a patterned electrode method. (This figure also appears on page 38.)

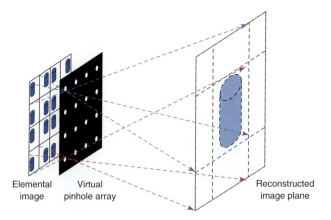

Figure 2.9 Concept of computational integral imaging reconstruction. (This figure also appears on page 46.)

Figure 2.10 Demonstrations of *Microsoft*'s Kinect for the Xbox 360. (This figure also appears on page 46.)

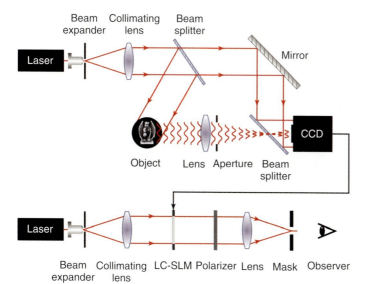

Figure 3.1 The optical scheme of electroholography including capture, transmission, and display based on a single CCD and transmissive LCSLM. (This figure also appears on page 60.)

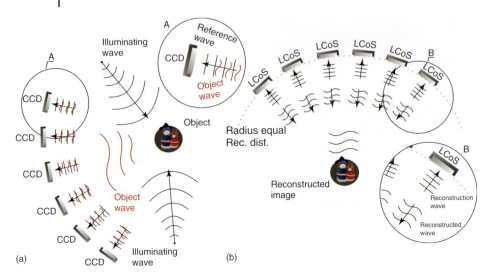

Figure 3.2 The general schemes of wide viewing angle capture and display systems based on (a) multiple CCDs and (b) multiple LCoS SLMs in circular configuration. (This figure also appears on page 61.)

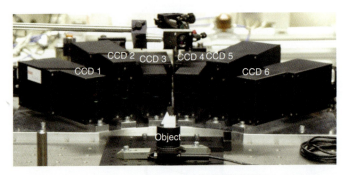

Figure 3.11 Photo of the setup used for capturing digital holographic videos – front view. (Source: Courtesy of BIAS [22].) (This figure also appears on page 73.)

Color Plates | 445

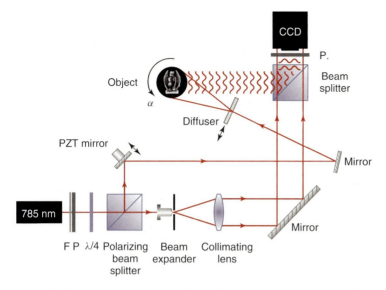

Figure 3.13 The holographic system for 360° capture of static 3D objects; P – polarizer, F – neutral density filter. (This figure also appears on page 76.)

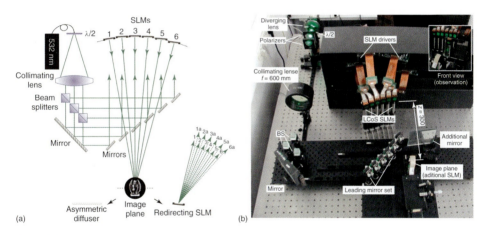

Figure 3.18 The general scheme of a holographic display (a) and the photo (b) of the display for "naked eye" observation mode. At the scheme there are two additional modules: asymmetric diffuser used for observation of images at large reconstruction distances (large FoV), redirecting SLM module for "naked eye" observation of close reconstruction (small FoV). (This figure also appears on page 82.)

446 | *Color Plates*

Figure 3.19 3D model of scene composed of chairs and the three views of holographic reconstruction. (This figure also appears on page 82.)

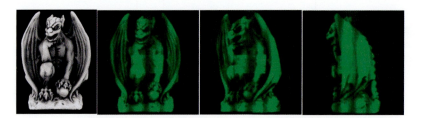

Figure 3.20 3D model of a Gargoyle statue and its reconstructed views (composed of six waves) as seen from three different perspectives. Reconstruction at an asymmetrical diffuser. (This figure also appears on page 83.)

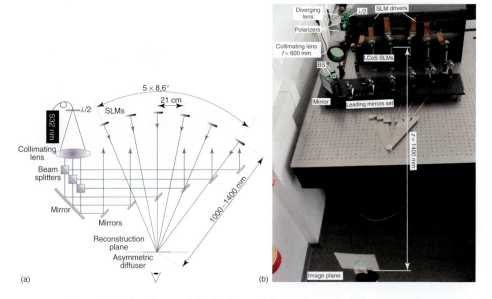

Figure 3.21 The scheme of the display used for displaying the holographic video captured by the multi-CCD holographic capture system shown in Figure 3.11. (This figure also appears on page 85.)

Color Plates | 447

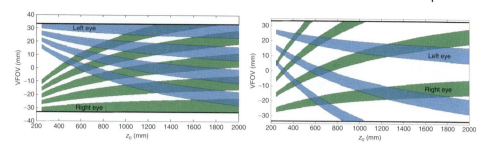

Figure 3.22 Representation of VFOV as a function of observation distance for reconstruction distance 1000 mm and on-axis binocular observation ($d_b = 65$ mm) for six SLM circular displays with (a) FF = 0.6 and (b) FF = 0.3. (This figure also appears on page 85.)

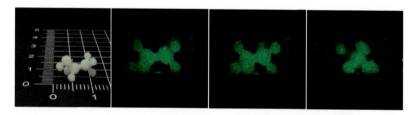

Figure 3.24 Printed model of ethanol molecule with 3D printer and views of optical reconstruction of digital holograms captured for this model in the system presented in Figure 3.17. (This figure also appears on page 89.)

Figure 3.25 Printed model of chairs scene with 3D printer and views of optical reconstruction of digital holograms captured for this model in the system presented in Figure 3.17. (This figure also appears on page 89.)

448 *Color Plates*

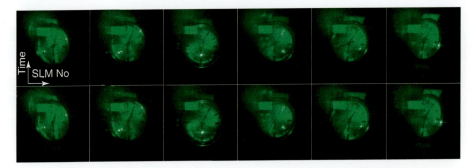

Figure 3.26 The exemplary reconstructions selected from a holographic video of the running watch. The photos in a row represent the images reconstructed by the sequential SLMs (1–6); the columns show two different states of the watch in time. (This figure also appears on page 90.)

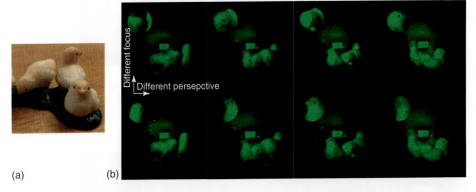

(a) (b)

Figure 3.27 The scene with three chicks at a table captured during rotation of the table: (a) the photo of the scene and (b) exemplary reconstructions selected from a holographic video of the scene. The photos represent reconstruction from the same SLM but captured (in a row) for different perspectives. In the columns, the photos are taken for two different focuses in order to show the depth of the scene. (This figure also appears on page 90.)

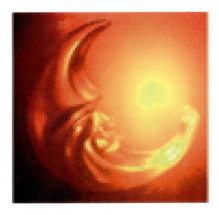

Figure 4.1 Art hologram, author Kalin Jivkov. (This figure also appears on page 111.)

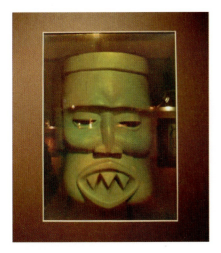

Figure 4.2 African mask, wood carving, 20th century A.D. (This figure also appears on page 112.)

Figure 4.3 Thracian mask, bronze, fourth to third century B.C. (This figure also appears on page 112.)

Figure 4.7 Three frames of the reconstructed 3D moving object from i-Lumogram; the total number of views is 640. (Source: With kind permission of Geola UAB, Vilnius, Lithuania.). (This figure also appears on page 116.)

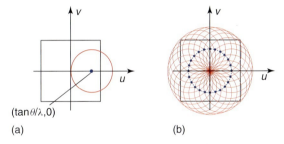

Figure 5.4 Extractive area on the $u - v$ plane from (a) one projection image and (b) a series of projection images. (This figure also appears on page 126.)

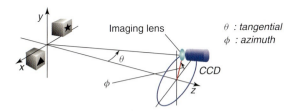

Figure 5.5 A recording optical system. (This figure also appears on page 126.)

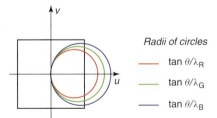

Figure 5.6 Adjustment of magnifications due to wavelength. (This figure also appears on page 127.)

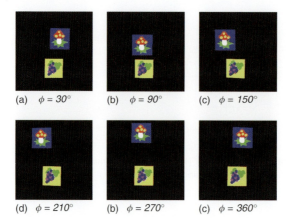

Figure 5.7 Color projection images at $\theta = 15°$. (This figure also appears on page 129.)

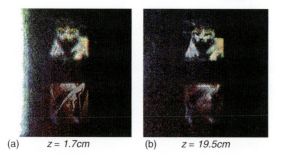

Figure 5.9 Experimental results of optical reconstruction. (This figure also appears on page 131.)

Color Plates | 453

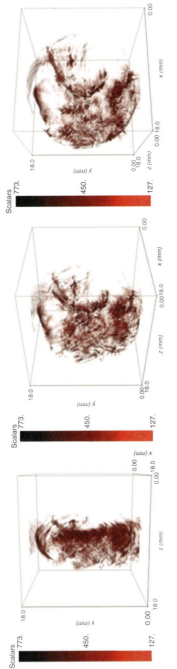

Figure 5.11 Reconstructed cancer images with different view angles. (This figure also appears on page 132.)

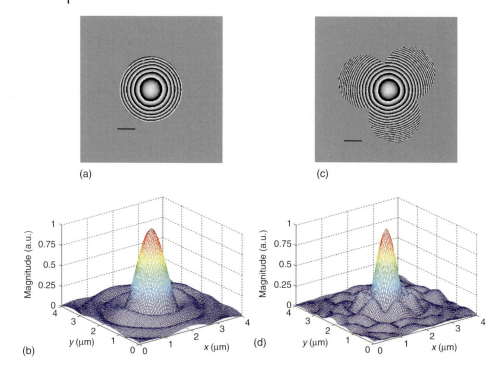

Figure 6.5 (a) Wrapped phase of the on-axis hologram of a 0.5 μm diameter pinhole. The scale bar is 10 μm. The phase distribution has a radius ~18 μm, a Fresnel number ~12, and a radius of curvature ~50 μm. (b) Amplitude of the reconstruction of the 0.5 μm pinhole using the online hologram. FWHM ~1.0 μm. (c) Wrapped phase of three off-axis holograms of the 0.5 μm pinhole illustrating the idea of pupil synthesis. The scale bar is 10 μm. (d) Amplitude of the reconstruction of the 0.5 μm pinhole using the composite off-axis holograms. FWHM ~0.7 μm. (This figure also appears on page 146.)

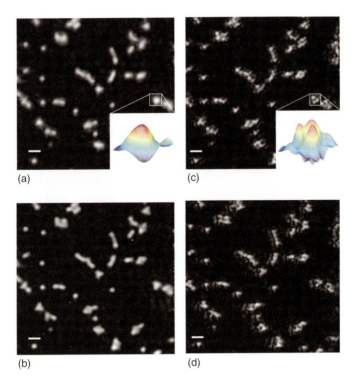

Figure 6.6 (a) Reconstruction of the on-axis hologram of a collection of ∼1.0 μm fluorescent beads at the "best focus" distance of 47.5 μm from the focal plane of the objective. The scale bar is 5 μm. Bead clusters are just barely resolved. (b) Same reconstruction at a focus distance of 49 μm. The two planes are within the Rayleigh range of the on-axis scanning FZP. (c) "Best focus" at 47.5 μm from the focal plane of the objective, where coherent sum of the complex amplitudes of the reconstructions of three off-axis holograms was recorded with off sets 120° apart. (d) Same reconstruction at a focus distance of 49 μm. The distance between the two planes is close to the Rayleigh range of the synthesized FZP, and different beads clusters are focused in different planes. (This figure also appears on page 147.)

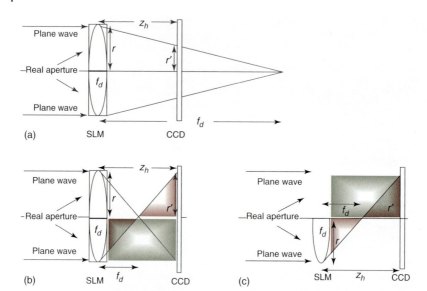

Figure 6.8 Possible configurations of recording holograms in the case of point-like object: (a) for FINCH where $f_d > z_h$. In this configuration, a hologram can be recorded, but, as indicated in the text, this hologram is suboptimal. (b) For FINCH where $f_d < z_h$. In this configuration, a hologram cannot be recorded because there is no interference between the plane and the spherical waves arriving from the same part of the SLM. (c) For T-SAFE, where $f_d < z_h$. In this configuration, the recorded hologram is optimal. The red and green areas indicate the spherical and plane waves, respectively. The rectangles in (a) and (b) symbolize the diffractive element of constant phase, where the lens symbol in all of the figures stands for the quadratic phase element, both the constant phase and quadratic phase elements are displayed on the same SLM. (This figure also appears on page 153.)

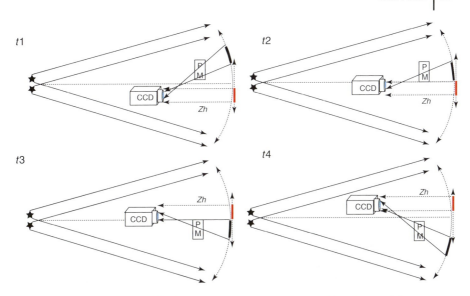

Figure 6.9 Proposed design of T-SAFE, which is based on spherical (black line) and flat (red line) mirrors rather than on SLMs. Four interfering steps, needed to obtain the synthetic aperture hologram, are shown. PM, phase modulator. (This figure also appears on page 154.)

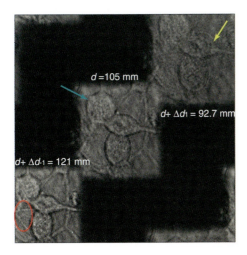

Figure 7.9 Amplitude reconstruction of a "cell" hologram at a distance $d = 105$ mm at which the cell indicated by the blue arrow is in focus. The $+1$ order correspond to a distance $d = 92.7$ mm at which the cell indicated by the yellow arrow is in good focus, while the -1 order correspond to a depth of $d = 121$ mm where the filaments are well visible, highlighted by the red ellipse. (This figure also appears on page 175.)

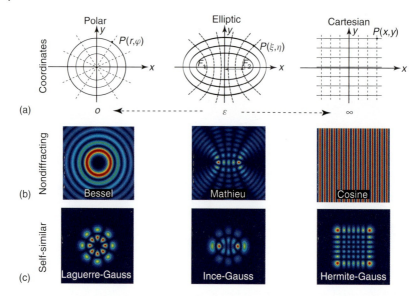

Figure 8.10 Nondiffracting and self-similar beams in polar coordinates (a), Cartesian coordinates (c), and elliptic coordinates (b), which include polar and Cartesian coordinates with an appropriately chosen ellipticity parameter ϵ. (This figure also appears on page 194.)

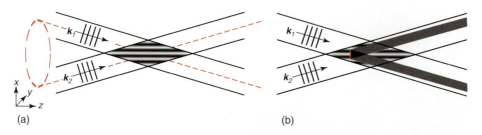

Figure 8.11 Nondiffracting cosine lattice. (a) Interference of two plane waves. (b) Self-healing of nondiffracting beams after disturbance of a small obstacle. (This figure also appears on page 194.)

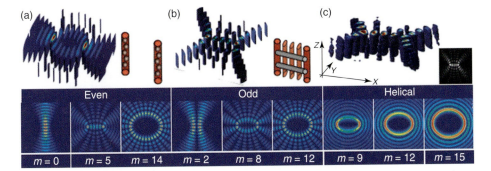

Figure 8.13 Nondiffracting Mathieu beams for three-dimensional particle manipulation. Cartoon and three-dimensional intensity distribution of (a) even Mathieu beam of seventh order enabling organization of microspheres and (b) even Mathieu beam of fourth order enabling orientation and organization of elongated particles. (c) Experimental intensity distributions of an even Mathieu beam of fourth order in the focal plane of the optical tweezers setup (left: three-dimensional measurement and right: transverse cut). Bottom row: selection of intensity distributions of even, odd, and helical Mathieu beams of different orders. (This figure also appears on page 197.)

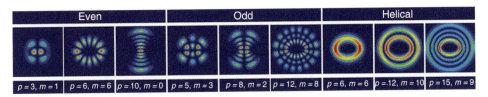

Figure 8.15 Selection of intensity distributions of even, odd, and helical self-similar Ince-Gaussian beams of different orders. (This figure also appears on page 199.)

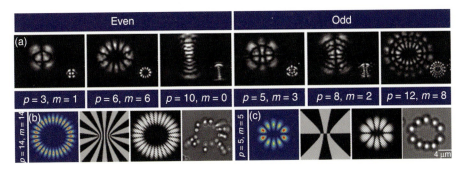

Figure 8.16 Experimental images of self-similar Ince-Gaussian beams. (a) Simultaneous imaging of both a conjugate and a Fourier plane of the trapping plane; the scaling between both planes can be seen clearly. (b,c) Intensity, phase, hologram, and particle configurations. In addition, in (b) the phenomenon of optical binding can be observed. (This figure also appears on page 199.)

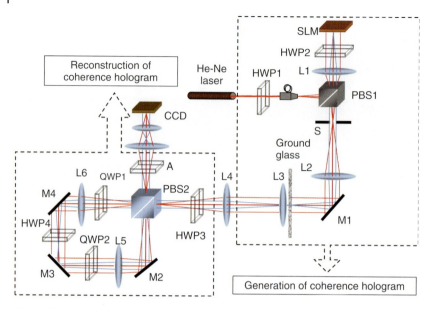

Figure 11.6 Experimental setup for generation and reconstruction of a Leith-type coherence hologram. (This figure also appears on page 244.)

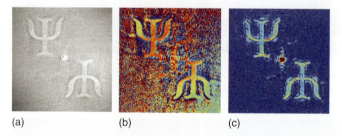

Figure 11.7 Reconstructed images of Leith-type coherence holography. (a) Raw intensity image resulted from shearing interference, (b) phase image, and (c) contrast image jointly representing the complex coherence function. (This figure also appears on page 246.)

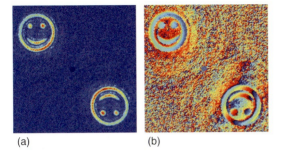

(a) (b)

Figure 11.10 Reconstructed images show (a) the fringe contrast and (b) the fringe phase, which jointly represent the complex coherence function. (This figure also appears on page 247.)

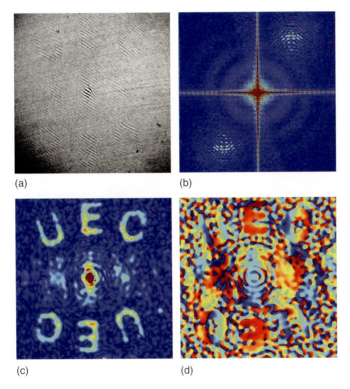

(a) (b)

(c) (d)

Figure 11.12 Reconstructed images. (a) Raw intensity image resulted from shearing interference in real time. Note that the region of high-contrast fringes represents the letters U, E, and C; (b) the Fourier spectrum of the interference image; and (c,d) are the contrast image and the phase image, respectively, which jointly represent the complex coherence function. (This figure also appears on page 249.)

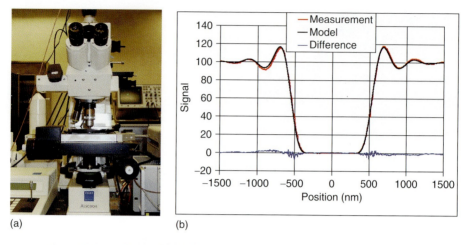

Figure 12.6 (a) Photo of the UV transmission microscope used at PTB for quantitative optical microscopy, primarily on photomasks, and (b) typical transmission intensity profile. (This figure also appears on page 264.)

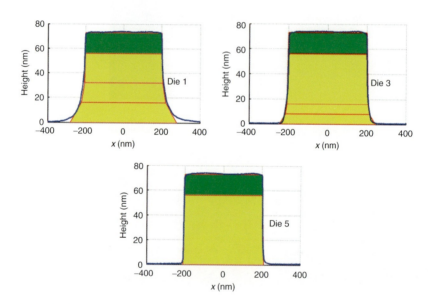

Figure 12.14 Approximation of the profiles used for the modeling of the UV microscopic images using a stack of four trapezoids. (This figure also appears on page 272.)

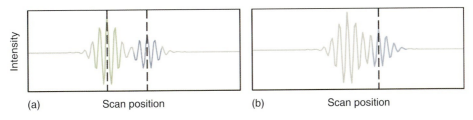

Figure 13.5 (a) Template signal matching for thick-film thickness measurement in CSI: the two best match signal positions provide an estimate of the optical thickness. (b) Template signal matching for topography measurement in the presence of a transparent film using a "leading-edge" template. (This figure also appears on page 293.)

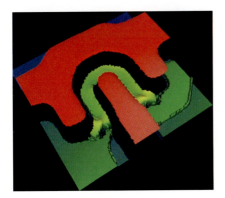

Figure 13.7 Inverse model film thickness measurement in CSI: map of photoresist thickness near a transistor gate on a TFT LCD panel. The horseshoe-shaped trench is nominally 400 nm thick and 5 μm wide at the bottom. (This figure also appears on page 294.)

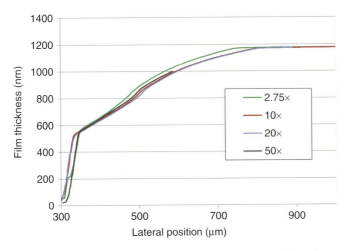

Figure 13.8 Inverse model film thickness measurement in CSI: thickness profiles of a silicon dioxide film on silicon measured at similar locations with Michelson and Mirau interference microscope objectives. (This figure also appears on page 294.)

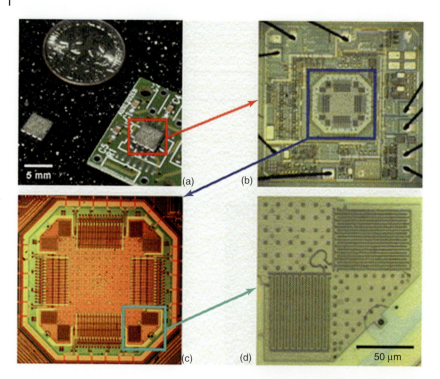

Figure 14.9 Dual-axes microaccelerometer: (a) overall view of a package, (b) the sensor – highlighted by the square in the center – and electronics integrated on a single chip, (c) the sensor consisting of a proof mass suspended by four pairs of folded springs – the square in the lower right corner highlights one of the spring pairs, and (d) view of the spring pair suspending the lower right corner of the proof mass. (This figure also appears on page 316.)

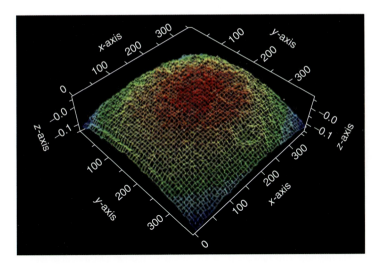

Figure 14.12 The out-of-plane 212 nm deformation component of the left shuttle based on the fringe patterns of Figure 14.11. (This figure also appears on page 318.)

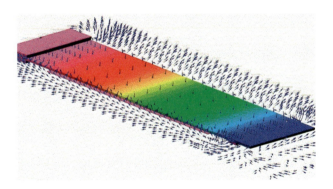

Figure 14.14 Computational multiphysics modeling of an RF MEMS switch closure at atmospheric conditions: 3D representation of air damping. (This figure also appears on page 321.)

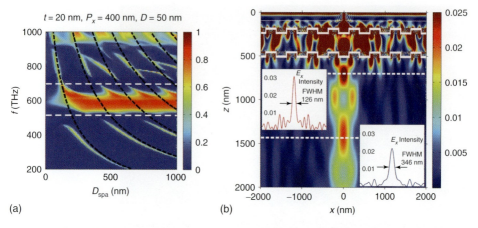

Figure 15.6 RCWA simulation of the transmission of a double meander structure as a function of the frequency f and the distance between the single structures D_{spa}. (a) The dashed black lines represent the predicted FP modes, while the dashed white lines indicate the SRSPP/LRSPP frequencies determined by dispersion diagram analysis (not shown). Ranging up to more than a micrometer, the passband is almost independent of D_{spa}. (b) Electric field intensity behind a slit with $w = 100$ nm and a double meander stack with $P_x = 400$ nm, $t = 20$ nm, $D = 50$ nm, and $D_{spa} = 200$ nm. The inset on the upper left side shows the electric field intensity along the x-axis at the focus plane. The inset on the lower right shows the electric field intensity at a distance of about two wavelengths ($\lambda = 500$ nm) behind the structure. (This figure also appears on page 343.)

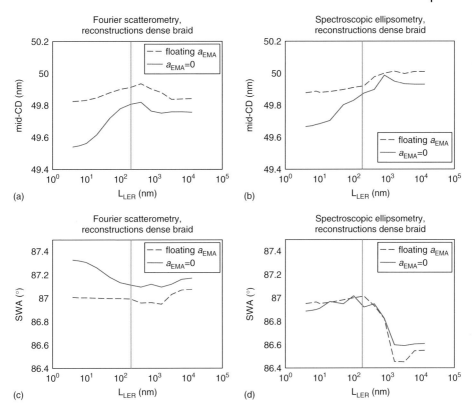

Figure 15.9 Reconstructions of CD and SWA using Fourier scatterometry and spectroscopic ellipsometry for vanishing and floating thickness of the EMA layer a_{EMA}. (a) Reconstructed CD (Fourier scatterometry), (b) reconstructed CD (spectroscopic ellipsometry), (c) reconstructed SWA (Fourier scatterometry), and (d) reconstructed SWA (spectroscopic ellipsometry). (This figure also appears on page 347.)

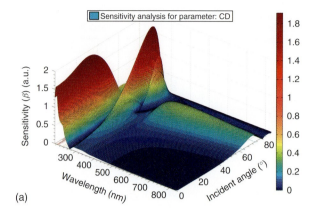

(a)

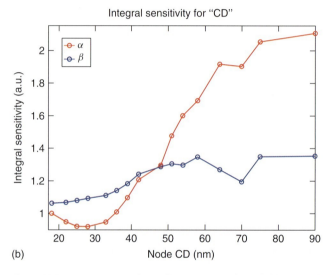

(b)

Figure 15.11 Sensitivity analysis of scatterometry tools. (a) Sensitivity toward the measurand β for a dense resist line grating (CD 48 nm) and the parameter CD dependent on the wavelength and the incident angle and (b) sensitivity trend (parameter CD, fixed incident angle) for STI line structures at different technology nodes. (This figure also appears on page 350.)

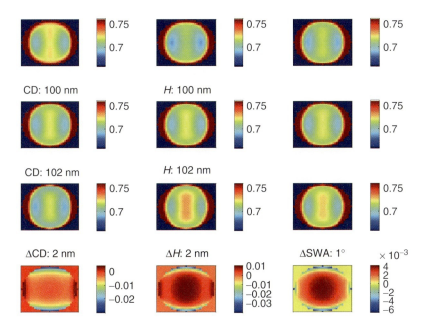

Figure 15.15 Simulated Fourier scatterometry pupil images for a photoresist line grating on silicon illuminated with 410 nm. The CD, height, and SWA are varied, and in the last row, the intensity difference is plotted to see the sensitivity toward that parameter. (This figure also appears on page 353.)

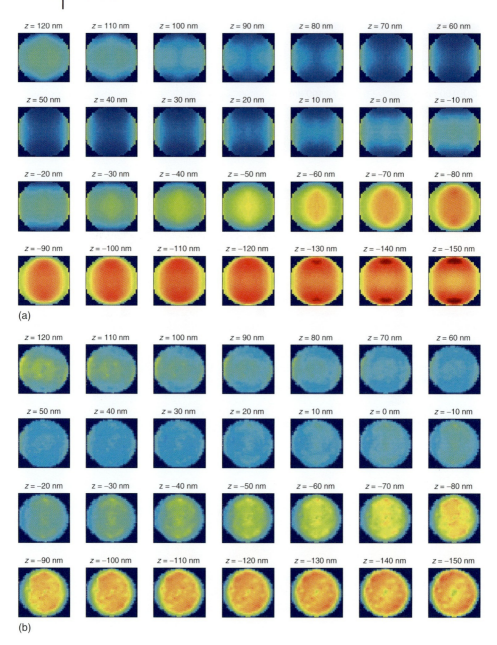

Figure 15.16 Complete scan of the reference mirror for an e-beam structured resist silicon grating of CD 200 and 400 nm pitch, with a height of 70 nm measured with white-light Fourier scatterometry: (a) measured results and (b) simulation results with the same parameters. (This figure also appears on page 356.)

Color Plates | 471

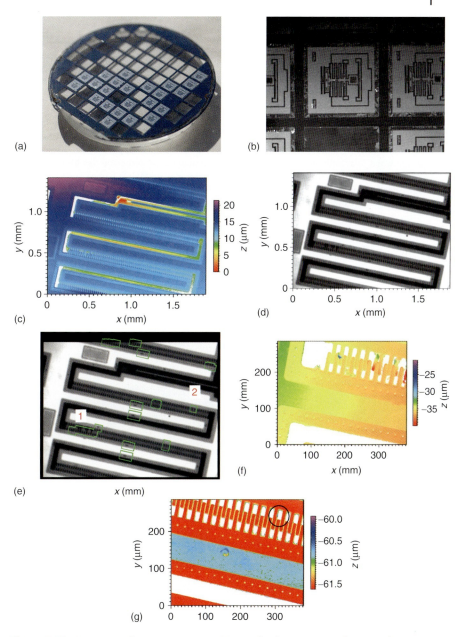

Figure 15.19 Inspection of a 6 in. wafer with in-plane microcalibration devices. (a) Photo of the wafer under test, (b) first scale measurement of MEMS structures made with a video microscope, (c) topography of the comb drive taken in the second scale with a 10× confocal microscope, (d) maximal intensity image of 10× confocal measurement, and (e) intensity image with highlighted region for possible defects, (f,g) third scale, high-resolution measurement with the 50× confocal microscope. (This figure also appears on page 361.)

  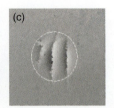

Figure 17.9 (a) Photograph of a wood inlay with a size of 15 cm × 15 cm. (b) Image of the wood inlay with shear of 2.5 mm in x- and y-direction. (c) Unwrapped phase image after filtering. Three cracks are identified, the encircled area indicates location and size of the delamination. (Source: Measurement at BIAS, sample supplied by M. Stefanaggi, Laboratorie de Recherche des Monuments Historique (LRMH), Champy sur Marne, France; see Ref. [49].) (This figure also appears on page 403.)

(a)

(b)

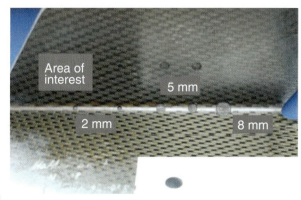

(c)

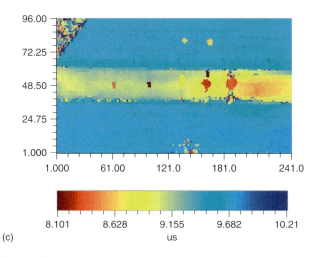

Figure 17.12 (a) Photograph of a CFRP component, (b) rear side with inserted flaws, and (c) C-scan, for explanation, see text. (Source: Taken from Ref. [16], first published in Ref. [64].) (This figure also appears on page 407.)

Index

a

analytical tools 307
ACES methodology 306–308
acousto-optics modulator (AOM) 43
active measurement 335–336
active stabilization 378–379, 390
active tiling system 59
adjusting beam 420–421
aliasing effect 70
aliasing images 64, 70
amplitude modulation 6, 9
analog drive scheme 6
analogy 170, 249, 251
analysis complex-valued 10–11, 153, 156, 164, 191
angular perspective 76
angular spectrum 43, 148, 168, 250, 275, 332, 336, 340
a-priori knowledge 266, 288, 293, 306, 331, 335–336
area related resolution 333, 337–338
assistance system 335, 338–339, 355
Association of University Technology Managers (AUTM) 222
asymmetrical diffuser 76, 81, 88
atmosphere 369–370, 373–374
atmosphere isolation 377–378
autostereoscopic displays 31, 33, 36–40, 91, 97

b

back projected reference beam 421
background illumination 100, 374–375
bacterial cells 189
bandlimited 333
bandwidth spatio-temporal 58, 183
Bayh-Dole Act 1980 217
beam, 10, 16, 40, 43–44, 58, 60, 67, 70, 76

beam hardening 225, 233–236
Bessel beams 195
bit depth 5, 7, 62
bottom-up strategy 335
bumps 360, 394
business 207–208, 210–212, 214–217, 219, 221–222

c

calibration devices 357–359
calibration procedure 88, 260
cantilever 314, 316–317
cantilever microcontact 319–322
carrier frequency 66–67, 248
cavities 191, 228–229, 394–395
charge-coupled device (CCD) 12, 14, 16, 60–62, 72–75, 80–81, 84–86, 123, 126–127, 130–131, 137, 148, 152–154, 157, 164, 166, 169, 172, 174, 196, 244–248, 265, 276–277, 288, 290, 309–310, 312, 351–352, 354, 397, 400, 402–403, 423–430
claim 219–220
coherence 42, 73, 76, 96, 101–102, 225, 256, 274
coherence holography 16, 239–252
Coherence Scanning Interferometry (CSI) 283, 294–296, 298
coherent 17, 40, 42, 95–96, 121–122, 135, 142, 146, 150, 164, 241–242, 248, 274, 295, 418
colloidal sciences 179, 181, 187–188
color reconstruction 59, 122, 126–127, 130
comparative deformation measurement 419
comparison measurements 257, 260, 268–271
complex-valued data 63
complex-valued modulation 10–11
compression lossless 62–63, 77

Optical Imaging and Metrology: Advanced Technologies, First Edition.
Edited by Wolfgang Osten and Nadya Reingand.
© 2012 Wiley-VCH Verlag GmbH & Co. KGaA. Published 2012 by Wiley-VCH Verlag GmbH & Co. KGaA.

compression lossy
computational integral imaging reconstruction (CIIR)   45–46
computational tools   308
compute unified device architecture (CUDA)   43
computed tomography   123, 225–237, 393
computer generated holograms   13, 43, 58–60, 80, 83, 96, 121, 179, 183, 239
computer-aided accuracy   233
confocal microscopy   225, 259–260, 339, 360
conjugate reference beam   419
contrast ratio   9
2D/3D conversion   47
copyright   208–210
cracks   359–360, 394–395, 400, 403
critical dimension (CD)   263, 298, 327, 340, 343
cross-modulation   11
cross-talk   33
crystalline structures   188
curvature of the microdisplay   4
cutoff frequency   142–144, 333

d

dark-beam scanning   340
dark-field illumination   279, 340
data capture   58–59, 334
data compression   62, 75
data decompression   62–63, 77
data processing   40, 62, 335, 339
deconvolution   334
decoupling of capture and display systems   63
defectoscopy   327
defects (Materials)   393–400, 402–404, 406
3D deformation field   432
delamination   227, 359–360, 401–403, 414
dents   394
depth cue   47, 57, 96
depth of focus   89–90, 163–166, 225, 327, 333
depthcube   40, 42
depth-fused display (DFD)   34, 44
3D depth information   45
depth-of-field   163, 172, 222
design Patent   208–209
dichromate gelatin   103–104
difference holographic interferometry (DHI)   418–423
differential interference contrast (DIC)   15
diffraction   10–12, 16, 43, 58–59, 61, 64, 66, 77–78, 95, 97–115, 122, 124, 128, 164–165, 168, 172–173, 180, 183–184, 193, 195, 225, 241, 245, 255–256, 262, 264–265, 268, 327, 331–333, 337, 341, 343–344, 347–348, 352, 384, 428
diffraction limited lateral resolution   327, 331–333
diffractive lens   149–150, 152, 156, 193
diffractive optical elements   99, 183–184
digital (many entries)
digital drive scheme   6–7
digital holograms in-line   95, 229
digital holograms reconstruction   11, 13, 19, 21–22, 45, 59–62, 64–70, 76, 80, 82–89, 90, 97–101, 104, 137, 140, 143–148, 150, 156, 164–172, 231, 240–244, 246–247, 290, 431–432
digital holograms recording   42–43, 57–58, 60, 88, 95–96, 99–102, 104–109, 111–115, 135–136, 150–151, 158, 160, 164, 240, 247, 418–421, 428–429, 431
digital holographic display   115–116,
digital holographic interferometry   383
digital holographic microscope   386–387
digital holography   11–12, 20, 59–60, 63, 115, 122, 135–159, 164, 168, 290, 340, 384, 386, 390, 413–433
Digital Micromirror Devices (DMD)   59
Digital Speckle Shearography   402–403
dipole approximation   180
direct and indirect solution strategies   305–322
direct problem   329, 331–332, 336
direct strategy   305–322
disclosure   208, 221–222
3D display techniques, recent   32, 44
dislplay with plane beam illumination   79
display 3D   19–20, 22, 31–48, 58, 115, 122, 350, 357
display autostereoscopic   31–33, 36–40, 97
display holographic   21, 31, 34, 40, 42–44, 58–59, 62–69, 77–85, 95,105, 108, 115–116
display holography   95–116
display stereoscopic   31–36, 40, 47–48, 57
display with circular configuration (geometry)   67–68
display with divergent (spherical) beam illumination   67, 79, 147
display with planar configuration (geometry)   66
display-independent processing   63
display-specific processing   63
double-exposure method   311
Digital Speckle Pattern Interferometry (DSPI)   369, 372, 382–384, 390, 414, 432

dual-axes accelerometer   314, 316
dual-use application   328
dynamic optical tweezers   181–183

e

effective medium approximation (EMA)   345
electrically controlled birefringence (ECB)   1
Electro-Holography   43, 59–60
Electronic Speckle Pattern Interferometer (ESPI)   382, 384–385, 390
ellipsometry   289, 291–292, 294–295, 299, 343–348
elliptical geometries   193
elongated object/non-spherical object   189
emerging technology (ET)   312
end-to-end 3D TV   62–63
european Patent Office Database   215
experimental tools   308
extended focus image (EFI)   163

f

far-field technique   328, 345
Feldkamp–Davis–Kress (FDK) method   231
feature size   269, 327–328, 341, 344
feedback loop   307, 335–337, 369
ferroelectric LC (FLC)   4
fiber optics sensors   388–389
field of view (FoV) monocular FoV (MFoV)   68
field of view binocular FoV (BFoV)   32, 47, 68, 84, 89
field stitching   345
fill factor capture   72, 75
fill factor display   75, 84, 89
filtered backprojection   231
final design   306–307
flicker noise   6–7
fluorescense imaging   334
fluorescent light   142
force measurement   187, 232
forward model   283–290
Fourier   15
Fourier hologram   12, 97, 136, 165, 170–172
Fourier inverse   124, 139, 248
Fourier transform   124, 139, 248
Fourier-scatterometry   343, 346–356
fractal dimension   355
frame rate of video   1–2, 5–7, 21, 35, 40, 59, 73, 75
Fresnel
Fresnel hologram   19, 22, 123, 136, 142, 147–148, 150–152, 155, 165–172, 242

Fresnel incoherent correlation holography (FINCH)   147–148, 150–153, 156
Fresnel zone plate (FZP)   142, 243, 249
fringe density   396, 418
fringe identifying   423
fringe integrator   424
fringe order   289, 311
fringe processing   374, 424
fringe projection   288, 332
fringe reflection technique (FRT)   399–402
fringe systems   419, 422–423, 425, 432
fringe-locus function   310–311
fringes   58, 86–87, 101, 104, 107, 150, 240–242, 244, 246, 248–249, 251, 289, 380, 400–401, 417–418, 420, 422, 425–429, 432
full-field-of-view (FFV)   308

g

genetic algorithms   344, 355
geometric transformation   166, 175
ghost imaging   23
ghost traps   184
google Patent   215–216
gradient force   180
graphics processing units (GPU)   43
grating interferometer   383
gratings and lenses approach   184
grooves   394
guide to the expression of uncertainty in measurement GUM   352

h

harsh agents   370–375
harsh environment   369–390
head mounted display, HMD   21, 49
head-up display   21
Helmholtz equation   195–196, 198
hermite-gaussian beams   198
heterodyne detection   142–143, 287
hierarchical supramolecular organization   190–192
hole drilling   369, 432–433
holocamera   414–415
holografika   33, 40, 42
holographic display   42–43, 58–59, 62–70, 77–83, 91, 96, 98, 105, 108, 115–116
hologram, several entries
hologram capture   13, 22, 63, 92
hologram classification - volume reflection   97, 99, 113
hologram computer generated   65, 80, 83, 121–133
hologram digital   140
hologram in-line   95

hologram phase shifting   63
hologram synthetic aperture   147–159
holographic 3D display   34
holographic illumination   419–421, 433
holographic interferometry (HI)   13, 288, 383, 413–433
holographic lithography   13–14
holographic optical tweezers   179–200
holographic sensor   12–13
holographic stereogram   43, 59, 98
holographic visualization   21–22
holography, several entries
holoprinter   97–98, 115–117
homodyne detection   142–143
human eye observation conditions   84
human vision   57
human visual system   32, 47–48
humidity   369, 371–373
hybrid approach   328

i

identification problem   329–331
Ill-posed   330–331, 341
Ill-posedness   328, 330
image conjugate   95, 140, 144, 182, 199, 241, 276, 419
image removal, DC image   62–63
image real   60, 65–68, 84, 87, 241, 419, 421, 427
image reconstruction   45, 61, 84, 334
image restoration   334
3D image scale error   233
image signal modeling   261–263
image synthesis   334, 340
image twin image removal   87, 136, 151, 155–156
imaging, several entries
imaging 2D   37–39, 45–47, 57–58, 60, 96–99, 116, 124–125, 135–136, 138, 230–231
imaging 3D   36–39, 44, 96, 122, 135, 164, 225–238
immersion technologies   341
Ince-Gaussian beams   198–199
incoherent, several entries
indicator   48, 50, 210, 338–339, 353, 360, 413
indirect problem   331
indirect strategy   306
inertial measurement unit (IMU)   305
inertial sensor   328
initial design   306–307
in-line digital Fresnel holography   73
in-line metrology   327

instrument transfer function   290, 336
integral imaging   33, 36, 39–41, 45–46
intellectual property IP   207–223
interference microscopy   260, 340, 351–352
interferogram   243, 248, 288–289, 310–311, 317–318, 374, 386, 413–421, 423, 426, 429, 432, 434
interferometer   4, 11, 73, 123, 135, 147, 240–246, 248, 251, 277, 284–285, 287–292, 309–310, 338, 354, 369–372, 374–386, 390, 404–405, 414–415
interferometry   13, 72, 96, 98, 239, 277, 288–289, 294, 297, 340, 369–389, 395–397, 413–433
international technology roadmap   327
invention   96, 207–208, 212, 214–223
invention disclosure form IDF   221
inverse model   283–284, 286–287, 290–301
inverse problem   283, 285, 328–331, 336–337
IP protection   207, 210–211, 217, 223
IP rights   211, 218
iterative fourier transformation algorithm (IFTA)   184

j

joint ownership   219
Jones matrix   8, 10, 12, 352
Joule heat   320

k

keyhole problem   58, 84
kissing bonds   394, 395

l

Laguerre-Gaussian beams   194, 198
laser ultrasonics (LUS, LAUS)   393–394, 398, 404–406, 408
length measurement error E   235–237
lenticular lens method   36–37
licensing   212, 220–222
line edge roughness LER   267, 270, 278, 344–348
line feature   264, 266–267
liquid crystal on silicon (LCoS)   1, 59, 187, 245
liquid crystal shutter glasses   32–36
liquid crystal spatial light modulator (LC SLM)   13, 40, 42, 60, 402
Lorentz force   180
Lorenz-Mie theory   180
low-pass filter   6, 332

m

macro interference pattern   429
magnification   86, 127–129, 150, 152
magnification of display angular   86
magnification of display longitudinal   86
magnification of display transverse   150, 152
mask work   209–210
master object   13, 419–422
Mathieu beams   190, 196–197, 200
maximum permissible error MPE_E   237
Maxwell-equations-solvers   262
Mccutchen theorem   248–249
measurand   260–261
measurement uncertainty   257, 260, 265–268
measuring limit   418
mechanical loading   395–396
mechanical shock   389
MEMS accelerometer   314, 319
MEMS gyroscope   314
MEMS inspection   305–323
MEMS switch   308, 317, 319–321
meta-material   341–343
Metrological traceability   257–260
micro interference pattern   429–430
microdevice   305, 321
microelectromechanical system, MEMS   1, 168, 187, 227, 283, 305, 328, 337, 358
microfluidics   181
microgyro   314, 318
microgyroscope   314, 317–319
microlens array   137
micromanipulation   179, 192, 195–199, 201
193 nm microscopy   273
microscopy applications   13–16, 22
microstructuring   13
microswitch   308, 314, 316–317, 320–322
microsystem   305, 308
milliscale to microscale ranges   322
minimum reconstruction distance   64–65, 70
mirror   1–3, 14, 18–20, 40–42, 44, 58–60, 76–77, 80–82, 85, 97, 101, 154–155, 182, 186–187, 242, 245–246, 277, 287, 351–352, 354, 379, 386–387, 396–397, 400–401, 421
model-based analysis   261, 269
model-based metrology   328, 334
multi CCD holographic capture system   72, 84–85
multi SLMs holographic display system   85
multicolor holographic recording   107, 111–115
multi-photon microscopy   13
multiple cameras   31, 45–46, 61, 741–75
multiple image planes   178
multiple patterning   332, 334, 341
multiple SLMs   42, 61
multiple viewpoint projections (MVP)   135
multiplexing spatial   11, 35, 40, 59, 79, 137, 154, 172
multiplexing temporal   40, 78–79
multi-region measurement   338
multiscale analysis   337–339
multi-scale sensor fusion   328

n

near real-time   322, 328
near field   328, 340, 342, 345
negative index materials   341
nominal-actual value comparison   232
Non-Destructive Testing (NDT)   229, 393–402
non-diffracting beams   194–197
noninvasive measurements   322
novelty   208, 214
numerical aperture   142, 150, 163, 180, 185–186, 256, 291, 293, 332, 358

o

3D objects   31, 40, 45, 58–59, 71, 75–76, 99, 121–126, 135, 137, 143, 147, 164, 226, 240–241
object size   73, 76, 127, 150, 229, 426
observation mode "asymmetrical diffuser"   76, 83, 88
observation mode "naked eye"   76, 83, 88
observation mode binocular   47, 68, 84–85, 89
off-axis   68, 87, 95, 97, 135–136, 142–147, 159, 173, 243–244, 340, 429
office of technology transfer, OTT   220
Ohmic-type MEMS switch   320
one shot interferometer   390
on-line   142, 146
optical angular momentum   181
optical assembly   190
optical binding   199
optical Coherence Tomography (OCT)   225, 239
optical comparison   418–419, 422
optical flow   45
optical microscope   185, 255–256, 260, 270–274, 278
optical non-destructive testing (optical NDT)   396–397
optical scanning   121–122, 135, 142–147
optical sorting   181
optical switching   19

Index

optical transfer function   290, 332
optical tweezers   179–200
optoelectronic holography (OEH)   309–312
optoelectronic laser interferometric microscope (OELIM) method   322, 328
optoelectronic laser interferometric microscope (OELIM) system   308, 312–313
optoelectronic methodology   309–312
orbital angular momentum   195–196, 198, 200

p

paraboloidal   154
parallax barrier method   36–37
paraxial approximation   66, 70–71, 143
patent   19, 39, 208–222
patent corporation treaty PCT   213–215
Pendry's perfect lens   342
3D perception cues   32, 47
phase change on reflection   285, 289
phase difference   169, 311, 372, 380, 418–421
phase modulation   4, 7, 9–11, 12–17, 19, 102, 104–107, 150, 185, 187,
phase shift procedure   245–247
phase shifting digital holography   246–247
phase shifting polarization interference microscopy   340
phase singularity   195, 243
phase space diagrams (Wigner charts)   65
Photoactivated Localization Microscopy, PALM   256, 341
photonic band gap materials   188
photopolymers   44, 99, 104–105, 108, 111–113
photoresist   105–106, 108
physiological cues   47
pixelated phase-mask interferometer   385
plant patent   208–209
point spread function (PSF)   13, 16–17, 136, 357
polarization modulation   9–10, 13, 16–18, 35
polarizing glasses method   32, 35–36
pores   191, 394–395
portable device   432
portable interferometer   375, 414–418
power spectral density   285, 290, 355
pressure   47, 179, 305, 313, 370–371, 373, 378, 389, 396
prior art   214–216
projection onto the constraint sets (POCS)   139

propagation algorithm between parallel planes   82
propagation algorithm between tilted planes   82
prototype design   306–307
PSF-engineering   13, 16–17
psychological cues   47
pulse code modulation, PCM   7
pulse shaping   19
pulse width modulation, PWM   6
puzzle read-out (PRo) technique   426–428

q

quadratic phase function   136–137, 140, 142, 149
quality factor (Q-factor)   323
quantitative microscopy   255–257, 260
quasi-common path interferometer   380–381, 387
quasicrystalline structures   188

r

radiation   98, 107, 113, 142, 179–180, 230, 234, 256, 264, 273, 370–371, 374–375, 378
radiation force   180
radiation isolation   378
Rayleigh approximation   165
Rayleigh criterion   101, 157, 322
Rayleigh resolution limit   144
rays optics approximation   180
real world objects and scenes   79
reconstruction artifacts   239
reconstruction distance   22, 64–67, 70, 81–87, 136–137, 150, 164, 167, 170, 172–173
reconstruction noise   139, 233, 237
reconstruction numerical   13, 88, 163–175, 431–432
reconstruction optical   101
reconstruction optoelectronic   62
reconstruction problem   330, 337
recording materials   43
reference surface   288, 432–433
reflectometry   343, 393–394, 396–398, 408
registered IP   210
regularization   330–331, 334, 340
reliability   105, 111, 307, 320, 322, 350, 354, 389
reliable electrical interconnection   322
residual stress   311, 369, 384–385, 431–433
resolution   332
resolution enhancement   256, 273, 276, 379, 334, 338–340
response time   7, 9, 11–12, 35, 43, 49

Index | 481

rewritable holographic stereogram 43
RF MEMS switch 308, 316, 319–321
rigid body motion compensation 433
rigorous coupled wave analysis (RCWA) 261–262, 297, 336
robust interferometer 380, 430
rod-shaped bacteria 1189–190
RSS-type uncertainty 311

s

Sagnac interferometer 243, 245, 248
sampling parameters mismatch in 86
scalability 322
scanning 142–147, 235–237, 350, 406, 426, 429–431
scanning error PF (form) 236
scanning error size (PS) 235–237
scanning white-light interference microscopy 350
scattering force 180–181
scatterometry 279, 295, 343–355
scenes dynamic 72
scenes static 43, 61
scratches 355, 359–360, 394
see real technology 19, 21, 59
Seereal 19, 21, 34, 43, 59
segmentation algorithm 87
self compensating interferometer 381
self-healing property 194–195
self-similar beams 197–200
sensitivity 269, 300, 348–350
sensitivity analysis 300, 348–350
sensor fusion 299, 328, 334, 335, 357–338
servicemark 210
shearography 13, 387–389
sidewall angle, SWA 258, 271, 295, 345
silver halide emulsions 105–108
simulator sickness questionnaire (SSQ) 48
single-beam-gradient trap 180
source trajectory 231, 233
space-bandwidth product 59, 66, 70, 76–78, 80, 86, 327–328, 333
spatial coherence comb 248–252
space spatial frequency analysis 3, 6, 10–11, 60, 67, 97–98, 101, 104, 127–128, 139–140, 142–145, 171, 249–251, 285, 290, 332–333, 340, 397
spatial light modulator (SLM), 1–23 (full chapter) 40, 57, 70, 77, 79–80, 83, 91, 172, 243, 245, 276, 328, 336, 379, 402–403
spatial light modulator (SLM) amplitude 9–11, 19–20, 23

spatial light modulator (SLM) liquid crystal (LCSLM) 58–60
spatial light modulator (SLM) phase only 70, 77, 80, 91
spatial light modulator (SLM) virtual 79–80, 83
spatial phase-only
spatial positions 45
spatial resolution 43, 58, 101, 128, 142, 159, 187, 232, 328, 385
spatially structured illumination 341
speckle contrast 332
speckle Multiframe 63
speckle noise 21, 59, 77, 424
speckle reduction 62–63
spectroscopic ellipsometry 343–344, 346–348
Spherical 64, 67, 70, 77–78, 80, 107–108, 142, 144, 147, 149, 153–154, 156, 165, 170, 180, 187, 189, 241, 420–422
start up 220–222
stereograms 59, 98
stereoscopic displays 32–36
stimulated emission depletion microscopy, STED microscopy 16, 256, 341, 363
stochastic optical reconstruction microscopy, STORM 256, 341
structured light 45
sub-wavelength feature 340–342, 345
sub-wavelength imaging 341–342
super multi-view (SMV) display 44
super-lens 341–343
superresolution 328, 334, 341, 363, 365
superresolution microscopy 16
superresolution structured illumination microscopy, SR-SIM 341
supramolecular organization 190–192
surface form
surface plasmon polaritons 341
surface profilometry 252, 350
surface roughness 236, 385, 321, 397
surface topography 285, 290, 292, 294, 301, 355
2D/3D switching display 38
synthetic aperture 136
synthetic aperture with fresnel elements (SAFE) 136, 147–159
synthetic diffraction grating 173
system aberrations 337

t

tailored light fields 193, 200
television (TV) 58
television 3D (3DTV) 32, 57, 60, 62–63, 91

television holographic   57–90
television real time 3DTV   64
temperature   7, 12, 102–103, 264, 313, 321, 369–372
temperature isolation   378
template matching   336
test object   229, 408, 414, 419–423
texture features   355
theoretical upper limit   425
thermal loading   387, 408, 422
thermal movement of X-ray source   237
thermography   393–394, 399–400, 408
thin film   84, 291–295
three-dimensional display   31–50
threshold-/gradient-based surface determination   235–237
tilted plane wave illumination of SLM   78, 81
tilted planes   82, 88, 168
time-of flight (TOF)   45
top-down strategy   335
total measurement uncertainty (TMU)   35–36
traceability   29, 57–58, 229, 257
trade secret   210–11
trademark   210–215
transmission, 76 entries
transmission and surface relief holograms   98
triangulation   226, 288, 332–333, 338
true 3D   57, 92, 344
two refractive index contouring   419
two-step and two-frequency upconversion   40
two-wavelength contouring   419, 421

u

ultrasound
uncertainty analysis   266, 307–308, 352
utility patent   209–210

v

van Cittert-Zernike theorem   241
vertically aligned nematic, VAN   3
vibration isolation   378–379
vibrations   59, 274, 313, 374–376, 378–381, 384, 385, 387–388
vibrometer   13, 405, 416
video 3D holographic   60, 84, 89–90
video electro-holographic
viewing angle of holographic display off-axis   68, 87
viewing angle of holographic display on-axis   65, 84–85
visual fatigue   32, 44, 47–50
visual perception   59, 62, 68, 70, 84, 92, 97–98, 116–117
volumetric displays   40–42

w

wave (several entries)
wave front sensing   290–291
well-posed   330
white-light   135, 136, 338, 340, 350–354, 356
white-light interference fourier-scatterometry   351, 353
Wigner distribution function (WDF)   64
world intellectual property organization WIPO   213

x

X-ray component alignment   234
X-ray cone-beam tomography   238
X-ray tomography (X-CT)   230–237

z

zcam   32, 45
zeolite crystals   191
zero order term   61